Modern Acoustics and Signal Processing

For further volumes:
http://www.springer.com/series/3754

The ASA Press

The ASA Press imprint represents a collaboration between the Acoustical Society of America and Springer dedicated to encouraging the publication of important new books in acoustics. Published titles are intended to reflect the full range of research in acoustics. ASA Press books can include all types of books published by Springer and may appear in any appropriate Springer book series.

ASA Press

Ning Xiang • Gerhard M. Sessler
Editors

Acoustics, Information, and Communication

Memorial Volume in Honor of Manfred R. Schroeder

ASA Press

Springer

Editors
Ning Xiang
The School of Architecture
Rensselaer Polytechnic Institute
Troy, NY, USA

Gerhard M. Sessler
Darmstadt University of Technology
Darmstadt, Germany

ISBN 978-3-319-05659-3 ISBN 978-3-319-05660-9 (eBook)
DOI 10.1007/978-3-319-05660-9
Springer Cham Heidelberg New York Dordrecht London

Library of Congress Control Number: 2014941939

Printed on acid-free paper

Springer is part of Springer Science+Business Media (www.springer.com)

Acoustical Society of America

The mission of the **Acoustical Society of America** (www.acousticalsociety.org) is to increase and diffuse the knowledge of acoustics and promote its practical applications. The ASA is recognized as the world's premier international scientific society in acoustics, and counts among its more than 7,000 members, professionals in the fields of bioacoustics, engineering, architecture, speech, music, oceanography, signal processing, sound and vibration, and noise control.

Since its first meeting in 1929, The Acoustical Society of America has enjoyed a healthy growth in membership and in stature. The present membership of approximately 7,500 includes leaders in acoustics in the United States of America and other countries. The Society has attracted members from various fields related to sound including engineering, physics, oceanography, life sciences, noise and noise control, architectural acoustics; psychological and physiological acoustics; applied acoustics; music and musical instruments; speech communication; ultrasonics, radiation, and scattering; mechanical vibrations and shock; underwater sound; aeroacoustics; macrosonics; acoustical signal processing; bioacoustics; and many more topics.

To assure adequate attention to these separate fields and to new ones that may develop, the Society establishes technical committees and technical groups charged with keeping abreast of developments and needs of the membership in their specialized fields. This diversity and the opportunity it provides for interchange of knowledge and points of view has become one of the strengths of the Society.

The Society's publishing program has historically included the *Journal of the Acoustical Society of America*, the magazine *Acoustics Today*, a newsletter, and various books authored by its members across the many topical areas of acoustics. In addition, ASA members are involved in the development of acoustical standards concerned with terminology, measurement procedures, and criteria for determining the effects of noise and vibration.

Series Preface for Modern Acoustics and Signal Processing

In the popular mind, the term "acoustics" refers to the properties of a room or other environment—the acoustics of a room are good or the acoustics are bad. But as understood in the professional acoustical societies of the world, such as the highly influential Acoustical Society of America, the concept of acoustics is much broader. Of course, it is concerned with the acoustical properties of concert halls, classrooms, offices, and factories—a topic generally known as architectural acoustics, but it is also concerned with vibrations and waves too high or too low to be audible. Acousticians employ ultrasound in probing the properties of materials, or in medicine for imaging, diagnosis, therapy, and surgery. Acoustics includes infrasound—the wind-driven motions of skyscrapers, the vibrations of the earth, and the macroscopic dynamics of the sun.

Acoustics studies the interaction of waves with structures, from the detection of submarines in the sea to the buffeting of spacecraft. The scope of acoustics ranges from the electronic recording of rock and roll and the control of noise in our environments to the inhomogeneous distribution of matter in the cosmos.

Acoustics extends to the production and reception of speech and to the songs of humans and animals. It is in music, from the generation of sounds by musical instruments to the emotional response of listeners. Along this path, acoustics encounters the complex processing in the auditory nervous system, its anatomy, genetics, and physiology—perception and behavior of living things.

Acoustics is a practical science, and modern acoustics is so tightly coupled to digital signal processing that the two fields have become inseparable. Signal processing is not only an indispensable tool for synthesis and analysis but it also informs many of our most fundamental models about how acoustical communication systems work.

Given the importance of acoustics to modern science, industry, and human welfare Springer presents this series of scientific literature, entitled Modern Acoustics and Signal Processing. This series of monographs and reference books is intended to cover all areas of today's acoustics as an interdisciplinary field. We expect that scientists, engineers, and graduate students will find the books in this series useful in their research, teaching, and studies.

July 2012 William M. Hartmann

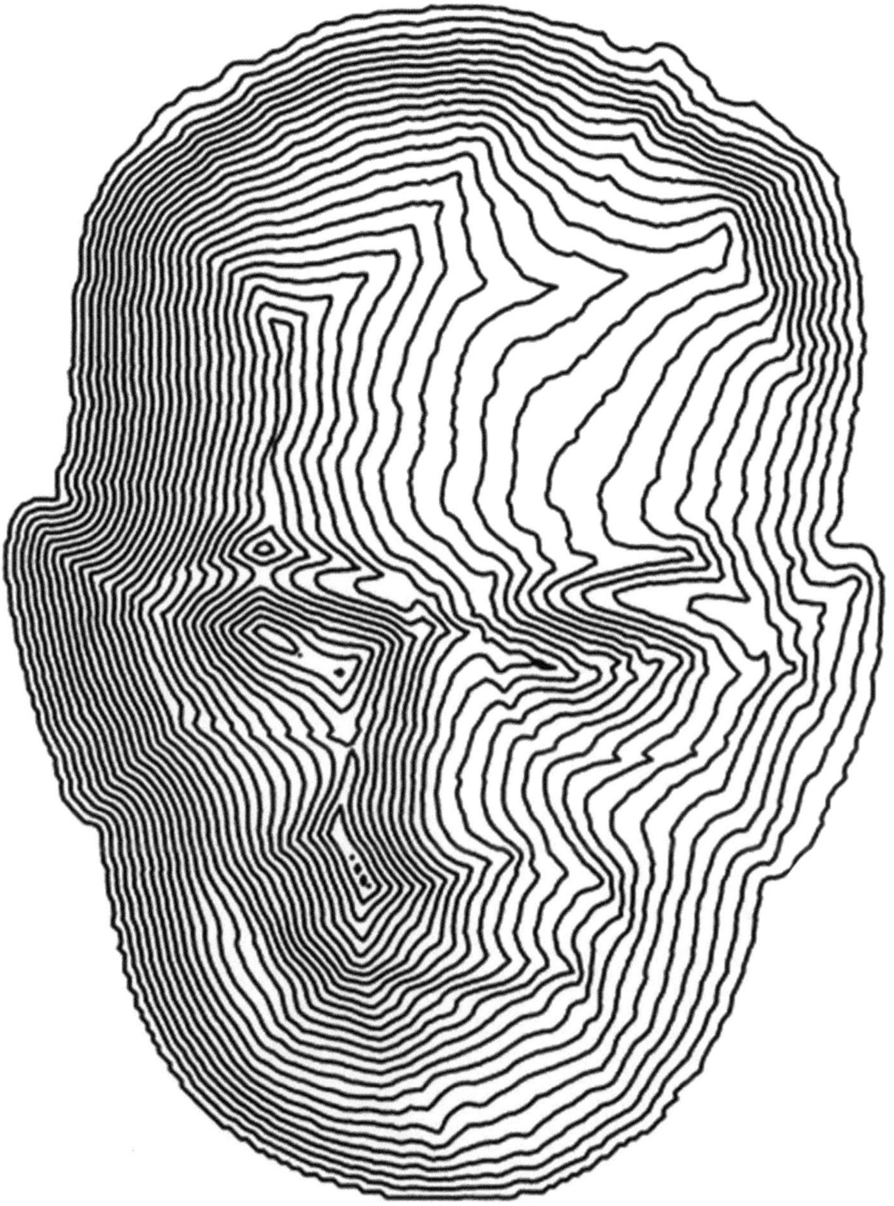

Manfred R. Schroeder's Portrait – Solution of Eikonal equation, programed by Wolfgang Moeller (see also Chapter 20, Fig. 20.11, see also at the beginning of Part II, Fig. 28 in photo collection on page 293) (Wolfgang Moeller is one of Schroeder's former students)

Preface

Manfred Schroeder was born in Ahlen, Westphalia, Germany. In Manfred's youth, his father, a mining engineer, and his mother encouraged an early interest in the beauty and utility of mathematics. His mathematical talent was already evident in his secondary schooling. Like many science-oriented youths of the time, he was attracted to radio technology: He built short-wave receivers and transmitters. He never considered himself a legitimate "radio amateur" ("ham") because of the difficulty of obtaining a government license, and the (not unusual) additional obstacle of Morse code. But, in all other respects, he was a "ham."

Also, like many other youths of his time, Manfred's continued education was interrupted by the hostilities of World War II. At age 16, he was drafted into the German air force, serving first in Poland and later in Holland. Because of his talent for electronics, he was trained in the early uses of radar for detecting range and direction. He thereby became part of an intensive technology race, an experience that no doubt contributed to his later remarkable zeal in acoustics research.

After the conclusion of hostilities, he returned home, rejoicing that his family was intact and unharmed in this turbulent time. Manfred continued his education and entered the University of Göttingen to study physics and mathematics. In due course, as his university studies advanced, he fell under the tutelage of Professor Erwin Meyer, director of the Drittes Physikalisches Institut (Third Physics Institute) and an internationally recognized authority in acoustics. Working towards his doctoral degree, Manfred conducted fundamental research on the distribution of microwave and acoustic normal modes in enclosures, relations that characterize the frequency response and spatial spread of electromagnetic and sound energy in structures such as waveguides, rooms, and auditoria. He received his Dr. rer. nat. degree as a student of Professor Meyer in 1954. Later that year he joined Bell Laboratories in Murray Hill, New Jersey. Manfred went on to be appointed Head of the Acoustics Research Department in 1958 and Director of the Acoustics and Speech Research Laboratory in 1963. Soon thereafter, he assumed responsibility for all areas of acoustics, speech, and mechanics research at Bell Labs. In 1969, while still maintaining some of his responsibilities at Bell, he was appointed Professor of Physics and Director of the Drittes Physikalisches Institut at the

University of Göttingen. After his retirements from Bell Labs and the University in 1987 and 1991, respectively, Manfred continued his activities as a scientist and teacher.

Manfred's scientific contributions are so numerous that it would require much more space than we have in this Preface to provide a complete introduction. One field about which Manfred was always enthusiastic and to which he made many fundamental and far-reaching contributions is room acoustics. As mentioned above, this area fascinated him as early as his thesis work in the early 1950s in Göttingen, where he developed his brilliant statistical theory of frequency responses of enclosures. Manfred built vacuum-tube-based devices for electromagnetic waves to validate his hypothesis. His "radio amateur" experience and early use of radar along with the dynamic research environment at Erwin Meyer's Institute prepared Manfred to be an excellent experimentalist. In a series of publications Manfred described the relations between the eigenmodes and the frequency response of a room. He explained the random character of the response above a certain frequency (today known as the Schroeder frequency). Other ingenious contributions followed, such as the use of frequency shifting to suppress feedback in public address systems and the integrated-tone-burst method for measuring sound energy decay functions of enclosures (today known as Schroeder curves via Schroeder's integration or backwards integration). His seminal paper in 1965 on backwards integration has led to numerous research activities in room acoustics (see Chap. 3 by Xiang). Manfred introduced binary maximum-length sequences in architectural acoustics in 1979—the so-called fast M-sequence transform for room impulse response measurements (see Chap. 6 by Xiang, Xie & Cox). Another outstanding achievement is Manfred's invention of pseudo-random surface structures based on number-theoretical schemes (Schroeder diffusers, see Chap. 6 by Xiang et al. and Chap. 9 by D'Antonio). Such devices are used to scatter sound waves in all directions, and hence improve the acoustics of concert halls by providing lateral sound reflections.

Of much public interest was Manfred's involvement in the activities to improve the acoustics of Philharmonic Hall in New York City. Soon after the opening of the Hall in 1962, grave deficiencies in its acoustics were detected, and Manfred together with a few acoustic consultants was invited to analyze and remedy the situation. He was asked to make an objective evaluation of the acoustics of the Hall and decided to use a novel measuring method based on digital signal generation and processing. It was in this period that Manfred first began to develop his backwards integration method. Preparation, performance, and evaluation of these experiments turned into a scientifically challenging project for Manfred and his coworkers. Later on, while at Göttingen, Manfred and his students evaluated 20 of the major concert halls of the world by subjective auditory comparison, thereby gaining significant information about subjective parameters of room acoustics, such as lateral reflections. In addition to these new digital measuring techniques, between 1968 and 1969 Manfred pioneered a ray-tracing method of computer simulation of large halls. This was shortly followed by active research followed in room-acoustics simulations, leading to many refined methods (see Chap. 2 by Krokstad et al.). These methods have gained great practical importance for the planning of new auditoria.

Manfred Schroeder was the first to recognize the importance of digital signal processing, not only in room acoustics but in many other areas as well, and also the first to actually employ these methods. Examples are the generation of artificial reverberation (Schroeder's all-pass filter) and artificial stereophony as well as digital signal processing applications in ultrasonics, speech, and electroacoustics. Manfred is also one of the pioneers of computer graphics: For his application of concepts from mathematics and physics to the creation of artistic works he was awarded the First Prize at the 1969 International Computer Art Competition.

Another area where Manfred made fundamental contributions is speech and hearing acoustics. Among his outstanding achievements in this field are the invention of the voice-excited vocoder in 1960 and the introduction of linear predictive coding (LPC) in 1967 during the exciting days of the digital revolution at Bell Labs. LPC remains an important technique for analysis and synthesis of speech and music signals. Also in 1967 Manfred used acoustic flectrometry to determine the human vocal tract. He also worked on early development of the concept of surround sound. His mathematical model of the inner ear has inspired much of the following subsequent work in auditory research.

Manfred's interest in mathematics and especially in number theory motivated him in 1984 to write his very successful treatise "Number Theory in Science and Communication," now in its fifth edition. This book differs from other texts on the topic by its intuitive approach and by emphasizing the applications of number theory to such diverse fields as cryptography, physics, computing, and self-similarity. Many amazing relationships are uncovered in this text which is not only stimulating to read but also a lot of fun to muse about. Similarly, Manfred's 1992 book "Fractals, Chaos, Power Laws: Minutes from an Infinite Paradise" illustrates the far-reaching influence of the concept of self-similarity on science, music, and the visual arts. Both books are marvellous examples of how to present difficult subjects transparently and in an amusing style. Manfred's third book, "Computer Speech," first published in 1999, spans an arch from speech recognition and synthesis to monaural and binaural hearing and modern signal analysis. This text summarizes Manfred's long involvement with speech and hearing research at Bell Labs. Apart from these monographs, Manfred Schroeder published more than 350 papers, many of them seminal communications, and holds 45 US patents.

As a professor and teacher at Göttingen University, Manfred was able to attract excellent students due to his inspiring lectures on such diverse topics as physics, acoustics, mathematics, and signal processing. Most of the over 45 Ph.D. candidates whom he supervised successfully continued their careers either in academia or in industry, many of them as professors, heads of companies, or chief scientists.

Manfred received numerous accolades and professional recognitions throughout his long and varied career. He was elected a member of the U.S. National Academy of Engineering and the Göttingen Academy of Science, and Fellow of the American Academy of Arts and Sciences and of the New York Academy of Sciences. As mentioned above, in 1969 he was awarded First Prize at the International Computer Art Competition for his application of concepts from mathematics and physics to the creation of artistic works. He was a founding member of the Institute de

Recherche Acoustique Musique of (IRCAM) in Paris. Manfred has also been recognized for his scientific achievements as the recipient of Gold Medals from the Audio Engineering Society in 1972 and the Acoustical Society of America in 1991. In addition, he was presented with the Rayleigh Medal from the British Institute of Acoustics in 1987, the Lower Saxonion State price in 1992, the Helmholtz-Medal of the German Acoustical Society in 1995, and the Technology prize of the Eduard Rhein-Foundation in 2004.

This introduction to Manfred Schroeder would be incomplete without a few words about Manfred's personality. Everybody who has met Manfred knows his kindness, his humor and cheerfulness, his ease of communication. Many of these attributes were demonstrated in his farewell lecture, presented at the occasion of his official "retirement," to an overflow audience in Göttingen's largest lecture hall: For 90 min, Manfred fascinated his listeners with a brilliant, lucid, and amusing demonstration of the most beautiful experiments from the fields of acoustics, optics, and chaos physics.

In December 2006, the Acoustical Society of America, at its Joint-Meeting with the Acoustical Society of Japan in Honolulu (ASA), organized an honorific session for Manfred, chaired by Ning Xiang, Juergen Schroeter, and Akiro Omoto. The ASA also organized a memorial session for Manfred, chaired by the present authors, at its Seattle meeting in May 2011. Among the invited speakers of these two Sessions were his former colleagues and students (in alphabetic order).

Joshua Atkins, Bishnu Atal, Peter Cariani (coauthor with Yoichi Ando), Peter D'Antonio, James Flanagan, Hiroya Fujisaki, Fumitada Itakura, Birger Kollmeier, Armin Kohlrausch, Roland Kruse, Heinrich Kuttruff, Max Mathews (deceased), Volker Mellert, Jean-Dominque Polack, Gerhard Sessler, Michael Vorlaender, James West, and Ning Xiang.

The first part of this book contains 13 chapters, written by colleagues and former students of Manfred, on a variety of topics. These range from architectural acoustics (chapters by James West and Joshua Atkins, Ning Xiang, Guillaume Defrance & Jean-Dominique Polack, Asbjørn Krokstad, Peter Svenson & Svein Strøm, Jean-Dominique Polack, Peter D'Antonio), psychological acoustics (Yoichi Ando & Peter Cariani, Armin Kohlrausch & Steven van de Par), and speech (Bishnu Atal) audio, acoustics measurements to electro acoustics (Roland Kruse & Volker Mellert, Gerhard Sessler, and Ning Xiang, Bosun Xie & Trevor J. Cox) and underwater acoustics (Dieter Guicking). The style of the chapters ranges from truly scientific to absolutely colloquial. Many chapters refer not only to Manfred's original work but also to new studies that originated from Manfred's studies. In all these papers, the relationship of the problems discussed to Manfred's own fields of interest is, in general, evident.

The second part of the book consists of memoirs which Manfred wrote over the last decade of his life. We are grateful to Manfred's wife Anny Schroeder and to Manfred's children Marion, Julian, and Alexander Schroeder for granting us permission to include the memoirs in this book. These recollections shed light on many aspects not only of Manfred's life but also on that of many of his colleagues, friends, and contemporaries. They also portray political, social, and scientific events during a time span that extends from the pre-war period to the present. These memoirs, written in an inimitable and witty style, are full of information, entertaining, and fun to read. We trust that the reader will find this volume useful and stimulating.

Troy, NY — Ning Xiang
Darmstadt, Germany — Gerhard M. Sessler

Contents

Part I Scientific Chapters

1 **Spatial Audio and Room Acoustics** 3
Joshua Atkins and James E. West

2 **The Early History of Ray Tracing in Acoustics** 15
Asbjørn Krokstad, U. Peter Svensson, and Svein Strøm

3 **Advanced Room-Acoustics Decay Analysis** 33
Ning Xiang

4 **Estimating the Crossover Time Within Room Impulse Responses** . . . 57
Guillaume Defrance and Jean-Dominique Polack

5 **Are Impulse Responses Gaussian Noises?** 77
Jean-Dominique Polack

6 **Recent Applications of Number-Theoretic Sequences in Audio and Acoustics** 93
Ning Xiang, Bosun Xie, and Trevor J. Cox

7 **Where Mathematics and Hearing Science Meet: Low Peak Factor Signals and Their Role in Hearing Research** 113
Armin Kohlrausch and Steven van de Par

8 **Neurally Based Acoustic and Visual Design** 145
Yoichi Ando and Peter Cariani

9 **Manfred R. Schroeder: A Personal Memoir, Optimizing the Reflection Phase Grating** 177
Peter D'Antonio

10 **Manfred R. Schroeder: Challenge the Present and There Is a Better Way** 199
Bishnu S. Atal

11 **Transducer Research at Bell Laboratories Under Manfred Schroeder** 213
Gerhard M. Sessler

12 **Manfred Schroeder and Acoustical Impedance** 229
Roland Kruse and Volker Mellert

13 **Research on Underwater Acoustics in Göttingen** 241
Dieter Guicking

Part II Manfred Schroeder's Memoirs

14 **Right Turns** 299
Manfred R. Schroeder

15 **School Years** 303
Manfred R. Schroeder

16 **July 1943** 325
Manfred R. Schroeder

17 **Radar** 329
Manfred R. Schroeder

18 **Lucky Breaks** 337
Manfred R. Schroeder

19 **The University of Göttingen** 343
Manfred R. Schroeder

20 **Bell Laboratories** 363
Manfred R. Schroeder

21 **The KGB** 401
Manfred R. Schroeder

22 **Professor of Physics** 407
Manfred R. Schroeder

23 **The German Language** 423
Manfred R. Schroeder

24 **Retirement and Beyond** 431
Manfred R. Schroeder

25 **The Year 2006: One of Our Best (So Far)** 437
Manfred R. Schroeder

Bibliography of Manfred R. Schroeder's Publications 441

Index 457

List of Contributors

Joshua Atkins Beats Electronics, LLC, Santa Monica, CA, USA

Yoichi Ando Graduate School of Science and Technology, Kobe University, Kobe, Japan

Bishnu Atal Department of Electrical Engineering, University of Washington, Seattle, WA, USA

Peter Cariani Harvard Medical School, Boston, MA, USA

Boston University, Boston, MA, USA

Trevor J. Cox Acoustics Research Centre, University of Salford, Salford, Greater Manchester, UK

Peter D'Antonio RPG Diffusor Systems, Inc., Upper Marlboro, MD, USA

Guillaume Defrance Acoustics and Lighting Dpt., CSTB, Saint-Martin-d'Heres, France

Dieter Guicking Drittes Physikalisches Institut der Universität Göttingen, Göttingen, Germany

Armin Kohlrausch Smart Sensing & Analysis Group, Philips Research Europe, Eindhoven, The Netherlands

Human-Technology Interaction, Eindhoven University of Technology, Eindhoven, The Netherlands

Asbjørn Krokstad Department of Electronics and Telecommunications, Norwegian University of Science and Technology, Trondheim, Norway

Roland Kruse Acoustics Group, Institute of Physics, Carl von Ossietzky Universität, Oldenburg, Germany

Volker Mellert Acoustics Group, Institute of Physics, Carl von Ossietzky Universität, Oldenburg, Germany

Steven van de Par Acoustics Group, Institute of Physics, Carl von Ossietzky Universität, Oldenburg, Germany

Jean-Dominique Polack Université Pierre et Marie Curie—Institut Jean Le Rond d'Alembert, M° Dupleix, France

Gerhard Sessler Department of Communications Technology, Technische Universität Darmstadt, Darmstadt, Germany

Svein Strøm COWI A/S, Trondheim, Norway

U. Peter Svensson Department of Electronics and Telecommunications, Norwegian University of Science and Technology, Trondheim, Norway

James E. West Department of Electrical and Computer Engineering, Johns Hopkins University, Baltimore, MD, USA

Ning Xiang Graduate Program in Architectural Acoustics, School of Architecture, Rensselaer Polytechnic Institute, Troy, NY, USA

Bosun Xie Physics, South China University of Technology, Guangzhou, China

Part I
Scientific Chapters

Chapter 1
Spatial Audio and Room Acoustics

Joshua Atkins and James E. West

1.1 Introduction

Manfred Schroeder's pioneering work in spatial hearing, speech coding, and room acoustics during his career at Bell Labs and the University of Göttingen paved the way for many teleconferencing advances that are still being researched today. Some of the current challenges, such as multichannel acoustic echo cancellation, spatial audio capture and reconstruction, noise reduction, and coding and compression, were originally studied by Dr. Schroeder. In this chapter we will discuss his contributions to the field and how current research continues to explore methods for building immersive communication systems.

1.2 Room Acoustics

Manfred Schroeder's interest in the behavior of sound in enclosures began with his Ph.D. thesis at the University of Göttingen under Professor Meyer and continued at Bell Labs. Work at Bell Labs was done mainly to support methods to improve communication between groups of people connected through the telecommunications network, i.e. teleconferencing. In a quote from the IEEE's oral history [1], he states:

> The third one was room acoustics, my old field from my Ph.D. thesis. That was reactivated in connection with work on Philharmonic Hall in 1963 at Lincoln Center for the Performing Arts in New York City. The acoustics there were severely criticized, and the management of Lincoln Center turned to AT&T. AT&T turned to Bell Labs, and then Bill Baker asked me to join a committee of four experts to look at the acoustics situation at Philharmonic Hall. The other three were consultants, but I, by the consent decree between Western

J. Atkins (✉)
Beats Electronics, LLC, 8600 Hayden Place Culver City, CA 90232, USA
e-mail: josh.atkins@beatsbydre.com

N. Xiang and G.M. Sessler (eds.), *Acoustics, Information, and Communication*, Modern Acoustics and Signal Processing, DOI 10.1007/978-3-319-05660-9_1,

Fig. 1.1 Manfred Schroeder (standing) and James West in Bell Labs anechoic chamber at Murray Hill, New Jersey, measuring the impulse response of two walls using a starter pistol

Electric and the government, was not allowed to do consulting. But I was permitted to make measurements and assess the situation.

Figure 1.1 shows Manfred Schroeder in the Philharmonic Hall at Lincoln Center (later renamed Avery Fisher Hall). At the time, tools for measuring properties of enclosures were mainly analog, and consequently inaccurate with respect to the properties that define clarity and good articulation of voices and instruments. Initial efforts using existing equipment and techniques were performed in the Bell Labs anechoic chamber. Walls were constructed to measure reflections and to determine the impulse response from just two perpendicular surfaces using a starter pistol as a sound source (shown in Fig. 1.2). This method, even with just two walls, presented many problems; poor signal-to-noise over a broad frequency range and variations in spectrum and energy from shot to shot were among the worst. So, we knew this would be a serious problem in larger enclosures. References to balloons, clappers, and other devices for generating an acoustical impulse were all lacking in some way, but the lack of energy presented the major problem. To solve the problem, we borrowed a small canon from Rutgers University that was used to celebrate home team football touchdowns. Outdoor measurements of the impulse from the canon were encouraging in energy, length of the burst, and repeatability. With the canon placed on the stage in Philharmonic Hall and microphones distributed throughout the audience, we fired the canon. The loud burst of energy sent sound waves throughout the hall that were picked up by all microphones. The loud sound caught the interest of the hall's manager who entered immediately and panicked because visibility was poor due to the smoke from the blast. The manager forbid any more shots and immediately began to try to remove the smoke and smell from the canon before the evening performance (Fig. 1.3).

So began the use of digital signal processing to more accurately determine the acoustical properties of enclosures [2, 3]. First, the proper stimulus was created in the form of a tone burst composed digitally and played back in the enclosure over a nearly omnidirectional loudspeaker. The response of the enclosure was picked up

Fig. 1.2 Manfred Schroeder on stage at Philharmonic Hall in 1963 at Lincoln Center

by strategically placed microphones and recorded for evaluation using a computer. Such measurements were repeated for different loudspeaker and microphone positions. This new method provided a much more accurate value for the reverberation time, derived from the integrated tone-burst decay. The digital method also allowed refined measurements that resolved the seat attenuation effect which explained why certain frequency bands were missing as sound propagated over the regularly spaced seats [4]. Hall raking, adding a gradual incline in seat height from the stage, can reduce this "seat effect."

The early reverberation decay, the first 50 ms, could be defined, as well as reverberation decay in other time domains. The digital method allowed measurement of other room properties that contributed to better understanding of how sound propagates and how it effects the impression of listeners [5]. One New York Times music critic found that the subjective impression in the hall differed greatly; the best seats were said to be in the upper balcony to the right of the stage. Standard analog measurements showed little difference between locations, but the digital

Fig. 1.3 Manfred Schroeder in the Bell Labs anechoic chamber surrounded by loudspeakers placed to represent specific echo response measured in Philharmonic Hall (now Avery Fisher Hall). The research group wanted to independently test the accuracy of the measurements by playing back the results obtained in Philharmonic Hall to see if the same subjective impression could be obtained just using early reflections

measurements of the impulse response showed many early reflections that supported better acoustics in the upper balcony.

Continuing along this line of work, a method for calculation of the sound energy decay from a room impulse response was developed, now called the Schroeder backward integration method [6]. This allowed for accurate estimation of the reverberation time. A method, based on number theory, was also created to measure the impulse response of a room without the need of an impulsive signal [7]. Based on maximum length sequences (MLS), the method relied on correlation of the input signal with the recorded signal and resulted in high SNR and the ability to take measurements of a concert hall during a performance to better understand the effect of occupied seats.

Today, room and concert hall measurements continue to be developed with similar techniques. It has been found that the MLS technique suffers from issues when the measured system has nonlinear behavior and a modified, correlation-based method based on a sine sweep is now widely used [8]. New microphone array systems are also enabling high resolution capture of the acoustic sound field as it propagates through a space [9–11]. Spherical arrays, treated as "acoustical cameras," can plot the sound energy decay in a room as a function of angle of incidence, and are leading to new methods for analysis of the complex sound field in large enclosures [12].

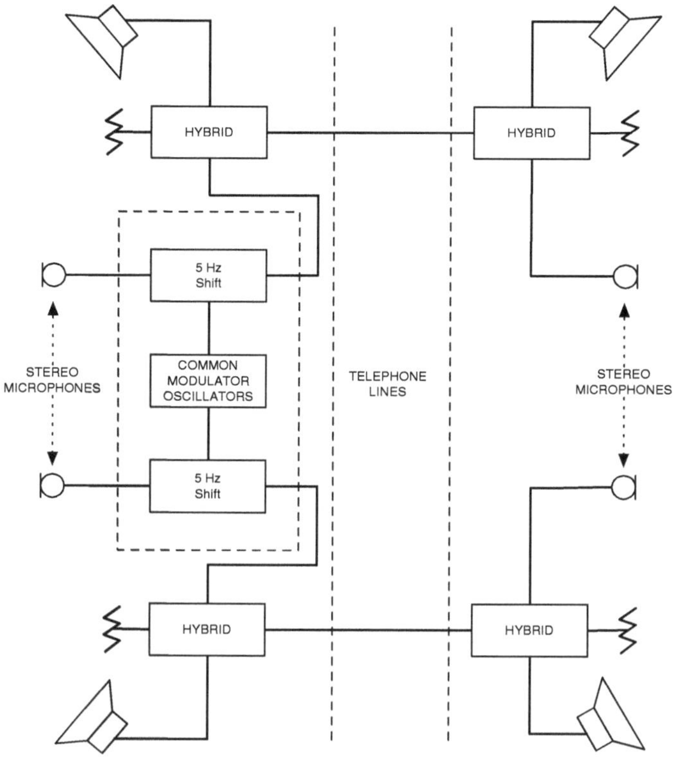

Fig. 1.4 The stereo teleconferencing system from [13]. The system shows a 5 Hz time-varying analog pitch shift that was used to minimize feedback on the far-end

1.3 Hands-Free Telephony

In the early 1960s, Bell Labs began pursuing research on conference telephony and speakerphones, which continues to be an active area of study even 50 years later. Early experiments in teleconferencing revealed the serious problem that half-duplex systems imposed on users; one-way communication simply did not work well, so efforts to reduce the echo return path began [13]. A solution was proposed using a simple frequency shifting circuit that altered the pitch of transmitted speech by 5 Hz to make the feedback path nonlinear [14]. Figure 1.4 shows a schematic of the system. This shift was nearly imperceptible on speech, but allowed for an additional 3 dB of gain in the system without "singing" artifacts.

It was later found that the single-channel acoustic echo cancellation problem can be solved efficiently using a digital system with an adaptive filter to estimate the room impulse response. However, as hands-free teleconferencing systems developed in the 1990s to include stereo and multichannel audio, a problem of correlation between the loudspeaker channels became apparent [15]. Here,

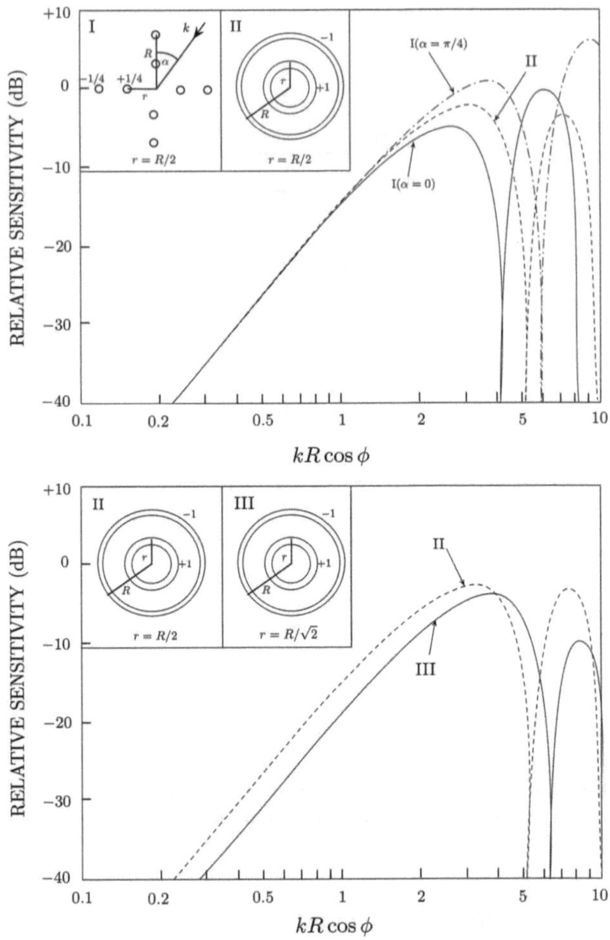

Fig. 1.5 Sensitivities of three toroidal microphones (point, line, and area sensors) from [18]. The specific microphone patterns were used to attenuate sound from a loudspeaker mounted above the microphone (in a null of the response pattern) while still capturing the voices of the teleconference participants

Schroeder's work on the problem became relevant again; in order to decorrelate the two signals, the method of adding a small frequency shift (with some slow time variation) to the speakers in a multichannel setup allows for the adaptive filters in the system to converge quickly [16]. Other methods of decorrelation, through nonlinearities or transform domain methods, have also been helpful in solving the problem [17].

The tools developed for concert hall measurements were also applied to problems incurred in small rooms for teleconferencing with considerable success. This leads to the need for microphones with special directional characteristics. The electret microphone opened the possibility to obtain small, inexpensive transducers with similar frequency and phase characteristics. This allows sensor arrays to form directional patterns to fit specific needs for reducing pickup of unwanted sounds such as noise.

Early work in this direction by Schroeder and others at Bell Labs led to the development of a differential microphone array with a constant toroidal pattern [18]. This allowed for the selective capture of speech from conference participants seated around an array and rejection of the audio from a loudspeaker mounted above the table. Figure 1.5 shows the measured sensitivities of three toroidal microphones constructed at Bell Labs. Further research at Bell Labs led to the development of other fixed directivity microphones with different polar patterns [19, 20].

At the time there was no possibility of digitally adjusting the multiple microphone signals, but today similar microphone array systems have evolved to be adaptable in real time to track multiple moving talkers and noise sources [21]. Through the use of beamforming, modern microphone arrays can remove noise without degrading speech quality, a problem in single-channel, statistical techniques.

Taking the differential array concept to its full extent, modern cylindrical and spherical microphone arrays are now under development [9]. Spherical arrays, such as the Eigenmike, can digitally control the microphone gains to create any beam shape (toroid, cardioid, hypercardioid, etc.) and steer the pattern in any direction [9]. Current implementations suffer from SNR issues at low frequencies due to mismatch between the array radius and wavelength, and aliasing issues at high frequencies due to microphone spacing larger than the wavelength [9, 22, 23].

1.4 Spatial Audio and Perception

Manfred Schroeder's interest in spatial audio and binaural perception began shortly after he joined Bell Labs in 1954. His work started with a series of papers examining a pseudo-stereophonic effect that could increase the spatial presence of a monaural signal by using a pair of comb filters [24]. This research later led to the development of a "colorless" artificial reverberation method still widely used in digital reverberation systems today [25].

Manfred's work on binaural perception continued with a research study on the subjective effects of transmission errors in two-channel stereo [26]. Shortly after this, Manfred and Bishnu Atal developed the crosstalk canceler, first described in a patent filed in 1962 [27]. A crosstalk canceler, shown in Fig. 1.6, is used to present binaural recordings (e.g., from a manikin head) for auralization over loudspeakers. Its aim is to transmit two independent signals from a set of loudspeakers to the two ears by applying filters that compensate for the acoustic crosstalk that happens when one loudspeaker's output is heard by the ear on the opposite side [28]. The crosstalk canceler was used in research systems through the 1970s. The most notable being Schroeder's comparative study of European concert halls at the University of Göttingen, where subjective preferences of many halls were correlated with objective measurements [29]. The result of this work showed that concert halls should be designed to provide dissimilar signals to each ear at every seat, necessitating diffusive surfaces on the ceiling in modern, wide halls.

Work on crosstalk cancellation systems continued at other research institutions with improvements for playback in non-anechoic spaces and extensions to systems with more than two loudspeakers [30–33]. Today, many home sound systems have this type of processing as a standard feature. While, binaural methods focus on presenting spatial sound for a single listener with a fixed head location, techniques

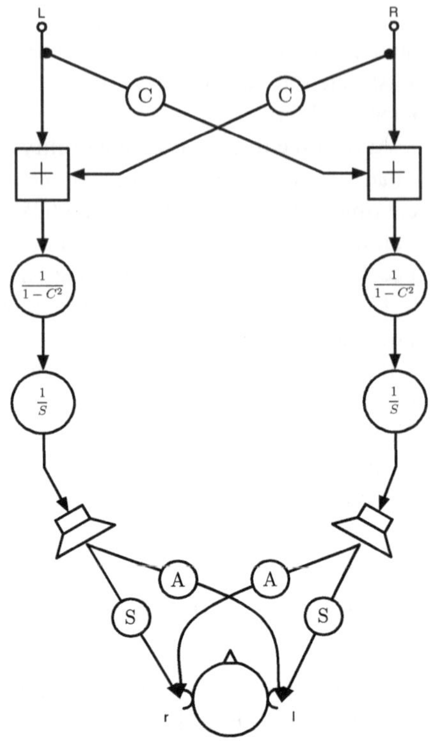

Fig. 1.6 The crosstalk canceler from [27]. The transfer functions A and S represent the ipsilateral and contralateral responses, respectively, and C represents the cancellation transfer function

for spatial sound reproduction for a group of listeners are also being developed under varying methodologies—Wave Field Synthesis [34], Ambisonics [35], and Vector Base Amplitude Panning [36]. Modern research in this area combines solutions to the wave equation in multiple dimensions with methods for solving inverse scattering problems. In many idealized cases the loudspeaker driving functions for an array can be solved analytically using simplifications to the Kirchhoff–Helmholtz integral [37] or a spatial harmonic solution to the wave equation [38]. However, whether solved numerically or analytically, attempts to recreate a given physical sound field for many listeners break down when the number of speakers is small. In many cases, only half of the system is being addressed; research on spatial hearing and psychoacoustics still has a large role to play in the future development of multichannel, multi-listener sound systems.

1.5 Summary

Today, we use many results from the early work on teleconferencing at Bell Labs in developing solutions for modern multichannel, immersive systems. Schroeder's work on modeling, simulating, and measuring room acoustics form the fundamentals of our understanding of sound behavior in rooms and how we develop and test new systems. His work on stereo reproduction and perception forms the basis for our motivation to explore multichannel systems that can present spatial auditory cues to participants. Early work on frequency shifting in public address systems was a first attempt at a problem that research is still focused on, multichannel acoustic echo cancellation. Directional microphone arrays, which are now digitally steerable, allow for a new direction to combat noise and echo in conferencing systems. While there is much work left to be done until immersive teleconferencing systems are equivalent face-to-face meetings, the systems we have today owe a lot to the focused work of Manfred Schroeder and many others at Bell Labs.

References

1. Schroeder, M., Nebeker, F.: Oral-History: Manfred Schroeder. IEEE History Center, Hoboken, NJ (1994)
2. Schroeder, M.R.: Digital-computer method of measuring concert-hall acoustics. J. Acoust. Soc. Am. **36**(5), 1011 (1964)
3. Atal, B.S., Schroeder, M.R., Sessler, G.M., West, J.E.: Evaluation of acoustic properties of enclosures by means of digital computers. J. Acoust. Soc. Am. **40**, 428 (1966)
4. Sessler, G.M., West, J.E.: Sound transmission over theatre seats. J. Acoust. Soc. Am. **36**(9), 1725–1732 (1964)
5. Atal, B.S., Schroeder, M.R.: Study of sound decay using ray-tracing techniques on a digital computer. J. Acoust. Soc. Am. **41**(6), 1598 (1967)
6. Schroeder, M.R.: New method of measuring reverberation time. J. Acoust. Soc. Am. **37**(6), 1187–1188 (1965)
7. Schroeder, M.: Integrated-impulse method measuring sound decay without using impulses. J. Acoust. Soc. Am. **66**(2), 497 (1979)
8. Farina, A.: Simultaneous Measurement of Impulse Response and Distortion with a Swept-Sine Technique. AES Convention, New York, NY (2000)
9. Meyer, J., Elko, G.: A highly scalable spherical microphone array based on an orthonormal decomposition of the soundfield. ICASSP **2002**(2), 1781–1784 (2002)
10. Merimaa, J.: Applications of a 3-D microphone array. In: AES Convention 112 (2002)
11. Gover, B.N., Ryan, J.G., Stinson, M.R.: Measurements of directional properties of reverberant sound fields in rooms using a spherical microphone array. J. Acoust. Soc. Am. **116**(4), 2138–2148 (2004)
12. O'Donovan, A., Duraiswami, R., Zotkin, D.: Imaging concert hall acoustics using visual and audio cameras. In: ICASSP 2008, pp. 5284–5287 (2008)
13. Harvey, F.K., Cisek, J.G., MacLean, D.J., Schroeder, M.R.: Some aspects of stereophony applicable to conference use. J. Audio Eng. Soc. **11**(3), 212–217 (1963)
14. Schroeder, M.R.: Improvement of acoustic-feedback stability by frequency shifting. J. Acoust. Soc. Am. **36**(9), 1718–1724 (1964)
15. Benesty, J., Morgan, D.R., Sondhi, M.M.: A better understanding and an improved solution to the problems of stereophonic acoustic echo cancellation. ICASSP **1997**(1), 303–306 (1997)
16. Herre, J., Buchner, H., Kellermann, W.: Acoustic echo cancellation for surround sound using perceptually motivated convergence enhancement. ICASSP **2007**(1), I-17–I-20 (2007)
17. Atkins, J.: Spatial Acoustic Signal Processing for Immersive Communication. Johns Hopkins University, Baltimore, MD (2011)
18. Sessler, G., West, J.E., Schroeder, M.R.: Toroidal microphones. J. Acoust. Soc. Am. **46**, 28 (1969)
19. Sessler, G.M., West, J.E., Kubli, R.A.: Unidirectional, second-order gradient microphone. J. Acoust. Soc. Am. **86**(6), 2063–2066 (1989)

20. Sessler, G.M., West, J.E.: Second-order gradient unidirectional microphones utilizing an electret transducer. J. Acoust. Soc. Am. **58**(1), 273–278 (1975)
21. Flanagan, J.L., Berkley, D.A., Elko, G.W., West, J.E., Sondhi, M.M.: Autodirective microphone systems. Acustica **73**(2), 58–71 (1991)
22. Atkins, J.: Optimal spatial sampling for spherical loudspeaker arrays. In: ICASSP 2010, pp. 97–100 (2010)
23. Atkins, J.: Robust beamforming and steering of arbitrary beam patterns using spherical arrays. In: IEEE WASPAA 2011 (2011)
24. Schroeder, M.R.: An artificial stereophonic effect obtained from using a single signal. In: AES Convention 9 (1957)
25. Schroeder, M.R., Logan, B.F.: 'Colorless' artificial reverberation. J. Audio Eng. Soc. **9**(3), 192–197 (1961)
26. Harvey, F.K., Schroeder, M.R.: Subjective evaluation of factors affecting 2-channel stereophony. J. Acoust. Soc. Am. **32**(11), 1521 (1960)
27. Atal, B.S., Schroeder, M.R.: Apparent sound source translator. US Patent 3,236,949, 1966
28. Schroeder, M.R.: Models of hearing. Proc. IEEE **63**(9), 1332–1350 (1975)
29. Schroeder, M.R., Gottlob, D., Siebrasse, K.F.: Comparative study of European concert halls: correlation of subjective preference with geometric and acoustic parameters. J. Acoust. Soc. Am. **56**, 1195 (1974)
30. Bauck, J., Cooper, D. H.: Generalized transaural stereo. In: AES Convention 93 (1992)
31. Ward, D.B., Elko, G.W.: Optimum loudspeaker spacing for robust crosstalk cancellation. ICASSP 1998 **6**, 3541–3544 (1998)
32. Kirkeby, O., Nelson, P.A., Hamada, H.: Local sound field reproduction using two closely spaced loudspeakers. J. Acoust. Soc. Am. **104**(4), 1973–1981 (1998)
33. Gardner, W.G.: 3-D Audio Using Loudspeakers. Massachusettes Institute of Technology, Cambridge, MA (1997)
34. Berkhout, A.J., de Vries, D., Vogel, P.: Acoustic control by wave field synthesis. J. Acoust. Soc. Am. **93**, 2764 (1993)
35. Gerzon, M.: Psychoacoustic decoders for multispeaker stereo and surround sound. In: AES 93rd Convention (1992)
36. Pulkki, V.: Virtual sound source positioning using vector base amplitude panning. J. Audio Eng. Soc. **45**(6), 456–466 (1997)
37. Spors, S., Rabenstein, R., Ahrens, J.: The theory of wave field synthesis revisited. In: AES Convention 124 (2008)
38. Poletti, M.A.: Three-dimensional surround sound systems based on spherical harmonics. J. Audio Eng. Soc. **53**(11), 1004–1025 (2005)

***

Biography

Joshua Atkins received the B.S. degree in Electrical Engineering with a minor in Computer Science from Johns Hopkins University in 2005. He joined the Electro-optics group at the Johns Hopkins University Applied Physics Lab and completed his M.S. in Electrical Engineering in 2006 with a dissertation on the polarization characteristics of fluorescence from biological aerosols. In 2011, he received his Ph.D. from the Department of Electrical and Computer Engineering at Johns Hopkins University for dissertation work on immersive communication systems. Since then, he has been working at Beats Electronics where he leads the speech and audio processing research group. His research interests include microphone and loudspeaker array design and signal processing, statistical and adaptive signal processing, models of human auditory perception, and musical and architectural acoustics.

James E. West is currently Research Professor at Johns Hopkins University, Department of Electrical and Computer Engineering and the Department of Mechanical Engineering. He was formally a Bell Laboratories Fellow, at Lucent Technologies.

Professor West holds more than 50 U.S. and 200 foreign patents on various microphones and techniques for making polymer electrets and transducers. He was inducted into The National Inventors Hall of Fame in 1999 for the invention of the electret microphone. West is a member of the National Academy of Engineering, a Fellow of IEEE and the Acoustical Society of America He is the recipient of the Acoustical Society of America's Silver Medal in Engineering Acoustics (1995), an honorary Doctor of Science degree from New Jersey Institute of Technology (1997). West was awarded the Mexican Institute of Acoustics' John William Strutt 3rd Baron of Raleigh Award (2003), the Acoustical Society of America's Gold Medal (2006), an honorary Doctor of Engineering from Michigan State University (2006), the National Medal of Technology (2006) and the Franklin Medal in Engineering (2010), an honorary Doctors degree from University of Pennsylvania(2013). Professor West has been active in creating and supporting programs aimed at improving diversity in the Science, Technology, Engineering and Mathematics (STEM) fields.

Chapter 2
The Early History of Ray Tracing in Acoustics

Asbjørn Krokstad, U. Peter Svensson, and Svein Strøm

Abstract Manfred Schroeder and Bishnu Atal are behind two seminal conference presentations on room acoustics, one in 1962 and the other in 1967. In the first, computerized auralization was outlined remarkably early, long before the term auralization was introduced. In the second, a ray tracing technique in room acoustics was presented. Independently, Asbjørn Krokstad and colleagues in Norway worked on the ray tracing technique, which lead to a journal paper in 1968. Here we describe some developments that lead up to these two breakthroughs for the use of computers in room acoustics.

2.1 Introduction

Bishnu Atal and Manfred Schroeder presented a paper titled "Study of sound decay using ray-tracing techniques on a digital computer" at the 73rd Meeting of the Acoustical Society of America in New York, April 1967, but only an abstract was published for this presentation [1]. In 1968, a similarly titled paper, "Calculating acoustical room response by use of a ray tracing technique" was published by Asbjørn Krokstad, Svein Strøm, and Svein Sørsdal in the relatively recent Journal of Sound and Vibration [2]. The latter was submitted in November in 1967, but no reference to the Atal and Schroeder presentation was made—since neither of the authors went to the ASA meeting. At the same time, Krokstad remembers the great inspiration from another conference paper by Schroeder, Atal, and Bird at the ICA congress in Copenhagen in 1962 on "Digital computers in room acoustics" [3]. Schroeder did eventually, in 1970, publish a journal paper titled "Digital simulation of sound transmission in reverberant spaces" in the Journal of the

A. Krokstad (✉)
Department of Electronics and Telecommunications, Norwegian University of Science and Technology, Trondheim, Norway
e-mail: asbjorn.krokstad@iet.ntnu.no

N. Xiang and G.M. Sessler (eds.), *Acoustics, Information, and Communication*, Modern Acoustics and Signal Processing, DOI 10.1007/978-3-319-05660-9_2,

Acoustical Society of America, submitted in February 1969, which described the ray-tracing technique but which did not cite the paper by Krokstad et al. [4]. Curious as this may seem to the researcher generation who has grown up with access to most publications right at the computer, in the times before efficient computer search engines it was quite a tedious work to look for all potentially relevant references.

Being the first published full paper, the paper by Krokstad et al. has become the standard work to cite for later studies on computerized prediction in room acoustics and has been a remarkably relevant paper to cite throughout more than 40 years. It was the first journal paper that presented the computerized ray-tracing technique for finding the impulse response, or rather echogram, in any three-dimensional model of a room. The reason that it has stayed relevant is that computerized prediction techniques still is an active research area, and such prediction still relies on the ray-tracing technique to a large extent. As an example, the two leading commercial softwares that are used by practitioners today, CATT Acoustic [5] and Odeon [6], use variants of the ray-tracing technique. In this paper, we will try to present some of the work in Trondheim and other places that lead up to the publishing of this paper. It will be a mix of a view from the inside (by the first and third author) and from the outside (by the second author). We will also see that the development was progressing in quite natural steps, yet at one point the development was perceived to have reached a mature enough stage that the "definitive" citation resulted. We will also see that in Trondheim there was an ongoing development of, and practical use for, the ray-tracing technique from the mid-1960s and onwards. The initial work by Atal and Schroeder, on the other hand, wasn't followed up to any larger degree, simply because there were several other highly significant developments going on, such as the linear prediction speech coder [7].

2.2 First Studies by Allred and Newhouse in 1958: Mean Free Path Calculations

In the late 1950s, researchers started to use computers quite widely for solving demanding numerical problems. The computers offered a new and more efficient tool to solve an existing problem, or to get more accurate results than had been possible before.

The field of room acoustics gives an interesting example of how the computers offered a straightforward improvement to existing numerical problems. Sabine's famous equation from 1898 states that the sound level in a room decays linearly with time, that is, an exponential decay for the sound pressure amplitude. The decay rate, or rather reverberation time, T_{60}, was determined by a very simple relationship, the first equality below,

$$T_{60} = 0.163\,\left[\mathrm{m}^{-1}\mathrm{s}\right]\frac{V}{\overline{\alpha}S} = 0.041\,\left[\mathrm{m}^{-1}\mathrm{s}\right]\frac{l_m}{\overline{\alpha}} \tag{2.1}$$

where V is the room volume (in m^3), $\overline{\alpha}$ is the average absorption coefficient, and S is the total surface area of the room (in m^2). Researchers have been kept busy ever

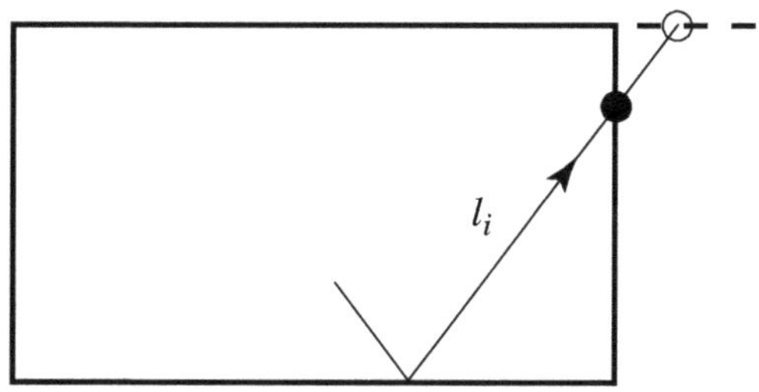

Fig. 2.1 Illustration of the Allred and Newhouse algorithm from 1958 [9]. Their paper studied three-dimensional shoe-box-shaped rooms; here a two-dimensional version is drawn. Each ray was reflected specularly and the next hit point was chosen as the closest of the three (two for the two-dimensional version) possible cross-points with other planes. The data sought were the lengths of the free paths, l_i

since studying to which degree this simple formula is true in various room geometry and absorption distribution cases. This relationship could also be written on a form which involves the so-called mean free path length in a room, that is, the average length that a sound wave can travel between wall hits, l_m (in m), and that is the second equality above.

This mean free path length had been established theoretically for some simple room shapes but novel approaches were used to find the mean free path in rooms of other shapes. In his book from 1932, *Architectural Acoustics*, Vern Knudsen described experiments where they used light rays in a scale model to figure out where sound waves would hit during several consecutive specular reflections [8]. By marking hit points on walls and measuring the corresponding lengths, he could collect data on these path lengths. Obviously, the accuracy was quite limited but he did find that quite different room shapes seemed to give the same average path length, or "mean free path." Such optical techniques in scale models continued to be in use, but also drawings, a pen and a ruler offered similar possibilities.

A first step towards computerized ray tracing was taken by collecting data on the path lengths and the collision probabilities of the walls using a computerized Monte Carlo technique by Allred and Newhouse in 1958 [9]. They studied rectangular/shoe-box-shaped rooms and Fig. 2.1 illustrates that the algorithm was quite straightforward. The algorithm generated rays in a number of randomized directions from one source position. For each ray, its path was followed through a succession of specular reflections. Data on path lengths and which reflecting walls that were hit were stored. Allred and Newhouse did not specify what kind of computer they used, or calculation times, but they could follow ten consecutive reflections for each ray, and they used 150 rays that were emitted in random directions.

This computerized calculation method gave accurate values for the mean free path in shoe-box rooms of different room dimension aspect ratios, and showed that there will be different reverberation times in rooms with different ratios. Two follow-up papers pointed out two errors in the original paper: the authors themselves found a programming error that gave around 10 % too long rev. times in the first paper [10]. Then Hunt, in 1964, showed that they had used a randomization of the source emission directions which did not correspond to an isotropic sampling of the sound field [11].

Allred and Newhouse did outline an extension of their work in [9]:

> The Monte Carlo method of machine computation is shown to be applicable to the evaluation of room acoustics. Its extension to the study of reverberation time is straightforward, and air absorption, as well as frequency dependence of air and absorbers, can also easily be taken into account. .... It would be a simple matter also to vary the absorption coefficient as a function of position, either discretely or continuously, ...
>
> The method is also applicable to the study of coupled rooms and rooms of irregular shape, such as auditoria. The complications are geometrical, and are of degree rather than kind as referred to our parallelopiped calculations.

Apparently, they did not follow this path of development, and the reverberation time was their focus also for future directions of the research.

2.3 Schroeder at ICA in 1962: The Path Towards Auralization

A visionary conference paper by Schroeder, Atal, and Bird was presented at the ICA in Copenhagen in 1962 titled "Digital computers in room acoustics" [3], as mentioned above. That paper was remarkable in that it laid out the methodology for what much later was to be called auralization, that is, the technique for creating audible computerized simulations of the sound in a room [12]. The auralization process was still in the early 1990s a computationally very demanding process but Schroeder and colleagues had identified the necessary steps:

The use of computerized convolution between a (simplified) impulse response and anechoic music.

The cross-talk cancellation technique for the presentation of binaural sound over two loudspeakers in an anechoic room, which can give the impression of sound incidence from any direction.

The development of a natural sounding reverberation unit without needing to carry out a convolution with a huge number of discrete reflections.

Interestingly, Schroeder mentioned that discrete echoes can be handled by discrete delay units but otherwise did not describe the computation of a detailed impulse response or how these discrete echoes should be found. Rather more focus was given to the use of computers for implementing Schroeder's earlier (1954) seminal work on the statistics of frequency response functions [13].

2.4 Developments in Oslo 1965–1967: Diffuseness of Reverberation Rooms with Hanging Reflectors

At the University of Oslo, Department of Physics, there were two professors who were both interested in room acoustics, but using different approaches: Johan Holtsmark, who published a paper titled "Reverberation as a stochastic process"

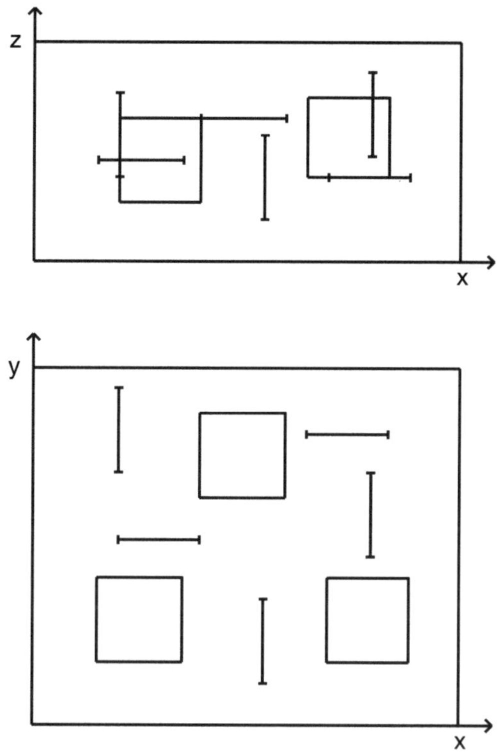

Fig. 2.2 Illustration of a shoe-box-shaped room with free-hanging reflectors. Reproduction from [15] where reflectors parallel to the walls were used

in 1966 [14]; his colleague Wilhelm Løchstøer was also interested in the reverberation process and supervised a Master's student Nikolai Stenseng who published his master thesis in January 1965 titled "The influence of diffusers on sound absorption measurements of a small volume" (translation of the Norwegian title) [15]. This thesis by Stenseng was a gradual development from the previous studies at that time in that it implemented a ray-tracing technique for a 3D room, albeit shoe-box-shaped, with free-hanging diffusors. These diffusers were always parallel to one of the walls, which made the computation of obstruction checks somewhat easier. Since reverberation rooms typically have smooth and flat walls and reflectors, only specular reflections were implemented and the need for something else was not discussed either. The program was written in the language Fortran IV for the computer UNIVAC 1107 (which had been delivered to the Norwegian Computing Center in Oslo in August 1963, after having been introduced in the USA in October 1962). An illustration from the thesis by Stenseng shows the parallel diffusers, see Fig. 2.2. Stenseng had spent a period at the NATO underwater research centre in La Spezia in Italy in the early 1960s, and remember being introduced to the power of the computers for the first time during that visit. When he returned to Oslo, he took one of the first programming courses offered, and when he started his master thesis, the idea to use a computer to simulate the propagation of light rays came quite naturally.

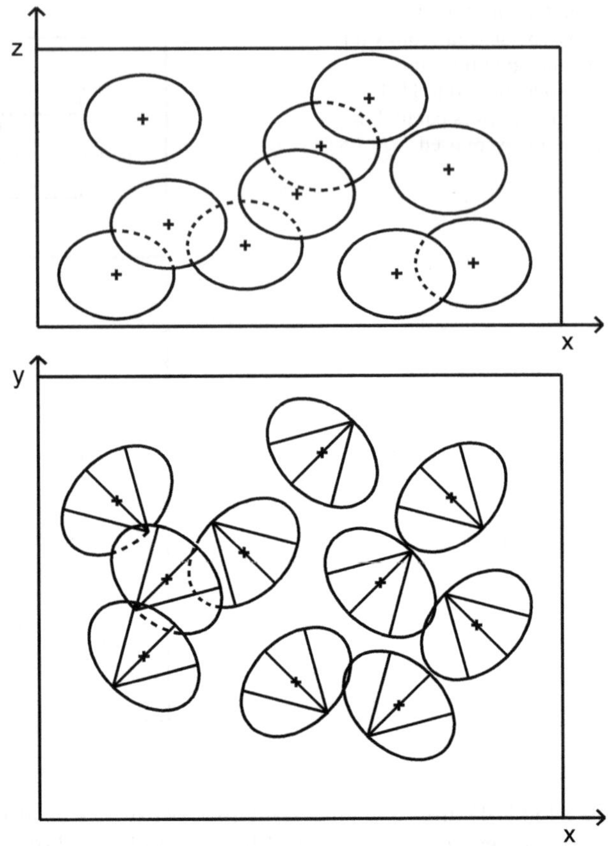

Fig. 2.3 Illustration of room with oblique reflectors (reproduced from [16])

An extension of this study by Stenseng was the master thesis by Svein Strøm in 1967, on "The diffusivity in small reverberation rooms with and without diffusers" (translation of the Norwegian title), also under Løchstøer in Oslo [16]. Strøm expanded Stenseng's study by implementing oblique reflectors, and also spherical reflectors. Furthermore, Strøm also computed actual decay curves, rather than collecting statistics data on path lengths. The purpose was still, however, to study the reverberation time. Figure 2.3 shows an example of a room with reflectors from [16].

2.5 Developments in Trondheim, Leading up to the Paper in 1968

Asbjørn Krokstad was hired as a research assistant at the Norwegian Technical Institute in Trondheim in 1956, and got the task to manage the specification of a new anechoic chamber, which was finished in 1958.[1] Soon after that, he got the further task to evaluate the performance of this new anechoic chamber. Since the chamber was rather small, compared to the recently finished one in Lyngby, Denmark, the classical free-field distance law criterion was difficult to apply, and Krokstad wanted to study the influence of the first-order reflections represented by image sources. The Institute got its first computer, a "GIER" from Denmark, in late 1962. Krokstad took the first course offered in Trondheim on the ALGOL programming language in 1963 (500 students and employees took this first course on computer programming!). During the course he started to use computer calculations to quantify the influence of reflections. The work on evaluating the anechoic chamber [17] later became part of his Ph.D. degree in 1963, which had Wilhelm Løchstøer and Gordon Flottorp, both from Oslo, as opponents (the supervisor of Krokstad was Reno Berg, associate professor in acoustics in Trondheim since 1935). As mentioned above Løchstøer was associate professor at the University of Oslo, Department of Physics, specializing on acoustics, whereas Flottorp was chief audiologist at the Oslo University Hospital. Curiously, this trio of supervisor and opponents were the three founders of the Acoustical Society of Norway in 1955.[2]

Krokstad got inspired by the successful application of the first-order image source method to the anechoic room and wanted to extend this to more general room shapes (and less absorbing ones too). Beranek had published his classical book "Music, acoustics & architecture" in 1962, and the book contained simplified plans for 100 of the most famous concert halls in the world [19]. This book gave the possibility to get input data for further studies, and furthermore, a possibly crucial event took place in Norway around this time: a new concert hall, "Grieghallen," was being planned in Bergen. An architectural bidding process was launched with a deadline in February 1965 and the winning design was made by the Danish architect Knud Munk. Krokstad naturally got very interested in studying the suggested design of the new hall.

The acoustics group in Trondheim, which had a brand new laboratory, opened in March 1965 [20], with Krokstad as director, hired Svein Sørsdal as a computer-savvy summer student, and the two of them tried to implement an image source

[1] The anechoic chamber in Trondheim still, in 2014, stands as it did in 1958 (but is rather ready for renovation). It has the quite unusual feature of graphite-covered wool wedges in order to improve the anechoic properties for electromagnetic waves.

[2] Reno Berg was also one of the founders of the *Nordic Acoustical Society*, founded on 19 June 1954 which happened to be the day of the 25th anniversary of the Acoustical Society of America. A telegram was sent by the "newly born son" to the "grandfather" ASA, as referred to in [18].

visibility test for general 3D geometries on the GIER computer that they both had experience with. This failed, however, since the visibility tests turned out to be too complex. They changed horses, so to say, aiming at using ray tracing instead and readily got in touch with Stenseng, and later Strøm, who were both in Oslo. At the same time, the Institute got their next type of computer, a UNIVAC 1107, in the end of 1965, after the GEIR computer had been running non-stop for a few years.

Strøm moved to Trondheim after he had finished his master thesis in early 1967 and joined the simulation project. His Fortran implementation of a 3D visibility test turned out to be exactly what was needed, and was combined with the ALGOL program developed earlier. Sørsdal implemented the main program structure, routines for reading geometry data, as well as for post-processing the result for each ray. A visualization technique was implemented where each ray's hit point in the audience area was plotted, with a little tail indicating the incidence angle onto a receiving surface. A Kingmatic plotter had been acquired by the Institute in 1966 and it made such visualizations possible.

Strøm carried out most of the actual calculations, and used the program as the Trondheim group were acoustics consultant in a sequence of projects: the "Hjertnes Kino og Kulturhus" in Sandefjord, opened in 1975; Grieghallen in Bergen, opened in 1978; and Oslo concert hall, opened in 1977. Prior to these projects, several of the halls in Beranek's book were studied and published in a report (in Norwegian) in 1971 [21]. The Grieghallen project had got delayed quite severely after the design competition had been settled in 1965. Locally in Bergen, Helmer Dahl, research director at the Christian Michelsen Research Institute, had got involved in the plans to build the new concert hall, and had tried to use the old lightbeam method of Vern Knudsen for studying different room shapes. These attempts had failed and Dahl therefore had contacted Krokstad who had the ray-tracing program up and running.

Svein Strøm describes that the small Trondheim group was hired as advisors for the Grieghallen project in Bergen [22]. The result was that the architect had to accept the suggestions from Trondheim for the orchestra enclosure, the ceiling shape, the sidewall reflectors, and for large sound reflecting objects that were hung from the ceiling along the sidewalls (Fig. 2.4). The exciting trial concert on the 12th of May, 1978, showed that the project was successful. Ten years later the local newspaper, *Bergens Tidende*, confirmed this in an article about the hall.

The journal paper was finished and submitted in November 1967. In order to develop a functioning engineering tool, they had taken several steps from the previous state of the art where ray length histories could be collected. In addition to the handling of arbitrary 3D geometries (admittedly with small numbers of polygons: up to 99 corners and 50 polygons per symmetric half of a room; and typically with around 2,000 emitted rays [21]), Krokstad et al. introduced some refinements [23]:

1. Some areas were marked as audience surfaces, and ray histories were collected for these. Also smaller sub-areas were introduced in order to study the sound field distribution over the audience area, see Fig. 2.5 (from [2]).

Fig. 2.4 Grieghallen in Bergen, one of the first concert halls where computerized ray tracing was used as a design tool

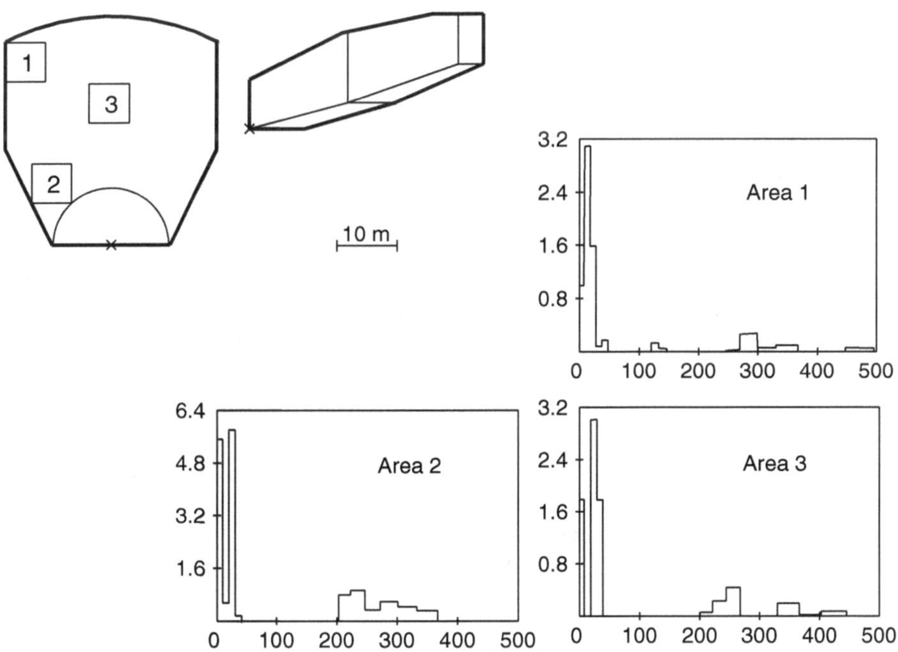

Fig. 2.5 Illustration of an irregularly shaped room, and computed echograms for three different receiver patches (reproduced from [2])

2. A kind of echogram was collected for each such audience surface, see Fig. 2.5. These resulted, however, simply from counting the number of hits per discrete time interval.
3. Different absorption coefficients were introduced—sort of. Surfaces were classified as totally reflecting, totally absorbing, or audience (which were also considered as totally absorbing). The motivation was that the room shape determined, to a large degree, the amount of early energy and its spatial distribution.

There were clearly some simplifications compared to what is common today, but some of these were commented by the authors in [2].

On the topic of specular reflections: "However, in the actual design of a hall, one may also wish to know the effects of changes in details, such as the introduction of diffusing wall elements. Even if such details can be included in the mathematical model of a hall without causing excessive computing time, the validity range of the results may be difficult to state." It was, however, not indicated how to handle diffuse reflections.

On the topic of absorption: "As mentioned in Sect. 2.2, the surfaces of the room are considered to be either totally reflective or totally absorbent. Other absorption characteristics of the surfaces, together with air damping, may readily be included in the computations. The authors believe that these factors are of minor interest in view of the building materials commonly used in concert halls."

2.6 Schroeder at the ASA Meeting in 1967: Computing Decays in Generally Shaped Convex Rooms

After the Allred and Newhouse study, and some subsequent theoretical discussion papers on mean free paths, other studies were looking into the computation of reverberation time (rather than mean free path and collision frequencies). At each reflection, the ray's energy was decreased by the wall reflection factor. Then, each ray gave a stepwisely decreasing contribution, and averaging across many of them gave a smooth exponentially decaying curve. This extension to the ray-tracing technique was presented by Atal and Schroeder at the ASA meeting paper in April 1967 [1], completely independently of the work in Norway. ASA meetings don't publish any papers, so all that is available is an abstract, plus some example results in two later papers by Schroeder in 1969 and 1970 [4, 24]. It did seem like Atal and Schroeder had a functioning program for general geometries, see Fig. 2.6, however, only for two dimensions.[3]

[3] In a paper by Wayman and Vanyo in 1977 [25], the quite straightforward extension of Schroeder's approach from two to three dimensions was presented, with no reference to the 1968 paper by Krokstad et al.

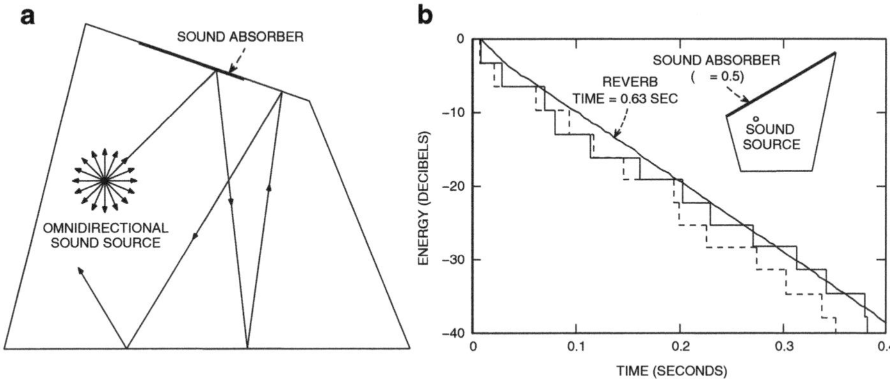

Fig. 2.6 Illustration of how the ray-tracing technique was used to gather individual ray-decay curves, which could be averaged to a smooth exponential decay (reproduced from [24])

The Atal and Schroeder algorithm was presented in the JASA paper in 1970 and their implementation clearly handled various degrees of absorption coefficients, as opposed to the approach by Krokstad et al. Furthermore, Atal and Schroeder specified the handling of diffuse reflections: "These rays are traced on the computer and are reflected from the walls either specularly or according to some specified random law." It seems to have been implemented as well, according to a further comment in the paper: "The discrepancies found for rooms with randomly reflecting walls and 'suspended' diffusing elements were generally, although not always, somewhat smaller."

On the other hand, Atal and Schroeder did not study the distribution of received sound across any audience surface; instead the global decay curve was studied, see Fig. 2.6. An interesting detail of the 1970 JASA paper by Schroeder is its reference number 16, to a paper by Atal and Schroeder in JASA which is marked as "To be published," but it never seems to have been published.

2.7 Later Developments of Ray Tracing in Room Acoustics

After the paper by Krokstad et al. was published in 1968, it appeared as if the paper was unknown for some time, at least to American researchers. In 1973, a paper published by Haviland and Thanedar outlined a method to compute the detailed response in one specific receiver location using ray tracing—but only for rectangular rooms [26]. This paper cited the Schroeder 1967 ASA abstract but not the 1968 paper by Krokstad et al. Another paper, published in 1977, that also seemed to be unaware of the Krokstad paper, has been mentioned above [25]. On the other hand, citations of the Krokstad paper appeared in 1971 [27], 1973 [28], and later. Krokstad et al. summarized the work in Trondheim in the paper titled "Fifteen years' experience with computerized ray tracing" [29].

Of special relevance is the paper by Kuttruff in 1971 [27]. That paper presented an integral equation formulation (later, independent versions of this became known as the radiosity method), and a discrete Monte Carlo-type solution where narrow beams where traced until they hit a wall, and then they were reflected in a direction generated by two random numbers so as to simulate a Lambert-like reflection. Probably, the beam was represented by a single center ray. From the paper: "In der so festgelegten Richtung läuft das Teilchen also wieder in den Raum, wird erneut an einer Wand gestreut usw., bis es schliesslich einmal auf die absorbierende Wand trifft, die seiner Wanderung ein Ende setzt.", or (our translation) "The narrow beam thus propagates through the room in the determined direction, hits another wall where it is scattered, etc., until it finally hits the absorbing wall, which will end its propagation."

A lot of research has been done to develop the ray-tracing technique further. Much work has studied accuracy issues since ray tracing is inherently a stochastic process (as long as there is at least one partly diffusing surface area in a room). In addition, there are surface-sampling issues for the deterministic part of the process. The so-called cone tracing was presented by Van Maercke et al. in 1986, replacing thin rays (hitting a receiver sphere) by propagating cones (hitting a receiver point) [30]. An important development was presented by Vorländer in 1989 with the so-called hybrid technique: ray tracing was used to find possible reflection paths but in a subsequent phase, the ray-tracing-identified specular reflection paths were replaced by their image source equivalents. An advantage was that the image source method gives exact reflection paths, which can be used for improved accuracy in the early path of the impulse response [31]. This linking between ray tracing and the image source methods was also explored through the beam-tracing technique presented by Walsh already in 1980, where ray bundles were treated as coherent "beams" [32]. This beam tracing was shown by Stephenson to facilitate the inclusion of diffraction and developments to avoid its computation time explosion [33]. Also, Funkhouser et al. have demonstrated very efficient implementations of beam tracing via special geometry structures [34].

A step towards avoiding the "either specular or diffuse reflection" approach in ray tracing was taken by Dalenbäck where a method let reflections generate both specular and diffuse reflections while avoiding an exponential growth in path number tracing as function of reflection order [35]. The ray-tracing technique was combined with another algorithm, radiosity, by Lewers [36]. Later, the Odeon software employed components from radiosity [37] in their implementation. In fact, ray tracing in a room with only diffusely reflecting walls, is a discretized/ Monte Carlo-type solution of the integral equation in radiosity, which was presented by Kuttruff in 1971 as described above [27].

2.8 Ray Tracing in Other Fields

So, was room acoustics the pioneer field for the computerized ray-tracing technique? Not really; as early as 1954 examples from optics were published. A paper titled "Ray tracing on the Manchester university electronic computing machine" by

Black was published in the British Proceedings of the Physics Society [38]. This was not so surprising—the term ray tracing does after all come from optics, and ray tracing in optics is typically used to study the refraction through optical lenses. Compared to room acoustics the "direct wave" is dominating, and studied in detail, while reflections might be ignored. On the other hand, the medium is refractive so that the rays do not travel along straight lines.

In underwater acoustics, early work was apparently done using ray tracing in the US Navy. The abstract in a 1956 report by Anderson and Peterson claims that "This report discusses: (1) an acoustic intensity program for estimating convergence zone propagation loss using ray theory" [39]. More publically, in an abstract from an ASA Meeting in November 1961, Norris claimed that "A ray-tracing program which has been developed for the IBM 704 and 7090 computers yields results which show excellent agreement with experimental data. The program permits ray computations for arbitrary source and receiver depths, bottom profiles, and velocity structures. The tabulated results give actual ray trajectories as well as the geometrical spreading associated with each path." [40].

In underwater acoustics, a central problem is that the medium is not homogeneous, which leads to that straight-line rays cannot be used (similar to in optics). On the other hand, geometries are practically always modeled as 2D, with a flat top surface (the sea surface) and a deterministically or stochastically shaped bottom profile. Multiple reflections are studied but with more restricted geometries and lower orders of reflection. Of curious interest is also early ray-tracing work on so-called analog computers—electronic circuits that performed calculations such as integrations and differentiations, by, e.g., Graber et al. in 1961 [41].

Ray tracing has more recently been developed heavily in computer graphics. A paper by Whitted in 1980 is considered as an important foundation of ray tracing in computer graphics [42], but an earlier paper from IBM in 1969 is considered as the original citation [43]. This kind of ray tracing was similar to the one in acoustics, except that rays were "shot" from a receiver's eye, through a grid of pixels that represent the display, reflected off surfaces, and finally reaching light sources. The more acoustics-like approach of emitting rays from (light) sources is usually called "global illumination" in graphics. Because of the generally large interest in computer graphics, many researchers are active in this field and the techniques are developed rapidly. In addition, the hardware for computer graphics generation (graphics processing units, or GPUs) has recently been developing at a faster pace than general CPUs for computers. This has lead to that so-called general-purpose GPUs (GPGPU) can be used as numerical coprocessors for many kinds of numerical calculations with processing power many times higher than CPUs [44]. Acoustical ray tracing has been demonstrated on such GPGPUs [45, 46].

A last field where ray tracing has become a common tool is the study of radio wave propagation, indoors as well as outdoors in city environments. This application applies specular reflections (as in room acoustics) as well as transmission, and of multiple orders. An early paper on this technique was published in 1991 [47]. It could be noted that diffraction over edges has been employed in radio propagation studies for a long time, either based on classical Fresnel diffraction, or based on the high-frequency asymptotic geometrical theory of diffraction (GTD) that was presented by Keller in 1962 [48].

2.9 Concluding Remarks

The longevity of the ray-tracing technique, and the number of fields that it is applied to, indicates a very flexible method which can handle many phenomena. Ray tracing is based on geometrical acoustics, and equivalent approximations to true wave fields. Consequently, there will always be wave-related aspects that are more or less difficult to represent but the concept of edge diffraction has been implemented to some degree.

This overview has clearly had a focus on the development in Norway, which we know the best, and on the early work by Atal and Schroeder. Early attempts that have been published through more or less easily available channels might certainly have slipped our attention, and we humbly apologize if this is the case. We think that the development of the ray-tracing technique in room acoustics is a nice example of quite a gradual development process where several steps have been taken by different researchers, at different places, and thereby reached a kind of breakthrough.

Acknowledgements The authors are grateful to discussions with Svein Sørsdal, Nikolai Stenseng, and Prof. Bishnu Atal for sharing their memories from their early work on ray tracing.

References

1. Atal, B., Schroeder, M.R.: Study of sound decay using ray-tracing techniques on a digital computer. J. Acoust. Soc. Am. **41**, 1598 (1967)
2. Krokstad, A., Strøm, S., Sørsdal, S.: Calculating the acoustical room response by the use of a ray tracing technique. J. Sound Vib. **8**, 118–125 (1968)
3. Schroeder, M.R., Atal, B.S., Bird, C.: Digital computers in room acoustics. In Proceedings of the 4th International Congress on Acoustics, Copenhagen (1962)
4. Schroeder, M.R.: Digital simulation of sound transmission in reverberant spaces. J. Acoust. Soc. Am. **47**, 424–431 (1970)
5. CATT Acoustic. www.catt.se
6. Odeon. www.odeon.dk
7. Atal, B.S.: The history of linear prediction. IEEE Signal Processing Magazine, Match 2006, pp. 154–161 (2006)
8. Knudsen, V.: Architectural Acoustics, pp. 133–141. Wiley, New York, NY (1932)
9. Allred, J., Newhouse, A.: Applications of the Monte Carlo method to architectural acoustics. J. Acoust. Soc. Am. **30**, 1–3 (1958)
10. Allred, J., Newhouse, A.: Applications of the Monte Carlo method to architectural acoustics. II. J. Acoust. Soc. Am. **30**, 903–904 (1958)
11. Hunt, F.V.: Remarks on mean free path problem. J. Acoust. Soc. Am. **36**, 556–564 (1964)

12. Kleiner, M., Dalenbäck, B.-I., Svensson, P.: Auralization—an overview. J. Audio Eng. Soc. **41**, 861–875 (1993)
13. Schroeder, M.R.: Die Statistichen Parameter der Frequenz Kurven von Grossen Räumen. Acustica **4**, 594–600 (1954)
14. Holtsmark, J.: Reverberation as a stochastic process. Phys. Norv. **2**, 63–70 (1966)
15. Stenseng, N.: Diffusorers innflytelse på lydabsorbsjonsmålinger i et klangrom med lite volum. M.Sc. Thesis, University of Oslo, Department of Physics (in Norwegian) (1965)
16. Strøm, S.: Diffusitet i små klangrom med og uten diffusorer. M.Sc. Thesis, University of Oslo, Department of Physics (in Norwegian) (1967)
17. Krokstad, A.: Design and construction of an anechoic chamber. Report. Department of Electrical Engineering, Technical University of Norway, Trondheim, (1964)
18. Firestone, F.A.: The twenty-fifth anniversary celebration. J. Acoust. Soc. Am. 26, 879 (1954)
19. Beranek, L.L.: Music, Acoustics and Architecture. Wiley, New York, NY (1962)
20. Nordahl, O.: Akustisk Laboratorium ved NTH/SINTEF. Master Thesis, Department of History, NTNU, Trondheim (in Norwegian) (2005)
21. Strøm, S.: Fordeling av reflektert lyd i modeller av eksisterende konsertsaler, Report LBA 381. Akustisk Laboratorium, SINTEF, Trondheim (1971) (in Norwegian)
22. Strøm, S.: A look back on 40 years of cooperation with Asbjørn on room acoustics consulting for multi-purpose halls. In: Svensson, P. (ed.) Reflections on Sound. In Honour of Professor Emeritus Asbjørn Krokstad. Norwegian University of Science and Technology, Department of Electronic and Telecommunication, Trondheim (2008)
23. Krokstad, A.: The hundred years cycle in room acoustic research and design. In: Svensson, P. (ed.) Reflections on Sound. In honour of Professor Emeritus Asbjørn Krokstad. Norwegian University of Science and Technology, Department of Electronics and Telecommunication, Trondheim (2008)
24. Schroeder, M.R.: Computers in acoustics: symbiosis of an old science and a new tool. J. Acoust. Soc. Am. **45**, 1077–1088 (1969)
25. Wayman, J., Vanyo, J.: Three-dimensional computer simulation of reverberation in an enclosure. J. Acoust. Soc. Am. **62**, 213–215 (1977)
26. Haviland, J., Thanedar, B.: Monte Carlo applications to acoustical field solutions. J. Acoust. Soc. Am. **54**, 1442–1448 (1973)
27. Kuttruff, H.: Simulierte Nachhallkurven in Rechteckräumen mit diffusem Schallfeld. Acustica **25**, 333–342 (1971)
28. Juricic, H., Santon, F.: Images et rayons sonores dans le calcul numerique des echogrammes. Acustica **28**, 77–89 (1973)
29. Krokstad, A., Strøm, S., Sørsdal, S.: Fifteen years' experience with computerized ray tracing. Appl. Acoust. **16**, 291–312 (1983)
30. Van Maercke, D.: Simulation of sound fields in time and frequency domain using a geometrical model. In Proceedings of 12th International Congress on Acoustics, Toronto, paper E11-7 (1986)
31. Vorländer, M.: Simulation of the transient and steady-state sound propagation in rooms using a new combined ray-tracing/image-source algorithm. J. Acoust. Soc. Am. **86**, 172–178 (1989)
32. Walsh, J.P.: The design of Godot: a system for computer-aided room acoustics modeling and simulation: In: Proceedings of 10th International Congress on Acoustics, Sydney, Paper E15.3 (1980)
33. Stephenson, U.: Quantized pyramidal beam tracing—a new algorithm for room acoustics and noise immission prognosis. Acta Acustica united with Acustica **82**, 517–525 (1996)
34. Funkhouser, T., Tsingos, N., Carlbom, I., Elko, F., Sondhi, M., West, J.E., Pingali, G., Min, P., Ngan, A.: A beam tracing method for interactive architectural acoustics. J. Acoust. Soc. Am. **115**, 739–756 (2004)
35. Dalenbäck, B.-I.: Room acoustic prediction based on a unified treatment of diffuse and specular reflection. J. Acoust. Soc. Am. **100**, 899–909 (1996)

36. Lewers, T.: A combined beam tracing and radiant exchange computer model of room acoustics. Appl. Acoust. **38**, 161–178 (1993)
37. Naylor, G.: Treatment of early and late reflections in a hybrid computer model for room acoustics. J. Acoust. Soc. Am. **92**, 2345 (1992)
38. Black, G.: Ray tracing on the Manchester university electronic computing machine. Proc. Phys. Soc. Sect. B **67**, 569–574 (1954)
39. Anderson, E.R., Pedersen, M.A.: Application of Computer Techniques to the Prediction of Underwater Sound Propagation, Report ADA066760. Navy Electronics Lab, San Diego, CA (1956)
40. Norris, A.: Ray-tracing program for high speed digital computers. J. Acoust. Soc. Am. **33**, 1659 (1961)
41. Graber, G., Rubin, A.: Underwater acoustic ray analysis. J. Acoust. Soc. Am. **33**, 840 (1961)
42. Whitted, T.: An improved illumination model for shaded display. Commun. ACM **23**, 343–349 (1980)
43. Appel, A.: Some techniques for shading machine renderings of solids. In: Proceedings of the April 30–May 2, 1968, Spring Joint Computer Conference, AFIPS '68, pp. 37–45 (1969)
44. www.gpgpu.org
45. Jedrzejewski, M., Marasek, K.: Computation of room acoustics using programmable video hardware. In: Wojciechowski, K. (ed.) Computer Vision and Graphics, Proceedings of International Conference ICCVG 2004, Warsaw, Poland, September 2004. Springer Netherlands, pp. 587–592 (2006)
46. Tsingos, N., Dachsbacher, C., Lefebvre, S., Dellepiane, M.: Instant sound scattering. In Kautz, J., Pattanaik, S. (eds.) Proceedings of the 18th Eurographics Symposium on Rendering 2007 (2007)
47. McKown, J.W., Hamilton, R., Lee, J.: Ray tracing as a design tool for radio networks. IEEE Netw. Mag. **5**, 27–30 (1991)
48. Keller, J.B.: Geometrical theory of diffraction. J. Opt. Soc. Am. **52**, 116–130 (1962)

***

Biography

Dr. **Asbjørn Krokstad** is a professor emeritus at the Norwegian University of Science and Technology, Trondheim, Norway, since 1998. He has a Ph.D. from the Norwegian Institute of Technology, Trondheim, in 1962, and was professor from 1970. He did research on computational room acoustics, with the first paper on ray tracing in 1968, reverberation enhancement systems, and was behind the development of one of the world's first digital hearing aid prototypes in 1989, and holds several patents related to this. He was president of the Norwegian acoustical society 1994–1998, and the general chairman of the ICA in Trondheim 1995. He has been an active musician and conductor throughout his career.

Dr. **U. Peter Svensson** is a professor of electro-acoustics at the Norwegian University of Science and Technology, Trondheim, Norway, since 1999. He has a Ph.D. from Chalmers University of Technology, Gothenburg, Sweden, in 1994. His research has dealt with auralization, especially computational methods involving diffraction modeling and loudspeaker reproduction techniques. He has also worked on beamforming techniques for microphone arrays, measurement techniques, reverberation enhancement, and interaction over Internet/video conferencing. He has been on the boards of the acoustical societies of Sweden, Norway, and the European Acoustics Association.

Svein Strøm is a senior adviser at COWI AS, Trondheim, Norway. He has an M.Sc. in physics from the University of Oslo in 1967. He worked at the ELAB/SINTEF in Trondheim with room acoustics research together with Dr. Krokstad until 1994. He did his M.Sc. thesis on the ray-tracing technique, which paved the way for the paper by Krokstad, Strøm, and Sørsdal in 1968. Strøm has worked with room acoustical design throughout Norway, including many examples of passive and active variable acoustics installations. Together with Prof. Krokstad, he was the acoustical consultant for the concert hall in Trondheim, Olavshallen, which was opened in 1992.

Chapter 3
Advanced Room-Acoustics Decay Analysis

Ning Xiang

Abstract Schroeder's integration method broke the new ground in classical architectural acoustics. Schroeder's integration method yields sound energy decay functions from room impulse responses. For reverberation time estimation based on a parametric model, a nonlinear regression method has been proposed. The nonlinear regression method yields reverberation time estimates, insensitive to background noise and the upper limit of integration. Recent interest in acoustically coupled-volume systems has prompted new challenges in analyzing sound energy decay characteristics, which are more complicated than just single-rate decays. This chapter will demonstrate a suitable framework for this room-acoustics application using Bayesian inference and the Schroeder's integration as the foundation of the advanced model-based energy decay analysis. Based on the Schroeder's integration, two levels of inference, decay order selection and decay parameter estimation, are discussed in details.

3.1 Introduction

In 1965, Manfred R. Schroeder published his seminal paper [1] on a new method of measuring reverberation time. The paper presents a novel method of obtaining sound energy decays in terms of integration of energy room impulse responses, *Schroeder integration*. The Schroeder integration method broke new ground in classical room-acoustics, and has since received wide acceptance in the architectural acoustics community. The sound energy decay function obtained by Schroeder integration is termed *Schroeder decay function*, or simply, *Schroeder curve* when

N. Xiang (✉)
Graduate Program in Architectural Acoustics School of Architecture, Rensselaer Polytechnic Institute, Troy, NY 12180, USA
e-mail: xiangn@rpi.edu

N. Xiang and G.M. Sessler (eds.), *Acoustics, Information, and Communication*, Modern Acoustics and Signal Processing, DOI 10.1007/978-3-319-05660-9_3,

presented graphically. This chapter discusses a parametric model derived from Schroeder integration, which has been proven to fully describe Schroeder decay functions. The parametric model plays a central role within the scope of advanced sound energy decay analysis since no characteristics can be inferred from experimental measurements for the relevant features without an appropriate model. Based on the Schroeder decay model, this chapter introduces a model-based Bayesian probabilistic inference in advanced room-acoustic energy decay analysis where both decay order (model) selection and decay parameter estimation problems are to be solved. A brief introduction to a model-based nonlinear regression approach for the specific case of reverberation time estimation is also given.

Sound energy decays in room-acoustics investigations are traditionally determined using random-noise excitations; once this signal has brought the sound field to its steady state, the noise excitation is interrupted. Normally sound pressures at receiver locations are recorded, and the squared pressures over time are termed steady-state *energy-time* functions. Schroeder considered a sound source-receiver arrangement in a room under investigation to be a linear time-invariant system. When the sound source is driven by the interrupted noise, the receiver receives a sound signal represented by convolution between the room impulse response and the interrupted noise. The steady-state energy decay functions can be derived by integration of the energy room impulse response, so-called Schroeder integration [1]. This method, from a single measurement of a room impulse response, produces an energy decay function which is theoretically equivalent to an ensemble average of a large number of steady-state energy decay functions when using interrupted noise. Following Schroeder, a number of investigations of this method [2–12] have substantiated its advantages. The Schroeder decay functions are not only free of unnecessary random fluctuations, but they contain significant single or multiple exponential processes as well.

3.2 Schroeder Decay and Model

A normalized stead-state sound energy decay function $d(t)$, via Schroeder integration [1, 13], can be expressed by its discrete version in digital domain as:

$$d(t_i) = \frac{1}{E}\sum_{\tau=t_i}^{t_K} h^2(\tau) = 1 - \frac{1}{E}\sum_{\tau=0}^{t_i} h^2(\tau), \quad 0 \leq i < K, \tag{3.1}$$

and

$$E = \sum_{\tau=0}^{t_K} h^2(\tau), \tag{3.2}$$

where variable t_i represents discrete time, i, K are integers with K being the total number of data points, so t_K represents the total time record length of *room impulse response* $h(t)$ as shown in Fig. 3.1a. The room impulse response $h(t)$ is defined between a sound source and a receiver in the room under investigation. In the following t_K is also termed *the upper limit of the integration*. E is the total energy contained in $h(t)$. Schroeder integration essentially establishes the relationship between the *energy room impulse response* $h^2(t)$ and the *steady-state energy decay function* $d(t)$ [7]. The right-hand side of Eq. (3.1) is of both theoretical and practical relevance, it represents the complementary relation between the energy decay and build-up [13], at the same time it is beneficial for numerical implementation due to a straightforward accumulative summation. In architectural acoustics practice, logarithmic energy room impulse responses $10\log_{10}[h^2(t)]$ often represented graphically over time, termed *energy-time curve* (ETC), as illustrated in Fig. 3.1b.

A parametric model (*Schroeder decay model*) with s exponential decay terms [14–16]

$$\mathbf{H}_s(\mathbf{A}_s, \mathbf{T}_s) = A_0(t_K - t_i) + \sum_{j=1}^{s} A_j \left[\exp\left(\frac{-13.8 \cdot t_i}{T_j}\right) - \exp\left(\frac{-13.8 \cdot t_K}{T_j}\right) \right], \quad (3.3)$$

can fully describe Schroeder decay function $d(t)$. Vector $\mathbf{A}_s = [A_0, A_1, \cdots A_s]$ contains $s + 1$ linear coefficients, while vector $\mathbf{T}_s = [T_0, T_1, \cdots T_s]$ contains s decay time parameters. The subscript s in the model in Eq. (3.3) is the decay order, denoting that the decay model contains s different exponential decay terms with s different decay times. The models in the present application are nested models—a higher order model embodies lower ones. Particularly, if the second-order model describes the data reasonably well, the first-order or the third-order models often also describe the data well. When $s = 1; 2; 3$, it is said that the sound energy decays are of single-, double- or triple-slope nature, respectively. For single-slope decays, the corresponding decay time is termed *reverberation time*, while in multiple-slope decay cases, multiple different *decay times* are expected. When t_K is selected large enough, the constant term $\exp(-13.8 \cdot t_K/T_j)$ in Eq. (3.3) is negligible.

There are two major issues reported in published literature since 1965; Schroeder integration yields energy decay functions from experimentally measured room impulse responses, which sensitively depend on the upper limit of the integration [2] and inevitable background noise [5]. The Schroeder decay model in Eq. (3.3) helps us understand the characteristic behavior of the Schroeder decay functions. A full understanding of this characteristic behavior, in turn, facilitates the discussion of the advanced analysis tools elaborated on later in this chapter. It further reveals that the two reported issues, associated with the upper limit of the integration and background noise, are actually redundant. Along with Fig. 3.1c we summarize the following characteristic features

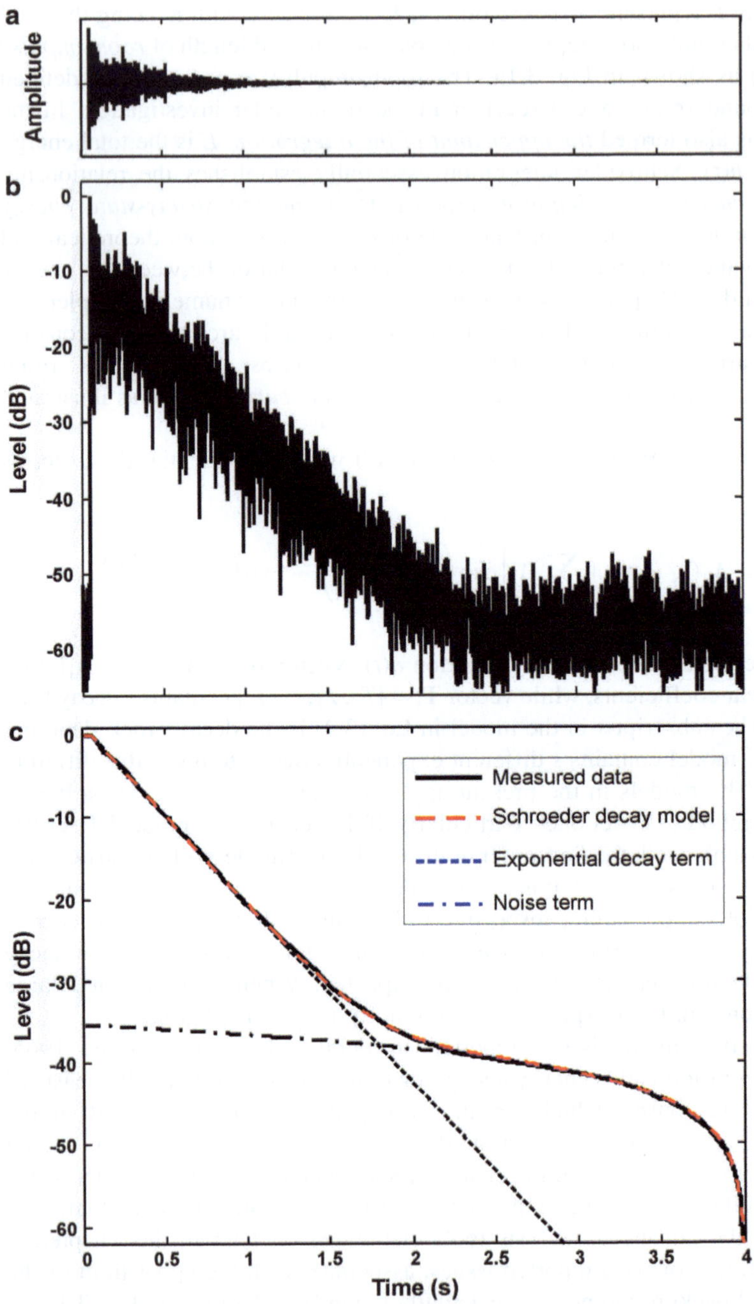

Fig. 3.1 Experimentally measured room impulse response, band-pass filtered at 1 kHz (Oct.), its corresponding energy-time curve and Schroeder decay curve. (**a**) Room impulse response. (**b**) Normalized energy-time curve (energy room impulse response) illustrated in logarithmic scale [level in (dB)], derived from the room impulse response shown in (**a**). (**c**) Normalized Schroeder decay curve illustrated in logarithmic scale [level in (dB)] derived from the energy room impulse response as shown in (**b**). The Schroeder decay model along with decomposed exponential decay term and the linear decay term is also illustrated for ease of comparison

1. The Schroeder decay curve will, generally (or very often), not exhibit a linearly decaying curve when graphically presented in a logarithmic scale [level in (dB)], even for the single-slope case. It can be characterized by three major portions.
   (a) *Beginning portion*: the decay curve decays within this portion with a slope or with slopes predominantly determined by room-acoustic properties, such as the reverberation time (decay times).
   (b) *Middle portion*: a curve often appears to decay with a somewhat slower slope in comparison to that in the beginning portion. Only when the upper limit of the integration is long enough this portion remains observable [see the curves between 1.5 and 2.5 s as shown in Fig. 3.1c], otherwise it can disappear so that the beginning portion transitions directly to the late portion.
   (c) *Late portion*: the decay curve ends up with a characteristic curvature towards the end of the time record of the room impulse response (the upper limit of the integration).
2. The linearly decreasing term $A_0(t_K - t_i)$ [see Eq. (3.3)] is predominantly responsible for the characteristic curvature of the late portion in the logarithmic scale, and it accounts for both the effect of the upper limit of the integration through parameter t_K [2] and the effect of background noise through parameter A_0; therefore, $A_0(t_K - t_i)$ is also termed the *noise term* in this chapter. It is actually a nuisance term, irrelevant to the room-acoustics problem at hand.
3. Schroeder decay functions become undefined at the upper limit of integration t_K [see Fig. 3.1c] when examined logarithmically since for $t_i \to t_K$, $d(t_i) \to 0$ and the parametric model accounts for this fact.

Luizard and Katz [6] recently proposed an alternative model to the envelope of the energy-time functions (energy room impulse responses)

$$\mathbf{H}_s(\boldsymbol{\alpha}_s, \mathbf{T}_s) = \alpha_0 + \sum_{j=1}^{s} \alpha_j \exp\left(\frac{-13.8 \cdot t_i}{T_j}\right). \tag{3.4}$$

The relationship between Schroeder decay functions and the energy time functions is already represented by Eq. (3.1) [1, 7]. Xiang et al. [8] also demonstrated the difference between the two types of decay functions using diffusion equation modeling results. The parameter estimates based on two different data models should not be directly compared without the conversion using Schroeder integration since Schroeder integration of the model in Eq. (3.4) between the time interval $[t_i, t_K]$ results in exactly the model in Eq. (3.3) with $A_j = \alpha_j T_j / 13.8$, when neglecting the constant term $\exp(-13.8 \cdot t_K / T_j)$ in Eq. (3.3). The model-based analysis discussed in this chapter, including nonlinear regression and particularly two levels of Bayesian inference, can also be applied to the smoothed energy-time functions using the alternative model in Eq. (3.4). The following discussions in the context of model-based decay analysis, however, remain focused on the Schroeder decay functions.

3.3 Linear and Nonlinear Regression

The characteristic features of Schroeder decay curves summarized above, along with the model in Eq. (3.3), help us understand the fact that the linear least-squares-fit approach to directly fit logarithmic Schroeder decay functions (curves) relies on an inappropriate model. A large number of investigations have been reported in major acoustics journals for decades after Schroeder published this method. The reported effort, among others, includes mitigation of the nonlinear decay curves [as shown in Fig. 3.1c] by careful selection of the upper limit of integration, the so-called truncation method [2, 9], or subtraction/elimination of background noise, the so-called subtraction method [5, 10, 11]. All these mitigation methods, however, employ linear (least-square) regression, implicitly relying on a linear decay model. The success of reverberation time estimation then depends critically on accurate estimation of the upper limit of the integration t_K, accurate estimation of A_0 associated with the background noise [14], or accurate elimination of the background noise [10]. In order to improve reverberation time estimates, iterative processes have also been proposed to accurately estimate the upper limit of the integration [12]. Separate multiple measurements at a single receiver location have also been proposed to reduce the effect of the background noise [10]. A close look at the model in Eq. (3.3) reveals that these methods depend critically on accurate estimations or removal of one or the other nuisance parameter (A_0, t_K).

In 1995, a nonlinear regression method [14] was introduced by this author[1] based on the nonlinear model in Eq. (3.3). For reverberation time estimates (single-slope case), the nonlinear regression removes the stringent requirement on accurate estimation of nuisance parameters, both the upper limit of the integration t_K or the parameter A_0 associated with the background noise. The nonlinear regression starts with a rough estimate of initial values of model parameters A_1, T_1, and A_0. It then iteratively improves the model parameters in terms of infinitesimal corrections of each individual parameter by solving a linear equation system. For the single-slope cases, it easily converges to the global minimum by properly estimating the initial values of A_1 and T_1, which can be straightforwardly estimated from a small segment at the beginning of a logarithmic Schroeder decay function. Accurate initial estimates of A_0 and t_K are uncritical, and t_K can also be set with ease to be somehow smaller or larger than T_1, such as one, two or even a multiple of the initial value of T_1 which has already been estimated [14], or simply set sufficiently large without any burden of accurate determination.

The nonlinear regression works efficiently for the reverberation time estimation (single-slope case) without the stringent requirement on parameters t_K and A_0 because it involves t_K and A_0 as adjustable model parameters in the regression process. The iterative process of the nonlinear regression will lead to convergence

[1] Dr. Wolfgang Ahnert was collaborative in the initial work [17, 18] by drawing this author's attention to a paper published in the *Computation Journal* in 1965 dealing with minimizing a sum of squares of nonlinear functions [14].

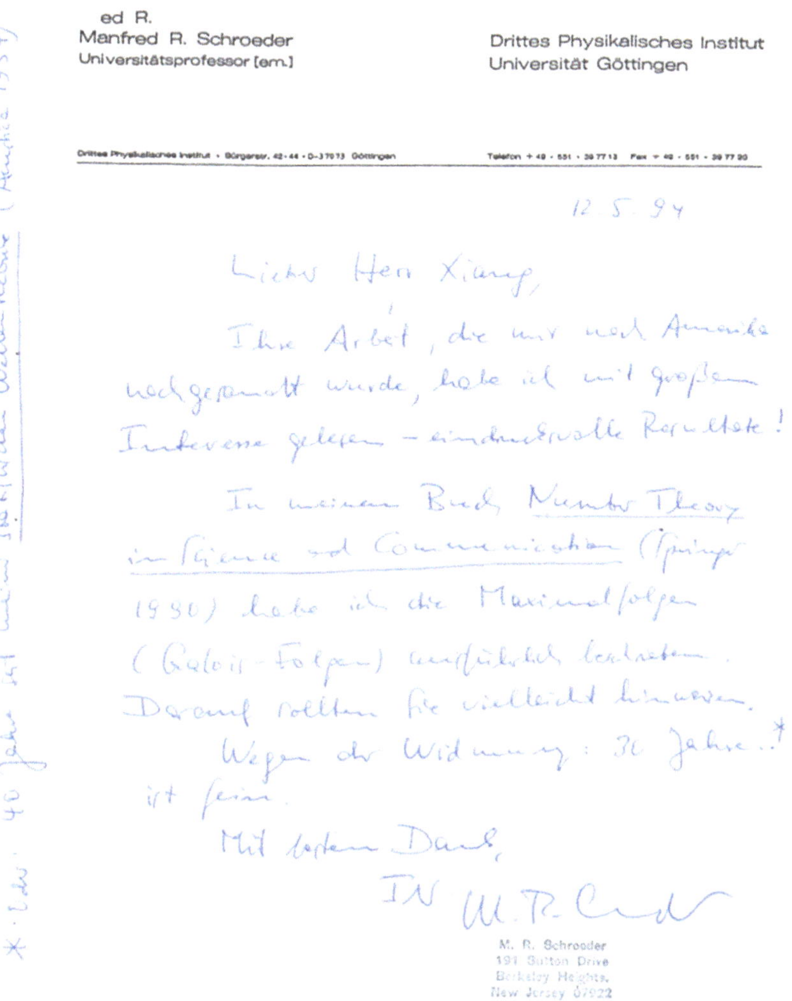

ed R.
Manfred R. Schroeder
Universitätsprofessor [em.]

Drittes Physikalisches Institut
Universität Göttingen

12.5.94

Lieber Herr Xiang,

Ihre Arbeit, die mir nach Amerika nachgesandt wurde, habe ich mit großem Interesse gelesen – eindrucksvolle Resultate!

In meinem Buch Number Theory in Science and Communication (Springer 1990) habe ich die Maximalfolgen (Galois-Folgen) ausführlich behandelt. Darauf sollten Sie vielleicht hinweisen.

Wegen der Widmung: 30 Jahre*! ist fein.

Mit bestem Dank,

Ihr M. R. Schroeder

M. R. Schroeder
191 Sutton Drive
Berkeley Heights,
New Jersey 07922

* Über 40 Jahre ist meine statistische Wellentheorie (Acustica 1954)

Fig. 3.2 Professor Manfred Schroeder's handwritten letter, shortly before publication of the nonlinear regression method [14] in the Journal of the Acoustical Society of America during the revision phase, the final draft was sent to Professor Manfred Schroeder for his critical comments. The first paragraph of the letter: "Your paper, relayed to me to America, I have read with great interest—impressive results!"

to the global minimum after only a few iterations even with rough estimates of initial values [14]. They require some added computational expenses, which are, on current PC-platforms, barely noticeable; on order of a few milliseconds. Professor Manfred Schroeder has shown a great interest in this work [see Fig. 3.2].

This author soon found out [19, 20], however, if the model order is higher than one (double- or multiple-slope cases), initial value estimation becomes too critical to allow convergence. For this reason, the model-based Bayesian method, elaborated in

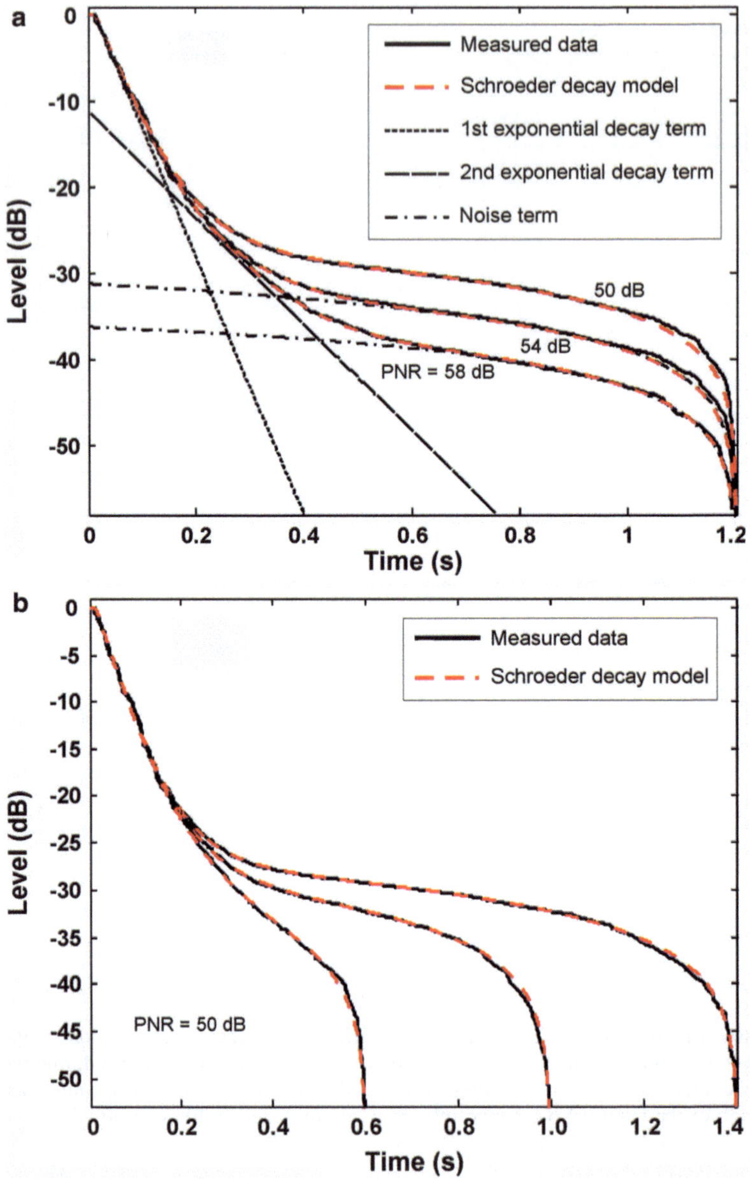

Fig. 3.3 Experimentally measured Schroeder decay curves at one specific source-receiver configuration. (**a**) Normalized Schroeder decay curves for a fixed upper limit of integration with its peak-to-noise ratio PNR = 50, 54, and 58 dB, respectively. Decomposed exponential terms and the noise term are also shown for ease of comparison. (**b**) Normalized Schroeder decay curves for a fixed peak-to-noise ratio (PNR = 50 dB) while changing the upper limit of integration $t_K = 0:6$ s; $1:0$ s, and $1:4$ s, respectively

the following, has emerged [20, 21]. Before describing Bayesian inference, a close look at experimentally measured data as illustrated in Fig. 3.3 motivates applications of advanced energy decay analysis methods. Figure 3.3a illustrates Schroeder decay

Table 3.1 Bayesian analysis of an experimentally measured Schroeder decay function in acoustic scale models as shown in Figs. 3.3 and 3.4, derived from a 1 kHz octave-band-pass-filtered room impulse response

	Single-slope	Double-slope	Triple-slope
BIC (neper)	−85.76	0.0	−81.08
A_0 (dB)	−62.40	(−58.5) −62.2 (−67.2)	−62.26
A_1 (dB)	−5.15	−5.15 (±3.4E-3)[a]	−5.26
T_1 (s)	0.483	0.411 (±9.3E-3)	0.39
A_2 (dB)	–	−14.91 (±1.3E-3)[a]	−14.24
T_2 (s)	–	0.765 (±9.7E-3)	0.768
A_3 (dB)	–	–	−20.43
T_3 (s)	–	–	1.22

Bayesian information criterion (BIC) along with decay parameters estimated using single-slope, double-slope, and triple-slope models are listed. The double-slope decay obtains the highest natural logarithmic BIC, by which the other BIC values are normalized. A_0 listed in column "Double-Slope" contains three different values corresponding to three different curves in Fig. 3.3a. The standard derivations dA_1, dT_1, dA_2, and dT_2 are also listed only for the double-slope case

[a] dA_1, dA_2 are listed linearly

curves, derived from a 1 kHz octave-band-pass-filtered room impulse response measured in a coupled-volume system at one specific receiver location, but with three different resulting peak-to-noise ratios (PNR = 50, 54, and 58 dB, approximately). The decay characteristics represented by these three curves are actually the same. According to Eq. (3.3), a model-based decay parameter estimation with $s = 2$ yields decay parameter values as listed in Table 3.1, (Column 3).

Using these estimated parameters, the decay model curves [see Eq. (3.3), plotted by dash lines in Fig. 3.3] agree well with the experimental data. The decay model curves consist of two exponential decay terms and one noise term [a linearly decaying term, $A_0(t_K - t_i)$]. The only parameter differing between the model curves is $10\log_{10}(A_0) = -58.5, -62.2, \text{and} -67.2$ dB for the three cases, respectively. Figure 3.3a also illustrates the noise term for the two highest peak-to-noise ratios. Figure 3.3b illustrates the decay curves for three different upper limits of integration with a fixed peak-to-noise ratio of 50 dB. The Schroeder decay models with the exact five decay parameters listed in Table 3.1 (Column 3), but with different upper limits of integration $t_K = 0.6, 1.0, \text{and } 1.4$ s, well describe the course of the experimental decay curves.

Some previous authors reported that Schroeder decay functions present difficulties when fitting the straight lines [2, 5, 9, 10, 12]. One may argue that corrections can be applied to make Schroeder decay curves look like linear decays by carefully selecting the upper limit of the integration or by subtraction/elimination of background noise, even resorting to iterative processes to optimize the corrections. The breaking mechanisms of the iterative process are conventionally set according to linearity of the resulting curves. From the viewpoint of the model-based analysis, these previous approaches have applied an inappropriate (linear) model for analyzing Schroeder decay data. As Fig. 3.3 demonstrates, however, if the data contain decay processes more complicated than a single-decay process, those correcting measures, implicitly assuming a straight-line decay function, will significantly destroy inherent decay characteristics in the data. More profound challenges are due to the fact that one cannot know prior to data analysis how many decay processes the data at hand contain, since in single-space rooms, multiple-slope decays may occur, whereas in coupled-volume systems, single-slope decays can often be found. The following section elaborates on an advanced method based on Bayesian inference, which will first determine how many slopes are in the data before using an appropriate model to analyze the data.

3.4 Bayesian Decay Analysis

The fact that the decay models in Eq. (3.3) describe Schroeder decay functions reasonably well, constitutes basic background information for the following discussion. The background information I encapsulates the fact that a number of models are competing with each other to describe the energy decay data reasonably well. In particular, the background information I restricts the number of models M to be finite, the models in set $\Phi = \{\mathbf{H}_1, \mathbf{H}_2, \cdots, \mathbf{H}_M\}$ are exhaustive and exclusive, and only one model is expected to be true (or useful) to explain the data. In practice, architectural acousticians are often expecting single-, double-, or triple-slope energy decays. All these propositions constitute the background information I which will be explicitly used in the subsequent formulation of model-based Bayesian decay analysis.

Professor Manfred Schroeder already found in his first computer room simulation based on ray tracing that non-exponential decays can be found if the simulated room contains one absorptive wall while all the others are highly reflective [22]. Multiple-slope decays are of auditory significance because they are believed to meet two competing auditory perceptual attributes simultaneously, both clarity and reverberance [20, 23]. Investigations on perceptual attributes, audibility of different decay profiles, or perceptual effects of multiple-slope decays over single-slope decays are beyond the scope of the current chapter, but the analysis methods described in this chapter will provide interested researchers with quantitative tools for the further studies.

3.4.1 *Two Levels of Bayesian Inference*

Bayesian probabilistic inference encompasses both the parameter estimation and the model selection problems by extensive use of Bayes' theorem. Bayesian inference applied to solving parameter estimation problems is referred to as the first level of inference, while solving model selection problems is referred to as the second level of inference [21, 24]. The following discussion begins with the second level of inference, model selection. This is also the scientifically logical approach;

one should determine which of the competing models is appropriate before the relevant model parameters are inferred.

3.4.1.1 Model Selection: The Second Level of Inference

Among a set of competing models, increased decay orders will always improve curve fitting but will often generalize poorly [24]. To avoid over-parameterization, Bayesian model selection applies Bayes' theorem to one of the competing models $\mathbf{H}_s$ given the background information I and the data $\mathbf{D}$, containing Schroeder decay function of K elements, as

$$p(\mathbf{H}_s|\mathbf{D}) = \frac{p(\mathbf{H}_s)p(\mathbf{D}|\mathbf{H}_s)}{p(\mathbf{D})}, \tag{3.5}$$

while pushing any interest in model parameter values into the background of the current problem. In Eq. (3.5), $p(\mathbf{H}_s)$ represents the *prior probability* of the model $\mathbf{H}_s$ given the background information I as defined previously at the beginning of this section (Sect. 3.4), but the conditional as "given the background information I" as noted by $p(\mathbf{H}_s|I)$ [or the other quantities, $p(\mathbf{D}|\mathbf{H}_s, I)$, $p(\mathbf{H}_s|\mathbf{D}, I)$], is omitted throughout this chapter for simplicity. Quantity $p(\mathbf{D}|\mathbf{H}_s)$ is termed *marginal likelihood*, also termed *evidence*. Quantity $p(\mathbf{D})$ is a constant, and $p(\mathbf{H}_s|\mathbf{D})$ represents the *posterior probability* of the model $\mathbf{H}_s$, given the data $\mathbf{D}$.

Model comparison between two different models $\{\mathbf{H}_j, \mathbf{H}_k\}$ relies on the posterior ratio,

$$\frac{p(\mathbf{H}_j|\mathbf{D})}{p(\mathbf{H}_k|\mathbf{D})} = \frac{p(\mathbf{H}_j)}{p(\mathbf{H}_k)}\frac{p(\mathbf{D}|\mathbf{H}_j)}{p(\mathbf{D}|\mathbf{H}_k)}, 1 \leq j, k \leq M; j \neq k. \tag{3.6}$$

While model selection exhaustively calculates every posterior probability $p(\mathbf{H}_s|\mathbf{D})$ among the finite model set, to determine which model leads to the maximum posterior probability, it can also sequentially evaluate Eq. (3.6) to rank the competing models.

The first ratio on the right-hand side in Eq. (3.6), termed *prior ratio*, represents the degree of one's initial belief in how much the model $\mathbf{H}_j$ would have been preferred over the model $\mathbf{H}_k$ before the data were acquired, while the second ratio on the right-hand side of Eq. (3.6), termed *likelihood ratio*, measures how much the data prefer the model $\mathbf{H}_j$ over $\mathbf{H}_k$. The prior ratio opens up the opportunity, if desirable, to incorporate a prior preference to $\mathbf{H}_j$, but it is discouraged. Assigning equal prior probability

$$p(\mathbf{H}_s) = \frac{1}{M}, \quad 1 \leq s \leq M, \tag{3.7}$$

implies no subjective preference to any of models, which results in

$$\frac{p(\mathbf{H}_j|\mathbf{D})}{p(\mathbf{H}_k|\mathbf{D})} = \frac{p(\mathbf{D}|\mathbf{H}_j)}{p(\mathbf{D}|\mathbf{H}_k)}. \tag{3.8}$$

This indicates that the marginal likelihood (or evidence) $p(\mathbf{D}|\mathbf{H}_s)$ plays a central role in Bayesian model selection.

3.4.1.2 Parameter Estimation: The First Level of Inference

For the purpose of parameter estimation, the first level of Bayesian inference applies Bayes' theorem

$$p(\mathbf{\Theta}_s|\mathbf{D},\mathbf{H}_s) = \frac{p(\mathbf{\Theta}_s|\mathbf{H}_s)p(\mathbf{D}|\mathbf{\Theta}_s,\mathbf{H}_s)}{p(\mathbf{D}|\mathbf{H}_s)}, \tag{3.9}$$

to the parameters $\mathbf{\Theta}_s$ containing $2 \cdot s + 1$ parameters and the data $\mathbf{D}$ once a specific model $\mathbf{H}_s(\mathbf{\Theta}_s)$ is determined by the model selection, where $p(\mathbf{\Theta}_s|\mathbf{H}_s)$ is the prior probability of parameters $\mathbf{\Theta}_s$, $p(\mathbf{D}|\mathbf{H}_s)$ is exactly the marginal likelihood in Eq. (3.5), and $p(\mathbf{D}|\mathbf{\Theta}_s, \mathbf{H}_s)$ is the likelihood function of the parameters, which is the probability of the residual error $\mathbf{e} = \mathbf{D} - \mathbf{H}_s$. Application of the principle of maximum entropy leads to an assignment of the likelihood function [25],

$$L(\mathbf{\Theta}_s) = p(\mathbf{D}|\mathbf{\Theta}_s,\mathbf{H}_s) \approx \frac{(2\pi E)^{-K/2}}{2}, \tag{3.10}$$

with

$$E = \frac{\mathbf{e}^{\mathrm{T}}\mathbf{e}}{2}, \tag{3.11}$$

where $(\cdot)^{\mathrm{T}}$ denotes matrix transpose.

The probability $p(\mathbf{\Theta}_s|\mathbf{D}, \mathbf{H}_s)$ on the left-hand side of Eq. (3.9) is the *posterior probability* of the parameters $\mathbf{\Theta}_s$, representing the updated knowledge about the parameters once the data become available. The integration over the entire parameter space $\int_{\mathbf{\Theta}_s}$ on both sides of Eq. (3.9), combined with the simplified notations $\pi(\mathbf{\Theta}_s) = p(\mathbf{\Theta}_s|\mathbf{H}_s)$ and $L(\mathbf{\Theta}_s) = p(\mathbf{D}|\mathbf{\Theta}_s, \mathbf{H}_s)$ in Eq. (3.10), yields

$$p(\mathbf{D}|\mathbf{H}_s) = Z_s = \int_{\mathbf{\Theta}_s} L(\mathbf{\Theta}_s)\pi(\mathbf{\Theta}_s)d\mathbf{\Theta}_s. \tag{3.12}$$

The quantity $p(\mathbf{D}|\mathbf{H}_s)$ in the denominator on the right-hand side of Eq. (3.9) is independent of the parameters $\mathbf{\Theta}_s$, and can be taken out of the integral. $p(\mathbf{D}|\mathbf{H}_s)$ determined by Eq. (3.12) is exactly the same as the marginal likelihood in Eqs. (3.5) and (3.8). It plays a central role in the model selection. Equation (3.9) can then be expressed as

$$\begin{array}{c} p(\mathbf{\Theta}_s|\mathbf{D},\mathbf{H}_s) \times Z_s = \pi(\mathbf{\Theta}_s) \times L(\mathbf{\Theta}_s). \\ \text{posterior} \times \text{evidence} = \text{prior} \times \text{likelihood} \end{array} \tag{3.13}$$

The quantity $Z_s = p(\mathbf{D}|\mathbf{H}_s)$ is often referred to as (Bayesian) *evidence* for model $\mathbf{H}_s$ [24]. Equation (3.13) states the logical relations among the quantities of Bayesian inference [28]: prior probability $\pi(\mathbf{\Theta}_s)$ and the likelihood function $L(\mathbf{\Theta}_s)$ are the inputs, while the posterior probability $p(\mathbf{\Theta}_s|\mathbf{D},\mathbf{H}_s)$ and the evidence Z_s are the outputs. Particularly, the posterior probability is the output for the first level of inference, the parameter estimation, while the evidence, Z_s, is the output for the second level of inference, the model selection. Bayesian evidence encapsulates the principle of parsimony, and quantitatively implements Occam's razor: *when two competing theories explain the data equally, the simpler one is preferred.* Using an asymptotic approximation, the next section elaborates on the quantitative implementation of Occam's razor.

3.5 Bayesian Information Criterion

The Bayesian information criterion (BIC) asymptotically approximates Bayesian evidence if a (multi-dimensional) Gaussian distribution can approximate the posterior probability distribution within a vicinity around the global extreme of the likelihood. Experimental results have shown that this is often the case for the energy decay analysis [16]. In this application, the (natural logarithmic) BIC for ranking a set of decay models $\mathbf{H}_1, \mathbf{H}_2, \mathbf{H}_3, \cdots$ is given by

$$\text{BIC} \approx 2 \cdot \ln\left[\widehat{L}\left(\widehat{\mathbf{\Theta}}_s\right)\right] - N_D \ln(K) \quad \text{[Neper]}, \tag{3.14}$$

with $N_D = 2 \cdot s + 1$ being the dimensionality, or the number of parameters involved in model $\mathbf{H}_s$. The quantity $\widehat{L}$ is the peak value of the likelihood whose location in the parameter space is denoted by $\widehat{\mathbf{\Theta}}_s$. The first term in Eq. (3.14) represents the degree of the model fit to the data, while the second term represents the penalty of over-parameterization. When increasing the decay model order, the first term may also increase, indicating a better fit to the data, but the BIC will dramatically decline because of the second term, which penalizes over-parameterization. It will outweigh the first term [16]. In the scope of energy decay analysis among a set of decay models $\mathbf{H}_1, \mathbf{H}_2, \mathbf{H}_3, \cdots$, the model yielding a significantly larger BIC value in Eq. (3.14) is the most concise model providing the best fit to the decay function data, while at the same time capturing the important exponentially decaying features evident in the data.

Taking the experimentally measured data (PNR = 54 dB, as shown in Fig. 3.3), Fig. 3.4 illustrates the Bayesian analysis results and shows the decomposed decay terms in the second- and third-order Schroeder decay models. The first-order decay model represents a strong misrepresentation of the data and is not shown here. Figure 3.4b illustrates the typical example of an over-parameterized model which may or may not improve the curve fitting significantly in comparison with that of the second-order model as shown in Fig. 3.4a, but Occam's razor, quantitatively implemented in Bayesian analysis, penalizes the over-parameterized model since it receives a significantly lower BIC value. Its BIC value is 81 nepers lower than that of the second-order decay model (see Table 3.1, last column).

3.6 Sampling Methods

The BIC critically relies on estimation of the global maximum of the posterior distribution Eq. (3.14); it does not actually calculate the integral of the likelihood volume/mass over the prior distribution Eq. (3.12). When the posterior distribution cannot be approximated by (multi-dimensional) Gaussian distributions within the vicinity at its global peak, the BIC will represent a poor approximation of the evidence. This section outlines calculation of the evidence, Z_s. Nested sampling, among other advanced sampling methods, most recently proposed by Skilling [26] provides a cohesive method for estimating the evidence Z_s as a primary quantity while obtaining the posterior samples $p(\boldsymbol{\Theta}_s|\mathbf{D}, \mathbf{H}_s)$ as subsidiary values useful for parameter estimations during the process of calculating Z_s. This becomes plausible by inspecting Eqs. (3.9) and (3.12) since Z_s is calculated via the integral of $L(\boldsymbol{\Theta}_s)\pi(\boldsymbol{\Theta}_s)$ over the entire parameter space $\widehat{\boldsymbol{\Theta}}_s$; at the same time, decay parameter estimation solely relies on calculating the posterior $p(\boldsymbol{\Theta}_s|\mathbf{D}, \mathbf{H}_s)$ through $L(\boldsymbol{\Theta}_s)\pi(\boldsymbol{\Theta}_s)/Z_s$ [see Eq. (3.9)]. The decay parameters can then be estimated in terms of probabilistic moment calculations [25] [see Eq. (3.23) in Sect. 3.6.3] or in terms of the maximum *a posteriori* (MAP) approach using these subsidiary posterior samples [25, 29].

3.6.1 Basic Formulation

This section outlines the nested sampling applied to both Bayesian model selection and parameter estimation tasks. More complete treatments can be found elsewhere [26–29]. The following discussion will use simplified notation so that Eq. (3.12) becomes

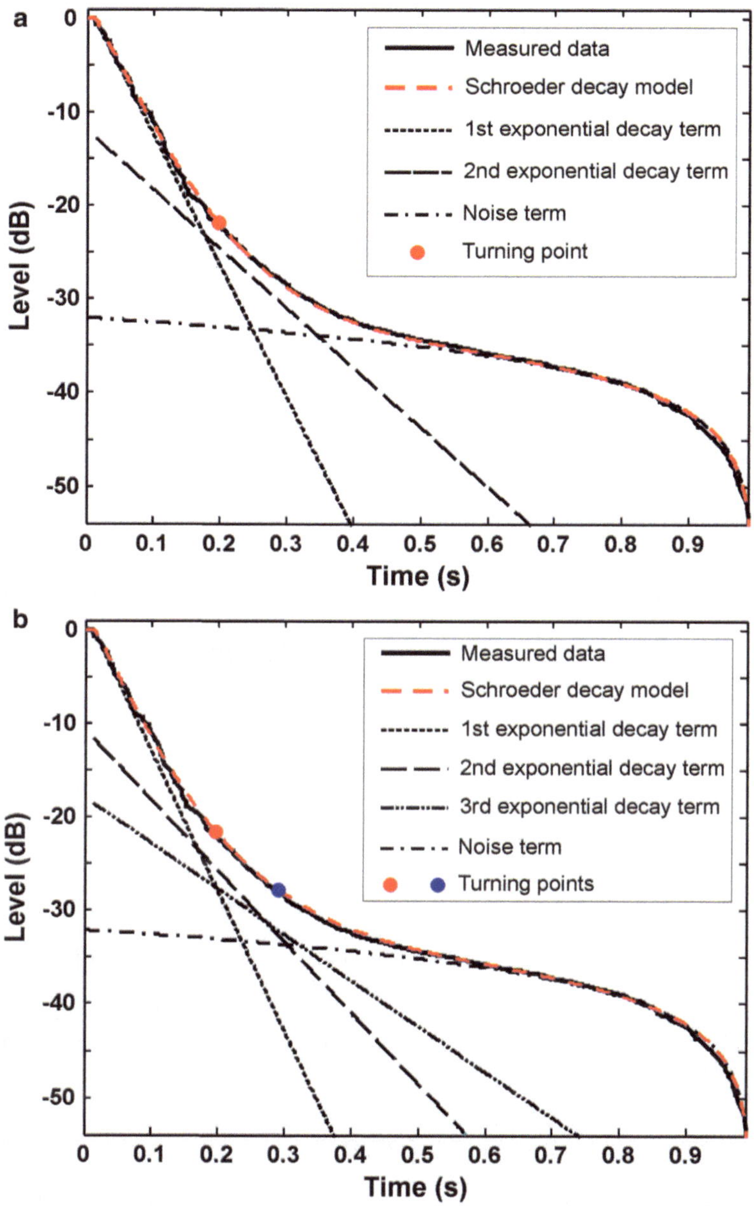

Fig. 3.4 Experimentally measured Schroeder decay curves as shown in Fig. 3.3 (the peak-to-noise ratio PNR = 54 dB). (**a**) Comparison with Bayesian model curve and decomposition using a double-slope model. (**b**) Comparison with Bayesian model curve and decomposition using a triple-slope model

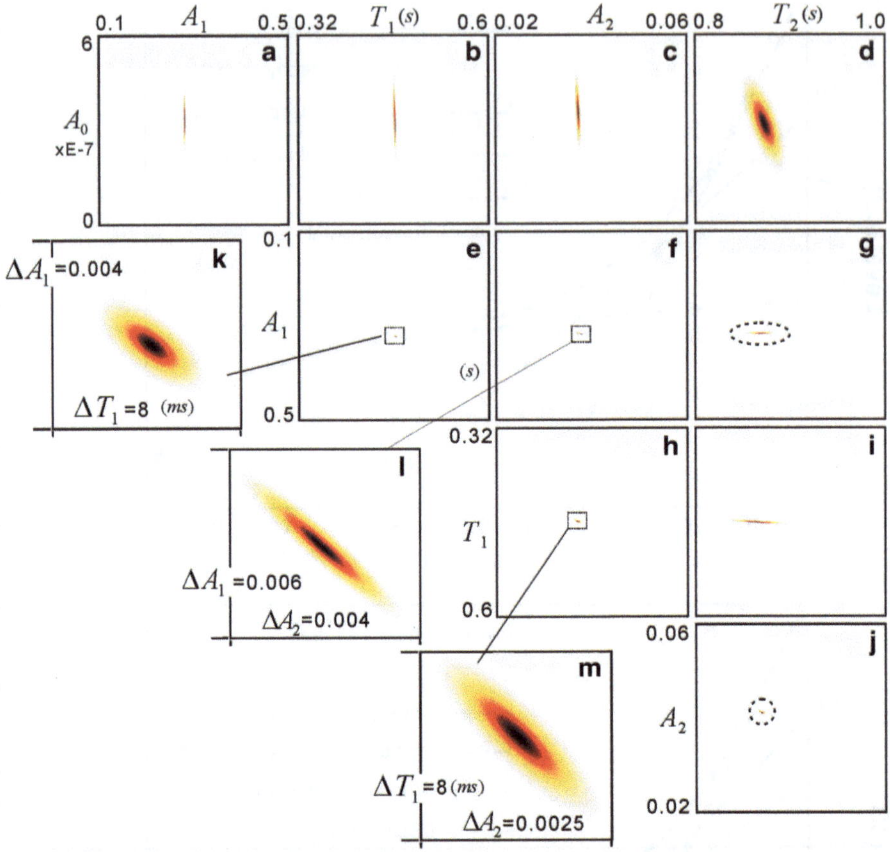

Fig. 3.5 Normalized (likelihood or) marginal posterior probability distributions (MPPDs) over two-dimensional (zoomed-in) parameter space from an experimentally measured data set in an acoustical scale model using the second-order decay model (double-slope model). (**a**) MPPD over $\{A_0, A_1\}$. (**b**) MPPD over $\{A_0, T_1\}$. (**c**) MPPD over $\{A_0, A_2\}$. (**d**) MPPD over $\{A_0, T_2\}$. (**e**) MPPD over $\{A_1, T_1\}$. (**f**) MPPD over $\{A_1, A_2\}$. (**g**) MPPD over $\{A_1, T_2\}$. (**h**) MPPD over $\{T_1, A_2\}$. (**i**) MPPD over $\{T_1, T_2\}$. (**j**) MPPD over $\{A_2, T_2\}$. A grid of 200×200 is used for (**a–d**), and (**k–m**), while a grid of 300×300 for (**e–j**)

$$Z = \int_{\boldsymbol{\Theta}} L(\boldsymbol{\Theta})\pi(\boldsymbol{\Theta})d\boldsymbol{\Theta}. \tag{3.15}$$

The subscript s is dropped, but still bear in mind that the parameters $\boldsymbol{\Theta}$ and the evidence Z are associated with a specific model $\mathbf{H}_s$ given the data $\mathbf{D}$ and the background information I.

This integral over the parameter space will become intractable even over a small dimensionality of the parameter space. In order to highlight this challenge, Fig. 3.5 illustrates posterior probability distributions $[L(\boldsymbol{\Theta}_s)\pi(\boldsymbol{\Theta}_s)]$ over a five-dimensional decay parameter space, often found in the decay analysis of room-acoustics practice. The figure shows one example experimentally measured in an acoustical scale model of coupled rooms. The measured room impulse response is first filtered at 1 kHz octave-band, then Schroeder-integrated to obtain the data vector **D**. In the following, the area/volume under the posterior probability distribution is termed *posterior mass*. The figure illustrates small two-dimensional subspaces of the five-dimensional parameter space while keeping the rest of other three parameters constant, being optimum-values determined from well-sampled estimates of the parameters. These two-dimensional distributions are referred to marginal posterior probability distributions (MPPDs). All the MPPDs as shown in Fig. 3.5 contain the concentrated posterior mass in order to show its shape and sharpness. In room-acoustics practice, prior knowledge of individual parameters is often vague. Without experimental data, decay times T_i can only be assumed to fall in a range of a few seconds (0.2 s–few seconds). Varied shapes and sizes of the posterior probability distribution, some extremely sharp, narrow, some very broad, already indicate that a direct integral over a few dimensions of relevant parameter space in terms of exhaustive sampling makes the problem hopelessly intractable.

The sampling method discussed in this section makes evaluating this integral a tractable task, Eq. (3.15) can be simply rewritten [26] as

$$Z = \int_0^1 L(\mu) d\mu, \tag{3.16}$$

where $d\mu(\mathbf{\Theta}) = \pi(\mathbf{\Theta}) d(\mathbf{\Theta})$ is the elementary hyper-volume under the prior probability distribution, with

$$\mu = \int \pi(\mathbf{\Theta}) d(\mathbf{\Theta}), \tag{3.17}$$

μ is termed *prior mass*. Figure 3.6 illustrates the concept of the likelihood function over the prior mass. Since it is a monotonic function, $L(\mu)$ is just the inverse of $\mu(L)$. A decreasing sequence of prior mass yields

$$1 = \mu_0 > \mu_1 > \cdots > \mu_i > \mu_{i+1} > \mu_\infty = 0, \tag{3.18}$$

when an increasing sequence of likelihood values

$$0 \approx L_{\min} = L_0 < L_1 < \cdots < L_i < L_{i+1} < \cdots < L_{\max} < \infty \tag{3.19}$$

can be determined. This is elaborated on in the following subsection. The elementary prior mass $d\mu$ is associated with the likelihood within $[L_i, L_{i+1}]$.

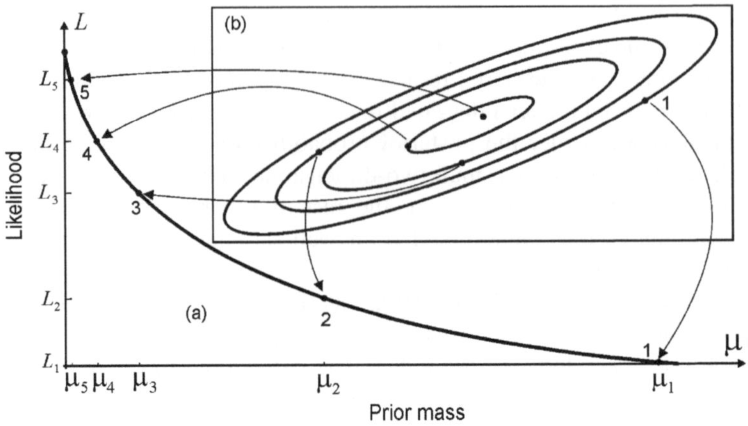

Fig. 3.6 Estimation of the Bayesian evidence by the nested sampling. (**a**) The area under the likelihood function curve over the prior mass between 0.0 and 1.0 represents the Bayesian evidence as defined in Eq. (3.16). A decreasing sequence of the prior mass along μ-axis corresponds to an increasing sequence of the likelihood values along L-axis. (b) The likelihood contours corresponding to different likelihood values

3.6.2 Numerical Implementation

In order to implement the above integration for Z in Eq. (3.16), one should start backwards from Eq. (3.19) to Eq. (3.15) by defining the prior mass sequence Eq. (3.18) based on constraints on likelihood values Eq. (3.19). The integral is processed sequentially by ordered elementary prior mass $d\mu$ confined to likelihood values within $[L_i, L_{i+1}]$. Each prior mass value μ_i corresponds to the likelihood value $L_i = L(\mu_i)$. In relating each L_i with its outward interval $[\mu_i, \mu_{i-1})$, the evidence can be approximated according to Eq. (3.16) by

$$Z \approx \sum_{i=1}^{\infty} L_i \Delta\mu_i, \quad \text{with} \quad \Delta\mu_i = \mu_{i-1} - \mu_i, \tag{3.20}$$

where $\mu_0 = 1$, $\mu_\infty = 0$, and $L_\infty = L_{\max} < \infty$.

Numerical implementation starts drawing a finite number N of random samples $[\boldsymbol{\Theta}^{(1)}, \cdots, \boldsymbol{\Theta}^{(N)}]$ from the prior probability $\pi(\boldsymbol{\Theta})$ with very low likelihood values. This is most easily achieved if outer bounds of the parameter ranges can be roughly specified *a priori*. The prior mass, according to Eq. (3.18), then corresponds to $\mu_0 \approx 1$. Among N samples as illustrated in Fig. 3.7, the next step is to identify and record the sample associated with the lowest likelihood value, denoted by $[\boldsymbol{\Theta}^{(1)}, L_1]$. The replacement of this lowest value by another new sample constrained by $L(\boldsymbol{\Theta}) > L_1$ keeps the total number of the samples to be N. At the ith iteration, a new sample associated with $[\boldsymbol{\Theta}^{(i)}, L_i]$ is recorded and replaced by the next new one constrained by $L(\boldsymbol{\Theta}) > L_i$.

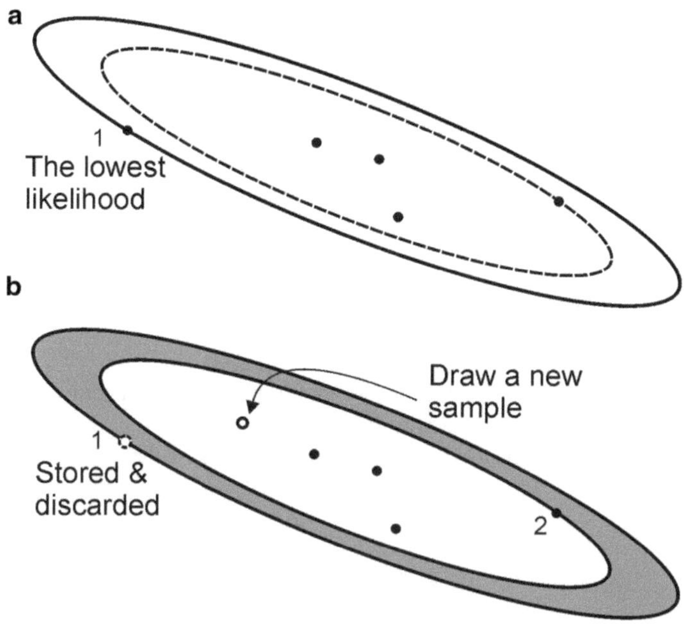

Fig. 3.7 Nested sampling algorithm using five random, uniformly distributed samples ($N = 5$) over the likelihood contours of a two-dimensional parameter space at each iteration. (**a**) The sample associated with the lowest likelihood value (L_1) is identified. (**b**) The sample (denoted by a *dashed circle*) associated with the lowest likelihood value is recorded and discarded. A new sample (denoted by a *circle*) is drawn constrained within the likelihood contour $> L_1$

Figure 3.7 conceptually illustrates this sampling procedure with $N = 5$. In Fig. 3.7a, among $N(=5)$ sampled likelihood values, the lowest one labeled by "1" is identified. In Fig. 3.7b this sample, labeled by a dotted circle, is recorded and discarded, upon which a new sample is drawn from the prior distribution constrained upon the condition $L_{\text{new}} > L_1$. After adding this new sample to the remaining $(N - 1)$ samples, another sample point is then identified to have the lowest likelihood value among the new N samples, labeled by "2" in Fig. 3.7b. In Fig. 3.7b the shaded area between two likelihood contours, when projecting onto the prior distribution, corresponds to the elementary prior mass $\Delta\mu_1$. It can be shown [26] that $\Delta\mu_i$ in Eq. (3.20) can be approximated statistically as

$$\Delta\mu_i \approx \left(\frac{1}{N+1}\right)\left(\frac{N}{N+1}\right)^{i-1}. \tag{3.21}$$

Accumulating Eq. (3.20) to infinity can be interrupted either by specifying termination after R iterations or by specifying a convergence criterion.

3.6.3 Posterior Samples for Parameter Estimation

While accumulating the evidence Z in Eq. (3.20), samples $\boldsymbol{\Theta}^{(i)}$ associated with the likelihood value L_i are recorded in R objects $[\boldsymbol{\Theta}^{(1)}, L_1]$, $[\boldsymbol{\Theta}^{(2)}, L_2], \ldots, [\boldsymbol{\Theta}^{(R)}, L_R]$. Posterior sample values are only of interest if the proper model preferred by the data has been selected via evidence estimation. According to Eqs. (3.13) and (3.21), the posterior samples are readily determined by

$$p_i(\boldsymbol{\Theta}|\mathbf{D}, \mathbf{H}) = \mu_i L_i / Z \approx \left(\frac{N}{N+1}\right)^i L_i / Z, \tag{3.22}$$

given the selected model without any further effort to sample them. For numerical convenience, Eq. (3.22) is evaluated logarithmically.

Probabilistic moment calculations [30] yield the mean values of the parameters $\langle \boldsymbol{\Theta} \rangle$ by

$$\langle \boldsymbol{\Theta} \rangle = \frac{1}{R} \sum_{i=1}^{R} \boldsymbol{\Theta}^{(i)}. \tag{3.23}$$

and bearing in mind that $\langle \boldsymbol{\Theta} \rangle$, given the model $\mathbf{H}_s$ of s slopes contains $2 \cdot s + 1$ parameters $\langle \boldsymbol{\Theta} \rangle = \{\langle \theta_1 \rangle, \ldots, \langle \theta_{2s+1} \rangle\}$ [see Eq. (3.3) with $\mathbf{A}_s$ and $\mathbf{T}_s$]. The expected element at jth row, kth column of the covariance matrix $\langle \mathbf{C} \rangle = [\langle C_{jk} \rangle]$ can be estimated by

$$\langle C_{jk} \rangle = \frac{1}{R} \sum_{i=1}^{R} \left(\theta_j^{(i)} - \langle \theta_j \rangle\right) \left(\theta_k^{(i)} - \langle \theta_k \rangle\right), 1 \leq j, k \leq 2 \cdot s + 1, \tag{3.24}$$

respectively, where each $\langle \theta_k \rangle$ is estimated by Eq. (3.23). The covariance matrix $\langle \mathbf{C} \rangle$ is a square matrix of dimension $2 \cdot s + 1$. The individual variance σ_j^2 and corresponding standard deviation σ_j of jth decay parameter can be estimated from the diagonal elements of the co-variance matrix $\langle \mathbf{C} \rangle$ as discussed in [30]. The inter-relationship between decay parameters is measured by the cross-correlation coefficient

$$ccc(j, k) = \langle C_{jk} \rangle / \sqrt{\langle C_{jj} \rangle \langle C_{kk} \rangle}. \tag{3.25}$$

If the mean values of decay parameters cannot meet the required accuracy based on the available posterior samples, the mean values in Eq. (3.23) and the covariance matrix in Eq. (3.24) estimated so far can guide a further effort to refined sampling, using e.g., importance sampling [25], where the proposal distribution, essential to the importance sampling, can be more accurately assigned.

3.7 Discussion

As an analysis method for quantifying energy decay characteristics, the methods reviewed in this chapter are model-based methods, which critically depend on the model involved. No decay characteristics can be inferred without an appropriate parametric model. Note that linear least-squares fitting is essentially a model-based method as well. The reason that correcting measures must be applied in previous analysis [2, 5, 9, 12] is that the linear model misrepresents Schroeder decay functions. Even Bayesian decay analysis can also provide misleading estimates if incorrect models are used. In this specific application, the decay order (number of slopes) must be determined before using the estimated parameters to explain the data. Fortunately, the Bayesian framework also embodies decay model selection, and quantitatively implements Occam's razor [24]. One should apply Bayesian model selection to a proper model set to determine which model among the competing models is appropriate before interpreting the decay parameters.

The model-based Bayesian decay analysis discussed in this chapter can be applied to both reverberation time estimation and to quantification of multiple-slope decay characteristics. Efficient sampling methods within the framework of Bayesian inference, particularly dedicated to multiple-slope decays, are still the on-going research of the author's team. Perceptual aspects associated with multiple-slope decays represent another potentially promising field of future research. Efficient implementation and successful applications of the model-based Bayesian decay analysis discussed in this chapter can benefit the future investigations since the quantification of acoustical conditions is inevitable in systematic psychoacoustic investigations.

3.8 Conclusions

Sound energy decay analysis in room-acoustics is of fundamental relevance and Schroeder integration is at its foundation. This chapter is mainly dedicated to sound energy decay analysis using the model-based Bayesian inference method. The parametric model derived from Schroeder integration is used for the task, and at the same time, it is also helpful to understand the behavior of Schroeder decay functions. In room-acoustics practice, Schroeder decay models of different orders are often competing with each other to explain or predict the data. The calculation of Bayesian evidence is then of central importance for both decay model selection and decay parameter estimation problems, more specifically in room-acoustic applications where reverberation times or decay characteristics in a room under investigation need to be quantified. This chapter outlines how the nested sampling method can be applied cohesively within the Bayesian framework to obtain both energy decay order selection and the decay parameter estimation.

Acknowledgments The author would like to dedicate this work to the memory of Manfred Robert Schroeder whose great interest in the author's work on room-acoustic energy decay analysis has inspired him to continue this endeavor. The author is thankful to Tomislav Jasa and Jonathan Botts for their insightful discussions and for implementing the nested sampling for this work, to Jean-Dominique Polack, Dejan Ciric, and Brian Katz for their critical comments on an early version of this chapter, and to Cameron Fackler and Paul Luizard for proofreading the final version of the chapter.

References

1. Schroeder, M.R.: New method of measuring reverberation time. J. Acoust. Soc. Am. **37**, 409–412 (1965)
2. Kuerer, R., Kurze, U.: Integrationsverfahren zur Nachhallauswertung. Acustica **19**, 314–322 (1967/1968)
3. Kuttruff, H., Jusofie, M.: Messungen des Nachhallverlaufs in meheren Räumen ausgefuehrt nach dem Verfahren der integrieten Impulseantworten (Energy decay measurements in rooms using the integrated impusle-response method). Acustica **11**, 1–9 (1969)
4. Bodlund, K.: On the use of the integrated impulse response method for laboratory reverberation measurements. J. Sound Vib. **56**, 341–362 (1978)
5. Chu, W.T.: Comparison of reverberation measurements using Schroeder's impulse method and decay-curve averaging method. J. Acoust. Soc. Am. **63**, 1444–1450 (1978)
6. Luizard, P., Katz, B.F.: Coupled volume multi-slope room impulse responses: a quantitative analysis method. In: Proceedings of IoA, Dublin (2011)
7. Cremer, L., Müller, H.A., Schultz, T.J.: Principles and Applications of Room Acoustics. Applied Science Publishers, London (1982)
8. Xiang, N., Jing, Y., Bockman, A.: Investigation of acoustically coupled enclosures using a diffusion equation model. J. Acoust. Soc. Am. **126**, 1187–1198 (2009)
9. Vorländer, M., Bietz, H.: Comparison of methods for measuring reverberation time. Acustica **80**, 205–215 (1994)
10. Hirata, Y.A.: A method of eliminating noise in power responses. J. Sound Vib. **82**, 593–595 (1982)
11. Xiang, N.: Two approaches to evaluation of reverberation times from Schroeders decay curves. In: Proceedings of 15th ICA, Trondheim, IV 175–178 (1995)
12. Lundeby, A., Vigran, T.E., Bietz, H., Vorländer, M.: Uncertainties of measurements in room acoustics. Acustica **81**, 344–355 (1995)
13. Schroeder, M.R.: Complementarity of sound buildup and decay. J. Acoust. Soc. Am. **40**, 549–551 (1966)
14. Xiang, N.: Evaluation of reverberation times using a nonlinear regression approach. J. Acoust. Soc. Am. **98**, 2112–2121 (1995)
15. Faiget, L., Legros, C., Ruiz, R.: Optimaization of the impulse response length: application to noisy and highly reverberant rooms. J. Audio Eng. Soc. **46**, 741–750 (1998)
16. Xiang, N., Goggans, P., Jasa, T., Robinson, P.: Bayesian characterization of multipleslope sound energy decays in coupled-volume systems. J. Acoust. Soc. Am. **129**, 741–752 (2011)
17. Xiang, N., Ahnert, W., Feistel, R.: A nonlinear regression approach to the evaluation of reverberation times using Schroeder's integrated impulse response method (A). J. Acoust. Soc. Am. **94**, 1846 (1993)
18. Xiang, N., Ahnert, W., Feistel, R.: Nachhallauswertung mittels nicht-linearer Regression (Reverberation time evaluation using nonlinear regression). In: Proceedings of Fortschritte der Akustik, DAGA93, pp. 255–258 (1993)
19. Xiang, N., Vorländer, M.: Using iterative regression for estimating reverberation times in two coupled spaces (A). J. Acoust. Soc. Am. **104**, 1763 (1998)

20. Xiang, N., Goggans, P.M.: Evaluation of decay times in coupled spaces: Bayesian parameter estimation. J. Acoust. Soc. Am. **110**, 1415–1424 (2001)
21. Xiang, N., Goggans, P.M.: Evaluation of decay times in coupled spaces: Bayesian decay model selection. J. Acoust. Soc. Am. **113**, 2685–2697 (2003)
22. Schroeder, M.R.: Digital simulation of sound transmission in reverberant spaces. J. Acoust. Soc. Am. **47**, 424–431 (1970)
23. Jaffe, J.C.: Innovative approaches to the design of symphony halls. Acoust. Sci. Tech. **26**, 240–243 (2005)
24. MacKay, D.: Information Theory, Inference and Learning Algorithms, Chapters 28, 29. Cambridge University Press, Cambridge (2002)
25. Jasa, T., Xiang, N.: Efficient estimation of decay parameters in acoustically coupled spaces using slice sampling. J. Acoust. Soc. Am. **126**, 1269–1279 (2009)
26. Skilling, J.: Nested sampling. In: Fisher, R. et al. (eds.) Bayesian Inference and Maximum Entropy Methods in Science and Engineering, American Institute of Physics, vol. 735, pp. 395–405 (2004)
27. Jasa, T., Xiang, N.: Using nested sampling in the analysis of multi-rate sound energy decay in acoustically coupled rooms. In: Knutz, K. et al. (eds.) Bayesian Inference and Maximum Entropy Methods in Science and Engineering, American Institute of Physics, vol. 803, pp. 189–196 (2005)
28. Sivia, D., Skilling, J.: Bayesian Data Analysis—A Bayesian Tutorial, 2nd edn. Oxford University Press, New York (2006)
29. Jasa, T., Xiang, N.: Nested sampling applied in the Bayesian room-acoustics decay analysis. J. Acoust. Soc. Am. **132**, 3251–3262 (2012)
30. Xiang, N., Goggans, P.M., Jasa, T., Kleiner, M.: Evaluation of decay times in coupled spaces: reliability analysis of Bayeisan decay time estimation. J. Acoust. Soc. Am. **117**, 3707–3715 (2005)

***

Biography

Prof. Dr. **Ning Xiang**, Director, Graduate Program in Architectural Acoustics, Rensselaer Polytechnic Institute, New York, USA. He obtained his Ph.D. (Dr.- Ing) in 1990 at BochumRuhr-University, Germany. He is a Fellow of the Acoustical Society of America, a Fellow of the Institute of Acoustics, UK, Associate Editor of the Journal of the Acoustical Society of America (JASA), and a book-series Editor with J. Ross Publishing. Inspired by Professor Manfred Schroeder, he has been working on applications of number-theoretic sequences since 1988. Among his long list of journal publications, he and Professor Manfred Schroeder have coauthored one JASA paper for simultaneous dual source measurements in 2003, a measurement technique being critical in outdoor sound propagation research. Since 1993 one of his research interests has been in room-acoustic energy decay analysis. Since 1999 he, along with his colleagues and students, has applied Bayesian inference in Schroeder's energy decay analysis. He is a recipient of Wallace Clement Sabine Medal from the Acoustical Society of America in 2014. [Photo: with Professor Manfred Schroeder in 2001, during the time he demonstrated initial success using Bayesian inference for advanced room-acoustics decay analysis.]

Chapter 4
Estimating the Crossover Time Within Room Impulse Responses

Guillaume Defrance and Jean-Dominique Polack

Abstract In 1954, M.R. Schroeder found that above a transition frequency, modes overlap and can even be superposed. This particular frequency has been called the Schroeder frequency. In the time domain, there also exists a time after which the density of sound rays, that have been emitted into a room, becomes homogeneous and isotropic. The purpose of the present chapter is to illustrate similar findings in the time domain, namely a domain where reflections are singled out, shortly after the direct sound, followed at longer times by superposed reflections that follow a different distribution in time. In previous publications, the transition time between the two regimes was called mixing time. Here, using two different approaches, it is experimentally shown that the estimated time is inconsistent with the theory of dynamical systems. Therefore, a more generic name is proposed for the transition from early reflections to late reverberation, the crossover time.

4.1 Introduction

In 1954, M.R. Schroeder experimentally investigated the distribution of eigenfrequencies measured in a room [1]. While, in 1912, Weyl had theoretically shown that this distribution is a quadratic function of the frequencies [2], Schroeder observed that above a certain frequency, the density of eigenmodes is apparently constant. In other words, he found that above a transition frequency, modes overlap and can even be superposed. This particular frequency has been called the Schroeder frequency. At low frequency, eigenmodes are discrete and are perceptively important since they participate to the sound coloration of the room. Above the Schroeder

G. Defrance (✉)
Laboratory of Signals and Systems, Dept. Electromagnetism, SUPELEC, 3 rue Joliot-Curie, 91192 Gif-sur-Yvette, France
e-mail: defrance.all@gmail.com

N. Xiang and G.M. Sessler (eds.), *Acoustics, Information, and Communication*, Modern Acoustics and Signal Processing, DOI 10.1007/978-3-319-05660-9_4,

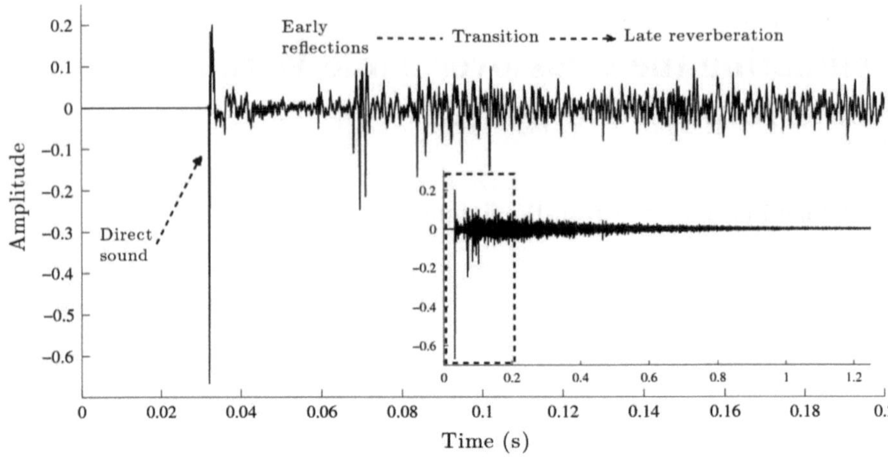

Fig. 4.1 Early times of an experimental RIR (encapsuled graphic), measured in Salle Pleyel

frequency, eigenmodes are statistically distributed [1]. This frequency depends on the reverberation time and the volume of the hall, and is given by:

$$f_{\mathrm{Sch}} = \sqrt{\frac{\pi c_0^3 T_{\mathrm{R}}}{4\pi V \ln(10)}} \tag{4.1}$$

$$\approx 2000\sqrt{\frac{T_{\mathrm{R}}}{V}} \tag{4.2}$$

where f_{Sch} is in Hz, T_{R} is the reverberation time of the room in s, c_0 is the speed of sound in m s^{-1}, and V is the volume of the hall in m^3.

In the time domain, there also exists a time after which the density of sound rays, that have been emitted into a room, becomes homogeneous and isotropic. We call that time the *crossover time*. This particular time stands for the transition between two regimes of room impulse responses: early reflections and late reverberation, also called the diffuse sound field. While at early times, reflections are discrete, the late reverberation is statistically distributed (i.e., superposition and overlap of sound rays) [3] (Fig. 4.1).

Therefore, the knowledge of the transition time can be particularly interesting when synthesizing room impulse responses (RIRs) since the computation of room impulse responses can be stopped at this time and synthesized by an exponentially decaying filtered Gaussian noise.

In the following, we shortly review statistical room acoustics in order to introduce the transition time in detail. Then, we show how to estimate the transition time within experimental RIRs using two different signal processing tools: Matching Pursuit and the eXtensible Fourier Transform. After comparing our results, we finally discuss the relationship between the transition time and dynamical properties of large rooms. In particular, we focus on the first name given to the crossover time, that is, the *mixing time*. By reviewing the meaning and definition of mixing and ergodicity, we justify the name of crossover time and explain its difference with the mixing time.

4.2 A Brief Review of Statistical Room Acoustics

A standard way to document the acoustics of a room is to measure a set of RIRs. The RIR is built by the superposition of *arrivals* [4], that is, modified versions of an original pulse emitted by the source and reaching the receiver after travelling through the room. Sound emitted by the source undergoes scattering and absorption (i.e., filtering) when encountering boundaries of the room. The source pulse is therefore divided in many wavelets that each follows a different *trajectory* within the hall. When these trajectories are closed, that is, when sound rays stick to the same path, eigenmodes are created.[1] RIRs are composed of all these trajectories with due consideration for their respective amplitude. In room acoustics, the time distribution of these arrivals plays an important role since it is directly linked to the acoustical quality of the room, as is demonstrated in numerous books [3, 5].

Since Weyl's path-breaking paper on eigenfrequency density in bounded rooms [2], it is well known that this density increases with the square of the frequency. However, when experimentally measuring this density in a box, Schroeder found out in 1954 [1] that it is not the case at high frequencies: the finite width of the modes, created by the losses, leads to a superposition of many modes at any frequency and to a constant density of peaks. Schroeder carried out further his investigation in order to determine the transition frequency between the two regimes, and introduced what is now called the Schroeder frequency, linked to volume and decay time, i.e., to mode width.

Weyl's estimation of eigenfrequency density was based on the approximation of the general solution to the Helmholtz equation in a bounded space by the Green function of free space. Thus, the notion of trajectory was implicitly introduced in mode counting, albeit in a crude way as represented by the Green function that links a source position to a receiver position. In Weyl's free space approximation, no reflections on the boundaries were taken into account, but later refinements to his theory took reflections into account [6].

As a consequence, Weyl's theory and its successors naturally induce researchers to look at the temporal succession of arrivals from trajectories linking the source to the receiver and taking into account reflections at the boundaries.

At short times, arrival density is not constant, but increases with time. A simple theory, derived for rectangular rooms [7], teaches us that the density $D_e(t)$ Eq. (4.3) is proportional to the square of the time elapsed since the sound was emitted by the

[1] In the semi-classical approach, we talk about periodic orbits.

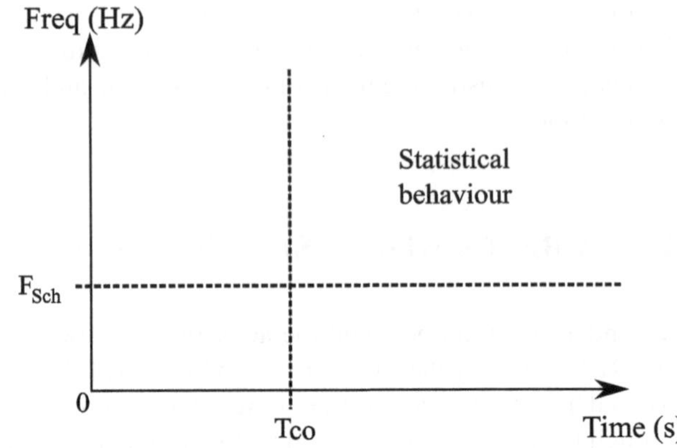

Fig. 4.2 Schematical representation of the Schroeder frequency and the crossover time

source (which is different from the time laps between the arrival and the direct sound):

$$D_e(t) = \frac{dN}{dt} = \frac{4\pi c_0^3}{V} t^2 \tag{4.3}$$

$$N(t) = \frac{4\pi V}{3c_0^3} t^3 \tag{4.4}$$

where $D_e(t)$ is in number of arrivals per second, N is the number of arrivals up to time t, c_0 is the speed of sound in m s^{-1}, and V is the volume of the room in m^3.

The present chapter investigates whether a transition time can experimentally be demonstrated between two domains: an initial time domain where many arrivals are discrete, and a late time domain where arrivals are superposed and therefore become statistically equidistributed, just as they are in the frequency domain (Fig. 4.2). Recall that the crossover time is measured from the time of emission of the impulse by the sound source.

4.3 Estimating the Crossover Time

4.3.1 Introduction

The first idea that comes to mind is to estimate the transition time by detecting the time at which the distribution of the RIR tends to be Gaussian. This idea arises from the report that if the spectrum of an RIR, measured in a large room, is Gaussian above the Schroeder frequency, then its inverse Fourier transform also provides a Gaussian distribution. However, the different statistical tests that one could use

(Kolmogorov–Smirnov, Lilliefors, etc.) are very dependent on the number of samples comprised within the analysis window, on the one hand, and on the threshold set for each test, on the other hand [8–10].

Therefore, estimating the crossover time must be carried out differently. Using the algorithm of Matching Pursuit (MP) [11], we show how the transition time can be estimated within RIRs by studying the temporal distribution of arrivals [12]. In the time–frequency domain, it is also possible to estimate the transition time as the time at which the Fourier coefficients of the spectrum of the RIR tends toward a Gaussian distribution. This is achieved using the eXtensible Fourier Transform [13].

4.3.2 Matching Pursuit

Assuming the hall to be a system of time invariant linear impulse responses, or a bunch of filters with delay, the source is expected to be filtered and translated in time along the RIR. Therefore, supposing a high correlation between the RIR and the pulses emitted by the source (i.e., the direct sound), with due consideration to the filtering of the room, MP appears to be well suited for this purpose. Matching Pursuit provides a sparse decomposition of a signal, that is, it decomposes any signal into a linear expansion of waveforms that belong to a dictionary. These waveforms are selected iteratively in order to best match the signal structures. Although Matching Pursuit is nonlinear [11], it maintains an energy conservation which guarantees its convergence. In practice the number of iterations must be finite, leading to only an approximate decomposition of the signal. In these experiments, the dictionary of waveforms (or atoms) is limited to the direct sound itself, translated in time. The determination of its exact temporal boundaries is of importance for a perfect match with the RIR, but is achieved within MP itself, as explored in [14].

Matching Pursuit can help understanding more deeply the architecture of an RIR since this algorithm introduced by [11] provides information, which can be seen as maxima of correlation Eq. (4.6) [15] between two signals, the RIR (x) and the direct sound (the mother-atom).

Matching Pursuit works as follows:

1. Initialization: $m=0$, $x_m=x_0=x$.
2. Computation of the correlations between the signal x_m and every atom γ of a dictionary φ using inner products:

$$\forall\gamma \in \varphi : \mathrm{CORR}(x_m, \gamma) = \left|\langle x_m, \gamma\rangle\right| \tag{4.5}$$

   The dictionary φ is a set of atoms γ, of the same length than x, constituted by the direct sound and translated in time, by step of one sample.
3. Search the most correlated atom, by searching for the maximum inner product:

$$\widetilde{\gamma}_m = \text{argmax}\{|\langle x_m, \gamma\rangle|\}_{\gamma \in \varphi} \tag{4.6}$$

4. Substracting the corresponding weighted atom $\widetilde{\alpha}_m \gamma_m$ from the signal x_m:

$$x_{m+1} = x_m - \alpha_m \widetilde{\gamma}_m \tag{4.7}$$

$$x_{\mathrm{R}}^{(m)} = \sum\nolimits_{k \le m} \alpha_k \cdot \widetilde{\gamma}_k \tag{4.8}$$

where $\alpha_m = \langle x_m, \widetilde{\gamma}_m\rangle$

5. (a) Stops if the desired level of accuracy is reached: $R = x_{m+1}$
   (b) Otherwise, re-iterate the pursuit to step 2:

$$m \leftarrow m + 1.$$

where x is the RIR, R the residual, γ the atom (here, the direct sound), φ the dictionary of atoms γ, and $x_{\mathrm{R}}^{(m)}$ the reconstructed signal.

In theory, any signal x can be perfectly decomposed in a set of atoms for an infinity of iterations. In practice, this number must be finite and a stopping criterium has to be set. The authors propose to use the signal/residual ratio (*SRR*) in dB of the norm L2 of x over the norm L2 of the residual (R) defined as:

$$SRR = 10\log_{10}\left(\frac{\|x_2\|}{\|R_2\|}\right) \tag{4.9}$$

The quality of the decomposition of x in atoms depends on the value of *SRR*, that is, the stopping criterium. On the one hand, for a too low *SRR*, the residual has an energy level too high, and the rebuilt signal $x_{\mathrm{R}}^{(m)}$ is an impoverished approximation of x. On the other hand, a too high *SRR* leads to a high number of iterations. In that case, it is not necessary to perform more iterations beyond a certain threshold of accuracy. In [12], we show that the best value of *SRR* is found when the mean differences of room acoustical indices (RT_{30}, RT_{20}, EDT, C_{80}, etc.) [16] between synthesized and original RIRs is below 5 %.

If the source impulse is unknown, it is possible to recover it from recorded RIRs running MP. The temporal boundaries of the direct sound that are thought to be the best correspond to the lowest number of iterations ran by MP onto experimental RIRs, with a stopping criterium of $SRR = 20$ dB to reach. For further details, please refer to [12, 14].

A key point of this method is the compensation of the energy decay of RIRs. Indeed, the probability to detect arrivals is directly linked to the local energy of the signal. As this latter decreases exponentially, one can expect the probability of detection to decrease exponentially too. Therefore, energy compensation, by making the signal stationary, ensures equal weight to all parts of the RIR, thus equiprobability of detecting arrivals. Note that the energy compensation is achieved

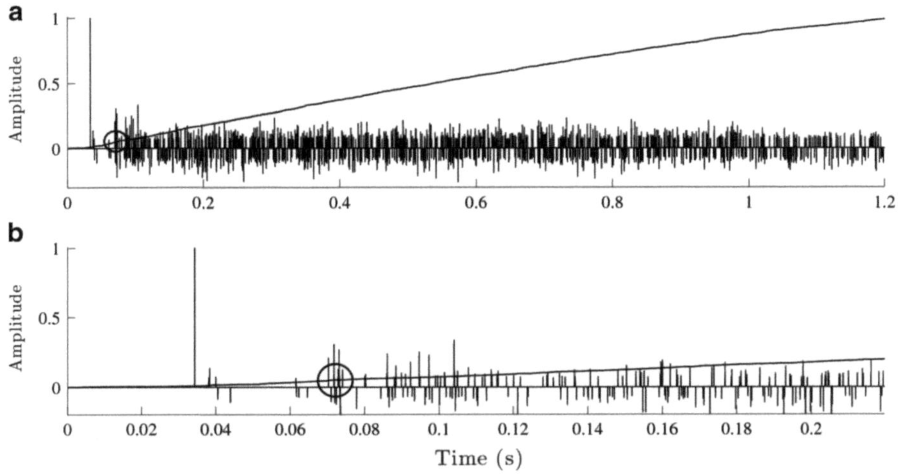

Fig. 4.3 (**a**) Crossover time (*circle*) detected on a *CDF* (*black bold line*) for an experimental RIR and the set of estimated arrivals (*grey*) (*SRR* = 5 dB). Plot (**b**) details the beginning of plot (**a**). Note the energy compensation applied to the signal

by applying an inverse exponential, whose argument is proportional to the reverberation time and to the mean absorption [3].

Matching Pursuit run on RIR provides a linear set of coefficients which can be seen as a temporal distribution of arrivals of the RIR, providing the knowledge of the cumulative distribution function (*CDF*) of arrivals (Fig. 4.3).

According to the theory of room acoustics, arrivals overlap for times greater than the transition time. We observe that MP detects only one arrival instead of two when the time delay between them is inferior or equal to the equivalent duration [17] of the source impulse (i.e., the direct sound). Therefore, the crossover time T_{co} is estimated by looking for:

$$T_{\text{co}} = \arg\min\left(t_{i+1} - t_i \leq \tilde{d}_{\text{Source}}\right) \tag{4.10}$$

where t_i is the time of occurrence of the ith arrival and $\tilde{d}_{\text{source}}$ is the equivalent duration of the source impulse.

At early times ($t < T_{\text{co}}$), a few arrivals are detected (Fig. 4.4). A cubic fit of this part of CDF curves permits to verify that the number of arrivals is a function of t^3, as seen in Eq. (4.4). The transition time marks the time at which CDF of estimated arrivals increases at a constant rate (linear part), which is coherent with the definition of late reverberation. In other words, for times greater than T_{co}, the RIR can be defined as behaving stochastically, rather than deterministically. In a large room like a concert halls, we observe that the evolution of the number of arrivals as a function of time (for $t \geq T_{\text{co}}$) is almost constant all over the room [12].

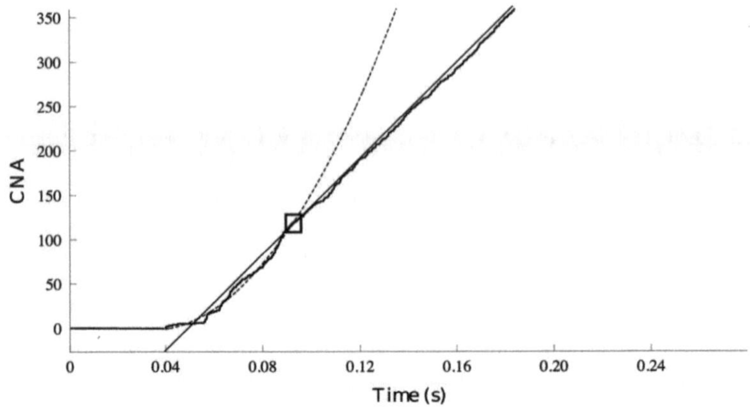

Fig. 4.4 Detail of a cubic ($t < T_{co}$) and a linear ($t > T_{co}$) fits made on a Cumulative Number of Arrivals (*CNA*). *Plain bold line*: *CNA*. *Dashed bold line*: cubic fit. *Dashed line*: linear fit. *Square*: crossover time

Remark If applying MP to RIRs is straightforward, this method presents disadvantages. The first outcome is the use of the dictionary of atoms φ, which is only constituted of the direct sound translated in time. The filtering of the room is thus not taken into account. Therefore, at each iteration of the algorithm, some spurious waveforms are created (due to Eq. (4.7)) that can be estimated as arrivals. This behavior of MP has been deeply studied by Sturm et al. [18].

4.3.3 eXtensible Fourier Transform

The eXtensible Fourier Transform [13], or XFT, is comparable to a memory process, that is, the time of occurrence of events (such as arrivals within RIRs) can be detected compared to what happened in the past (e.g., time of occurrence of the direct sound).

The XFT consists in computing the Discrete Fourier Transform of a signal using a window that grows from a given starting time (Fig. 4.5). For a discrete signal $x(n)$ of length N samples, its XFT is defined as:

$$X(p,f) = \int_{n=0}^{N-1} w\left(\frac{n}{N}\frac{M}{p}\right) x(n) e^{-2j\pi n f/N} \tag{4.11}$$

with $w(t)$ the analysis window, which is a continuous function of time:

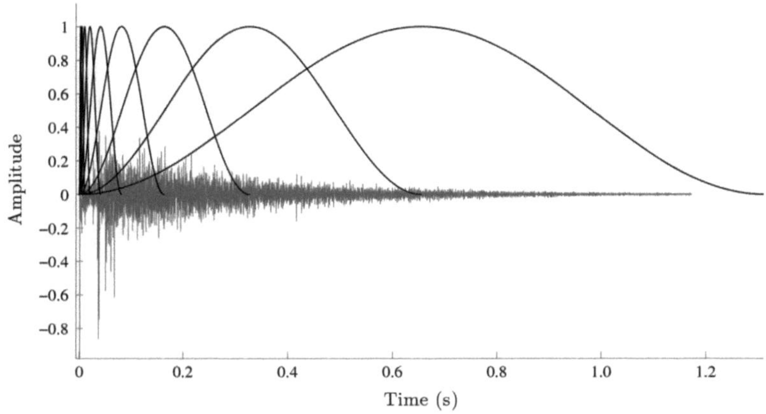

Fig. 4.5 Scheme of the principle of the XFT. For each window, a Discrete Fourier Transform of the signal x is computed. In that case, the window w is a Hann window, which starts from the time of arrival of the direct sound. The latter is detected as described in [19]

$$\mathrm{supp}(w(t)) = [0, 1], \tag{4.12}$$

where $X(p,f)$ is the complex matrix of the XFT, $x(n)$ a signal of N samples length, w the analysis window that grows with a constant rate while p increases (the first window has a width L_0), f is the discrete frequency in Hertz ($f \in Z$), M is the maximum number of windows w that the signal x can admit ($M = N/L_0$), p is the number of windows ($p = 1, 2, \ldots, M$), and $\mathrm{supp}(w(t))$ is the support of the function $w(t)$.

Equation (4.11) can be written differently:

$$X(p,f) = \int_{m=1}^{p} \int_{n=0}^{L_0-1} w\left(\frac{(m-1)L_0 + n}{pL_0}\right) x(n) e^{-2j\pi fn/N}. \tag{4.13}$$

Equation (4.13) emphasizes the fact that the XFT is a sum in the time domain of spectral components of the signal. Furthermore, the XFT can be seen as a DFT computed at different times of the signal but keeping the phase shift information.

The XFT is calculated for the same total number of samples (using zero padding) in order to keep the same nominal frequency resolution for each window. The number of samples (N_{DFT}) for computing each Discrete Fourier Transform is calculated as the highest power of 2: $N_{\mathrm{DFT}} = 2^{\lceil(\log(N)/\log(2))\rceil}$, where $\lceil\cdot\rceil$ is the ceiling symbol and N is the number of samples of the signal x.

Interpreting the magnitude of the XFT is not straightforward, although it is related to Schroeder's work. Indeed, it permits to visualize the progressive establishment of modes within the room at a given position. Therefore, we concentrate on the phase of the XFT, noted $\Phi(p,f)$.

The time evolution of $\Phi(p,f)$ can be understood as follows. As long as $t_{k+1} - t_k \gg 0$ (i.e., p is small), the function *unwrap*($\Phi(p,f)$) is approximately a linear function of frequency f, which is more or less robust. While for larger p ($\lim t_{k+1} - t_k \rightarrow 0$), if the number of arrivals is large enough, then the coefficients of the Fourier series expansion become independent and normally distributed [20, 21]. The XFT does not permit one to claim if a signal is normally distributed, but it does allow the visualization of the evolution of the system toward a random behavior. The time evolution of the XFT phase embodies the changing statistics of a signal. We propose calculating the sum of the differences between the unwrapped phases in the first window ($p = 1$) and successive ones ($p = 2, 3, 4, \ldots M$). The phase differences are calculated at each frequency f as follows:

$$\delta\Phi(p) = \int_{f=0}^{N-1} \mathrm{unwrap}[< X(1,f)] - \mathrm{unwrap}[< X(p,f)]. \tag{4.14}$$

Equation (4.14) can be seen as an estimate of the distance between the phases of $X(1,f)$ and $X(p,f)$. Therefore, the smaller the quantity $\delta\Phi(p)$, the more similar are the phase spectra. Overall, Eq. (4.14) is an estimate of the time of occurrence of arrivals, regarding the time origin given by the first window ($p = 1$).[2]

Figure 4.6 shows several graphs of the unwrapped phase of the XFT computed on an experimental RIR. As expected [3], the phase wraps faster between $\pm\pi$ with time, denoting a gradual changing of behavior in the signal, due for instance to an increasing number of events (such as arrivals and diffusion). This confirms findings from earlier investigations [12, 22, 23]. Estimating the crossover time is thus equivalent to estimating the time at which an important change of behavior occurs in the phases.

As seen in Figs. 4.7 and 4.8, $\delta\Phi(p)$ is an increasing function of time, up to a certain time, and comprises one main inflexion point. The latter is a breaking point in the statistics of the signal, and separates two states of behavior of the impulse response. The first state is deterministic, that is, arrivals can be discriminated, and their time of occurrence can be predicted (Sect. 4.2). The function that best fits this part of the curve $\delta\Phi(p)$ is a cubic function of time (Fig. 4.7), which confirms Eq. (4.4). The second state of behavior of the curve $\delta\Phi(p)$, which has a constant slope until a certain time, denotes a statistical behavior: arrivals are statistically distributed in time.

For a proper estimate of the crossover time, the first window must comprise the direct sound. The detection of the onset of RIRs and the estimation of the direct sound duration can be found in [14, 19]. Furthermore, as in Sect. 4.3.2, the energy decay of the RIR needs to be compensated in order to give the same weight to all arrivals within the signal, especially those at large times. In that case, the

---

[2] One may link this to [12, 22] and Sect. 4.2, in which the statistics of RIRs are investigated by looking at the cross-correlations in the time domain between the direct sound and the rest of the signal.

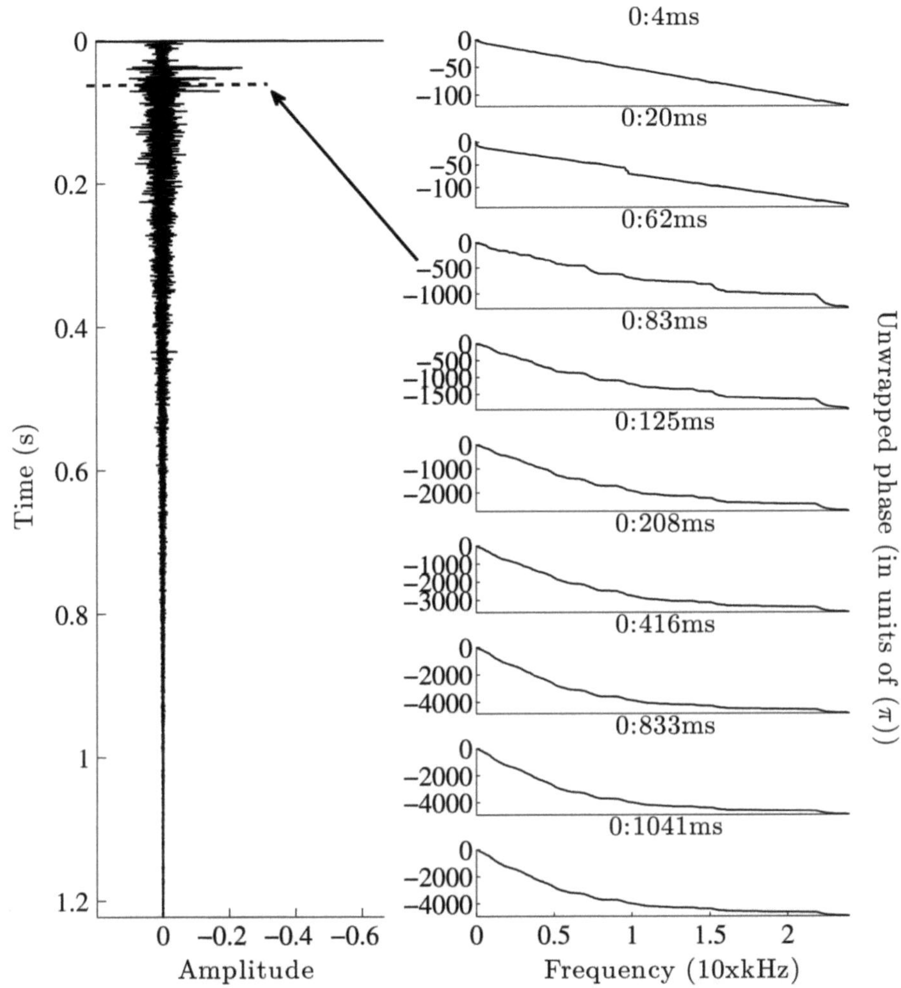

Fig. 4.6 Visualization of the unwrapped phases of the XFT computed on a RIR measured in Salle Pleyel (sound source: pistol shot). Note that the compensation of the energy decay of the RIR is applied, but not shown for a better readability. Furthermore, the time of propagation between the source and the receiver positions is not shown here. First window width: $L_0 = 128$ samples. The estimated crossover time is materialized by a *dashed line* superposed to the RIR

compensation of energy decay leads to use a rectangular window w Eq. (4.11) in order to avoid any other modifications of the amplitude of the signal x.

We propose to estimate the crossover time by searching the time of occurrence of the maximum of the curve $\delta^2\Phi(p)$, which is defined as:

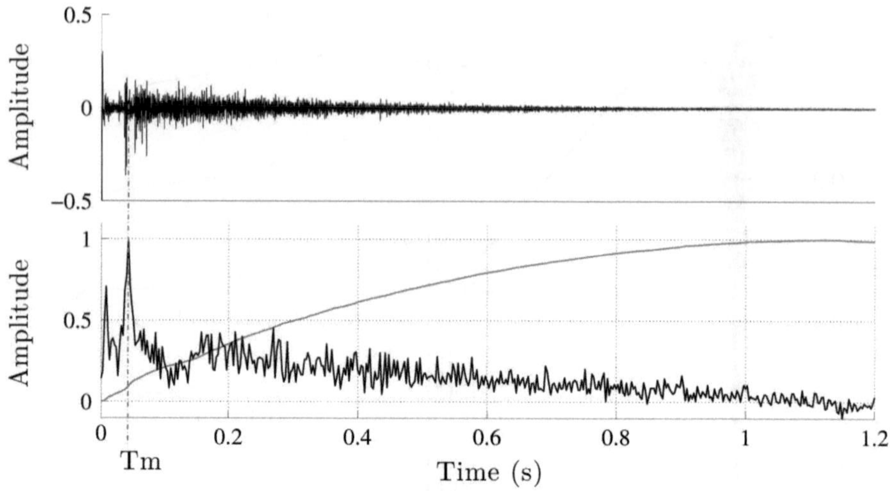

Fig. 4.7 *Grey*—experimental RIR (the energy decay compensation is applied, but not shown for a better readability). *Plain black line*—curve $\delta\Phi(p)$ Eq. (4.4) (normalized). *Black bold dashed line*—cubic fit. Note that $\delta\Phi$ is over sampled for a better readability. The graphic included shows the details of the selected region within the *dashed rectangle*

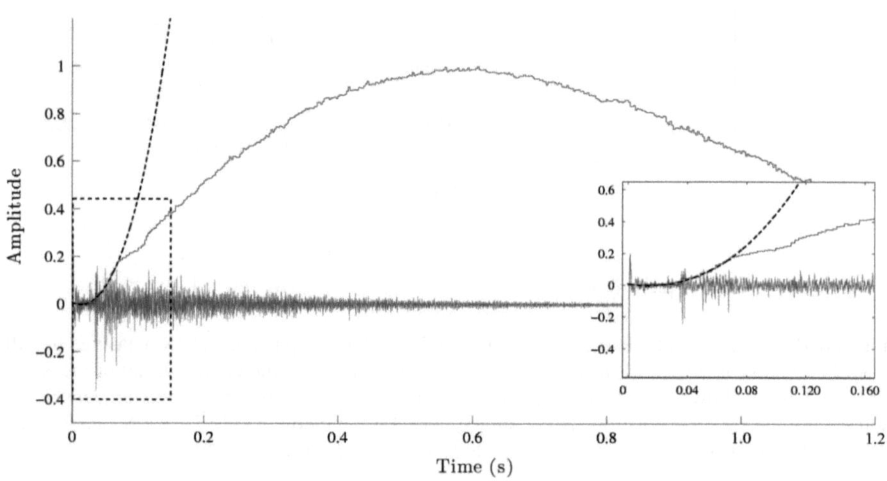

Fig. 4.8 *Top*—an experimental RIR (the energy decay compensation is applied, but not shown for a better readability). *Bottom*—Eq. (4.14) in *grey*; Eq. (4.15) in *black*. The breaking point (i.e., the crossover time) in the statistics of the RIR is materialized by a *dashed line*. Note that bottom graphic is over sampled for a better readability. Notice the other breaking points after the crossover time (other peaks). The time of propagation between the source and the receiver positions is not shown here. On *bottom graphic*, curves are normalized for an easier readability

$$\delta^2\Phi(p) = \delta\Phi(p+1) - \delta\Phi(p) \tag{4.15}$$

$$T_{\mathrm{co}} = \mathrm{argmax}_p\left(\delta^2\Phi(p)\right) * L_0/f_S \tag{4.16}$$

where $p = 1, 2, \ldots, M-1$, L_0 is the width of the first window w of analysis in samples, f_S is the sampling frequency in Hz, and T_{co} is the crossover time in second.

4.3.4 Comparing MP and XFT

In this section, we present crossover times estimated using MP and XFT[3] on experimental RIRs measured in Salle Pleyel (Paris, France) [25]. Salle Pleyel is a concert hall of 19,000 m^3 and 1,900 seats. We carried out a set of 21 measurements using pistol shots as sound sources and an omnidirectional microphone. Figure 4.9 shows crossover times estimated using MP and XFT.

For both estimators, transition times are a function of the source/receiver distance: the slope of each linear regression is proportional to the speed of sound. If several values estimated with MP and the XFT are the same, most of them are different. These differences can be explained as follows.

When using MP, transition times are a function of the stopping criterion (Sect. 4.3.2). The higher the SRR, the larger the number of estimated arrivals, and thus the larger the probability to detect the transition times at early times. Moreover, as the dictionary of MP is only composed of the direct sound of each RIR, artifacts of estimation are created at each iteration. Therefore, the number of arrivals can be overestimated. Finally, for MP and the XFT, the estimation of the direct sound can be erroneous when it is immediately followed by an arrival. And when using the XFT, the estimation of transition times also depends on the window width of analysis.

An interesting result is the values of transition times when the source and the receiver positions are the same. In such a case, values returned by the XFT predict that the time needed in order to reach the diffuse sound field is 0.044 s, that is, the mean free path of Salle Pleyel ($\widetilde{\ell} = 14\,\mathrm{m}$). The linear regression performed on the values given by MP predicts approximately half of the mean free path. It is difficult to conclude about this report. However, it seems realistic to assume that the diffuse sound field can only be established once sound rays have travelled the mean distance between two reflections within the hall.

[3] Both estimators, MP and XFT, have been validated on thousands of synthetic RIRs [12, 13], created by a stochastic model [24].

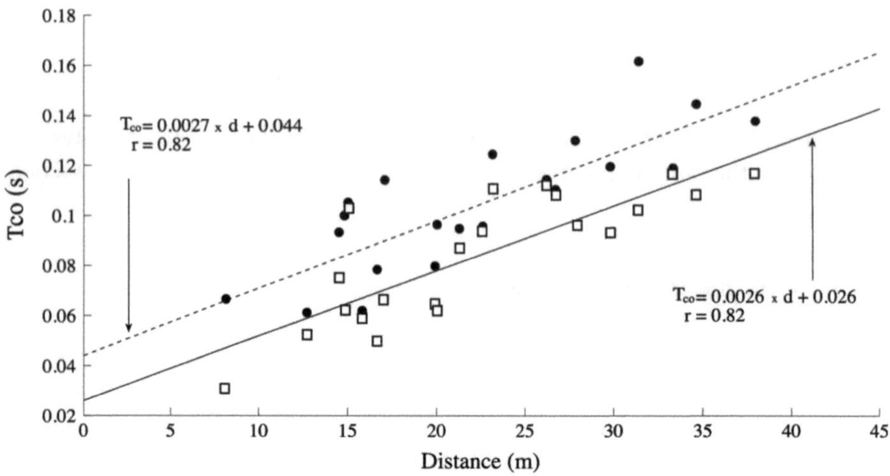

Fig. 4.9 Crossover times estimated within experimental RIRs. *Circles*: estimated with XFT. *Stars*: estimated with MP

4.4 Discussion Around the Crossover Time and the Mixing Time

The introduction of ergodic dynamical systems, made by Joyce [26], was able to justify Schroeder's spectral observations in the frequency domain. Joyce assumed large halls to be ergodic and mixing. In fact, he theoretically showed that the expression of the reverberation time given by Sabine is recovered if the room is mixing—thus, ergodic—and if absorption is weak and uniformly distributed within the space [26]. Since the publication of that article, mixing has been assimilated to late reverberation. In particular, the transition time from early reflections to the diffuse sound field was called the *mixing time* in the literature of room acoustics.

After briefly reviewing ergodicity and mixing, we discuss our methods and results regarding the theory of dynamical systems. We finally propose to use the general term *crossover time* instead of mixing time.

4.4.1 Ergodicity and Mixing

The hypothesis of ergodicity was first introduced by Boltzmann in 1871 for his kinetic theory of gases. But it is only in the 1930s that it has been theoretically proven [27]. The validity of the equidistribution of energy relies on the hypothesis of ergodicity. In fact, this hypothesis is validated a posteriori based on its relevant predictions, but it is still an hypothesis. Considering an ergodic system (at the thermodynamical equilibrium), all the states of the phase space have the same probability to be occupied. This means that the energy is equidistributed. In the theory of room acoustics, when the energy is equidistributed, that is, after a certain number of reflections, the diffuse sound field is fully established within the hall.

Referring to Krylov's theory of mixing [28], we learn that most dynamical systems gradually loose memory of their history with time. Thus, trajectories become independent of their origin, so that the probability of reaching any phase point, at any time, along the trajectory becomes equivalent (ergodicity). In such a case, the arrival density, at any receiver position, becomes constant and independent of the receiver position. The mixing character of a room is associated to the (exponential) divergence of sound rays emitted by a sound source, that were initially adjacents when emitted. Divergence of trajectories are created by obstacles (seats, balconies, etc.) or even by asperities of walls. Once the system is mixed, sound rays density is homogeneous and isotropic.[4]

Shannon's theory of information states that if all the states of a system are equiprobables, then its entropy is maximum [29]. The entropy is related to the information contained within the system. When the entropy is the highest, the system has lost all its initial information, that is, there is no memory of its initial state. In room acoustics, the loss of information is created by the numerous filtering that sound rays undergo on their trajectory (reflections on obstacles, dissipation, etc.). According to Shannon's theory, as long as the entropy of a system is not maximum, the system keeps evolving until it reaches its final state. For instance, a mixing system provides a maximum entropy.

It is worthwhile to recall that ergodicity has been experimentally proven only for a few systems (like Sinai's billiard or the Bunimovich stadium [30, 31]) but never for large halls, except using numerical simulations [32–34]. This explains why some can doubt about the relevance of this hypothesis within the context of room acoustics [35].

In fact, the *mixing time* is a naive concept. Indeed, a mixing system tends progressively (i.e., asymptotically) toward equilibrium, but does not suddenly reach this state. The hypothesis of ergodic rooms is useful to room acousticians since it allows one to assume that the energy is uniformly distributed within the hall after a sufficient number of reflections [26, 35, 36]. If experimental observations show that the rate of decay of the sound energy is statistically the same within the hall [37, 38], none of the studies prove that the energy is uniformly distributed within the phase space. Moreover, this equipartition of energy, that we assume in room acoustics, is only valid up to a certain distance of the boundaries of the hall, that is, where the migration of energy cannot be observed.

Several definitions of the mixing time can be found in the literature. The first one comes from the theory of dynamical systems, the second one can be found in the room acoustics literature (i.e., that transition time between early reflections and the diffuse sound field), and the third one is related to psychoacoustics. The latter has been given by Polack [39] based on Cremer's work [3]. Polack proposed to define the mixing time such as the time at which ten arrivals overlap within a window of 24 ms. Therefore, using $N = 10$ and $\Delta t = 24$ ms in Eq. (4.4), we obtain a heuristical formulation of the mixing time that is now well known:

[4] Note that a mixing system is ergodic but the inverse is not necessarily true.

$$T_M \approx \sqrt{V} \tag{4.17}$$

where T_M is expressed in ms and V is the volume in m^3.

The existence of these three definitions makes sometimes difficult the comprehension of the concept of mixing and mixing time. To estimate the mixing time according to its dynamical definition, one must experimentally observe the time evolution of the system in the phase space. While this approach requires to follow, within the room, the divergence of trajectories of sound particles, the two other definitions focus on a physical and a perceptual analysis of the system at a given position within the hall. It is thus obvious that the dynamical and the perceptual definitions are not related since mixing, in the perceptual sense, has to be understood as the inability to discriminate more than ten events in a given period of time. In this work, we do not discuss this alternative definition of mixing, especially because the comparison between the time distribution of arrivals estimated by MP and Eq. (4.17) is inconclusive [8].

4.4.2 Why the Crossover Time Cannot be Called the Mixing Time

Recent work have experimentally attempted to estimate the mixing time within room impulse responses [8–10, 12, 13]. While tests of Gaussianity, Matching Pursuit or the eXtensible Fourier Transform point out several phenomena such as the progressive apparition of diffusion, the growth of entropy, and the change of distribution of arrivals, the hypothesis of ergodicity and mixing is still not verified.

Running MP on RIRs measured in Salle Pleyel, we show that, after the crossover time, the total number of sound rays received (until the reverberation time) is not the same according to the receiver position in the hall. Nevertheless, the time evolution of this number is approximately constant in the hall (1,140 arrivals/s ± 7 %) [8, 12]. If this result suggests that the energy is uniformly distributed over the room, one cannot conclude that the system is mixing. But this result depends on the stopping criterion used for estimating arrivals within RIRs using MP (Sect. 4.3.2). Indeed, the higher the stopping criterion, the larger the number of estimated arrivals. Later measurements of RIRs carried out in Salle Pleyel with a B-format microphone have shown that there exists some privileged directions of propagation of sound rays at high frequencies. This is in contradiction with the assumed equipartition of energy.

On the other hand, the XFT and the phase difference are closer to the dynamical definition of mixing than MP. Indeed, the XFT is an estimate of the loss of memory of the system, that is, a progressive fall of correlation of the system regarding its initial state, in agreement with Krylov's and Shannon's theories. Moreover, as detailed in [13], the use of the XFT allows one to visualize when the distribution of a RIR tends to become Gaussian. This is to relate to Schroeder's work and his crossover frequency, but this has nothing to do with the dynamical definition of mixing.

Applying either MP or the XFT to experimental RIRs (Sect. 4.3.4), we find an obvious relationship between the transition time and the source/receiver distance, as well as with the sound source. However, the theory of dynamical systems clearly states the contrary [40]: mixing occurs approximately at the same time within the system, and is not a function of the source excitation. Moreover, as none of these methods do consider the trajectories of sound particles within the space, we cannot conclude to the equivalence between late reverberation and the dynamical definition of mixing. Therefore, the estimated transition time cannot be called the mixing time. We propose instead to call it the *crossover time*.

Proposition The crossover time defines the time above which arrivals are statistically distributed at a given position in the hall. In other words, the crossover time indicates when the sound field is locally diffuse, but does not mean it is diffuse elsewhere within the hall.[5]

Finally, in the hypothesis of a mixing hall, the mixing time should be reached later than the crossover time. In addition to develop a system for measuring the divergence of sound particles, the question is now to know whether RIRs measured in concert halls last enough time to estimate the dynamical properties of such systems.

References

1. Schroeder, M.R.: Statistical parameters of the frequency response curves of large rooms. J. Acoust. Eng. Soc. **35**(5), 307–316 (1987)
2. Weyl, H.: Das asymptotische Verteilungsgesetz der Eigenwerte linearer partieller Differentialgleichungen. Math. Ann. **71**, 441–479 (1912)
3. Cremer, L., Muller, H.A., Schultz, T.J.: Principles and Applications of Room Acoustics, vol. 1. Applied Science Publishers Ltd., London (1982)
4. Polack, J.-D.: Modifying chambers to play billiards: the foundations of reverberation theory. Acustica **76**, 257–272 (1992)
5. Beranek, L.: Concert and Opera Halls How They Sound. Acoustical Society of America, New York (1996)
6. Balian, R., Bloch, C.: Distribution of eigenfrequencies for the wave equation in a finite domain: I. Three-dimensional problem with smooth boundary surfaces. Ann. Phys. **60**, 401–447 (1970)
7. Bolt, R.H., Doak, P.E., Westervelt, P.J.: Pulse statistics analysis of room acoustics. J. Acoust. Soc. Am. **22**(3), 328–340 (1950)
8. Defrance, G.: Caractérisation du mélange dans les réponses impulsionnelles de salles. Application à la détermination expérimentale du temps de mélange (Characterization of mixing within room impulse responses. Application to the experimental estimation of the mixing time). Ph.D. thesis, Université Pierre et Marie Curie (UPMC—Paris VI) (2009)
9. Stewart, R., Sandler, M.: Statistical measures of early reflections of room impulse responses. In: Proceedings of the 10th International Conference on Digital Audio Effects (DAFx-07), Bordeaux, France, pp. 59–62 (2007)

[5] Here, the term *diffuse* means that sound rays come with the same probability from any region of the space.

10. Defrance, G., Polack, J.-D.: Measuring mixing time in auditoria. J. Acoust. Soc. Am. **123**(5), 3499 (2008)
11. Mallat, S., Zhang, Z.: Matching pursuit with time-frequency dictionnaries. IEEE Trans. Signal Process. **40**(12), 3397–3415 (1993)
12. Defrance, G., Daudet, L., Polack, J.-D.: Using matching pursuit for estimating mixing time within room impulse responses. Acta Acust. Acust. **95**(6), 1082–1092 (2009)
13. Defrance, G., Polack, J.-D.: Estimating the mixing time of concert halls using the eXtensible Fourier Transform. Appl. Acoust. **71**, 777–792 (2010)
14. Defrance, G., Daudetm, L., Polack, J.-D.: Characterizing sound sources for room acoustical measurements. In: Proceedings of the Institute of Acoustics, vol. 30 (2008)
15. Krstulovic, S., Gribonval, R.: Mptk: matching pursuit made tractable. In: ICASSP'06, Toulouse, France (2006)
16. ISO 3382: Acoustics-measurements of the reverberation time of rooms with reference to other acoustical parameters (1997)
17. Bendat, J.S., Piersol, A.G.: Random Data: Analysis and Measurements Procedures. Wiley, New York (1971)
18. Sturm, B.L., Shynk, J.J., Daudet, L., Roads, C.: Dark energy in sparse atomic estimations. IEEE Trans. Audio Speech Lang. Processing **16**(3), 671–676 (2008)
19. Defrance, G., Daudet, L., Polack, J.-D.: Finding the onset of a room impulse response: straightforward? J. Acoust. Soc. Am. **124**, EL248 (2008)
20. Hurwitz Jr., H., Kac, M.: Statistical analysis of certain types of random functions. Ann. Math. Stat. **15**(2), 173–181 (1944)
21. Rice, S.O.: Mathematical analysis of random noise. Bell Syst. Tech. J. **23**, 13–29 (1944)
22. Defrance, G., Daudet, L., Polack, J.-D.: Detecting arrivals within room impulse responses using matching pursuit. In: Proceedings of the 11th International Conference on Digital Audio Effects (DAFx-08), Espoo, Finland (2008)
23. Schodder, G.R.: Uber Die Verteilung Der Energiereicheren Schallruckwurfe In Salen. Acustica **6**, 445–465 (1956)
24. Polack, J.-D.: Playing billiards in the concert hall: the Mathematical Foundations of Geometrical Room Acoustics. Appl. Acoust. **38**, 235–244 (1993)
25. Defrance, G., Polack, J.-D., Katz, B.-F.G.: Measurements in the new Salle Pleyel. In: Proc. Int. Symp. Room Ac., Sevilla, Sept 2007
26. Joyce, W.B.: Sabine's reverberation time and ergodic auditoriums. J. Acoust. Soc. Am. **58**(3), 643–655 (1975)
27. Birkhoff, G.D.: Proof of the Ergodic theorem. Proc. Natl. Acad. Sci. **17**, 656–660 (1931)
28. Krylov, N.S.: Works on Foundations of Statistical Physics. Princeton University Press, Princeton, NJ (1979) (translated by A.B. Migdal and Ya. G. Sinai and Yu. L. Zeeman)
29. Shannon, C.E.: The mathematical theory of communication. Bell Syst. Tech. J. **27**, 379–423 (1948) and 623–656
30. Sinai, Y.G.: Dynamical systems with elastic reflections. Russ. Math. Surv. **25**, 141 (1972)
31. Bunimovich, L.A.: Conditions of stochasticity of two-dimensional billiards. Chaos **1**(2), 187–193 (1991)
32. Mortessagne, F.: Dynamique et interférences géométriques dans les billards chaotiques. Ph.D. thesis, Univ. Paris 7, Denis Diderot (1994)
33. Kawabe, T., Aono, K., Shin-ya, M.: Acoustic ray chaos and billiard system in Hamiltonian formalism. J. Acoust. Soc. Am. **113**(2), 701–704 (2003)
34. Koyanagi, S., Nakano, T., Kawabe, T.: Application of Hamiltonian of ray motion to room acoustics. J. Acoust. Soc. Am. **124**(2), 719–722 (2008)
35. Lee, J.B.: Why concert halls are not ergodic? J. Acoust. Soc. Am. **85**(6), 2680–2681 (1989)
36. Sabine, W.C.: Collected Papers on Acoustics. Harvard University Press, Cambidge (1964) (reprinted by Dover, New York)
37. Kuttruff, H.: Room Acoustics, 3rd edn. Elsevier Applied Science, New York (1991)
38. Penna, P.: Some statistical properties of reverberation process in diffuse sound fields. M.S. thesis, The acoustics laboratory, Denmark tekniske hojskole (1993)

39. Polack, J.-D.: La transmission de l'énergie sonore dans les salles. Ph.D. thesis, Thèse de doctorat d'Etat, Université du Maine, Le Mans, France (1988)
40. Katok, A., Hasselblatt, B.: Introduction to the modern theory of dynamical systems. Cambridge University Press, Cambridge (1996)

***

Biography

Guillaume Defrance received the French equivalent of a B.S. degree in Physical Measurements and in Mechanical Engineering from Université du Maine, Le Mans in 2003 and 2004, respectively, and finally the M.S. in Acoustics and Physics and Ph.D. degrees (Electrical and Mechanical Engineering) at Université Pierre et Marie Curie (UPMC), Paris, France, in 2006 and 2009, respectively. Dr. Defrance specializes in statistical room acoustics, late reverberation and diffuse sound field, stochastic and random processes, using algorithms for sparse approximation and time–frequency signal processing tools. In 2010, Dr. Defrance was a Marie Curie Research Fellow at the University of Sheffield (UK) where he carried out a research project on the diffuseness of urban sound fields. In 2011, he worked on the spatial repartition of energy at low frequencies in reverberant chambers at UPMC. As a research fellow at Echosens (Paris), he currently works on the characterization of disordered media using ultrasounds. He is currently a research fellow in the department of electromagnetism of the Laboratory of Signals and Systems at SUPELEC.

Dr. **Jean-Dominique Polack** received his Ph.D. in 1882 in Göttingen. He joined CNRS as Research Fellow in 1983, then was elected Professor of Acoustics at University Pierre et Marie Curie in Paris in 1997. From 1999 to 2001, he was appointed Professor of Electroacoustics at the Technical University of Denmark. From 2002 to 2006, he was Head of the Laboratoire d'Acoustique Musicale, now merged into Institut d'Alembert. He is presently in charge of the Graduate School at UPMC. Dr. Polack's areas of expertise cover loudspeaker models, room-acoustic measurements, ergodic theory of sound fields, and sound quality of soundscapes.

Chapter 5
Are Impulse Responses Gaussian Noises?

Jean-Dominique Polack

Abstract Many years ago, Manfred Schroeder proved that transfer functions in rooms follow complex Gaussian distributions. This property was extended to impulse responses by the author, following a suggestion by Moorer for simulating impulse responses. More recently, several authors have checked again the later property with modern signal analysis tools. They obtained mixed success, with results that strongly depend on the length of the analysis window. In order to understand these unstable results, the present paper takes a closer look at the statistical distributions of both impulse responses and transfer functions. It shows that an accurate model of both the impulse response and the transfer function is necessary in order to test the distributions. Further, it presents several sources of artefacts that skew the distributions, and show that they can be ascribed to the signal processing methods used to extract the responses. Finally, the conditions for obtaining true Gaussian distributions are specified.

5.1 Introduction

In 1954, Manfred Schroeder published two seminal papers [1, 2] on sound distribution in rooms. In the first paper, in refutation of the accepted theory of the time, he was able to show that resonance frequency in rooms are random if one introduces in the room small objects of dimensions equivalent to the wave length of interest. Up to then, practitioners believed that big diffusing objects were necessary to

J.-D. Polack (✉)
Institut Jean Le Rond d'Alembert, Sorbonne Universités, UPMC Univ Paris 06, UMR 7190, Paris F-75005, France

Institut Jean Le Rond d'Alembert, CNRS, UMR 7190, Paris F-75005, France
e-mail: jean-dominique.polack@upmc.fr

N. Xiang and G.M. Sessler (eds.), *Acoustics, Information, and Communication*, Modern Acoustics and Signal Processing, DOI 10.1007/978-3-319-05660-9_5,

randomize the modal distribution. In the second paper, he carried on with statistical properties of such randomized sound field, using advanced statistics recently developed by Rice. Thus, he could show that, above a cut-off frequency dependent of volume and reverberation time, modes overlap and combine so that average distance between peaks, between troughs, and even the standard deviation, was predicted by the theory of random noise.

Several years later in 1962, he and his collaborators went one step further [3]. They simulated the random distribution of sound fields in rooms by means of a powerful tool of random process theory: Monte-Carlo simulation. In analogy with the hazard game of roulette, which made Monte-Carlo casino world famous, Monte-Carlo technique simply simulates complex processes by choosing values at random, but following a given probability distribution, and combining them according to the properties of the process to be simulated. Using this powerful technique, Schroeder and his collaborators were thus able to simulate the frequency response of a room, showing that it displayed the same characteristics as real measurements.

The assumptions upon which Schroeder based his simulation are now well known. They are twofold. Firstly, the resonance frequencies of the modes are distributed randomly, that is, they locally follow a uniform distribution. Secondly, the transfer functions of the modes at their resonant frequency follow exponential distributions with imaginary random arguments. Thus, the transfer function at any frequency can be obtained by superposition of the different modes that respond at the frequency. When modes overlap, the transfer function becomes Gauss distributed by virtue of the Central Limit Theorem of probability. Of course, this is only valid at high frequencies, when many modes overlap, but the approximation is considered satisfactory as soon as ten modes respond at a given frequency.

The present paper develop this idea and extend it to the time domain, as was suggested by Moorer [4] when he wrote that an impulse response is similar to white noise exponentially decreasing with time. Indeed, the properties of the Fourier transform ensure that the time domain response, that is, the impulse response of a room, also is Gauss distributed. Several researchers [5, 6] have independently tried to check this property on measured impulse responses, with mixed success. We shall review some attempts, and stress the reason for their failure: the absence of a proper model of impulse response. Indeed, statistics teaches us that it is almost impossible to prove statistical properties ex post. Only the existence of a model, that is, analysis ex ante, can prove the statistical properties of room response.

As a consequence, the next step is a review of Schroeder's model for the transfer function, and its translation in the time domain, generalizing Moorer's model. The paper then carries on with the analysis of a measured impulse response, arbitrarily selected among some 200 obtained during a recent campaign in Paris [7]. It stresses the necessity of compensating for the decay, in order to reconstruct a stationary signal, before carrying out a detailed statistical analysis in both the time and frequency domains. It concludes with the necessity of improving the deconvolution algorithms currently used in room acoustics to obtain impulse response from sine sweep.

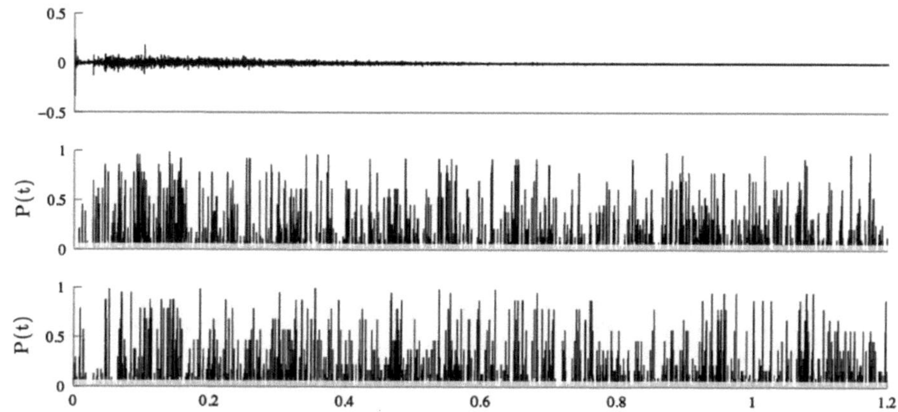

Fig. 5.1 Impulse response from Salle Pleyel (*top*) and the probability that samples are distributed according to Gaussian law of same variance and mean value (Kolmogorov-Smirnov test)—*middle curve*: using 200 samples; *bottom curve*: using 120 samples—gray values correspond to intervals where the test fails

5.2 Results from Previous Research

In 1989, the present author offered a generalization of Schroeder's model [8] that gave the theoretical framework of Jot's approach to digital reverberation filters [9]. As a consequence, several authors [5, 6] have tried recently to check the validity of the model. They used different statistical tests, but all of them rely on some statistical estimator: the ratio of the kurtosis to the variance [5], or Kolmogorov test [6]. We briefly present the results of Defrance [6].

Defrance used a set of impulse responses measured in Salle Pleyel with pistol shots (Fig. 5.1), and checked whether it was distributed according to a Gaussian distribution of same variance and mean value (null hypothesis). He used Kolmogorov-Smirnov test, that is, he compared the distribution of the experimental values to the empirical distribution of data obtained with a Gaussian random generator fed with the same mean value and the same variance. Most of the time, the impulse response passes the test (probability $P(t) > 0.05$, black values in Fig. 5.1), although the probability remains low, but every now and then the null hypothesis is rejected, meaning that the two distributions are not the same. A striking feature of Fig. 5.1 is the fact that the results of the test depend on the length of the window used (200 or 120 samples), with slightly better results for longer windows. Other tests give similar results [6], leading to the rejection of the hypothesis that impulse responses are distributed according to Gaussian laws (see also [5]).

In order to check the correctness of the procedure, Defrance also evaluated simulated impulse responses constructed by weighting Gaussian random noise with an exponential window. The results of Kolmogorov-Smirnov test are given in Fig. 5.2. This time, the impulse response passes the test most of the time, with

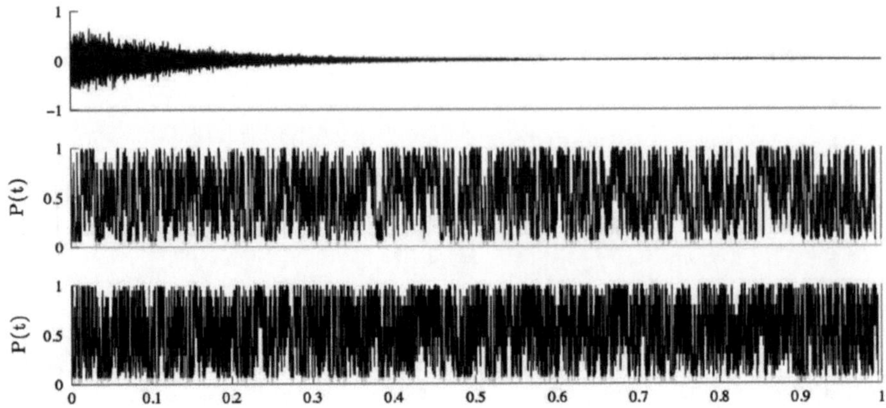

Fig. 5.2 Gaussian impulse response (*top*) and the probability that samples are distributed according to Gaussian law of same variance and mean value (Kolmogorov-Smirnov test)—*middle curve*: using 200 samples; *bottom curve*: using 120 samples—gray values correspond to intervals where the test fails

rather high probability. But every now and then, it fails the test. Once again, the results depend on the length of the analysis window used for the test.

The conclusion of this survey of previous research is that the methodology they use is not appropriate. Indeed, they are not constructed on a theoretical framework, like Schroeder's model, but only attempt to test a property: that impulse responses are Gaussian. In other words, they want to prove statistical properties ex post, whereas an analysis ex ante, making use of the properties of a model, is necessary to prove the statistical properties of room responses.

5.3 Generalizing Schroeder's Model to Time Domain

As stressed in the previous section, no proof of the statistical properties of impulse responses can be achieved without a proper model of impulse responses. This model relies on Schroeder assumptions:

- The resonance frequencies of the modes locally follow a uniform distribution.
- The transfer functions of the modes at their resonant frequency follow exponential distributions with imaginary random arguments.

At arbitrary frequencies, therefore, one must take into account the bandwidth of each mode, and superpose the modes accordingly (Fig. 5.3). The result is a random walk.

As shown in Fig. 5.3, it results in a complex transfer function, which is best decomposed into its real part $X(\omega)$ and its imaginary part $Y(\omega)$:

$$H(\omega) = X(\omega) + jY(\omega) \tag{5.1}$$

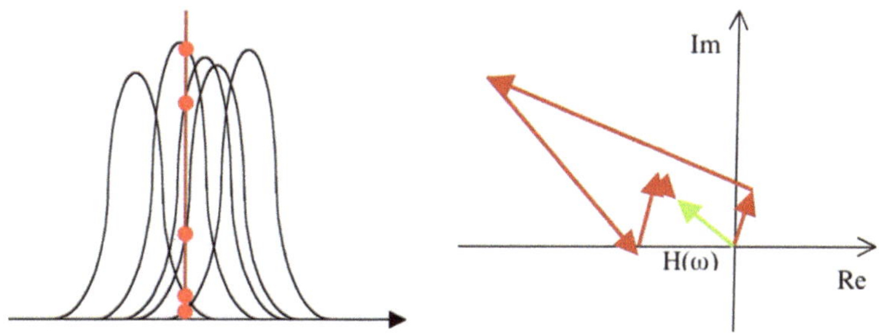

Fig. 5.3 Superposition of modes with random initial phases, taking into account their bandwidth (*right*)—it results in a random walk (*right*)

The theory of random walk then predicts that both the real and imaginary parts follow centered Gaussian distribution of same variance:

$$\begin{gathered}\langle X(\omega)\rangle = \big\langle Y(\omega)\big\rangle = 0 \\ \big\langle X^2(\omega)\big\rangle = \big\langle Y^2(\omega)\big\rangle\end{gathered} \tag{5.2}$$

In other words, the real and imaginary parts are equidistributed. Further, they are decorrelated [3]:

$$\langle X(\omega)Y(\omega)\rangle = 0 \tag{5.3}$$

Now, Schroeder model is defined in the frequency domain, and applies to modes taken individually, with due consideration of their bandwidth, that is, of their decaying nature. As a consequence, when translated into the time domain, it means that each mode will be exponentially decaying with random initial phase. By summing up all the modal contributions, one obtains an exponentially decaying random noise. However, since modes last the whole duration of the impulse responses, the random distribution of their superposition must be checked over the whole duration of the impulse response. Locally at some instant, nothing stops the impulse response from deviating from a Gaussian distribution.

In the following, we keep in mind the long-term characteristics of room impulse responses. As a consequence, we compensate for time variation, using the property of exponential decay. In a similar way, we must take into account the spectrum of the source when checking the Gaussian distribution of the transfer function; therefore, the next section is devoted to the presentation of room impulse responses.

5.4 Raw Analysis of Impulse Responses

For the purpose of illustrating the properties of room impulse responses, we arbitrarily chose an impulse response measured at Opéra Garnier in Paris during a recent campaign in 16 Parisian theaters and concert halls [7]. This response,

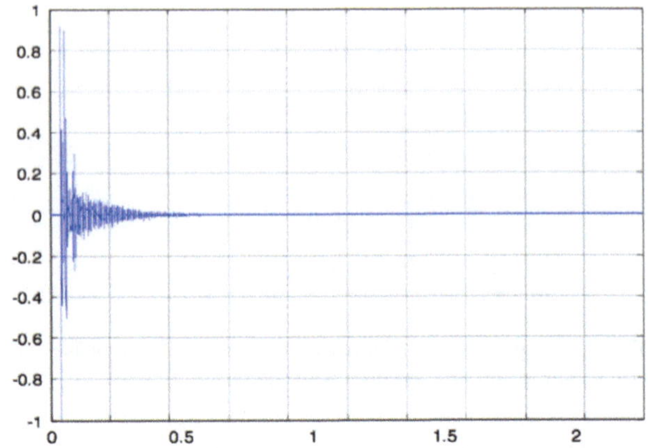

Fig. 5.4 Impulse response measured at Opéra Garnier in Paris

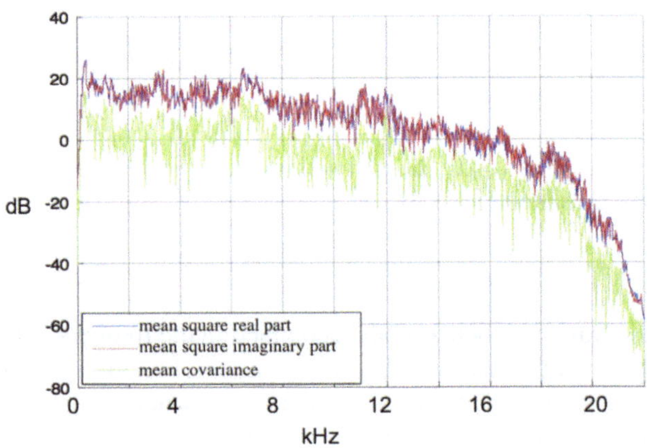

Fig. 5.5 Transfer function measured at Opéra Garnier. *Blue*: real part; *red*: imaginary part; *green*: cross-correlation

sampled at 44.1 kHz, is presented in Fig. 5.4. It was obtained by deconvolution of a logarithmic sine sweep of 30 s duration, using the Aurora suite developed by Farina [10].

From this response, we computed the transfer function by Fourier transformation, and estimated the running variances and cross-correlation according to Eqs. (5.2) and (5.3). These estimates were obtained by averaging 44 adjacent values of the squared real part, of the squared imaginary part, and of the product of the two. These estimates are presented in Fig. 5.5, where the real part is traced in blue, the imaginary part in red, and the cross-correlation in green. It is evident from Fig. 5.5 that the averaged squared real and imaginary parts coincide, since the blue curve is almost completely hidden behind the red one. On the other hand, the cross-correlation does not vanish, as predicted by Eq. (5.3). Proper estimation of the empirical process corresponding to averaging 44 adjacent values of the product of the real and imaginary parts leads to the prediction that the green curve should lay 16.5 dB below the two others [8], a value which agrees with Fig. 5.5.

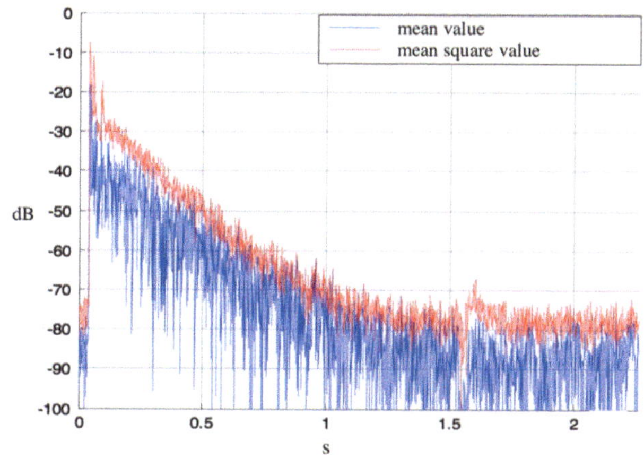

Fig. 5.6 Logarithmic plot of the impulse response. *Blue*: mean value; *red*: quadratic value

It is also instructive to have a look at the logarithmic display of the impulse response. Once again, we estimated the running variance and the running average on 44 adjacent values. They are displayed in Fig. 5.6, where the running average is traced in blue, and the running variance in red. This time, due to correlation between successive values, the difference between the two curves is much less than 16 dB [8].

It can be seen from Figs. 5.5 and 5.6 that Schroeder's model needs adaptation before any Gaussian distribution can be checked. First, the decay in time must be compensated, so that the impulse response approximates a stationary signal; then the spectrum must be compensated, to account for the spectrum of the source. These two issues are successively addressed in the next sections.

5.5 Compensating for Decay

The compensation procedure for the decay is illustrated in Fig. 5.7. Firstly, it is necessary to detect the background noise, or noise floor to which the impulse decays at larger times. This background noise is traced in dark blue for the mean square value, and in red for the mean value. Only the mean square value is of interest, and the noise floor is estimated by linear fit (green line in Fig. 5.7). By convention, any value within 10 dB of the noise floor is considered as background noise [11]. Therefore, the impulse response is reduced to the light blue part for its mean square, and the brown part for its mean.

The next step consists in estimating the decay by a linear fit of the mean square—magenta line in Fig. 5.7. Notice that the linear fit of the mean decay (yellow line) does not run parallel to the magenta line. This difference will come to light later on.

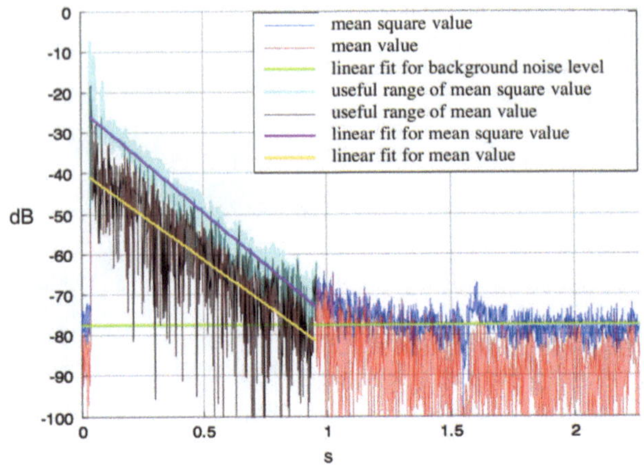

Fig. 5.7 The different linear fits used in the decay compensation process

Once the mean decay is obtained by linear fit, it is straightforward to compensate for the decay by multiplying the impulse response by the exponential of the logarithmic decay, while windowing it from 0.04 to 0.96 s. However, inspection of the compensated mean square (Fig. 5.8) reveals that the procedure is not sufficient. There remain abnormal values at the onset of the response, corresponding to coherent reflections. In a similar fashion, the mean square value increases toward the end of the window, revealing that some background noise is influencing the response. A shorter window must be used.

Figure 5.9 presents the mean compensated decay over the same window extending from 0.04 to 0.96 s. Now, the trend is different, with a positive mean slope of the curve. This corresponds to the fact that the yellow fit curve is not parallel to the magenta one in Fig. 5.7, having a smaller slope. In fact, it simply traduces the fact that computing mean values amount to low-pass filtering, albeit a very rough one, and that reverberation times at low frequencies usually are longer than at higher frequencies.

As a consequence, it is not sufficient to look at mean and mean square values to check for that compensated impulse responses follow Gaussian distribution, as was originally proposed in [8]. One must refine the analysis.

5.6 Detailed Analysis

The Artefact (cf. Oxford Dictionary) observed in the previous section led us to slightly amend the compensation procedure by selecting a more conservative time window in its last step. Figure 5.10 presents the thus selected portion of the impulse response used in this section.

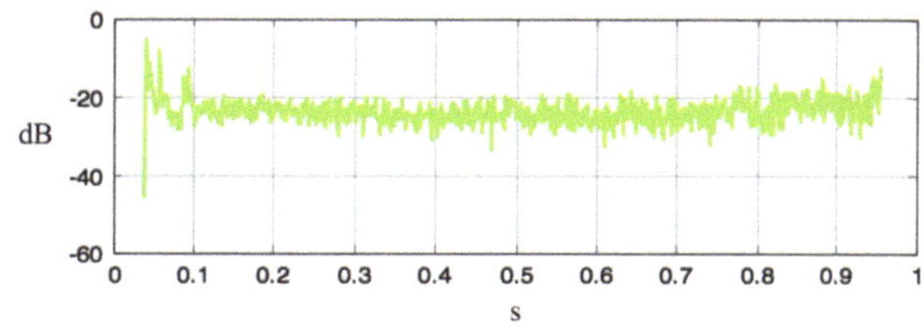

Fig. 5.8 Mean square of compensated impulse response

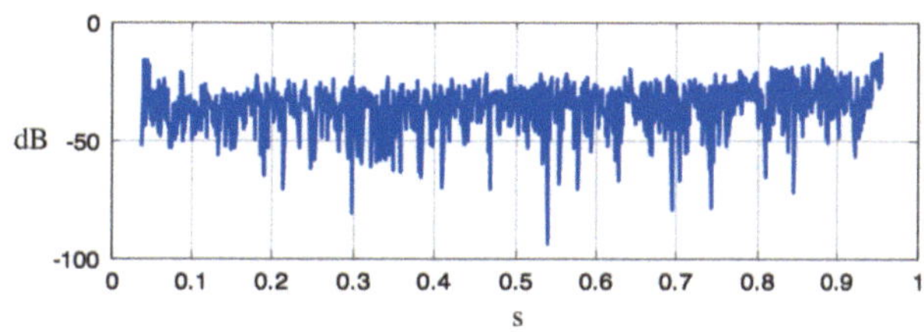

Fig. 5.9 Mean compensated impulse response

5.6.1 Impulse Response

Visual inspection of Fig. 5.10 reveals that the selected portion of the compensated impulse response looks indeed like white noise. It is therefore meaningful to compare its properties to the properties of white noise, for example by looking at the histogram of the values it takes (Fig. 5.11).

Figure 5.11 compares the histogram of the compensated impulse response, traced in red, to the theoretical histogram of a Gaussian distribution with zero mean and the same variance, traced in blue, computed for the same number of samples as contained in the impulse response. The two traces look similar. Therefore, we decided to check the distribution with statistical test. We used the Kolmogorov-Smirnov test, which compares the empirical distribution to a sample of the same number of random values that follow the theoretical distribution. We repeated the test several times with different simulated samples of the Gaussian distribution, and the compensated impulse response always passed Kolmogorov-Smirnov test.

However, a zoom on the histogram around the peak of the distribution reveals a slight skew of the impulse response. Indeed, its distribution does not top at zero value, as expected, but slightly below it (Fig. 5.12). We interpret it as a misalignment of the deconvolution filter that recovers the impulse response from the sine sweep measurement. Great care must be taken in the alignment of the filter in order to ensure Gaussian distribution, as well as the right amount of background noise.

5.6.2 Transfer Function

In order to check the distribution of the transfer function, it is also necessary to carry out some compensation of the source spectrum. Indeed, the transfer function

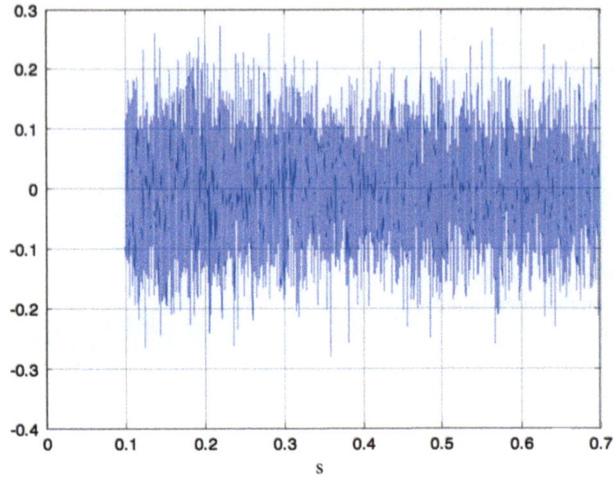

Fig. 5.10 Selected portion of compensated impulse response

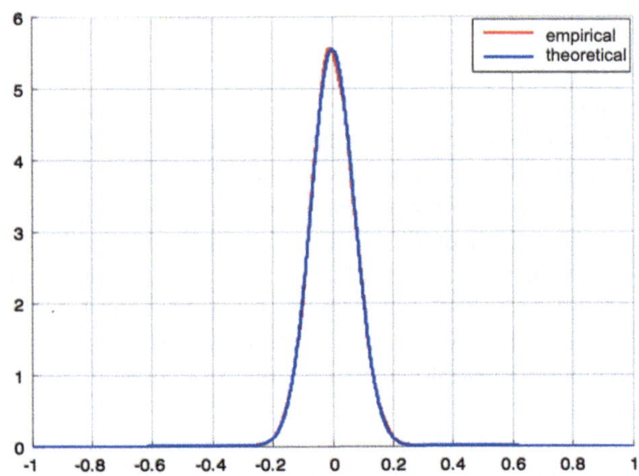

Fig. 5.11 Histogram of compensated impulse response

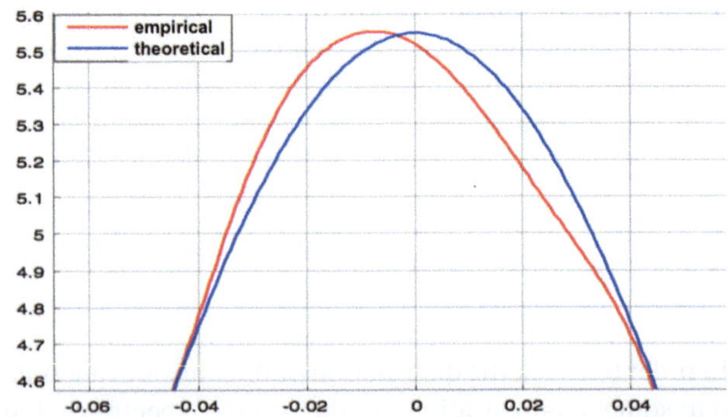

Fig. 5.12 Zoom on the histogram revealing misalignment of the deconvolution filter

computed from the impulse response by Fourier transformation is far from being flat, as shown in Fig. 5.5, but only the compensated part of the impulse response must be taken into account.

Figure 5.13 presents the raw spectrum of the compensated part of the impulse response (red). This estimate of the transfer function is far from being flat, and need smoothing before any compensation can be envisaged. Therefore, Fig. 5.13 also display several methods for smoothing the spectrum, from Welch spectrum computed with window lengths of 64 samples and 50 % overlap (green), to Burg spectral estimation based on 32nd order autoregressive process (light blue), and Yule-Walker spectral estimator with the order of the autoregressive model set to 16 (dark blue).

As little difference subsists between the Burg and Yule-Walker estimators, we decided to carry out the spectral compensation using the 16th order Yule-Walker estimator. In a similar way to the decay compensation of the impulse response, the complex spectrum computed from the compensated impulse response is then divided by the Yule-Walker estimator, yielding real and imaginary parts of the spectrum (Fig. 5.14) that look very similar after truncation of the central part of the spectrum—between 200 Hz and 14 kHz—and once again similar to white noise. Thus, proper distribution analysis can now be carried out.

Figure 5.15 presents the histogram of the real part of the compensated transfer function, traced in blue, and compares it to the theoretical histogram of a Gaussian distribution with zero mean and the same variance, traced in red, computed for the same number of samples as contained in the transfer function. The two traces look similar. Therefore, we decided to check the distribution with statistical test. Once again, we used the Kolmogorov-Smirnov test, and repeated the test several times with different simulated samples of the Gaussian distribution, and the compensated transfer function only passed Kolmogorov-Smirnov test some of the times.

This time, a zoom on the histogram around the peak of the distribution reveals a skew of both the empirical and theoretical distributions. None of them tops at zero value, as expected, but slightly below it for the empirical distribution, and slightly above it for the theoretical distribution (Fig. 5.16). But this time, an interpretation of this discrepancy is less evident, although it is obvious from Fig. 5.14 that the variance of the distribution slightly decreases with frequency, probably explaining why the Kolmogorov-Smirnov test sometimes fails.

Figure 5.17 presents the histogram of the imaginary part of the compensated transfer function, traced in blue, and compares it to the theoretical histogram of a Gaussian distribution with zero mean and the same variance, traced in red, computed for the same number of samples. The two traces look similar, except around the center of the distribution where the empirical distribution visibly has a higher peak than the theoretical one. As a consequence, several repetitions of the Kolmogorov-Smirnov test with different simulated samples of the Gaussian distribution always failed.

A zoom on the histogram around the peak of the distribution confirmed that the empirical distribution has a higher, and simultaneously narrower, peak than the theoretical one. Further, they both top at the same slightly negative value, but not at zero as expected (Fig. 5.18). This behavior points to a non-constant variance over

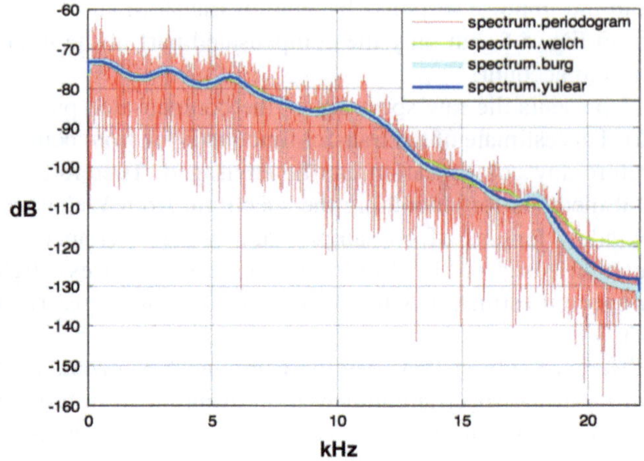

Fig. 5.13 Spectrum of compensated part of impulse response, and its estimation according to several procedures

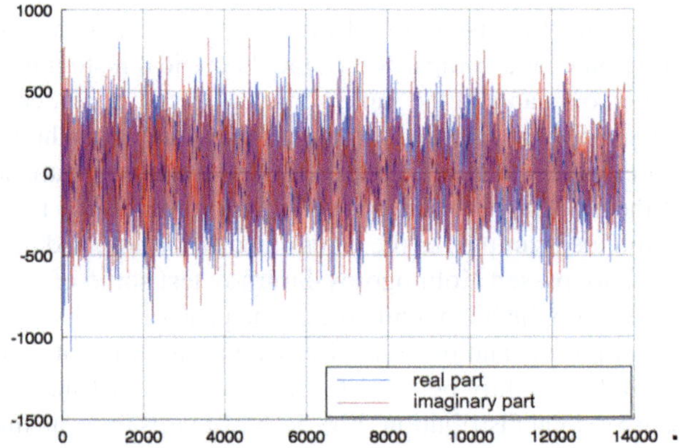

Fig. 5.14 Compensated complex spectrum. *Blue*: real part; *red*: imaginary part

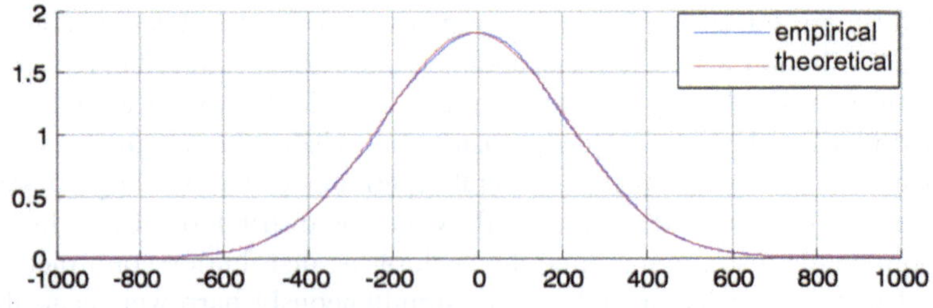

Fig. 5.15 Histogram of real part of compensated transfer function

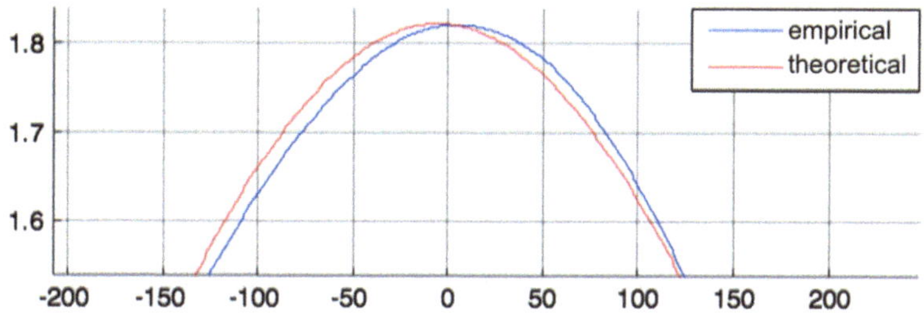

Fig. 5.16 Zoom on the histogram revealing abnormal values of both curves

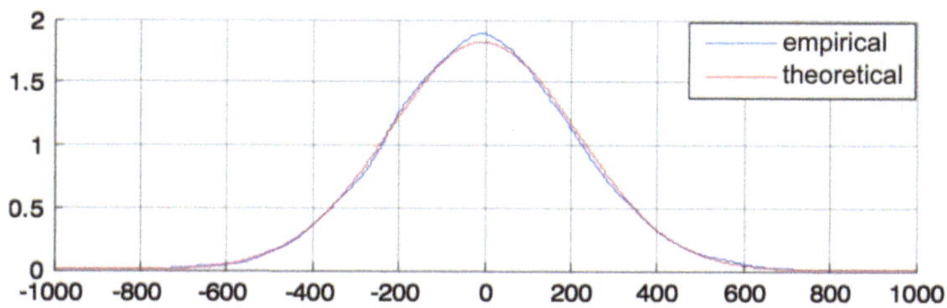

Fig. 5.17 Histogram of imaginary part of compensated transfer function

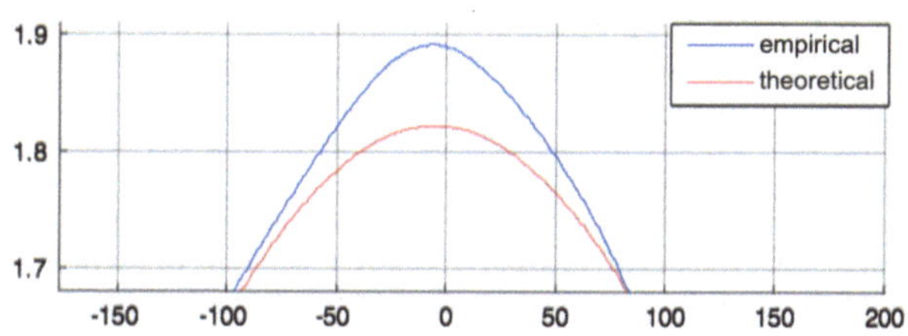

Fig. 5.18 Zoom on the histogram revealing abnormal peak of empirical distribution

the whole frequency range, as is visible in Fig. 5.14: the variance of the distribution decreases with frequency. This is why the Kolmogorov-Smirnov test always fails.

Indeed, complete analysis of the impulse response [8] reveals that the variance of the frequency response at a given frequency is proportional to the reverberation time at that frequency. Since reverberation times at high frequencies are always shorter, this is why the variance at high frequency is also smaller. Proper compensation of the spectrum should therefore take this property into account.

5.7 Conclusion

In this paper, we hope to have convinced the reader that, despite some unsuccessful previous attempts to prove it, impulse responses are Gaussian process, *provided* that global analysis is carried out on hand of a proper model of impulse responses. In this process, three points are essential: discard the early part with strong reflections, and the late part which simply is background noise; accurately compensate for the decay; and compensate for the source spectrum. Further, we have shown that such an analysis can reveal the shortcomings of the measurement procedure, especially of the inverse filtering used to recover the impulse from the measurement signal: care must be taken that it accurately provides zero mean.

Moreover, the analysis has also revealed that Schroeder's simple model is not sufficient, especially in the frequency domain. There remains a frequency dependent variance which Schroeder's model does not account for. As a consequence, a more complex time-frequency compensation is needed to improve the model so that it passes the statistical tests. This sets the goal for further research in the domain.

Acknowledgement The author wants to thank Xi Zhang, who carried out all the computations used in this paper.

References

1. Schroeder, M.R.: Eigenfrequenzstatistik und Anregungsstatistik in Raümen. Acustica **4**, 456–465 (1954)
2. Schroeder, M.R.: Die statistischen Parameter der Frequenzkurven von grossen Raümen. Acustica **4**, 594–600 (1954); reprinted in M.R. Schroeder, "Statistical parameters of the frequency response curves of large rooms", J. Acoust. Eng. Soc. 35 (1987) 307–316
3. Schroeder, M.R., Kuttruff, H.: On frequency response curves in rooms. Comparison of experimental, theoretical and Monte Carlo results for the average frequency spacing between maxima. J. Acoust. Soc. Am. **34**, 76–80 (1962)
4. Moorer, J.A.: About this reverberation business. Comput. Music J. **3**, 13–18 (1979)
5. Stewart, R., Sandler, M.: Statistical measures of early reflections of room impulse responses. In: Proceedings of the 10th International Conference on Digital Audio Effects (DAFx-07), Bordeaux, France, pp. 59–62 (2007)
6. Defrance, G., Polack, J.D.: Measuring mixing time in auditoria. J. Acoust. Soc. Am. **123**, 3499 (2008)
7. Figueiredo, F.L.: Indices acoustiques et leurs rapports statistiques: vérification objective et subjective pour un ensemble de salles de spectacles. Ph.D. thesis, Université Pierre et Marie Curie, Paris (2011)
8. Polack, J.D.: La transmission de l'énergie sonore dans les salles. Thèse d'Etat—Université du Maine, Le Mans (1988)
9. Jot, J.M.: Etude et realisation d'un spatialisateur de sons par modeles physiques et perceptifs. Ph.D. thesis, Telecom Paris (1992)
10. Farina, A.: Advancements in impulse response measurements by sine sweeps. In: AES 122nd Convention, Vienna, Austria (2007)
11. ISO 3382: Acoustics - Measurement of the reverberation time of rooms with reference to other acoustical parameters, (1997)

***

Biography

Dr. Jean-Dominique Polack received his Ph.D. in 1882 in Göttingen. He joined CNRS as Research Fellow in 1983, then was elected Professor of Acoustics at University Pierre et Marie Curie in Paris in 1997. From 1999 to 2001, he was appointed Professor of Electroacoustics at the Technical University of Denmark. From 2002 to 2006, he was head of the Laboratoire d'Acoustique Musicale, now merged into Institut d'Alembert. He is presently in charge of the Graduate School at UPMC. Dr. Polack's areas of expertise cover loudspeaker models, room-acoustic measurements, ergodic theory of sound fields, and sound quality of soundscapes.

Chapter 6
Recent Applications of Number-Theoretic Sequences in Audio and Acoustics

Ning Xiang, Bosun Xie, and Trevor J. Cox

Abstract Manfred R. Schroeder made influential contributions to acoustics by applying number theory. He introduced number-theory sequences in room impulse response measurements, applied number-theoretical sequences to shape wall surfaces for diffuse sound reflections. Another area Manfred Schroeder made broad impact is invention of an artificial reverberator using all-pass transfer-function properties to create colorless reverberation using a simple algorithm. Vast research activities and engineering developments over past decades have extended Schroeder's work involving artificial reverberators, artificial stereo and applications of maximum-length sequences. This chapter briefs a number of recent applications of maximum-length and quadratic-residue sequences, such as simultaneous dual/multiple sources measurements in acoustical tomography, in artificial pseudo-binaural reverberation, in decorrelation for spatial audio signals, and room acoustic diffuser design.

6.1 Introduction

In 1961, Manfred R. Schroeder, along with Ben Logan [1, 2] invented an artificial reverberator using all-pass transfer-function properties while having an exponentially decaying impulse response, thereby creating colorless reverberation using a simple algorithm. It has been widely accepted both in analog and digital audio devices for processing audio signals. One can now find colorless reverberators and artificial stereo, including simulations of virtual sound sources and auditory

N. Xiang (✉)
Graduate Program in Architectural Acoustics, School of Architecture, Rensselaer Polytechnic Institute, Troy, NY 12180, USA
e-mail: xiangn@rpi.edu

N. Xiang and G.M. Sessler (eds.), *Acoustics, Information, and Communication*, Modern Acoustics and Signal Processing, DOI 10.1007/978-3-319-05660-9_6,

environments in various stereophonic and surround sound reproduction formats, in practically all electronic music instruments (See Schroeder's Memoirs in Part II).

In the 1970s, Manfred Schroeder made another influential contribution to acoustics with the introduction of number-theory sequences, applying bipolar, binary maximum-length sequences (MLSs) to applications such as shaping wall surfaces for diffuse sound reflections [3], and acoustics system identification to enable room impulse response measurements based on correlation techniques [4]. Extending the wall surface work to exploit nonbinary sequences, led to Schroeder devising the quadratic residue diffuser [5] and other designs widely used in studios and performance spaces to improve acoustics [6] (see also Chap. 9).

Vast research activities and engineering developments over past decades have extended Schroeder's work involving artificial reverberators, artificial stereo and applications of MLSs. This chapter is dedicated to the memory of Manfred Robert Schroeder in briefing a number of recent applications of maximum-length and quadratic residue sequences, such as simultaneous dual/multiple sources measurements in acoustical tomography, in artificial pseudo-binaural reverberation, in decorrelation for spatial audio signals, and room acoustic diffuser design.

6.2 Properties of Maximum-Length Related Sequences

MLSs, also termed pseudorandom sequences/noises, offer excellent pseudorandom properties. Therefore, they are used in communication, acoustics, and many other relevant applications. They possess properties similar to those of random noise, but are periodically deterministic and have a strict time structure within their periods. In the following, a brief review of some basic properties pertaining to acoustic applications is given. Detailed description and definitions can be found in [7–9]. In particular, Xiang [9] has most recently summarized these properties succinctly.

An n-stage linear shift-register device with proper feedback taps can generate periodic binary MLSs $\{a_i\}$ with $a_i \in (0, 1)$. Binary MLSs can also be generated numerically using a simple recurrence in the digital domain

$$a_i = \sum_{k=1}^{n} c_k a_{i-k}, \quad 0 \le i < L, \tag{6.1}$$

with $c_k \in (0, 1)$ and $c_0 = c_n = 1$, where the summation and indices are calculated, *modulo* 2 and the positive integer n is said to be the *degree* of the linear feedback registers or of the corresponding MLS. Both analog generation using linear feedback registers and numerical generation using the linear recurrence are subject to a nonzero *initial state* $\{a_{-n}, a_{-n+1}, a_{-n+2}, \ldots a_{-2}, a_{-1}\}$ denoted as $\{\overleftarrow{a}_i\} \neq \{0\}$. The proper feedbacks are expressed by feedback coefficients $c_k \in (0, 1)$; when resulting in a periodic sequence of length $L = 2^n - 1$, the sequence arrives at its maximum possible length, therefore, termed MLS. Under this MLS condition, the proper set of feedback coefficients can also be expressed in polynomial form, $f(x) = \sum_{k=0}^{n} c_k x^k$, termed the *primitive polynomial*. One primitive polynomial given degree n corresponds uniquely to one MLS. The feedback coefficients c_k and the binary sequence element a_i are defined over Galois field GF(2) while the sequence $\{a_i\}$ is defined over Galois field $GF(2^n)$. The underlying mathematical calculus is a substantial subject of number theory [7].

Among a number of number-theoretic properties, the most important property of MLS is that the normalized periodic autocorrelation function of a bipolar MLS within one period is a two-valued function [8]

$$\phi(i) = \frac{L+1}{L}\delta(i) - \frac{1}{L}. \tag{6.2}$$

With large enough period length $L = 2^n - 1$, the periodic autocorrelation function approximates the *unit sample sequence* $\phi(i) \approx \delta(i)$. Therefore the power spectrum of any MLS is flat, expect for a "dip" at zero-frequency.

An MLS $\{b_i\}$ can always be derived from a given one $\{a_i\}$ in terms of a circular phase shift such that $b_i = a_{i+\tau}$ so that an invariant decimation $b_i = b_{2i}$ can be satisfied. Here all the indices are calculated modulo L. This specific MLS $\{b_i\}$ is designated as *characteristic* MLS and *self-similar* MLS [9]. This invariant decimation holds only for characteristic MLSs with decimation factors of 2^m with m being a positive integer [9]. There exists a unique characteristic MLS for any given primitive polynomial and a unique initial state $\{\overleftarrow{a}_i\}$ of the shift register to generate it. An algorithm deriving the required initial state $\{\overleftarrow{a}_i\}$ given the primitive polynomial for generating the characteristic MLS has recently been described by Xiang [10]. The characteristic MLSs with the invariant decimation are of practical significance, they are also relevant to the following applications.

Furthermore, *a variant decimation* is also of practical significance. When the decimation factor d is properly chosen, the decimation will lead to a pair of MLSs $(\{a_i\}, \{b_i\})$ with $\{b_i\} = \{a_{di}\}$, whose periodic cross-correlation function (PCCF) is of clearly lower value than the peak of the periodic autocorrelation function of either sequence. In this context, all the indices are calculated modulo L. One of the straightforward variant-decimation factors is $d = L - 1$. This decimation $\{b_i\} = \{a_{(L-1)i}\} = \{a_{-i}\}$ simply indicates that a time reversal of MLS $\{a_i\}$ always leads to another MLS $\{b_i\}$. This pair of MLSs $(\{a_i\}, \{b_i\})$ derived from decimating $\{a_i\}$ using factor $d = L - 1$ or equivalently from time reversing $\{a_i\}$ is termed reciprocal MLSs due to the fact that their corresponding primitive polynomials are reciprocal to each other [10]. The most relevant correlation properties pertaining to the applications described in this chapter are that one can find MLS pairs or sets of sequences such that the PCCFs between them are of lower values, and the absolute bounds of these low values are deterministically predictable. For example, the bound value of the reciprocal MLS pair of degree n has been reported [8, 11]

$$l_r(n) = \frac{2^{(n+2)/2} - 1}{2^n - 1}. \tag{6.3}$$

For MLSs whose degrees are not a multiple of 4, some decimation factors are of the form $d = 2^k + 1$ or $d = 2^{2k} - 2^k + 1$, where k is chosen such that $n/\text{gcd}(n, k)$ is odd, with gcd(,) being the greatest common divider. A decimation using these factors leads to an MLS pair $(\{a_i\}, \{b_i\})$, with $\{b_i\} = \{a_{di}\}$. For MLSs whose degrees are a multiple of 4, a decimation factor $d = 2^{(n+2)/2} - 1$ leads to a MLS pair. The cross-correlation between the original MLS $\{a_i\}$, and the decimated one $\{b_i\}$ results in small values bounded [8] by

$$l_p(n) = \frac{2^{\lfloor (n+2)/2 \rfloor} - 1}{2^n - 1}, \tag{6.4}$$

where $\lfloor \alpha \rfloor$ denotes the integer part of the real number α.

Besides the reciprocal MLS pairs, this chapter will refer to MLS pairs derived from above-mentioned decimation factors with low-valued PCCFs as *preferred MLS pairs* (see Xiang [8] for a detailed summary). In addition, one can derive a large number of sequences by combining the preferred or reciprocal MLS pairs $(\{a_i\}, \{b_i\})$, as $G_\tau(i) = a_i \oplus b_{i+\tau}$ with $\oplus$ denoting addition modulo 2, so-called Gold sequences $\{G_\tau(i)\}$, whose correlation bound values between each other are the same as those of either preferred MLS pairs or reciprocal pairs [8]. Similarly a decimation from an MLS $\{a_i\}$ of even-numbered degree with a decimation factor $d = 2^{n/2} + 1$ leads to a pair $(\{a_i\}, \{e_i\})$, with $e_i = a_{di}$. The combination $K_\tau(i) = a_i \oplus e_{i+\tau}$ leads to a binary Kasami sequence $\{K_\tau(i)\}$, and their low-valued PCCFs represent even lower bounds,

$$l_K(n) = \frac{2^{n/2} + 1}{2^n - 1}, \tag{6.5}$$

approximately half of those of preferred pairs and Gold sequences.

Figure 6.1 illustrates the bounds of the PCCF of preferred pair, reciprocal pairs, and Kasami sequences. Gold sequences have the same bound values as those of preferred pairs and reciprocal pairs for even-numbered degrees. The preferred pairs possess approx. 3 dB lower bound values than those of reciprocal pairs for odd-numbered degrees. And the bound values of Kasami sequences, existing only for even-numbered degrees, are as 6 ~ dB lower than those of the others. Figure 6.2 shows auto- and cross-correlation functions of Kasami sequences of degree 14 to illustrate the excellent correlation properties of MLS-related sequences.

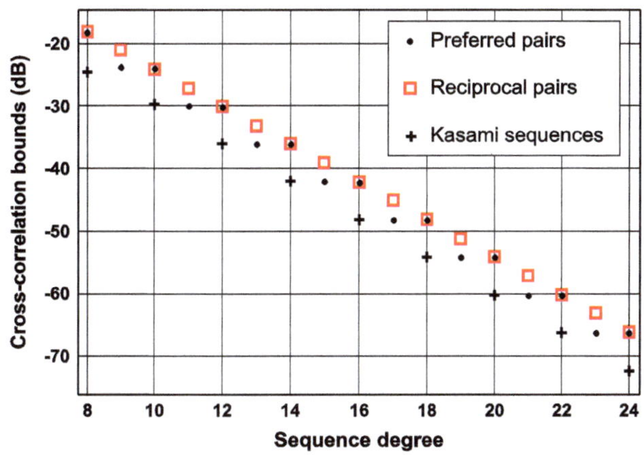

Fig. 6.1 Bound values of the cross-correlation functions of preferred, reciprocal MLS pairs and of Kasami sequences. The bound values are expressed in dB relative to the peak value of their normalized autocorrelation function [11]. [Reproduced from Xiang, N., Daigle, J. N., and Kleiner, M.: Simultaneous acoustic channel measurement via maximal-length-related sequences, *J. Acoust. Soc. Am.*, 117, 2005, pp. 1889–1894. Copyright 2005, Acoustical Society of America.]

6.3 Acoustical Measurements Using Simultaneous Sound Sources

Recent research in outdoor sound propagation for acoustic atmospheric tomography [12] calls for a critical measurement exploiting these correlation properties of MLSs and related coded signals. In tomographical applications, a number of sound sources have to be excited at the same time, within the same frequency range. This simultaneous multiple acoustic source measurement (SMASM) can be used in acoustic delay-time tomography to investigate temperature distributions and wind profiles near the ground surface in outdoor environments. With simultaneous excitations of multiple sound sources and one or multiple sound receivers, the SMASM considers the acoustic system under test as a linear time-invariant multiple-inputs and multiple-outputs (MIMO) system, at least approximately during the excitation period. Using the excellent correlation properties of the coded signals (sequences) briefly discussed above, a large number of coded signals can be straightforwardly derived for the SMASM technique.

Figure 6.3 illustrates a SMASM scheme, where the vector $\mathbf{s} = [s_1, \ldots s_n]^T$ stands for the multiple coded signals used as system excitations, and $\mathbf{r} = [r_1, \ldots r_p]^T$ denotes the system's responses to these excitations, with $[,]^T$ standing for matrix transpose; n is the number of simultaneous sources, while p is the number of receivers.

With properly designed excitations, the system identification task is to determine the impulse response matrix $\mathbf{h} = [h_{ij}]$ for $1 \leq i \leq p$, $1 \leq j \leq n$, determined by

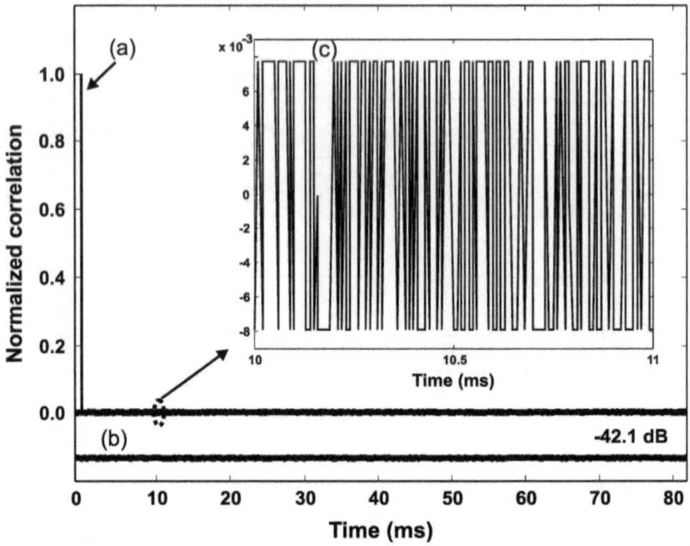

Fig. 6.2 Normalized correlation functions of 14-degree Kasami sequences [9]. (a) Periodic autocorrelation function (PACF) of individual sequences. (b) Periodic cross-correlation function (PCCF) between two Kasami sequences (shifted downwards beneath the autocorrelation function for a convenient comparison). The peak value of the PCCF is 42.1 dB lower than that of the delta-like PACF. (c) Magnified presentation of a segment from (a). Their peak values in the side-lobe of the PACF are the same as those of the PCCF in (b). [Reproduced from Xiang, N., Daigle, J. N., and Kleiner, M.: Simultaneous acoustic channel measurement via maximal-length-related sequences, *J. Acoust. Soc. Am.*, 117, 2005, pp. 1889–1894. Copyright 2005, Acoustical Society of America.]

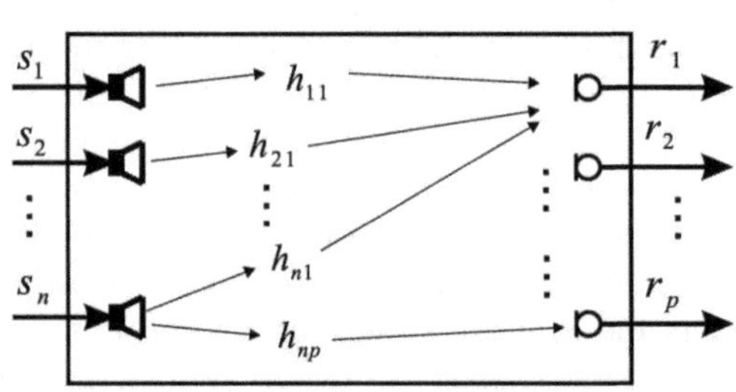

Fig. 6.3 An acoustical linear time-invariant system under investigation with n number of sound sources and p number of sound receivers. An $p \times n$ impulse response matrix **h** describes the entire system under investigation

$$\mathbf{h} \approx \mathbf{r} \otimes \mathbf{s}^T, \tag{6.6}$$

with $\otimes$ denoting periodic cross-correlation and the correlation properties discussed above leading to this concise form [11]. The physical meaning of Eq. (6.6) is that the cross-correlation between the receiver (column) vector $\mathbf{r} = [r_1, \dots r_p]^T$ and the source (row) vector $\mathbf{s}^T = [s_1, \dots s_n]$ approximately results in the channel impulse response matrix **h**. The approximation is due to the fact that the cross-correlation functions among individual source signals in the form of coded sequences (Kasami sequences, preferred and reciprocal MLS pairs) are of finite low-values rather than zeros. In calculating the individual cross-correlations as expressed in Eq. (6.6), particularly for $n > 2$, a specialized algorithm [13] can be used for efficient calculations. In the case of $n = 2$, reciprocal MLS pairs can be exploited with a dedicated fast MLS transform algorithm recently published by Xiang and Schroeder [10].

6.4 Artificial Reverberations Using Reciprocal MLS Pairs and Related Sequences

Following Schroeder [1, 2], there have been many developments of all-pass filter type artificial reverberators. The most recent overviews on all-pass-filter artificial reverberators can be found in [14, 15]. Another line of development is reverberators using finite-impulse-response (FIR) filters, probably due to Moorer [16]. He proposed the use of exponentially decaying random noise for the late part of room impulse responses, so that when convolved with anechoic sound materials, artificial reverberation with a desirable degree of reverberance is created. With the rapid advent of digital audio processing on personal-computer and DSP platforms, FIR filtering via linear convolutions have reached multiples of real-time speed at low cost, so that even a large number of multiple audio channels can be filtered in real-time with sufficiently long FIR-filter coefficients (room impulse responses). FIR-filtering algorithms have also emerged that ensure sufficiently short latency for long room impulse responses [17].

There is a need for rendering the FIR-filter-based artificial reverberations binaurally, such as in binaural room-acoustic simulations [18], where the tails of binaural room impulse responses (see Fig. 6.4) are replaced by exponentially decaying random noise. The decay rates (reverberation times) are determined via statistical room-acoustic principles. Alternately, other room-acoustic information is first extracted from the early part of detailed room-acoustic simulations, such as ray-tracing, image-source approaches, or hybrid approaches [18]. Such schemes have recently been used for psychoacoustics studies investigating speech intelligibility, as well as perceptual aspects of acoustically coupled-volume systems [19].

In principle, an artificial reverberation for a binaural rendering can be achieved with a spatial envelopment, so-called *listening envelopment* when the two random-noise late reverberation tails are incoherent. This avoids using geometrical room-acoustic simulation (e.g., ray-tracing) to simulate the late reverberation tails, saving extremely time-consuming processing given the current technology in numerical simulations. Reciprocal MLS pairs have been used [19, 20] for this purpose, since the cross-correlation between pairs, given the MLS degree, is of low-value (see Fig. 6.2). The intriguing aspect of using MLS pairs, in contrast with other methods

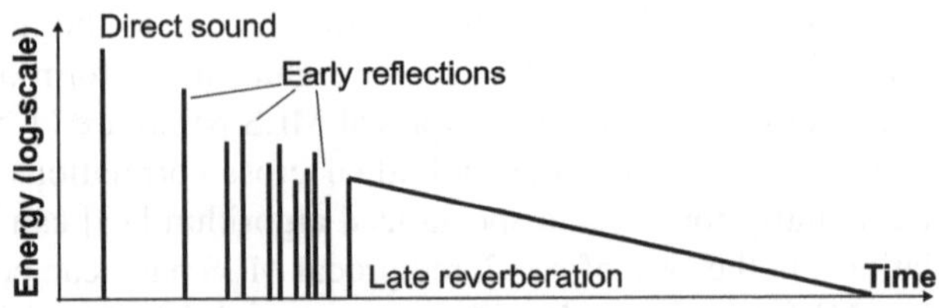

Fig. 6.4 Conceptual echogram containing the direct sound, early reflections, and the late reverberation tail

used in previous work [18] are that MLS pairs (and their related coded sequences) possess deterministically predictable, low values of cross-correlation. Hardly any other random noise signals can be found as low as these, Kasami sequences being the lowest. Lower values correspond to a high degree of enveloping spaciousness in the reverberance. In addition to the highest achievable degrees of spatial envelopment, one can use mix-networks [20] to control the cross-correlation values of the mixed MLS pairs for binaural channels, so that the degree of spatial envelopment can also be adjusted. This is of practical value, since different enclosure conditions will provide different spaciousness, in addition to the reverberance. As recent psychoacoustical investigations [21, 22] have demonstrated, simply shaping the decaying slopes of random noise in the late reverberation tails for achieving artificial reverberations using FIR-filtering technique is not sufficient in terms of naturalness and spatial envelopment of the artificial reverberation. Our recent work [21, 22] adjusts the decay slope (for controlled reverberance) in individual octave/third-octave bands. For each band, the target-reverberation time shapes the exponentially decaying envelope of the reciprocal MLS pair. In addition, interaural decorrelation coefficients which are complimentary to interaural cross-correlation coefficients (IACC) of the binaural reverberation tails

$$\mathrm{IADC} = 1 - \mathrm{IACC}, \tag{6.7}$$

in the individual bands also need to be adjusted accordingly in order to achieve targeted degrees of spatial envelopment. Adjusting only the decay slopes and the interaural decorrelation coefficients so that they are similar to those often observed in experimentally measured binaural room impulse responses creates natural sounding, artificial reverberance with the desired (or targeted) spatial envelopment. Figure 6.5 illustrates the procedure of the binaural artificial reverberation with controllable spatial envelopment and reverberance. A reciprocal MLS pair is first octave-band filtered. Two channels of band-pass filtered pseudorandom noise are mixed with an attenuation factor α_k with $0 \le \alpha_k \le 1$ and k being the octave-band index running from 63 Hz to 8 kHz at each octave-band step.

In Fig. 6.5 between "A" and "B" the two channels of the band-pass filtered, reciprocal MLS-pair are mixed to obtain the desired "interaural decorrelation." The band-pass filtered pseudorandom noise is then multiplied by an exponentially decaying function $E = \exp(-6.9 \cdot t/T_k)$, with the desired reverberation time T_k within octave band k. For broadband resulting "binaural" reverberation tails (late portion of room impulse responses)

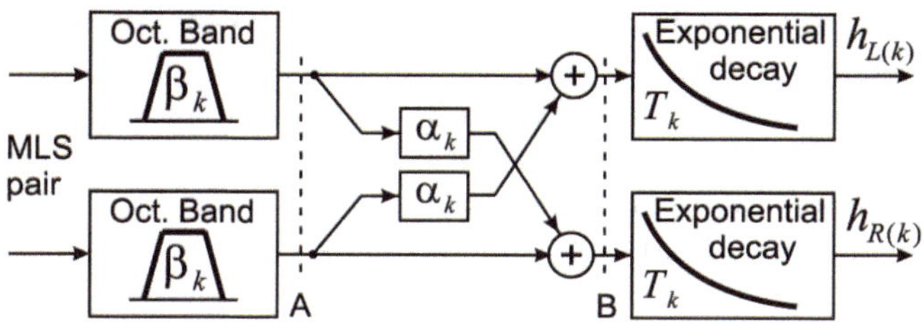

Fig. 6.5 Generation of single-band "binaural" reverberation tails with controllable interaural decorrelation coefficients and the reverberation times within octave band (k)

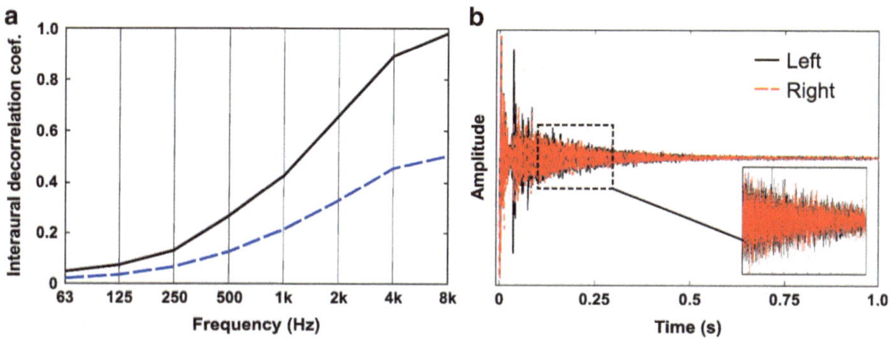

Fig. 6.6 Spatial profiles and the reverberation tails. (a) Two spatial profiles (large spaciousness, middle spaciousness) adjusted for two different degrees of listening envelopment. Interaural decorrelation is defined as 1 – IACC. (b) "Binaural" reverberation tails, properly scaled in terms of amplitude, it will be appended to the early portion of a pair of "binaural room impulse responses at time instance at 90 ms

$$h_{\mathrm{L,R}} = \sum_{k=1}^{M} h_{\mathrm{L,R}(k)}, \tag{6.8}$$

with $k = 1$ for octave band 63 Hz, $k = M$ for octave band 8 kHz, other k values correspond octave bands between 125 Hz and 4 kHz. In order to create naturally sounding, enveloping reverberance, in addition to the desired reverberation time profile, the interaural decorrelation profile also has to be adjusted. Figure 6.6a illustrates two different profiles of interaural decorrelation obtained using Fig. 6.5, inspired from analysis of experimentally measured binaural room impulse responses in existing real concert halls. Figure 6.6b illustrates one (binaural) pair of late-portion artificial "reverberation tails."

Advantages of this artificial enveloping reverberation scheme using reciprocal MLS-pairs are

- MLSs of degree 16–19 with a length between $2^{16} - 1$ and $2^{19} - 1$ are typical lengths used for the artificial reverberation application. At standard audio sampling frequency, the reciprocal MLS pairs are easily generated, and they

provide sufficient reverberation density when sampled using standard audio sampling frequency.
- Spectral flatness of each individual MLS ensures a colorless reverberance.
- Low cross-correlation values of reciprocal MLS pairs (see Fig. 6.2) ensure a high degree of listener envelopment. Reduced degrees of listener envelopment can be straightforwardly achieved using a mixing network [20].

6.5 Decorrelation of Audio Signals Using Reciprocal MLS Pairs [23]

Audio signal decorrelation is a technique that creates two or more replicas of an input signal, which have different waveforms but are perceived similarly to the original signals in most aspects except in some spatial auditory effects [24]. This is applicable to a variety of spatial audio processing effects, such as broadening the auditory source width, enhancing the listener envelopment in surround sound reproduction, and the externalization of auditory events in headphone representation, among others.

In order to create decorrelated audio signals without altering perceived timbre, decorrelation processing should change the signal waveform but leave the magnitude or power spectra of signals intact. A straightforward method is to filter the input signal with a pair of all-pass digital filters with unit magnitude and random phase responses ranging from $-180°$ to $180°$ at every discrete frequency. However, the degree of decorrelation resulting from this method is uncontrollable and unrepeatable.

As stated above, bipolar MLSs are pseudorandom sequences with deterministic and periodic structures, but possess characteristics similar to random noise. In particular, the favorable characteristics of nearly uniform power spectra and deterministic, low-valued cross-correlation functions between each reciprocal MLS pair make it an excellent candidate for the design of all-pass digital filters for audio signal decorrelation. In addition, taking advantage of the deterministic and periodic characteristics of MLS, the design of MLS-based decorrelation filters is controllable and repeatable. This is the advantage of MLS-based decorrelation filters over conventional all-pass filter with random phase. We can also use Kasami sequences or other coded pairs for the decorrelation algorithm, but reciprocal MLS pairs are easier to generate.

The design steps of reciprocal MLS decorrelation filters are outlined as follows:

1. Create a pair of n-degree or $L = 2^n - 1$ points reciprocal bipolar MLSs $\{a_i\} = \{a_0, a_1, \ldots a_{L-2}, a_{L-1}\}$ and $\{b_i\} = \{b_0, b_1, \ldots b_{L-2}, b_{L-1}\} = \{a_{L-1}, a_{L-2}, \ldots a_1, a_0\}$.
2. Sequences $\{a_i\}$ and $\{b_i\}$ are respectively used as the coefficients of a decorrelation FIR-filters pair.

As shown in Eq. (6.3) and Fig. 6.1, the bounds of the normalized periodic cross-correlation value of a reciprocal MLS pair, and thereby that of the reciprocal MLS-based filter coefficients, decreases with increasing degree or length of the MLS. A high degree or long length leads to nearly zero cross-correlation between the filter coefficients. The normalized cross-correlation between two filtered output signals is related to the normalized cross-correlation of filter coefficients and normalized autocorrelation $\Phi_0(n)$ of the input signal by:

$$\Phi_{\mathrm{out}} = \Phi_{ab}(n) * \Phi_0(n), \tag{6.9}$$

where $\Phi_{ab}(n)$ is the normalized PCCF of the reciprocal MLS pair. Therefore, the lower the values of normalized cross-correlation between filter coefficients, the more obvious the decorrelation effect will be. For a better decorrelation effect, a pair of high degree or long length reciprocal MLS filters is preferred. This makes design of real-time devices complicated, however. In implementing real-time devices, the specialized convolution algorithm discussed in Daigle and Xiang is of highly practical significance [25]. Moreover, filters with excessive length may distort the transient properties of signals. In practice, the length of reciprocal MLS filters is selected as a compromise between decorrelation effect and simplicity.

Because the perceived effect of decorrelation is greater for frequencies below 1 kHz than for high frequency above 3 kHz, the perceived performance of decorrelation filters at a given length can be optimized or improved by properly adjusting the filter coefficients. In fact, a cyclic time shift of inverse-order MLS $\{a_{L-1}, a_{L-2}, \ldots a_1, a_0\}$ yields L new sequences $\{a_0, a_{L-1}, a_{L-2}, \ldots a_1)$, $\{a_1, a_0, a_{L-1}, a_{L-2}, \ldots a_2\}$, and so on. The cyclic time shift of an inverse-order MLS is equivalent to adjusting its phase characteristics. The original sequence $\{a_0, a_1, \ldots a_{L-2}, a_{L-1}\}$ and each of these time-shifted inverse-order MLS exhibit auto- and cross-correlation characters similar to those in Eq. (6.3) and Fig. 6.1. We can select the one from these time-shifted MLSs that makes the cross-correlation of filters output lowest for low-pass input signal (such as below 1.5 kHz).

Similar to the case of artificial reverberation, a pair of output signals with controllable but higher values of cross-correlation can be approximately obtained by an appropriate mixture of the reciprocal MLS pair.

The algorithm of reciprocal MLS-based audio signal decorrelation is applicable to broadening or controlling the auditory source width (ASW) in spatial sound reproduction. The ASW is an important attribute of hall spatial auditory perception and closely related to early lateral reflections in the hall [26]. The early IACC, which is derived from the interaural cross-correlation within the first 80 ms, can be used as an objective index to evaluate ASW. In stereophonic or multichannel surround reproduction, the cross-correlation between/among the loudspeaker signals can also be used to evaluate the degree of the decorrelation effect. Taking conventional stereophonic reproduction as an example, results from psychoacoustic experiments show that a pair of 2,047-point reciprocal MLS filters (at 44.1 kHz sampling frequency) yields a relatively natural decorrelation effect in terms of left-right symmetry of broadened auditory events as well as less timbre coloration.

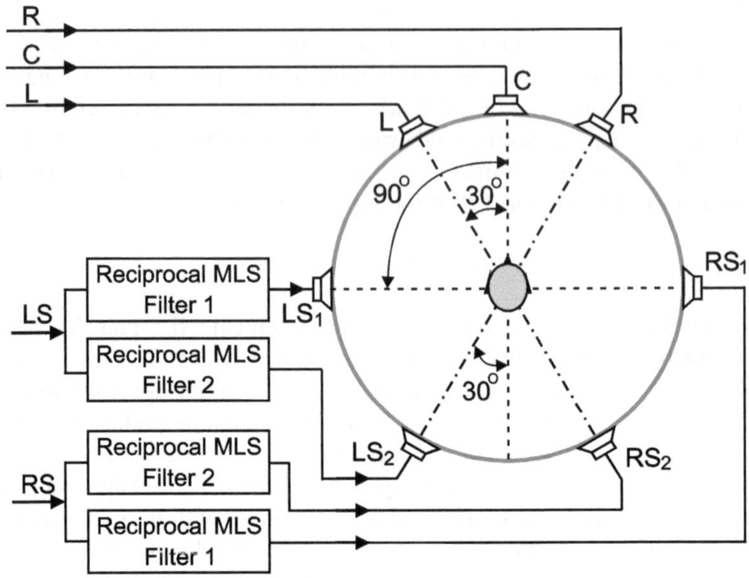

Fig. 6.7 Block diagram of converting 5.1 channel surround sound signals to 7.1 channel surround sound signals

The perceived performance of 511-point filters can be improved using the optimizing scheme mentioned above so that it is perceptually better than the original 1,023-point filters.

Another application of the algorithm of reciprocal MLS-based audio signal decorrelation is enhancing the listener envelopment in surround sound reproduction. It is well known that reproducing decorrelated signals via a series of surround loudspeakers results in the experience of better listener envelopment. An example is converting the 5.1 channel surround sound (music) signals to 7.1 channel reproduction. Figure 6.7 shows the block diagram of such an application. There are five independent full audio bandwidth channels in a 5.1-channel surround sound system, including left (L), center (C), right (R), left-surround (LS), and right-surround (RS), plus an optional low frequency effect channel (called .1 channel, which has been omitted in the figure). In the figure, the original L, C, and R signals are directly reproduced by three corresponding loudspeakers. The original LS and RS signals are filtered by two pairs of reciprocal MLS filters to yield four decorrelated surround signals LS_1, LS_2, LS_3, and LS_4, respectively, and then reproduced by four surround loudspeakers. The two reciprocal MLS pairs are derived from one pair of preferred MLSs. The length of reciprocal MLS filters is 511 points at a 48-kHz sampling frequency, and the optimizing scheme has been incorporated in the filter design. Here, a left-right symmetric decorrelation processing is adopted. That is, the reciprocal MLS filter 1 for signal LS is identical to that for signal RS, and so is the reciprocal MLS filter 2. For two pairs of resultant signals, LS_1 and RS_1 as well as LS_2 and RS_2, when the original signals LS and RS

Fig. 6.8 Gateway Mastering, Portland, ME showing a fractal diffusing rear wall (Diffractal®). (Photo courtesy of Gateway Mastering + DVD)

are decorrelated, the four signals are also decorrelated. While original LS and RS are correlated to each other, the signals within each pair are correlated, and the signals among different pairs are decorrelated. A preliminary subjective experiment shows that the algorithm of reciprocal MLS filter-based decorrelation improves listener envelopment in surround reproduction.

6.6 Diffuser Sequences

Figure 6.8 shows a diffuser applied to the rear wall of a studio, a fractal design which exploited the devices invented by Schroeder in the 1970s. Schroeder originally devised slatted wall surfaces based on MLSs [3], but these devices only operate across about an octave centered around the design frequency. Consequently, Schroeder sought out nonbinary number sequences such as the quadratic residue and primitive root sequences to form diffusers that had wells with many different depths that then operate over a wider bandwidth.

A quadratic residue sequence a_i is given by:

$$a_i = i^2 \bmod N; \quad 0 \leq i < N \tag{6.10}$$

where mod indicates the least nonnegative remainder; N is the number generator that must be an odd prime number and is also the number of wells per period. For example, one period of an $N = 7$ sequence is $\{0,1,4,2,2,4,1\}$.

For room diffusers, it is a complex exponential R_i incorporating the sequence that is crucial to how the structure reflects sound:

$$R_i = \exp\left(\frac{2\pi j a_i}{N}\right) \tag{6.11}$$

where j is $\sqrt{-1}$. Schroeder noted that R_i had an "astounding property" [5]. He was referring to the fact that the magnitude of the discrete Fourier Transform was constant and consequently the periodic autocorrelation function is the unit sample sequence.

The diffuser shown in Fig. 6.8 causes scattering in the horizontal plane. To diffuse sound vertically requires two-dimensional number sequences. To do this for a quadratic residue diffuser requires two number sequences, one for the horizontal (x-direction), one for the vertical (z-direction). Then the z-sequence is used to amplitude modulate the x sequence. For a quadratic residue sequence, this can be expressed as [5]:

$$a_{i,k} = \left(i^2 + k^2\right) \bmod N; \quad 0 \leq i < N; \; 0 \leq k < N \tag{6.12}$$

where i and k are the integers that index the sequence in the x and z direction respectively.

Another common method for making multidimensional phase grating diffusers is to use the Chinese remainder theorem [27]. This folds a one-dimensional sequence of length $N \cdot M$ into a two-dimensional array of size $N \times M$ while preserving the ideal autocorrelation and Fourier properties of the one-dimensional sequence. To use this method N and M must be co-prime. By co-prime, it is meant that the only common factor for the two numbers is 1. A quadratic residue sequence cannot be folded using the Chinese remainder theorem because it has a prime number of elements, and so an alternative sequence is needed, and the primitive root sequence is one possibility.

A primitive root sequence a_i is defined as:

$$a_i = r^i \bmod N; \quad 1 \leq i < N \tag{6.13}$$

where N is an odd prime, r is the primitive root of N, and the sequence has $N-1$ elements per period. A primitive root is an integer that yields a sequence a_i for $i=1$, 2, ... $N-1$ that are all unique. For example, $N=13$ has a primitive root of 2, so $a_i = \{2,4,8,3,6,12,11,9,5,10,7,1\}$, which generates every integer from 1 to $N-1$. Primitive roots can be found by a process of trial and error, alternatively, tables can be found in texts such as [28]. The autocorrelation of the primitive root sequence placed in an exponential using Eq. (6.11) is a two-valued function.

Consider taking the length 12 primitive root sequence and wrapping it into a 3×4 array. The one-dimensional sequence is written down the diagonal of the array, and as it is periodic, every time the edge of the array is reached, the position is folded back into the base period. The process is illustrated in Fig. 6.9.

a_i mod 3 \ a_i mod 4				
	2			3
		4		
			8	

a_i mod 3 \ a_i mod 4				
	2	10	11	3
	6	4	7	9
	5	12	8	1

Fig. 6.9 *Top*: The first four elements of the length 12 primitive root sequence {2,4,8,3,6,12,11,9,5,10,7,1} placed in a 4 × 3 array using the Chinese Remainder Theorem. *Bottom*: the final complete table for one period

This sequence folding technique still maintains the good autocorrelation properties of the original one-dimensional sequence. In the case of the primitive root sequence, the two-dimensional autocorrelation will be two-valued and the power spectrum is flat, expect for a decrease at zero-frequency.

6.7 Diffusers

The sequences with good autocorrelation properties were turned into wall corrugations by Schroeder. When sound encounters a phase grating diffuser (see Fig. 6.10), provided that half a wavelength is larger than the width between the dividers, then plane waves will result within the wells. Plane waves propagate down each well, reflect from the bottom and return to the mouth of the diffuser. For the ith well with depth d_i, for sound with a wavelength of λ, the sound undergoes a phase change of $2jd_i(2\pi/\lambda)$ while propagating in the wells and the pressure reflection coefficient of the well at the front face of the diffuser is:

$$R_i = \exp\left(\frac{4\pi j d_i}{\lambda}\right). \tag{6.14}$$

Comparing Eqs. (6.14) and (6.11) shows that when:

$$\frac{2d_i}{\lambda} = \frac{a_i}{N}. \tag{6.15}$$

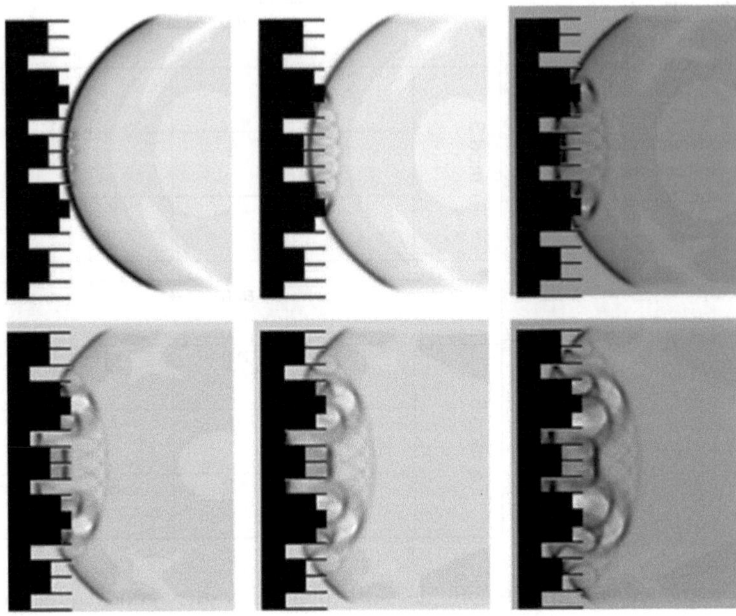

Fig. 6.10 Cylindrical wave reflected from a Schroeder diffuser calculated using a Finite Difference Time Domain (FDTD) model (after Cox and D'Antonio [6])

The sequence of sound waves reflected from the diffuser surface has the same astounding Fourier property that Schroeder identified. A rearranged version of Eq. (6.15) is therefore used to turn the number sequences into physical depths for a particular design wavelength.

What is heard some distance from the diffuser is a complicated interference pattern caused by the waves that emerged from each well before propagating to the listener. It is well-known from classical optics, that a Fourier transform of a wavefront passing through an aperture gives the far field diffraction pattern. Translating this rule for acoustic diffusers, it can be shown that the Fourier transform of the surface reflection coefficients gives the sound distribution in the far field. When there are many diffusers stacked side-by-side, the spatial periodicity causes energy to be preferentially reflected in certain directions creating so-called grating lobes. When a diffuser is formed from a quadratic residue sequence, at the design frequency these grating lobes all have the same energy. (See Fig. 9.11 in Chap. 9) This happens because the autocorrelation of the reflection coefficients is ideal.

In the 40 years since Schroeder pioneered modern diffusers for performance spaces, designs have been refined and improved as detailed in [6]. The autocorrelation of a quadratic residue sequence might be ideal at most multiples of the design frequency, but when translated into a real diffuser operating over many octaves, there are weaknesses in performance that allow room for further improvement. Researchers tried new number sequences, sequences were used in combination to remove periodicity and fractal constructions were developed. All these designs are still, however, recognizable descendants of Schroeder's original diffusers.

One of the most ingenious features of Schroeder's work was the concept of using wells to create a set of known surface reflection coefficients. But this locked diffusers into a geometry that gave an appearance that was not liked by all architects. This drove researchers to come up with new shapes, like curves, designed using numerical optimization. While numerical optimization allows diffusers to be designed with a visual appearance to better match an architect's concept, the design method lacks the elegant simplicity of Schroeder's original ideas. Despite the complex underpinning of Schroeder's designs with number theory, the end design process was just a case of applying a few simple equations on a calculator.

References

1. Schroeder, M.R.: Improved quasi-stereophony and "Colorless" artificial reverberation. J. Acoust. Soc. Am. **33**, 1061–1064 (1961)
2. Schroeder, M.R.: Colorless artificial reverberation. J. Audio Eng. Soc. **9**, 192–197 (1961)
3. Schroeder, M.R.: Diffuse sound reflection by maximum-length sequences. J. Acoust. Soc. Am. **57**, 149–150 (1975)
4. Schroeder, M.R.: Integrated-impulse method measuring sound decay without using impulses. J. Acoust. Soc. Am. **66**, 497–500 (1979)
5. Schroeder, M.R.: Binaural dissimilarity and optimum ceilings for concert halls: more lateral sound diffusion. J. Acoust. Soc. Am. **65**, 958–963 (1979)
6. Cox, T.J., D'Antonio, P.: Acoustic Absorbers and Diffusers, 2nd edn. Taylor & Francis, London (2009)
7. Schroeder, M.R.: Number Theory in Science and Communication. Springer, Berlin (1991). 2nd enl. edn
8. Xiang, N.: Digital sequences. In: Havelock, D., Kuwano, S., Vorländer, M., et al. (eds.) Handbook of Signal Processing in Acoustics, pp. 87–106. Springer, Berlin (2008)
9. Xiang, N., Genuit, K.: Characteristic maximum-length sequences for the interleaved sampling method. Acustica **82**, 905–907 (1996)
10. Xiang, N., Schroeder, M.R.: Reciprocal maximum-length sequence pairs for acoustical dual source measurements. J. Acoust. Soc. Am. **113**, 2754–2761 (2003)
11. Xiang, N., Daigle, J.N., Kleiner, M.: Simultaneous acoustic channel measurement via maximal-length-related sequences. J. Acoust. Soc. Am. **117**, 1889–1894 (2005)
12. Ziemann, A., Arnold, K., Raabe, A.: Acoustic tomography as a method to identify small-scale land surface characteristics. Acta Acust. Acust. **87**, 731–737 (2001)
13. Xiang, N., Li, S.: A pseudo-inverse algorithm for simultaneous measurements using multiple acoustical sources (L). J. Acoust. Soc. Am. **121**, 1299–1302 (2007)
14. Gardner, W.C.: Reverberation Algorithms. Kluwer Academic Publishers, Dordrecht (1998)
15. Blesser, B.: An interdisciplinary synthesis of reverberation viewpoints. J. Audio Eng. Soc. **49**, 867–903 (2001)
16. Moorer, J.: About this reverberation business. Comput. Music J. **3**, 13–28 (1979)
17. Gardner, W.G.: Efficient convolution without input-output delay. J. Audio Eng. Soc. **43**, 127–136 (1995)
18. Van Maercke, D., Martin, J., Vian, J.-P.: Binaural simulation of concert halls: a new approach for the binaural reverberation process. J. Acoust. Soc. Am. **94**, 3255–3264 (1993)

19. Olson, S.: Understanding the perception of double-sloped energy decays: a multidimensional scaling approach. M.S. thesis, School of Architecture, Rensselaer Polytechnic Institute (2008)
20. Blauert, J.: Spatial Hearing. MIT Press, Cambridge, MA (2001)
21. Trivedi, U., Dieckman, E., Xiang, N.: Reciprocal maximum-length and related sequences for artificial spatial enveloping reverberation (A). J. Acoust. Soc. Am. **125**, 2735 (2009)
22. Trivedi, U., Xiang, N.: Utilizing reciprocal maximum length sequences within a multichannel context to generate a natural, spatial sounding reverberation (A). J. Acoust. Soc. Am. **126**, 2155 (2009)
23. Xie, B.-S., Shi, B., Xiang, N.: Audio signal decorrelation based on reciprocal maximal-length sequence filters and its applications to spatial sound. Preprint 8805, the 133rd convention of the Audio Eng. Soc., San Francisco, CA, 2012
24. Kendall, G.S.: The decorrelation of audio signals and its impact on spatial imagery. Comput. Music J. **19**, 71 (1995)
25. Daigle, J.N., Xiang, N.: A specialized cross-correlation algorithm for acoustical measurements using coded sequences. J. Acoust. Soc. Am. **119**, 330 (2006)
26. Beranek, L.: Concert Hall and Opera Houses, 2nd edn. Springer, New York (2004)
27. Schroeder, M.R.: Chaos, Power Laws: Minutes from an Infinite Paradise. W.H. Freeman & Co., San Francisco, CA (1991)
28. Fan, P., Darnell, M.: Sequence Design for Communications Applications. Wiley, New York (1996)

***

Biography

Prof. Ning Xiang earned his Ph.D. in 1990 from the Department of Electrical Engineering at the Ruhr-University Bochum (RUB) in Bochum, Germany. Since Aug 2003 he has been with the Department of Electrical, Computer, and Systems Engineering and the School of Architecture, Rensselaer Polytechnic Institute, Troy, New York, USA. Since 2005, Dr. Xiang has been the chair of the Graduate Program in Architectural Acoustics. Dr. Xiang is a Fellow of the Acoustical Society of America, a Fellow of the Institute of Acoustics (United Kingdom), and a Member of the German Acoustical Society since 1991, a Member of the Audio Engineering Society, a Member of INCE-USA, the Institute of Noise Control Engineering-USA. Dr. Xiang is currently an Associate Editor of the *Journal of the Acoustical Society of America* for the section Architectural Acoustics. In the Acoustical Society of America, he is now serving the Technical Committee Chair on Signal Processing in Acoustics. He is a recipient of Wallace Clement Sabine Medal from the Acoustical Society of America in 2014.

Prof. Bosun Xie received a Bachelor of Science degree in physics and a Master of Science degree in acoustics at South China University of Technology. In 1998, he received a Doctor of Science degree in acoustics at Tongji University. Since 1982, he has been working at South China University of Technology and is currently a professor at Acoustic Lab., School of Science. His research fields include binaural hearing, spatial sound, and sound signal processing. He has published a book named "head-related transfer function and virtual auditory display" and over 150 papers in these fields (most in Chinese). He is a vice-chairman of China Audio Engineering Society and a committee member of Acoustical Society of China.

Trevor Cox is Professor of Acoustic Engineering at the University of Salford, UK, a former Senior Media Fellow funded by the Engineering and Physical Sciences Research Council, and a past president of the UK's Institute of Acoustics (IOA). One major strand of his research is room acoustics for intelligible speech and quality music production and reproduction. Trevor's diffuser designs can be found in rooms around the world. He has co-authored a research book entitled "acoustic absorbers and diffusers." He was awarded the IOA's Tyndall Medal in 2004. Trevor has a long track record of communicating acoustic engineering to the public and has been involved in engagement projects worth over £1M. He was given the IOA award for promoting acoustics to the public in 2009. He has developed and presented science shows to 15,000 pupils including performing at the Royal Albert Hall, Purcell Rooms, and the Royal Institution. Trevor has presented 18 documentaries for BBC radio including: Life's soundtrack, Save our Sounds and Science vs. the Strad. He authored The Sound Book for W.W. Norton (entitled Sonic Wonderlands in the UK).

Chapter 7
Where Mathematics and Hearing Science Meet: Low Peak Factor Signals and Their Role in Hearing Research

Armin Kohlrausch and Steven van de Par

Abstract In his scientific work, Manfred Schroeder touched many different areas within acoustics. Two disciplines repeatedly show up when his contributions are characterized: his strong interest in mathematics and his interest in the perceptual side of acoustics. In this chapter, we focus on the latter. We will first give a compressed account of Schroeder's direct contributions to psychoacoustics, and emphasize the relation with other acoustics disciplines like speech processing and room acoustics. In the main part of the chapter we will then describe psychoacoustic work being based on or inspired by ideas from Manfred Schroeder. Due to Schroeder's success in securing a modern online computer for the Drittes Physikalisches Institut after returning to Göttingen in 1969, his research students had a head start in using digital signal processing in room acoustics for digital sound field synthesis and in introducing digital computers into experimental and theoretical hearing research. Since then, the freedom to construct and use specific acoustic stimuli in behavioral and also physiological research has grown steadily, making it possible to test many of Schroeder's early ideas in behavioral experiments and applications. In parallel, computer models of auditory perception allowed users to analyze and predict how specific properties of acoustic stimuli influence the perception of a listener. As in

This chapter is an adapted version of a chapter which the authors contributed to a book published on the occasion of the 60th anniversary of the Drittes Physikalisches Institut (DPI) at the Georg-August Universität Göttingen [37].

A. Kohlrausch (✉)
Behavior, Cognition, & Perception Group, Philips Research Europe, High Tech Campus 36 (MS WO 02), Eindhoven 5656AE, The Netherlands

Human-Technology Interaction, Eindhoven University of Technology, Eindhoven, The Netherlands
e-mail: armin.kohlrausch@philips.com

N. Xiang and G.M. Sessler (eds.), *Acoustics, Information, and Communication*, Modern Acoustics and Signal Processing, DOI 10.1007/978-3-319-05660-9_7,

other fields of physics, the close interplay between experimental tests and quantitative models has been shown to be essential in advancing our understanding of human hearing.

7.1 Introduction

One of the scientific areas that was close to the heart of Manfred Schroeder was psychoacoustics. During his first period in Göttingen, the focus of his work was on the physical and statistical side of acoustics. On moving to the Bell Laboratories in 1954, he came into an environment with a long history in hearing and speech research, started by Harvey Fletcher in the 1910s. In his autobiographic chapter, Schroeder describes a number of examples how he got interested in the perceptual side of acoustics, both from his research in room acoustics and in speech processing. During his time as professor at the University of Göttingen, the number of diploma and Ph.D. students focusing on hearing related studies increased continuously. This increased interest was supported by a rebuilding of the central space in the "Halle," in front of the "Reflexionsarmer Raum," where the control panels for the loudspeaker dome (well known to acousticians from a photo in Blauert's book on spatial hearing, see Fig. 3.50 in [2]) were dismantled and spaces for two listening booths were created—unofficially "owned" by the two research groups around Birger Kollmeier and the first author. In parallel with the acoustic spaces, the computer infrastructure for controlling listening experiments and generating acoustic stimuli also grew steadily. This experimental infrastructure and a growing group of young scientists were essential for the increasing level of sophistication of hearing research at the Drittes Physikalisches Institut (DPI).

In this chapter, we want to briefly summarize Schroeder's contributions to the psychoacoustic literature. In the major part of this chapter, we will describe how the particularly close link between hearing research and another of Manfred Schroeder's scientific interests, defining signals with very specific properties, like low peak factors, has influenced the psychoacoustic community. Some aspects of these developments are also closely interlinked with other facets of Schroeder's life, as will become clear by comparing the present text with the corresponding passages from his autobiographic chapter.

The advanced digital signal processing capabilities at the DPI, quite unusual for an institute of physics in the 1970s and strongly inspired by Schroeder's experience at the Bell Laboratories, allowed a high level of creativity in constructing signals with specific spectral and temporal properties in both listening experiments and model simulations. These innovative approaches, which started at the DPI, spread to other places like Eindhoven and Oldenburg, and later on to Lyngby/Kopenhagen, and have influenced research paradigms in hearing research groups all over the world.

7.2 Schroeder's Major Contributions to Hearing Research

Schroeder's interest in psychoacoustics had a number of roots. Through his work on efficient speech coding and the naturalness of synthesized speech, he developed an interest in the role of the signal envelope and fine structure on timbre. A specific manifestation was his work on "monaural phase effects." Furthermore, he used concepts of perceptual masking and perceptual distance measures in speech coding algorithms to improve the trade-off between perceptual quality and bit rate. Through his work on concert hall acoustics and digital reverberators, he recognized the great importance of spatial parameters such as the interaural cross-correlation, for good perceptual quality of a reverberant environment; this had led to work on binaural modeling. Actually, stimulated by his research in concert hall acoustics in the 1970s, binaural psychoacoustics evolved in the 1980s into one of the strongholds of his research group at the DPI in Göttingen. And finally, he had a genuine interest in developing and improving models of the auditory periphery—models that could help to understand specific perceptual phenomena, like the level dependence of difference tones. An excellent overview about how deeply he thought about these various aspects of hearing science can be found in his overview article from 1975: Models of Hearing [59].

Probably the psychoacoustic topic that interested Schroeder the earliest was the question of the extent to which the human hearing system could decode the phase spectrum of a signal. His first publication on this topic was an abstract from the 58th meeting, in 1959, of the Acoustical Society of America with the title: New results concerning monaural phase sensitivity [54]. Schroeder describes the perceptual consequences of changing the component phase in harmonic complex signals comprising up to 31 components. Aspects mentioned are the strong influence of the peak factor on the signal timbre, the absence of timbre changes for signals with identical envelopes, and the possibility to create strong and varying pitch phenomena allowing one to play melodies, just by changing phases of individual components.

The emphasis on the relation between timbre and waveform had a direct link to the work on vocoder quality, where, on the synthesis side, the excitation signal for voiced speech was composed of harmonic complex tones for which one could choose the relative phases freely. Schroeder and colleagues had observed that the excitation function strongly influenced the quality of vocoder speech (see, e.g., [55]). Schroeder described this early work, including waveform examples for different phase choices, in his 1975 IEEE paper [59] where he also referred to an earlier, less-known account by R. L. Pierce on this work in the American Scientist from 1960 [50]. In the course of his research on monaural phase effects, he also formulated closed solutions for generating periodic signals with low peak factors [57]. This short mathematical paper from 1970 remained relatively unknown for nearly 20 years, receiving only 29 citations until 1989. After this phase rule was introduced into hearing research in 1986 in [69] (see next section), the number of citations exploded and grew to a total of 312 (as of October 2013), making it, together with the paper on measuring reverberation time, by far the most cited of his scientific publications.

Through his work on efficient speech coding, eventually resulting in code-excited linear prediction, Schroeder became acquainted with the phenomenon of masking and the great value of perceptually based distance metrics (see also the Chap. 10 in this book). The term "subjective error criteria" appeared first in the title of a conference contribution by Atal and Schroeder in 1978 [1]. As stated in the abstract of that paper, the human ear does not use simple RMS error measures when judging distortions introduced through coding. The new approach by Atal and Schroeder was to minimize the subjective distortion by shaping the spectrum of the resulting prediction error to be optimally masked by the speech spectrum. This approach does not reduce the total power of the error signal, which is determined by the quantizer, but redistributes its power along the frequency axis to have a more constant signal-to-noise ratio at all frequencies.

The concept of masking and its application in speech coding appeared in a follow-up article in 1979, jointly written with Atal and J.L. Hall [62]. The paper contains an extensive description of how to transform a short-term power spectrum into an excitation pattern, including transformation through outer and middle ear, calculation of critical-band densities, and transforming these into an excitation pattern from which the loudness was finally computed. The masked thresholds due to a masking signal, which in this application was the speech signal, were computed by multiplying the signal (speech) excitation function by a sensitivity function. In this way, the authors were able to predict whether, for a given short-term spectrum of the speech signal, a specific noise would fall below this masking function and was therefore inaudible. In addition to predicting the masked threshold, they derived, for suprathreshold noise levels, an objective degradation scale by relating the speech and the noise loudness (which, in modern terms, was calculated as partial noise loudness in the presence of the speech signal). This approach, in which state-of-the-art perception science was functionally integrated in a real-life signal-processing application, was later on also successful in the development of perceptual audio coding starting in the mid-1980s. An early account of this audio coding work, referring back to the earlier work by Schroeder et al. [62], is the paper by Johnston: Transform coding of audio signals using perceptual noise criteria [28].

An early account of spatial perceptual aspects in concert hall acoustics is given in the context of the acoustic measurements in the Philharmonic Hall [63]. One of the analyzed parameters which varied strongly between listener positions was the directional distribution of the early energy within 50 ms of the direct sound. After the acoustic changes in the hall had been finished, the variability of this parameter was much reduced. In their conclusion, the authors state, "It is therefore tentatively concluded that the directional distribution of early reflections is a significant contributing factor to acoustical quality" (p. 440 in [63]). This recognition of the role of spatial dissimilarity for good concert hall acoustics was further supported by the large comparative study of European concert halls, performed together with Gottlob and Siebrasse in Göttingen [64]. From their multidimensional analysis of paired comparison data from many listeners, the interaural coherence turned out to be an independent dimension, being negatively correlated with subjective preference. This article addresses spatial hearing also from a different perspective. It describes the setup in the anechoic chamber in Göttingen, which permitted the reproduction of two-channel binaural recordings via two loudspeakers in an anechoic environment. The critical element in the reproduction chain was a cross talk cancellation computation, as it had been described and demonstrated 10 years earlier by Schroeder and Atal [61]. With cross talk cancellation the sound transmission from two loudspeakers could be directed to the two ears as otherwise only possible by headphone reproduction.

In his paper "New viewpoints in binaural interaction" at the fourth International Symposium on Hearing in Keele, 1977 [60], Schroeder made an original contribution to theories of binaural hearing, inspired by his work on basilar membrane characteristics. The core of his proposal used cochlear delays as a basis for the analysis of interaural delays. The computation of waveform delays between right and left ear would, instead of being enabled by interaural neural delay lines (as they are included in many types of binaural models), be realized by comparing interaurally the activities between different places on each basilar membrane. Such a comparison was possible and meaningful, because the basilar membrane activity has, for a given signal, a systematic relation between place and delay, and the *basilar membrane delay* values are of the magnitude necessary to analyze realistic *interaural delay* values. This proposal was later implemented and evaluated by Shamma and colleagues [67] who coined the term "stereausis" for this scheme. A consequence of this way of thinking about interaural delay, as pointed out by Joris and colleagues at the 13th International Symposium on Hearing in Dourdan, France [29], was that the traditional delay-line model using neural delay lines needed to have a very exact anatomical link between equivalent basilar membrane positions in the two cochleae. Only small spatial mismatches would introduce considerable offsets in the interaural delay values, particularly at low frequencies.

A final topic which we want to mention is Schroeder's research on basilar membrane modeling and its relation to specific psychoacoustic phenomena. Schroeder's interest in the relation between basilar membrane and hair cell properties on one side and perceptual observations on the other might have its roots in his attempts to relate cubic difference tone (CDT) data to the critical-band concept and basilar membrane mechanics. In his first paper from 1969 [56], he analyzed CDT data from Goldstein, and added basilar membrane simulation and additional CDT phase measurements performed at Bell Laboratories. As in similar earlier work by Zwicker, this established a close relation between CDT generation and auditory excitation. In addition, Schroeder concluded that a "purely mechanical model of the ear" was insufficient to explain the amplitude and phase characteristics of the $2f_1 - f_2$ difference tone. An additional amplitude nonlinearity was required, and Schroeder concluded, based on the sign of the resulting CDT, that it had to be an amplitude-*limiting* nonlinearity.

In the following years, Schroeder, often together with J.L. Hall, contributed papers on basilar membrane models [58] and a model for mechanical to neural transduction in the inner hair cell [65], both of which have been highly influential in the development of more realistic models for the auditory periphery. How useful fast and realistic time-domain basilar membrane models can be for the interpretation of psychoacoustics results will be demonstrated in the next section.

7.3 Harmonic Complex Tone Stimuli and the Role of Phase Spectra

A particular class of auditory stimuli are signals with a periodic waveform. They can be analyzed in terms of their temporal properties (e.g., the perceptual influence of a slight deviation from temporal regularity in an otherwise regular series of clicks, see [43]) or in the context of their spectral properties (consider, for example, the role of the vocal tract filter on the resulting vowel quality). Of course, from a mathematical point of view, time domain and spectral descriptions are fully equivalent, as long as the complex spectrum, including phase, is considered. Historically, however, temporal and spectral views were quite distinct, mainly because of the influence of signal analysis systems that represented the power spectrum but not the phase spectrum. Also, the psychoacoustic paradigm of critical bands and auditory filters, which for a long time were only defined in terms of their overall bandwidth and their amplitude characteristics (see, e.g., [19]), made it difficult to bring the temporal and spectral views closer together. Again, the increasing use of computer programs to generate acoustic stimuli and to perform time-domain modeling of perceptual processes emphasized the role of the phase spectrum on the perceptual quality of periodic signals (e.g., [16]). In addition, for us students in Göttingen, there was the strong interest of Schroeder in phase effects which prepared us more than other colleagues at that time to think in terms of signal waveforms, envelopes, and the time dimension (e.g., [66]). In the following, we will focus on one specific type of complex tones, tones with so-called Schroeder phases.

7.3.1 Schroeder-Phase Harmonic Complex Tones

7.3.1.1 Definition

The term "Schroeder phase" refers to a short paper by Schroeder from 1970 [57]. In this paper, he addressed the problem how the peak-to-peak amplitude of a periodic waveform can be minimized. His proposed solution lies in a phase choice which gives the signal an FM-like property. The general solution derives the individual phase values without any restriction on the amplitude spectrum. A specific case is that of a harmonic complex with a flat power spectrum having a band-pass characteristic. The solution for the individual phase values of such a complex is as follows:

$$\phi_n = -\pi n(n-1)/N, \tag{7.1}$$

with N the total number of components in the complex. The important term in this equation is the quadratic relation between component number, n, and component phase, ϕ_n, which leads to an approximately linear increase in instantaneous frequency over time. The normalization with N creates a signal, for which the instantaneous frequency sweeps once per period from the frequency of the lowest to that of the highest component in the complex. In fact, the instantaneous frequency has a periodic sawtooth-like course for such Schroeder-phase signals.

It is obvious that reversing the initial sign in (7.1) has no influence on the peak factor of the resulting signal, but it will invert the direction of the linear frequency sweep. Because these two versions of a Schroeder-phase signal lead to substantially different percepts, a convention has been introduced to distinguish them. A *negative* Schroeder-phase signal is a signal were the phases of individual components are chosen as in (7.1). In contrast, a *positive* Schroeder-phase signal has phase values with a positive sign in front of the fraction. One can memorize this relation by using the fact that the sign of the phase is opposite to the direction of change in instantaneous frequency.

7.3.1.2 Acoustic Properties

By construction, Schroeder-phase stimuli have a relatively flat temporal envelope and the peak factor, defined as the ratio between envelope maximum and the RMS-value of the signal, is much lower than for other phase choices. This is demonstrated in Fig. 7.1, which compares the waveforms for harmonic complexes composed of 19 equal-amplitude harmonics of a 100-Hz fundamental from 200 to 2,000 Hz. There are three different choices of the component phases: positive Schroeder phase, negative Schroeder phase, and, in the top part, a zero-phase stimulus. For this latter stimulus, the energy is concentrated at very short instances within each period, leading to a much higher peak factor. The spectro-temporal properties of these signals can be seen more clearly in a short-time spectral representation.

Figure 7.2 shows the spectra of the three signals from Fig. 7.1 calculated using a moving 5-ms Hanning window. The sawtooth-like frequency modulation of the two Schroeder-phase complexes is pronounced in this representation. In addition, the plot for the zero-phase complex shows ridges at the spectral edges of 200 and 2,000 Hz. These relative spectral maxima can be perceived as pitch, superimposed on the 100-Hz virtual pitch of the complexes, and the presence of these pitch percepts has been used as a measure of the internal representation of such harmonic complexes [34, 35]. The visibility of the temporal structures in the short-time

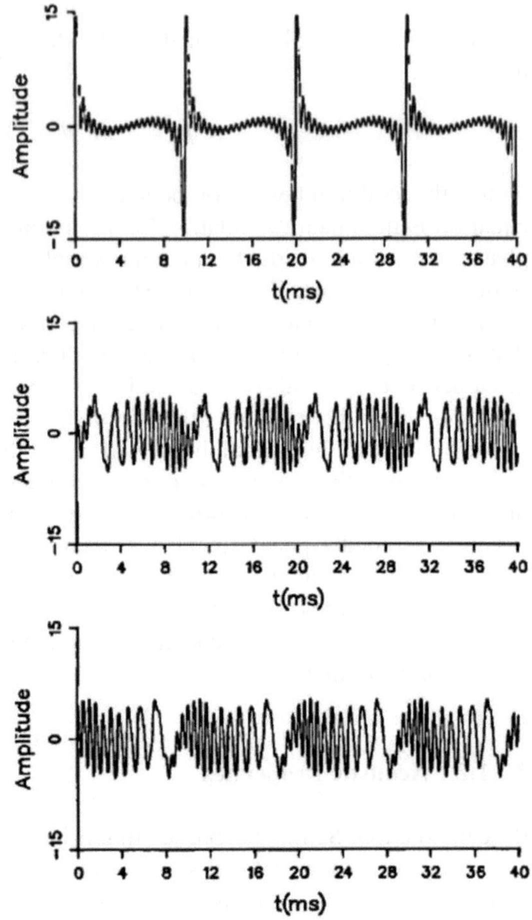

Fig. 7.1 Time functions of harmonic complexes for three different choices of the component's starting phases. *Top*: $\phi_n = 0$ zero-phase complex, *middle*: $\phi_n = -\pi n(n-1)/N$, negative Schroeder-phase complex, *bottom*: $\phi_n = +\pi n(n-1)/N$, positive Schroeder-phase complex. All complexes are composed of the equal-amplitude harmonics 2 to 20 of fundamental frequency 100 Hz. For this plot, the amplitude of an individual harmonic was set at 1. Reprinted from [36]. Copyright (1995) Acoustical Society of America

spectrum depends critically on the duration of the temporal window, relative to the period of the sound. The shorter the window, the more visible are temporal changes; the longer the window, the more the spectral properties are emphasized.

7.3.2 Role in Hearing Research and Perceptual Insights

The first paper in which the Schroeder-phase formula was used in hearing experiments was published by Mehrgardt and Schroeder in the proceedings of the 6th International Symposium on Hearing, 1983 [41]. In this paper, the quadratic phase formula from (7.1) was combined with an additional scaling factor, which permitted control of the spread of signal energy throughout the period. The spectrum of the harmonic complex was, however, not flat as in most later

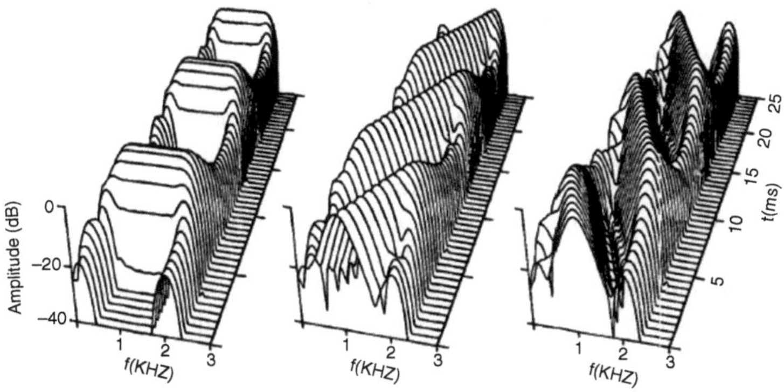

Fig. 7.2 Short-time spectral representation of the signals from Fig. 7.1 using a Hanning window with length 5 ms. The *left* panel shows the zero-phase signal, the *middle* panel the negative Schroeder phase and the *right* panel the positive Schroeder-phase complex. Reprinted from [36]. Copyright (1995) Acoustical Society of America

investigations, but the individual components had Hanning-weighted amplitudes. This paper emphasized the influence of the masker's temporal waveform on the observed masking behavior and showed how strongly the acoustic waveform and the resulting masked thresholds can vary by just varying the phase spectrum.

The great potential of Schroeder-phase signals to expose the phase characteristic of an auditory filter was discovered quite accidentally. During his master thesis research, Bennett Smith, traveling back and forth between Göttingen and Paris, where he continued to work as an engineer at the IRCAM, was interested in acoustic figure-ground phenomena, translating spatial orientation in the visual domain into linear frequency modulation in the auditory domain [68]. In the construction of his acoustic background stimuli, he made use of the Schroeder-phase formula. He did, unfortunately, not find any effect of acoustic figure-ground orientation on audibility, but made instead another observation: The audibility of a sinusoidal stimulus in a complex-tone masker depended strongly on the sign in the phase formula: When the background was constructed with a positive sign, the masked thresholds were lower by up to 20 dB, compared to the situation in which the phase sign was negative.

This large threshold difference formed a considerable scientific puzzle which kept the participants of Schroeder's weekly seminar busy for a prolonged period: Both masker versions had a similarly flat temporal envelope, so there was no reason to assume a difference in masking potential for simultaneously presented sinusoids. In addition, no temporal asymmetry, as in backward versus forward masking, could play a role given the simultaneous presentation of masker and test signal. We even considered the effects of FM to AM conversion of the masker waveform that should occur when the linear frequency modulation of the masker interacts with the shallow slopes of the auditory filter, without arriving at a satisfactory explanation. The effect was finally understood on the basis of computer simulations with a time-domain basilar membrane model which had been realized by Hans-Werner Strube

[70]. This model was, at that time, the first computer model with a realistic phase characteristic for the basilar membrane, which was able to compute the basilar membrane output in the time domain while being sufficiently time efficient. Strube observed that the two versions of the Schroeder-phase complex led to very different waveforms at the output of the various model elements that each represented a specific place on the basilar membrane. One waveform was clearly more modulated, and the existence of valleys in the masker waveform fitted nicely with the observation by Smith that this masker also led to lower masked thresholds: The target signal could be detected easier, i.e., at a lower level, within these masker valleys—a phenomenon for which the term “listening in the valleys” is used in the psychoacoustic literature. On the basis of the consistent explanation made possible by these simulations, we dared to summarize our experimental data and submit the manuscript to JASA. For me (AK), this was to become my first internationally peer-reviewed publication, and actually the only one jointly co-authored by Manfred Schroeder [69].

In the next years, a great many further experiments and model simulations were performed in Göttingen [30, 36], which led to the following insights: The clue to understanding the differences between positive and negative Schroeder-phase stimuli lies in the phase characteristic of the auditory filter, in combination with some general rules in masking of narrowband signals. If subjects have to detect a narrowband stimulus in a broadband masker, only the frequency region around the target frequency is of interest. In order to simulate the transformations in the auditory periphery for such an experiment, we have to compute the waveforms of the acoustic stimuli at the output of the auditory filter that is centered on the target frequency. For random noise maskers, only the amplitude characteristic of the filter is relevant, in line with the early “critical-band” concepts. For periodic signals as considered here, however, this filtered waveform, and its masking behavior, will depend critically on the amplitude and the *phase* characteristic of the auditory filter.

The most important conclusion was that, for the right choice of stimulus parameters, the phase characteristic of a Schroeder-phase stimulus matches quite closely the phase characteristic of the auditory filter, at least in the spectral region of maximum transfer of the filter. Since Schroeder-phase stimuli come in two flavors, one version, the negative Schroeder-phase stimulus, will have the same phase characteristic as the auditory filter, while the positive Schroeder-phase stimulus has a phase spectrum that is *opposite* to that of the filter. In the transfer through a filter, the input phase spectrum and the filter phase spectrum add to give the phase spectrum of the output signal. For the positive Schroeder phase, we thus have the situation of phase compensation, and the resulting filtered signal at the output of the auditory filter has a nearly constant phase of the strongest components, those with a frequency at the center of the corresponding basilar membrane filter, resulting in a highly peaked signal. Conceptually, this situation is quite similar to pulse compression through frequency modulation, as it is used in radar and sonar technology. In a way, the positive Schroeder-phase stimuli are matched in their phase characteristic to the auditory filters as they are realized mechanically in the inner ear.

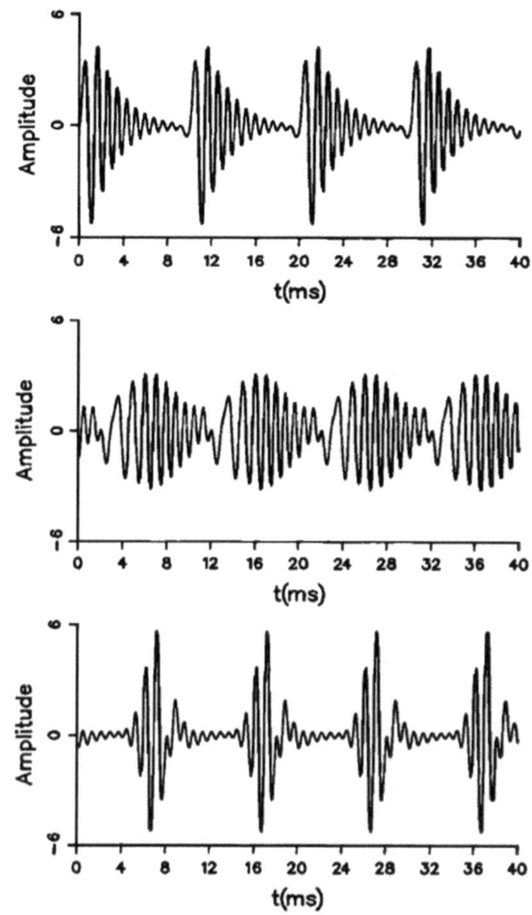

Fig. 7.3 Responses of a linear basilar membrane model at resonance frequency 1,100 Hz to the three harmonic maskers with fundamental frequency 100 Hz shown in Fig. 7.1. The *top* panel shows the zero-phase signal, the *middle* panel the negative Schroeder phase and the *bottom* panel the positive Schroeder-phase complex. Reprinted from [36]. Copyright (1995) Acoustical Society of America

This interpretation leads to some interesting, counterintuitive predictions: If a specific Schroeder-phase stimulus is optimally matched in its phase to the inner-ear filter at a certain frequency, then such a signal should have a higher peak-factor after filtering than a zero-phase input stimulus. This prediction is analyzed in Fig. 7.3, which shows waveforms of the three stimuli from Fig. 7.1 at the output of a linear basilar membrane filter tuned to 1,100 Hz. The two panels at the bottom show the waveforms of the two Schroeder-phase complexes. While the broadband input signals to the filter have a flat temporal envelope (cf. Fig. 7.1), both waveforms have a clear amplitude modulation after filtering, which follows the 10-ms periodicity of the stimulus. But the depth of modulation is quite different for the two signals, the positive Schroeder-phase stimulus at the bottom has a much higher peak factor and a longer period of small envelope values than the negative Schroeder-phase complex in the middle. This simulation reflects the initial observation made by Strube in [70].

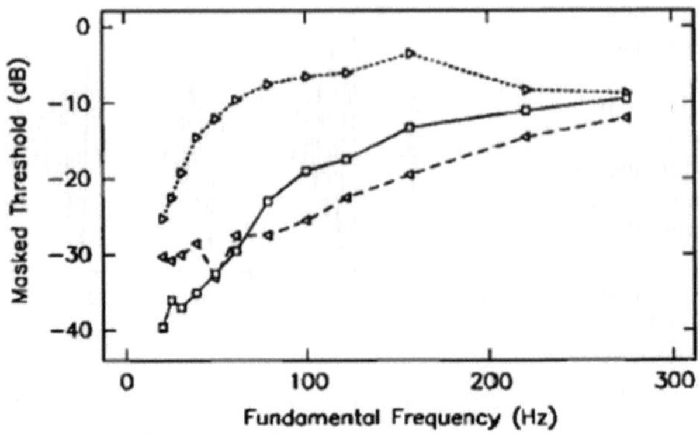

Fig. 7.4 Simultaneous masked thresholds of a 260-ms, 1,100-Hz signal as a function of the fundamental frequency f_0 of the harmonic complex masker. Thresholds are expressed relative to the level of a single masker component. The maskers were presented at a level of 75 dB SPL. *Squares*: zero-phase complex; *right-pointing triangles*: negative Schroeder-phase complex; *left-pointing triangles*: positive Schroeder-phase complex. Reprinted from [36]. Copyright (1995) Acoustical Society of America

The top panel shows the filtered version of the zero-phase complex, which has a highly peaked input waveform. The filtered waveform reflects, within each period, the impulse response of the auditory filter, because the zero-phase complex is similar to a periodic sequence of pulses. By careful analysis one can even see some properties of the filter's frequency-dependent group delay: around each peak, low-frequency components with a long waveform period occur first, followed later on by higher-frequency components. Comparing the top and the bottom panel reveals that, indeed, the positive Schroeder-phase complex has a somewhat higher peak factor than the zero-phase complex, and its energy is more concentrated in time, as expected based on the pulse-compression analogon.

The relation between zero-phase and positive-Schroeder-phase stimuli also formed the key to estimating the phase properties of a specific point on the basilar membrane. If we were able to determine the phase curvature, for which the match between stimulus and filter phase is "optimal," then this value would be an indication of the phase curvature (or the frequency-dependent group delay) of the filter. The clue to such an analysis is given by comparing perceptual thresholds for positive Schroeder-phase maskers with those for zero-phase maskers. For maskers, for which this difference is largest (and for which the Schroeder-phase stimulus as masker gives lower masked thresholds), the phase curvature at the signal frequency is an estimate of the filter phase. Figure 7.4 shows data from [36] which were used for such a computation.

In the region of f_0 values between 100 and 150 Hz, the differences in thresholds obtained with the positive Schroeder-phase complex (left-pointing triangles) and with the zero-phase complex (squares) are largest. For theses complexes, the second derivative of the phase-versus-frequency relation, which indicates the phase curvature has values between $1.05 \times 10^{-5}\pi/\text{Hz}^2$ and $0.74 \times 10^{-5}\pi/\text{Hz}^2$. We can conclude that the curvature of the phase characteristic for the basilar membrane filter centered at 1,100 Hz should be in the range of these two values. A similar conclusion about the phase curvature can be derived from the parameters of those complexes, for which positive Schroeder-phase and zero-phase complexes lead to approximately the same threshold. In this case, the internal envelope modulation of the two complexes after filtering on the basilar membrane should be approximately equal. As explained in detail in [36], the phase curvature in the Schroeder-phase stimuli should be half the value of the filter curvature, and this is reached for fundamental frequencies of 50–75 Hz. And exactly in this region, the two lower curves in Fig. 7.4 cross each other.

This consideration allowed a first computation of the auditory filter phase for one frequency, 1,100 Hz. In [36], additional threshold measurements were included for frequencies 550, 2,200 and 4,400 Hz, thus covering a range of 3 octaves. It is often assumed that the auditory filter has a nearly constant quality factor across the range of audible frequencies. If this scaling relation were to hold also for the phase characteristic, then the results obtained at 1,100 Hz would allow a direct prediction for the phase characteristic in the range of 3 octaves around 1,100 Hz. The comparison with the results at 550 Hz indeed revealed the expected relation, while towards higher frequencies, the curvature changed somewhat less with center frequency than expected for a system in which the amplitude and phase characteristics of the filters remain constant on a logarithmic frequency scale.

One final important observation from these initial Schroeder-phase studies needs to be mentioned. In the 1980s, the view on the shape of auditory filters was strongly influenced by the work of Patterson, Moore and colleagues, who had used the notched-noise technique to estimate the amplitude characteristic of the auditory filter. The best characterization was possible with a so-called rounded exponential filter shape (see, e.g., [49]). A time-domain implementation of a filter with such an amplitude characteristic was possible based on so-called gamma-tone filters [27, 48]. Because of the many studies supporting this concept of auditory filters, we were interested to analyze the Schroeder-phase stimuli with such a filter.

Figure 7.5 presents, in a format similar to Figs. 7.1 and 7.3, four periods of the waveform for harmonic complexes with fundamental frequency 100 Hz. The analysis shows the output of the gamma-tone filter tuned to 1,100 Hz, and the three subpanels are for the three different phase choices. It is apparent that this filter does not lead to differences in the modulation depth between the three stimuli, and based on this simulation one would expect quite similar masking behavior of all three complex tones, in contrast to the experimental data. The major reason for the similar treatment of the two Schroeder-phase maskers by the gamma-tone filter is its antisymmetric phase characteristic close to its resonance frequency. The curvature of the filter phase changes its sign at the resonance frequency from negative to positive. A filter with such a phase characteristic can never flatten out the phase of a Schroeder-phase complex that changes over the full range of its passband, and should therefore be used with caution in experiments where signal phase matters. As has been shown later by Lentz and Leek [40], this shortcoming of

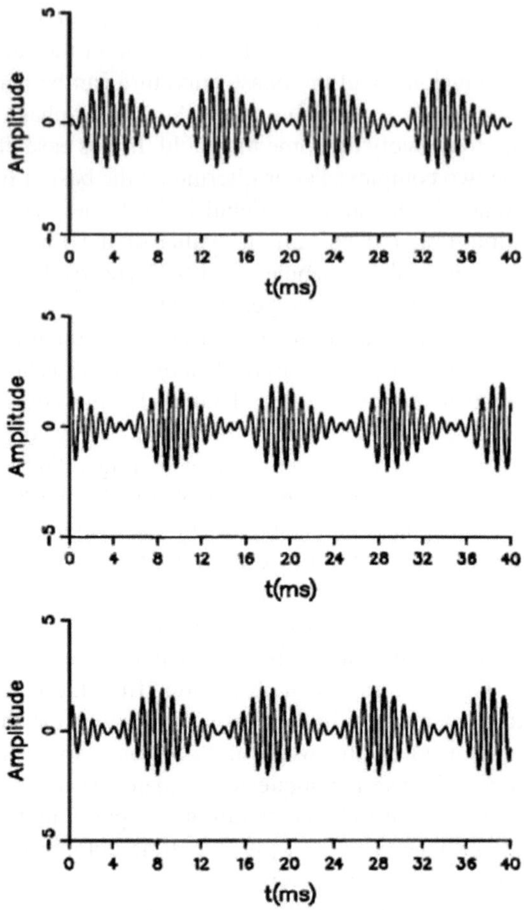

Fig. 7.5 Responses of a linear, fourth-order gamma-tone model at resonance frequency 1,100 Hz to the three harmonic maskers with fundamental frequency 100 Hz shown in Fig. 7.1. The *top* panel shows the zero-phase signal, the *middle* panel the negative Schroeder phase and the *bottom* panel the positive Schroeder-phase complex. Reprinted from [36]. Copyright (1995) Acoustical Society of America

the gamma-tone filter can be overcome by using a nonlinear extension of this filter, the gammachirp filter proposed by Irino and Patterson [26].

7.3.3 Later Developments

Although the first paper on Schroeder-phase stimuli was already published in 1986 ([69]), the paradigm was only widely adopted after publication of our second paper in 1995 [36]. The first papers that used the term "Schroeder phase" in their title were published in 1997 [9, 10]. Many authors related psychoacoustic findings with Schroeder-phase stimuli to the properties of the basilar membrane. Differences that were found between normal-hearing and hearing-impaired subjects and also influences of the overall presentation level indicated some role of active processes in creating large differences between positive and negative Schroeder-phase stimuli [10, 71, 72]. Based on the results of these studies, Summers concluded: "The current results showed large differences in the effectiveness of positive and negative Schroeder-phase maskers under test conditions associated with nonlinear cochlear processing. The two maskers were more nearly equal in effectiveness for conditions associated with more linear processing (high levels, hearing-impaired listeners). A number of factors linked to the cochlear amplifier, including possible suppressive effects and level-dependent changes in the phase and magnitude response of effective filtering, may have contributed to these differences." [[71], p. 2316].

The analysis of the phase characteristics of auditory filters was further refined by Oxenham and Dau [44, 45]. They varied the phase curvature of Schroeder-phase complexes by using a scalar multiplier in front of (7.1), very similar to the use of the Schroeder-phase formula by [41]. They concluded that the scaling invariance of filter phase with filter center-frequency, as expected for a set of filters with constant quality factor, might hold for frequencies above 1 kHz, but not for lower frequencies.

Schroeder-phase stimuli have also been used in physiological experiments, which allowed a direct test of the basic hypothesis of the role of peripheral filtering on modulation depth of the waveform, as published in 1985 by Strube. In 2000, Recio and Rhode [52] measured the basilar membrane response in the chinchilla for positive and negative Schroeder-phase stimuli and also for clicks, thus using types of stimuli very similar to the early psychoacoustic studies. They concluded: "The behavior of BM responses to positive and negative Schroeder complexes is consistent with the theoretical analysis performed by Kohlrausch and Sander in 1995, in which the curvature, i.e., the second derivative of the phase versus frequency curve of the BM was used to account for the differences in the response to each of the two Schroeder phases... Hence, phase characteristics of basilar membrane responses to positive Schroeder-phase stimuli show reduced curvatures (relative to the stimulus), and, as a result, peaked waveforms (Kohlrausch and Sander, 1995)" [[52], p. 2296].

Given the relevance of signal phase in Schroeder's work and in hearing research, the first author chose this topic in his contribution to the session "Honoring Manfred R. Schroeder, his contributions and his life in acoustics" at the 4th joint meeting of the Acoustical Society of America and the Acoustical Society of Japan in Honolulu, Hawaii (2006) in his talk: "Schroeder's phase in psychoacoustics." This session was held in the period that the DPI building at the address Bürgerstr. 42–44 was being dismantled and rebuilt for non-academic use. In order to preserve some memories, I (AK) was able to physically remove and to hand over to Schroeder a sign from the institute's parking lot, indicating the place where he and his family used to park their car during their weekend strolls through downtown Göttingen (see Fig. 7.6). The plate had been signed by all speakers of this memorial session.

Fig. 7.6 Photo from the 4th joint ASA and ASJ meeting, Honolulu, 2006, showing the first author in local outfit handing over to Manfred Schroeder a sign that used to indicate the parking spot reserved for goods delivery from the parking lot of the Drittes Physikalische Institut in Göttingen. Photo courtesy of session co-chair Akiro Omoto

7.4 Noise Signals with Non-Gaussian Statistics

In the last part of our chapter we want to demonstrate that the use of the very traditional auditory stimulus known as "noise" has also been influenced by advanced signal processing possibilities. Again, the role of mathematics, in this case of amplitude and envelope statistics and of the amount of intrinsic fluctuations and their spectral composition, is evident and has enabled major steps forward in understanding working principles of the human auditory system. Although the meaning of the term noise is wider in daily use, in hearing sciences, it refers to signals that are inherently random. When we consider for example white Gaussian noise, samples taken from its temporal waveform are randomly distributed according to a Gaussian distribution, and samples taken at subsequent moments in time are uncorrelated. The frequency-domain representation of white Gaussian noise shows a complex spectrum where the real and imaginary parts are also normally distributed.

Noise signals are often subjected to some kind of spectral filtering. Although this influences the spectral envelope of the signal, the Gaussian distributions of the time-domain samples and the complex spectral components are not influenced. However, the correlation across samples taken at different moments in time is influenced. This is reflected in the auto-correlation function. For a white noise signal, the autocorrelation function has a peak at lag zero and is zero at all other lags, in line with the idea that samples are mutually uncorrelated. For a filtered noise, however, there will be correlation across samples which is reflected in the autocorrelation function being unequal to zero at non-zero time lags.

Noise signals have a long history in hearing research because they are relatively easy to generate by analog means. Also, the most often used modifications, spectral filtering and temporal windowing, could be realized using basic analog electronics. Noise stimuli have been used extensively to study auditory masking where noise often serves as a masker. For example, in the early experiments related to critical bandwidth by Hawkins and Stevens [23], white, Gaussian-noise maskers were used to measure the frequency dependence of masked thresholds of a tonal signal. The preference for using white noise signals can be attributed to properties such as uniform energy distribution across time and frequency and the fact that the signals can be described by using only a few parameters.

The inherently stochastic nature of noise has strong implications for its masking behavior. This was demonstrated in studies that employed reproducible noise for which the stochastic uncertainties in the noise are effectively removed. Generally, reproducible noise produces lower masked thresholds than running noise (e.g., [39, 77]).

The probability distribution of noise amplitudes and envelopes can also have an influence on the masking properties of a noise signal. As long as stimuli for hearing research were generated by analog means, noise signals typically had a Gaussian statistic. One early exception to this was so-called multiplied noise (also called multiplication, or regular zero-crossing noise), generated by direct multiplication of a sinusoid and a lowpass noise [21]. It was a very convenient way to generate band-pass noises with tunable center frequency, and these noises were therefore quite useful in spectral masking experiments in which noise maskers with variable center frequencies and steep spectral cutoffs were required (e.g., [46, 47]). The fact that the envelope statistics of this stimulus differed significantly from that of Gaussian noise was only recognized after some time, once it was shown that these properties influence the outcome of listening experiments (see, as an example, [76]). One particularly relevant detail, often overlooked, is that the addition of two multiplied noise signals, for which the two sinusoidal center components are uncorrelated, results in a noise with Gaussian statistics.

With the advance of digital computers, additional noise types have been developed, often with the goal to vary the amount and nature of envelope fluctuations in a controlled way. Transposed stimuli were constructed to realize in the envelope of a high-frequency sinusoidal carrier the same waveform that resulted from auditory filtering and half-wave rectification of noise signals with a low center frequency [73]. "Sparse noise" was generated to have a white noise with a controllable, high amount of envelope fluctuations [25]. The opposite in terms of envelope fluctuations was "low-noise noise," a term coined by Pumplin in 1985 [51]. Due to its resemblance with periodic low-peak signals, we will focus in the remaining part of the chapter on this stimulus.

7.4.1 Low-Noise Noise

7.4.1.1 Definition

In the previous section, we noted that a filtering operation on white Gaussian noise causes the auto-correlation function to change from a delta function at lag zero, to a pattern that reflects predictability of successive time samples of such noise. This predictability is reflected in a smooth development of the envelope of the time-domain noise waveform.[1] The rate of fluctuation in the temporal envelope is proportional to the bandwidth of the filtered noise signal. The spectrum of the envelope has a large DC component and a downward tilting slope that leaves very little spectral power beyond frequencies equal to the bandwidth of the band-pass noise. Interestingly, the *amount* of fluctuation, i.e., the ratio between the AC and the DC part of the envelope, is independent of the bandwidth for Gaussian noise, a property which is reflected in the Rayleigh distribution of the temporal envelope values. Thus, there is an inherently high degree of fluctuation in Gaussian noise.

The inherent fluctuations that are present in Gaussian noise have prompted the development of so-called low-noise noise [51]. This special type of noise has the same spectral envelope as Gaussian noise, but a much lower degree of inherent fluctuations in its temporal envelope, hence the name *low-noise* noise. This stimulus allowed the study of the contribution of envelope fluctuations to auditory masking phenomena by comparing the masking effect of Gaussian and low-noise noise. The first authors to pursue this idea were Hartmann and Pumplin [22].

7.4.1.2 Stimulus Generation

The original means of generating low-noise, such as promoted by [51] was via a special optimization algorithm. First, a band-pass noise was digitally generated in the frequency domain by setting amplitudes in a restricted spectral range to some specific values, e.g., one constant value, and randomizing the phases. Such a noise will approximate all the properties of a band-pass Gaussian noise when the product of duration and bandwidth is sufficiently large. Via a steepest-descent algorithm, the phase spectrum was modified step-by-step in the direction that made the temporal envelope more flat, according to some statistical measure of envelope fluctuation.[2] After a sequence of iterations, a low-noise noise waveform was obtained with a rather flat temporal envelope and the initial amplitude spectrum. Summarizing, the Pumplin's method obtained low-noise noise by modifying the phase spectrum in a special way.

---

[1] There are alternative ways to determine the envelope of a signal which lead to somewhat different envelopes. We will consider here the Hilbert envelope.

[2] In this case, the normalized fourth moment of the waveform.

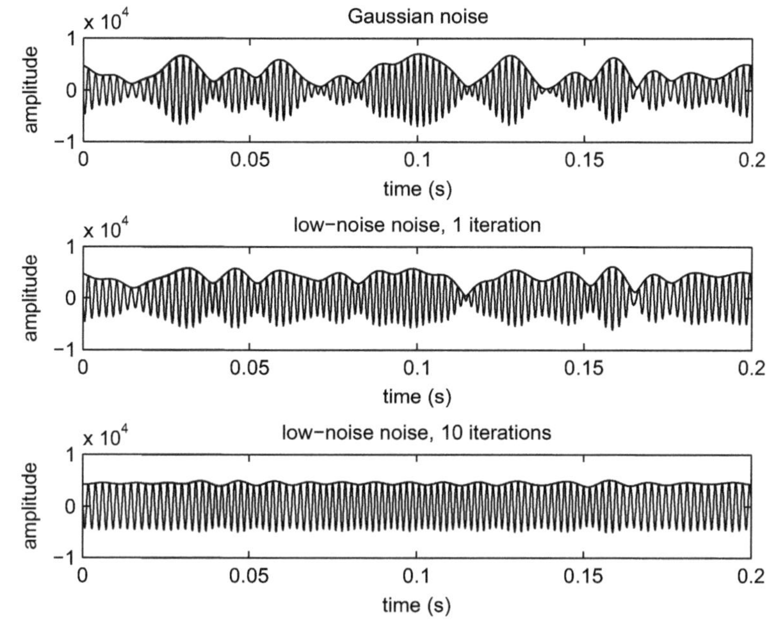

Fig. 7.7 Illustration of the low-noise noise generation. The *top* panel shows the time-domain Gaussian noise at the start of the iterative process, the *middle* panel the low-noise noise after one iteration, the *lower* panel the low-noise noise after 10 iterations. All waveforms are shown with their respective envelopes

Later on, in a publication dedicated to Manfred Schroeder on the occasion of his 70th birthday, several alternative methods of generating low-noise noise were proposed and evaluated by Kohlrausch et al. [33]. We will here describe the method that led to the lowest degree of fluctuation in the temporal envelope. The method consists of an iterative process that is initiated by generating a time-discrete Gaussian band-pass noise. The iterative process then consists of a sequence of straightforward steps.

First, the Hilbert envelope of the noise is calculated. Secondly, the noise waveform is divided by its Hilbert envelope on a sample-by-sample basis in the time domain. For the rare occasions that the Hilbert envelope is equal to zero, the resulting division is set to zero. In the third step, a band-pass filtering is applied to remove the new spectral components outside of the specified band-pass range that were introduced by the division operation in the previous step. By repeating these steps several times, a much flatter envelope is obtained.

After the first two steps, i.e., after calculating the Hilbert envelope and dividing the noise waveform by this envelope, the resulting waveform will have a flat temporal envelope. The spectrum, however, will also be modified considerably. The division by the Hilbert envelope can be seen as a time-domain multiplication by the reciprocal Hilbert envelope. In the frequency domain, this is equivalent to a convolution of the band-pass noise signal with the spectrum of the reciprocal Hilbert envelope. Due to the large DC component present in the Hilbert envelope,

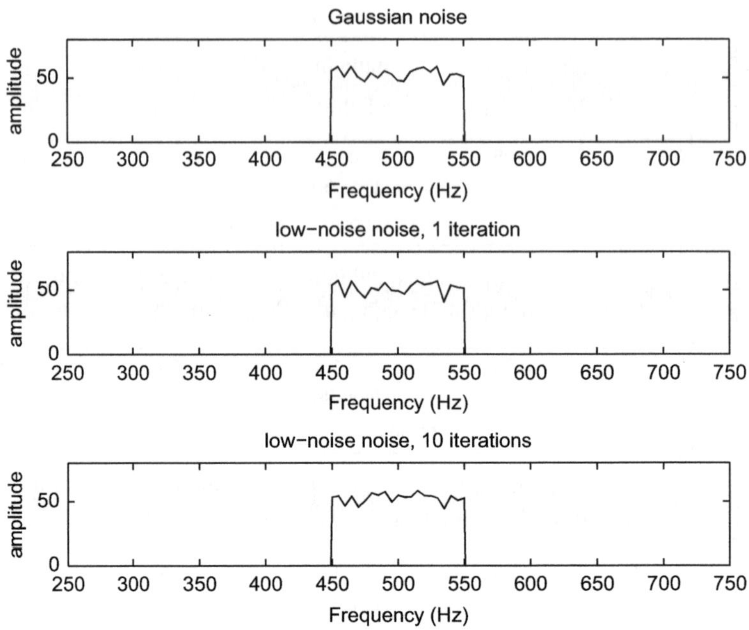

Fig. 7.8 Illustration of the low-noise noise generation. The *top* panel shows the power spectrum of the Gaussian noise at the start of the iterative process, the *middle* panel the spectrum of the low-noise noise after one iteration, the *lower* panel the spectrum of the low-noise noise after 10 iterations

the reciprocal Hilbert envelope will also have a large DC component. Thus, the convolution in the frequency domain will be dominated by this DC component and as a consequence, the spectrum of the band-pass noise will remain largely intact. However, there will be additional, new spectral components that are outside the band-pass range of the original band-pass noise.

Therefore, in the third step, band-pass filtering is applied to remove the new spectral components outside of the specified band-pass range that were introduced by the division operation in the previous step. Considering the argumentation given above, only a relatively small amount of signal power is removed by this operation. Nevertheless, after filtering, the temporal envelope will not be flat anymore. In Fig. 7.7, the temporal waveforms are shown for the original 100-Hz wide Gaussian band-pass noise centered at 500 Hz, that was input to the iterative process (top panel), and after the first iteration of our algorithm (middle panel). As can be seen, the degree of envelope fluctuation is reduced considerably. By repeating the iterative steps 10 times, a much flatter envelope is obtained (lower panel). Convergence is assumed to be obtained due to the DC component in the Hilbert envelope becoming more dominant over the higher spectral components after each iteration.

In Fig. 7.8, the same signals are shown, only now represented in the frequency domain. As can be seen, the original, band-pass Gaussian noise has a uniform spectral envelope. The spectrum of the low-noise noise signal is, even after

Table 7.1 Normalized fourth moment of low-noise noise as a function of the number of iterations

Number of iterations	Normalized fourth moment
0	3.030
1	1.845
2	1.701
4	1.591
6	1.552
8	1.535
10	1.526
Hartmann and Pumplin	1.580

The bottom row gives the value from [22] for comparison

10 iterations, quite similar to the spectrum of the Gaussian signal. There is, however, a tendency for the spectrum to have a somewhat lower level towards the edges of the band-pass range.

As a measure of envelope fluctuation, Table 7.1 shows the normalized fourth moment for different numbers of iterations of our algorithm. The value obtained by [22] is shown at the bottom of the table. As can be seen, already after 6 iterations, we obtain a lower degree of envelope fluctuation than the method of Hartmann and Pumplin. After 10 iterations, the normalized fourth moment is 1.526, close to the theoretical minimum of 1.5 for a sinusoidal signal.

In summary, the iterative method is able to create a low-noise noise by modifying both the phase and the amplitude spectrum. The specific ordering of spectral components in the passband causes the flat envelope that is seen in the lower panel of Fig. 7.7. Due to the very specific arrangement of phase and amplitude values throughout the noise spectrum, any modification of this spectral ordering will affect the flatness of the temporal envelope. In Fig. 7.9, the low-noise noise signal of Fig. 7.7, which was centered at 500 Hz, and had a bandwidth of 100 Hz, is shown after being filtered with a 78-Hz-wide gamma-tone filter centered at 500 Hz. As can be seen, the degree of envelope fluctuation has increased considerably. Since the gamma-tone filter used here is a reasonable first-order approximation of auditory peripheral filtering, this figure demonstrates that the properties that are present in the external stimulus should not be taken to be representative for the manner in which the stimulus is represented within the auditory system (see also Figs. 6 and 7 in [33]).

7.4.2 Role in Hearing Research and Perceptual Insights

As discussed before, Gaussian noise is frequently used as a stimulus in experiments investigating auditory masking. Early experiments by Fletcher [18] used noise signals of various bandwidths to determine detection thresholds of sinusoidal signals centered in the band-pass noise maskers. In these experiments, it was found that only the masker energy that was spectrally close to the sinusoidal target signal contributed to the masking effect of the noise. This led to the concept of the

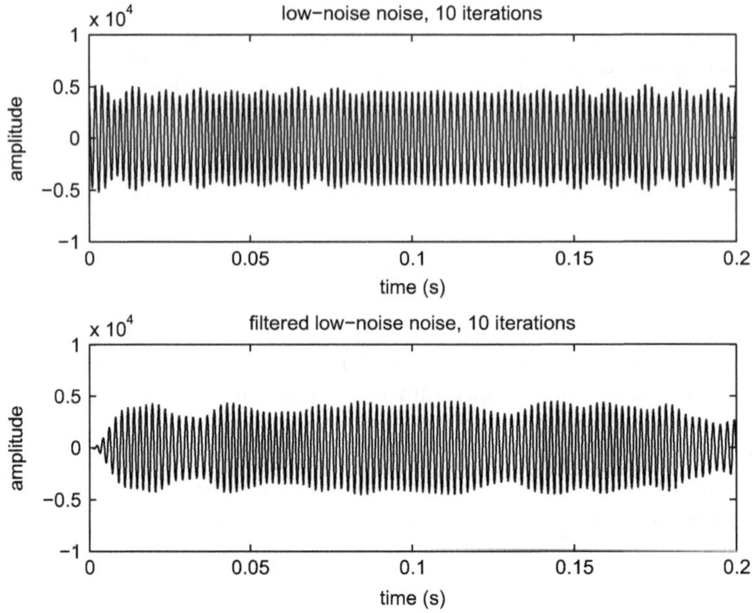

Fig. 7.9 Illustration of the low-noise noise generation. The *top* panel shows a low-noise noise with a bandwidth of 100 Hz after 10 iterations, the *lower* panel shows the same waveform after peripheral filtering with a gamma-tone filter of 78-Hz width

critical band, which indicates the spectral range that contributes to the masking effect on the sinusoidal signal. The integrated intensity of the masker within this range determines the masked threshold.

This purely intensity-based account of masking does not provide insights into the reasons for observing quite different masked thresholds when narrow-band noises or sinusoidal signals of equal level are used as a masker. When the bandwidth of the noise is smaller than the critical bandwidth, there is no difference in the masker intensity within that critical band for both masker types. Thus if overall masker intensity determines masking, thresholds should be the same. Typically, however, thresholds for tonal maskers are about 20 dB lower than for narrowband Gaussian-noise maskers (e.g., [42]).

One of the factors that is believed to contribute to the different masking strengths of these signals is the difference in the inherent envelope fluctuations. A tonal masker has no inherent envelope fluctuations, and the addition of the target tone will introduce a beating pattern that may be an effective cue for detecting the presence of the target. A noise masker, however, will have a high degree of fluctuation of its own. Addition of the sinusoidal signal does not alter the properties of the envelope fluctuations by a significant degree, and therefore, changes in the temporal envelope pattern may be a less salient cue for a noise masker.

Low-noise noise maskers provide an elegant stimulus to verify that the inherent fluctuations in Gaussian noise are an important factor contributing to its strong masking effect. Such an experiment had been done by Hartmann and Pumplin [22], but the difference in masked thresholds that they found was only 5 dB. This difference is considerably smaller than the 20-dB difference in masking found for Gaussian-noise maskers and tonal signals. A complicating factor may be that the inherent fluctuations in the low-noise noise that was used by Pumplin and Hartmann were still strong enough to cause a significant masking effect. Furthermore, the bandwidth of their low-noise noise stimulus was 100 Hz around a center frequency of 500 Hz. Although such a bandwidth agrees approximately with the estimates of auditory filter bandwidth at this frequency, peripheral filtering may have caused a significant reduction of the flatness of their low-noise noise stimulus, as we demonstrated in Fig. 7.9, where a low-noise noise signal with their spectral properties was filtered with a 1-ERB wide filter.

A more recent experiment by Kohlrausch et al. [33] used low-noise noise created by the iterative method outlined in the previous section resulting in an even lower degree of inherent fluctuation. In addition, the experiment of Kohlrausch et al. measured masked thresholds as a function of center frequency of the low-noise noise masker while keeping the target tone always spectrally centered within the 100-Hz wide noise masker. The highest center frequency in their experiment was 10 kHz, a frequency where the peripheral filter bandwidth is considerably larger than the masker bandwidth. As a result it can be assumed that peripheral filtering will only have a marginal effect on the temporal envelope flatness of the low-noise noise. Thus in these conditions, the difference in masking thresholds between Gaussian noise and low-noise noise should be about the same size as the difference seen for Gaussian-noise and tonal maskers.

One of the results from [33] is shown in Fig. 7.10. As can be seen, the masked thresholds for Gaussian noise (triangles) are constant for center frequencies of 500 Hz and above. This is in line with the fact that auditory filtering does not reduce masker intensity for a 100-Hz wide masker in this frequency range, and that the degree of inherent envelope fluctuations does not vary as a function of center frequency. For low-noise noise (circles), however, we see a clear dependence of thresholds on the center frequency. Although low-noise noise thresholds were lower than Gaussian-noise thresholds already at a center frequency of 1 kHz, for 10 kHz we see a much larger difference of more than 15 dB, which is much more similar to the difference observed for sinusoidal and Gaussian-noise maskers. The higher thresholds for lower frequencies are well in line with the idea that peripheral filtering affects the temporal envelope flatness of low-noise noise.

A variant of the experiment by Kohlrausch et al. [33] investigated the effect of a frequency offset between masker and sinusoidal target signal with the target always higher in frequency than the masker [75]. The addition of the sinusoidal target to the masker band introduces modulations with a rate that is characterized by the frequency difference between target and masker. When the target is centered within the masker, the newly introduced modulations will have a rate comparable to those already present within the masker alone and will therefore be difficult to detect. When the target is sufficiently remote from the masker band, the modulations that will be introduced due to the addition of the target will be of considerably higher

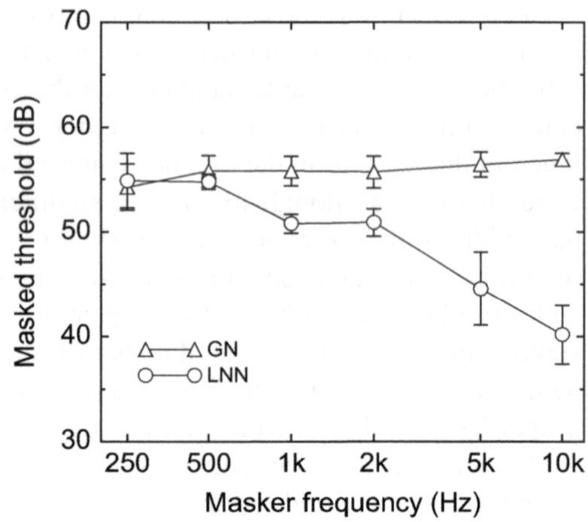

Fig. 7.10 Masked thresholds for 100-Hz wide Gaussian-noise (*triangles*) and low-noise noise maskers (*circles*) as a function center frequency. Reprinted from [33] with permission from Hirzel Verlag and the European Acoustics Association

rate than those already present within the masker and may be much easier to detect. Regarding the perceptual processing of stimuli with various modulation rates, there is evidence for frequency selectivity associated with the processing of temporal envelope fluctuations, which led to the modulation filterbank model proposed by [11]. Low-noise noise is a very suitable stimulus to test some of the non-intuitive consequences of this modulation filter bank concept.

In Fig. 7.11, results of the experiment in [75] are summarized. Masked thresholds are shown for Gaussian noise (dashed lines) and low-noise noise (solid lines) maskers centered at 10 kHz for bandwidths of 10 and 100 Hz, indicated by circles and squares, respectively. The abscissa gives the frequency of the sinusoidal target in terms of the frequency offset of the target relative to the masker center frequency. For an abscissa value of 0 Hz, the target is placed at the center frequency of the noise masker. The values at 0 Hz offset represent the most common on-frequency masking conditions. Note that for the two Gaussian-noise maskers (dashed curves), thresholds increase by about 5 dB, when the masker bandwidth is reduced by a factor of 10 from 100 to 10 Hz. This corresponds to a change of 1.5 dB/oct, exactly the value that is predicted by the notion that for this condition the detection process is dominated by an energy cue in combination with increasing variability of this cue for decreasing degrees of freedom in the masker [3, 20, 74].

The corresponding threshold values for low-noise noise are considerably lower, in line with the results from [31], replotted in Fig. 7.10. Again, thresholds for the 10-Hz masker are higher than those for 100 Hz. This can best be explained by the notion of frequency specific modulation perception. The total modulation energy is the same for the two masker bandwidths, but it is concentrated in a much smaller frequency range (basically, within a 10-Hz range of modulation rates) for the 10-Hz masker than for the 100-Hz masker. Thus, the amount of masker modulation within the lowest modulation filter, ranging up to 10 Hz, will be very high for the 10-Hz masker, making it difficult to detect the extra modulations introduced by adding the target (for this argument, see also Figs. 3 and 9 in [31]).

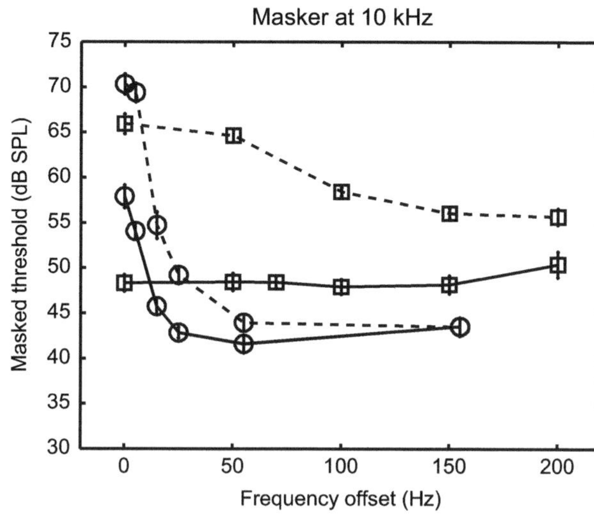

Fig. 7.11 Masked thresholds for Gaussian-noise (*dashed lines*) and low-noise noise maskers (*solid lines*) as a function of frequency offset between the masker and target. Squares show 100-Hz wide maskers and circles 10-Hz wide maskers both presented at 70 dB SPL

When the target frequency offset is increased, very different patterns are observed for the two different masker bandwidths, but also for the two maskers types that have very different spectra of their intrinsic envelope modulations. For 10-Hz maskers (circles), thresholds start to drop sharply once the offset exceeds 10 Hz. This drop is best explained by the introduction of masker-target modulations with a frequency range exceeding that of the masker modulations. These can be detected the easier, the further they are removed from the masker fluctuations. For the largest offset of 150 Hz, the two maskers lead to identical thresholds, an indication that detection is no longer limited by the intrinsic masker fluctuations ("external noise"), but by the hearing system's limited resolution ("internal noise"). For this condition, the target threshold is about 27 dB below the masker level. This value is in excellent agreement with data for a *sinusoidal* masker which amounted to a threshold about 29 dB below the masker level for a sinusoidal masker at 10 kHz and a sinusoidal target 150 Hz above the masker frequency (see Fig. 8 in [32]).

As can be seen from the squares indicating the 100-Hz-wide maskers, the low-noise noise thresholds (squares with solid lines) are roughly independent of frequency offset. They are about 5 dB higher than the values for the 10-Hz masker, so there is a small, but significant remaining influence of the 100-Hz masker. This can be understood by looking at the envelope spectrum. For a 100-Hz wide masker, the dominant rate of internal envelope fluctuations is around 100 Hz, and this corresponds exactly to the frequency difference between upper edge of the masker and the target for an offset of 150 Hz. Thus, even for frequency offsets of 150 Hz, the relatively small amount of masker envelope fluctuation prevents an optimal detection of the target based on temporal cues. In contrast, the Gaussian-noise thresholds for the same bandwidth (squares with dashed lines) show a clear dependence on the frequency offset, and thresholds are generally higher than the low-noise noise thresholds, in line with the idea that the inherent fluctuations in the

Gaussian masker prohibit the detection of the modulations introduced through the addition of the sinusoidal target signal. At larger frequency offsets, however, the modulations introduced by the target signal become higher in rate, and thresholds decrease systematically. This decrease reflects the decrease of the masker envelope spectrum for a Gaussian noise (cf. Fig. 8 in [15]).

One should note that the frequency offsets discussed here are considerably smaller than the peripheral filter bandwidth at the masker frequency of 10 kHz, which amounts to more than 1,000 Hz. Thus the patterns of thresholds observed in Fig. 7.11 cannot reflect peripheral *spectral* filtering. The dependence of thresholds on target frequency offset that is observed here is a reflection of the processing of temporal envelope fluctuations and not of spectral resolution.

7.4.3 Outlook

We have seen that the use of low-noise noise as masker does lead to different thresholds compared to Gaussian noise. This supports the idea that temporal fluctuations in Gaussian noise are a significant factor in auditory masking. Thus, low-noise noise may be an interesting stimulus also in the future to study the contribution of envelope fluctuations to masking.

Low-noise noise may also be of value for acoustical measurement techniques because it is a signal that couples a low crest factor with a continuous spectrum. Whereas in hearing experiments, the bandwidth of low-noise noise is usually limited to, at most, that of one critical bandwidth to prevent peripheral filtering from re-introducing fluctuations in the envelope, for physical measurements this restriction may not exist and wideband low-noise noise may be used to put maximum wideband power into a system that is somehow restricted in its maximum amplitude.

Various studies have applied low-noise noise for a number of different purposes. Dau et al. [15] have used low-noise noise, together with a number of different noise types to study the spectral processing of envelope fluctuations. Due to its flat temporal envelope, low-noise noise has a markedly different envelope spectrum as compared to Gaussian and multiplied noise. The envelope spectral content seems to govern the degree of masking that is observed. Buss et al. [8] have presented low-noise noise in their studies on comodulation masking release, where low-noise-noise provided a masker with little fluctuation but similar bandwidth as a Gaussian-noise masker. Low-noise noise has also been used to measure the minimal level to mask tinnitus [53] because it lacks excessive peaks and troughs. A final application for low-noise noise comes from speech perception research. Healy and Bacon synthesized artificial speech by amplitude modulating bands of low-noise noise and used this type of speech to measure the critical band for speech [24].

7.5 Conclusion

The topics in this chapter are naturally biased towards examples, to which we have ourselves contributed, and which were inspired by ideas and publications of Manfred Schroeder. In this way we hope to demonstrate how influential his way of thinking and his early emphasis on using computers in acoustics has been, using the example of psychoacoustics. The great potential of mathematically well-defined acoustic stimulus types lies in the possibility that they can equally well be used in behavioral as well as in physiological experiments, and that they can serve as input to those types of perception models which allow the processing of arbitrary signal waveforms (e.g., [5–7, 11–14]). This close interplay between psychoacoustics, physiology and modeling is one of the central themes of a conference series, the International Symposia on Hearing, which has been mentioned at many places in this chapter. This series was initiated in 1969 by, among others, Manfred R. Schroeder and Jan Schouten, the founding director of the Institute for Perception Research (IPO) in Eindhoven, where the present authors met each other some 20 years ago. The last of the symposia in which Schroeder participated was held 1988 in Paterswolde near Groningen, and Schroeder, with his love for the Dutch people, language, and country was happy to deliver the after dinner address at the conference social outing (see pp. vi–vii in [17]). This symposium never took place in Göttingen, but it can be seen as a late echo of the psychoacoustic research tradition at the DPI that two editions of these symposia, the 12th in 2000 [4] and the 14th in 2006 [38], were co-organized by scientists who received their initial academic training and were shaped in their scientific interests by Manfred Schroeder.

Acknowledgments The work described in this chapter reflects the close and fruitful cooperations that I (AK) had over the past more than 30 years with friends and colleagues at the DPI in Göttingen and later on in Eindhoven where I was joined at the IPO by SvdP in 1992. We both would like to thank all our colleagues for the creative atmosphere, never-ending curiosity and great fun, which helped to generate some interesting scientific insights. A part of the Göttingen atmosphere was in 1991 exported to Eindhoven, and as can be seen from the reference list, this has become a similarly fruitful period.

References

1. Atal, B.S., Schroeder, M.R.: Predictive coding of speech signals and subjective error criteria. In Proceedings of the IEEE International Conference on Acoustics, Speech, and Signal Processing, ICASSP 1978, pp. 573–576 (1978)
2. Blauert, J.: Spatial Hearing. The Psychophysics of Human Sound Localization. The MIT Press, Cambridge, MA (1997)
3. Bos, C.E., de Boer, E.: Masking and discrimination. J. Acoust. Soc. Am. **39**, 708–715 (1966)
4. Breebaart, J., Houtsma, A., Kohlrausch, A., Prijs, V., Schoonhoven, R.: Physiological and Psychophysical Bases of Auditory Function. Shaker Publishers, Maastricht (2001)

5. Breebaart, J., van de Par, S., Kohlrausch, A.: Binaural processing model based on contralateral inhibition. I. Model structure. J. Acoust. Soc. Am. **110**, 1074–1088 (2001)
6. Breebaart, J., van de Par, S., Kohlrausch, A.: Binaural processing model based on contralateral inhibition. II. Predictions as a function of spectral stimulus parameters. J. Acoust. Soc. Am. **110**, 1089–1104 (2001)
7. Breebaart, J., van de Par, S., Kohlrausch, A.: Binaural processing model based on contralateral inhibition. III. Predictions as a function of temporal stimulus parameters. J. Acoust. Soc. Am. **110**, 1105–1117 (2001)
8. Buss, E., Grose, J.H., Hall III, J.W.: Features of across-frequency envelope coherence critical for comodulation masking release. J. Acoust. Soc. Am. **126**, 2455–2466 (2009)
9. Carlyon, R.P., Datta, A.J.: Excitation produced by Schroeder-phase complexes: evidence for fast-acting compression in the auditory system. J. Acoust. Soc. Am. **101**, 3636–3647 (1997)
10. Carlyon, R.P., Datta, A.J.: Masking period patterns of Schroeder-phase complexes: effects of level, number of components, and phase of flanking components. J. Acoust. Soc. Am. **101**, 3648–3657 (1997)
11. Dau, T., Kollmeier, B., Kohlrausch, A.: Modeling auditory processing of amplitude modulation: I. Detection and masking with narrowband carriers. J. Acoust. Soc. Am. **102**, 2892–2905 (1997)
12. Dau, T., Kollmeier, B., Kohlrausch, A.: Modeling auditory processing of amplitude modulation: II. Spectral and temporal integration. J. Acoust. Soc. Am. **102**, 2906–2919 (1997)
13. Dau, T., Püschel, D., Kohlrausch, A.: A quantitative model of the 'effective' signal processing in the auditory system: I. Model structure. J. Acoust. Soc. Am. **99**, 3615–3622 (1996)
14. Dau, T., Püschel, D., Kohlrausch, A.: A quantitative model of the 'effective' signal processing in the auditory system: II. Simulations and measurements. J. Acoust. Soc. Am. **99**, 3623–3631 (1996)
15. Dau, T., Verhey, J., Kohlrausch, A.: Intrinsic envelope fluctuations and modulation-detection thresholds for narrowband noise carriers. J. Acoust. Soc. Am. **106**, 2752–2760 (1999)
16. Duifhuis, H.: Audibility of high harmonics in a periodic pulse. J. Acoust. Soc. Am. **48**, 888–893 (1970)
17. Duifhuis, H., Horst, J.W., Wit, H.P. (eds.): Basic Issues in Hearing. Academic Press, London (1988)
18. Fletcher, H.: Auditory patterns. Rev. Mod. Phys. **12**, 47–65 (1940)
19. Glasberg, B.R., Moore, B.C.J.: Derivation of auditory filter shapes from notched-noise data. Hear. Res. **47**, 103–138 (1990)
20. Green, D.M., Swets, J.A.: Signal Detection Theory and Psychophysics. Wiley, New York, NY (1974). reprinted by Krieger
21. Greenwood, D.D.: Auditory masking and the critical band. J. Acoust. Soc. Am. **33**, 484–502 (1961)
22. Hartmann, W.M., Pumplin, J.: Noise power fluctuation and the masking of sine signals. J. Acoust. Soc. Am. **83**, 2277–2289 (1988)
23. Hawkins, J.E.J., Stevens, S.S.: The masking of pure tones and of speech by white noise. J. Acoust. Soc. Am. **22**, 6–13 (1950)
24. Healy, E.W., Bacon, S.P.: Measuring the critical band for speech. J. Acoust. Soc. Am. **119**, 1083–1091 (2006)
25. Hübner, M., Wiegrebe, L.: The effect of temporal structure on rustling-sound detection in the gleaning bat, Megaderma lyra. J. Comp. Physiol. A **189**, 337–346 (2003)
26. Irino, T., Patterson, R.D.: A time-domain, level-dependent auditory filter: the gammachirp. J. Acoust. Soc. Am. **101**, 412–419 (1997)
27. Johannesma, P.I.M.: The pre-response stimulus ensemble of neurons in the cochlear nucleus. In: Cardozo, B.L., de Boer, E., Plomp, R. (eds.) Proceedings of the 2nd International Symposium on Hearing, pp. 58–69. Institute for Perception Research, Eindhoven (1972)
28. Johnston, J.D.: Transform coding of audio signals using perceptual noise criteria. IEEE J. Selected Areas Commun. **6**, 314–323 (1988)

29. Joris, P.X., van der Heijden, M., Louage, D.H., de Sande, B.V., Kerckhoven, C.V.: Dependence of binaural and cochlear "best delays" on characteristic frequency. In: Pressnitzer, D., de Cheveigné, A., McAdams, S., Collet, L. (eds.) Auditory Signal Processing: Physiology. Psychoacoustics, and Models, pp. 477–483. Springer, New York (2005)
30. Kohlrausch, A.: Masking patterns of harmonic complex tone maskers and the role of the inner ear transfer function. In: Duifhuis, H., Horst, J., Wit, H. (eds.) Basic Issues in Hearing, pp. 339–350. Academic Press, Harcourt Brace Jovanovich. Publ, London (1988)
31. Kohlrausch, A., Fassel, R.: Binaural masking level differences in nonsimultaneous masking. In: Gilkey, R.H., Anderson, T. (eds.) Binaural and Spatial Hearing in Real and Virtual Environments, Chapter 9, pp. 169–190. Lawrence Erlbaum Ass, London (1997)
32. Kohlrausch, A., Fassel, R., Dau, T.: The influence of carrier level and frequency on modulation and beat-detection thresholds for sinusoidal carriers. J. Acoust. Soc. Am. **108**, 723–734 (2000)
33. Kohlrausch, A., Fassel, R., van der Heijden, M., Kortekaas, R., van de Par, S., Oxenham, A.J., Püschel, D.: Detection of tones in low-noise noise: further evidence for the role of envelope fluctuations. Acustica United Acta Acustica **83**, 659–669 (1997)
34. Kohlrausch, A., Houtsma, A.J.M.: Edge pitch of harmonic complex tones. IPO Annual Progress Report **26**, 39–49 (1991)
35. Kohlrausch, A., Houtsma, A.J.M.: Pitch related to spectral edges of broadband signals. Phil. Trans. R. Soc. Lond. B **336**, 375–382 (1992)
36. Kohlrausch, A., Sander, A.: Phase effects in masking related to dispersion in the inner ear. II. Masking period patterns of short targets. J. Acoust. Soc. Am. **97**, 1817–1829 (1995)
37. Kohlrausch, A., van de Par, S.: On the use of specific signal types in hearing research. In: Kurtz, T., Parlitz, U., Kaatze, U. (eds.) Oscillations, Waves, and Interactions. Sixty Years Drittes Physikalisches Institut. A Festschrift, pp. 37–71. Universitätsverlag Göttingen, Göttingen (2007)
38. Kollmeier, A., Klump, G., Hohmann, V., Langemann, U., Mauermann, M., Verhey, J. (eds.): Hearing—From Sensory Processing to Perception. Proceedings of the 14th International Symposium on Hearing. Springer, Berlin (2007)
39. Langhans, A., Kohlrausch, A.: Differences in auditory performance between monaural and diotic conditions. I: Masked thresholds in frozen noise. J. Acoust. Soc. Am. **91**, 3456–3470 (1992)
40. Lentz, J.J., Leek, M.R.: Psychophysical estimates of cochlear phase response: masking by harmonic complexes. J. Assn. Res. Otolaryngol. **2**, 408–422 (2001)
41. Mehrgardt, S., Schroeder, M.R.: Monaural phase effects in masking with multicomponent signals. In: Klinke, R., Hartmann, R. (eds.) Hearing—Physiological Bases and Psychophysics, pp. 289–295. Springer, New York (1983)
42. Moore, C., Alcántara, J.I., Dau, T.: Masking patterns for sinusoidal and narrow-band noise maskers. J. Acoust. Soc. Am. **104**, 1023–1038 (1998)
43. Münkner, S., Kohlrausch, A., Püschel, D.: Influence of fine structure and envelope variability on gap-duration discrimination thresholds. J. Acoust. Soc. Am. **99**, 3126–3137 (1996)
44. Oxenham, A.J., Dau, T.: Reconciling frequency selectivity and phase effects in masking. J. Acoust. Soc. Am. **110**, 1525–1538 (2001)
45. Oxenham, J., Dau, T.: Towards a measure of auditory-filter phase response. J. Acoust. Soc. Am. **110**, 3169–3178 (2001)
46. Patterson, R.D.: Auditory filter shape. J. Acoust. Soc. Am. **55**, 802–809 (1974)
47. Patterson, R.D.: Auditory filter shapes derived with noise stimuli. J. Acoust. Soc. Am. **59**, 640–654 (1976)
48. Patterson, R.D., Allerhand, M.H., Giguère, C.: Time-domain modelling of peripheral auditory processing: a modular architecture and a software platform. J. Acoust. Soc. Am. **98**, 1890–1894 (1995)
49. Patterson, R.D., Nimmo-Smith, I., Weber, D.L., Milroy, R.: The deterioration of hearing with age: frequency selectivity, the critical ratio, the audiogram, and speech threshold. J. Acoust. Soc. Am. **72**, 1788–1803 (1982)

50. Pierce, J.R.: Some work on hearing. Am. Scientist **48**, 40–45 (1960)
51. Pumplin, J.: Low-noise noise. J. Acoust. Soc. Am. **78**, 100–104 (1985)
52. Recio, A., Rhode, W.S.: Basilar membrane responses to broadband stimuli. J. Acoust. Soc. Am. **108**, 2281–2298 (2000)
53. Schneider, P., Andermann, M., Wengenroth, M., Goebel, R., Flor, H., Rupp, A., Diesch, E.: Reduced volume of Heschl's gyrus in tinnitus. Neuroimage **45**, 927–927 (2009)
54. Schroeder, M.R.: New results concerning monaural phase sensitivity. J. Acoust. Soc. Am. **31**, 1579 (1959)
55. Schroeder, M.R.: Vocoders: analysis and synthesis speech. Proc. IEEE **54**, 720–734 (1966)
56. Schroeder, M.R.: Relation between critical bands in hearing and the phase characteristics of the cubic difference tones. J. Acoust. Soc. Am. **46**, 1488–1492 (1969)
57. Schroeder, M.R.: Synthesis of low-peak-factor signals and binary sequences with low autocorrelation. IEEE Transact. Inf. Theor. **16**, 85–89 (1970)
58. Schroeder, M.R.: An integrable model for the basilar membrane. J. Acoust. Soc. Am. **53**, 429–434 (1973)
59. Schroeder, M.R.: Models of hearing. Proc. IEEE **63**, 1332–1350 (1975)
60. Schroeder, M.R.: New viewpoints in binaural interaction. In: Evans, E.F., Wilson, J.P. (eds.) Psychophysics and Physiology of Hearing, pp. 455–456. Academic Press, New York (1977)
61. Schroeder, M.R., Atal, B.S.: Computer simulation of sound transmission in rooms. IEEE Int. Convention Record **11**, 150–155 (1963)
62. Schroeder, M.R., Atal, B.S., Hall, J.: Optimizing digital speech coders by exploiting masking properties of the human ear. J. Acoust. Soc. Am. **66**, 1647–1652 (1979)
63. Schroeder, M.R., Atal, B.S., Sessler, G.M., West, J.E.: Acoustical measurements in Philharmonic Hall (New York). J. Acoust. Soc. Am. **40**, 434–440 (1966)
64. Schroeder, M.R., Gottlob, D., Siebrasse, K.F.: Comparative study of European concert halls. J. Acoust. Soc. Am. **56**, 1195–1201 (1974)
65. Schroeder, M.R., Hall, J.L.: Model for mechanical to neural transduction in the auditory receptor. J. Acoust. Soc. Am. **55**, 1055–1060 (1974)
66. Schroeder, M.R., Strube, H.W.: Flat-spectrum speech. J. Acoust. Soc. Am. **79**, 1580–1583 (1986)
67. Shamma, S.A., Shen, N., Gopalaswamy, P.: Stereausis: Binaural processing without neural delays. J. Acoust. Soc. Am. **86**, 989–1006 (1989)
68. Smith. Der Einfluß des Vorzeichens des Momentanfrequenzverlaufes von Maskierern auf die Detektion von reinen Tönen (The influence of the sign of the change in the instantaneous frequency of a masker on the detectability of pure tones). Master's Thesis, Georg-August-Universität Göttingen (1984)
69. Smith, K., Sieben, U., Kohlrausch, A., Schroeder, M.R.: Phase effects in masking related to dispersion in the inner ear. J. Acoust. Soc. Am. **80**, 1631–1637 (1986)
70. Strube, H.W.: A computationally efficient basilar-membrane model. Acustica **58**, 207–214 (1985)
71. Summers, V.: Effects of hearing impairment and presentation level on masking period patterns for Schroeder-phase harmonic complexes. J. Acoust. Soc. Am. **108**, 2307–2317 (2000)
72. Summers, V., Leek, M.R.: Masking of tones and speech by Schroeder-phase harmonic complexes in normally hearing and hearing-impaired listeners. Hear. Res. **118**, 139–150 (1998)
73. van de Par, S., Kohlrausch, A.: A new approach to comparing binaural masking level differences at low and high frequencies. J. Acoust. Soc. Am. **101**, 1671–1680 (1997)
74. van de Par, S., Kohlrausch, A.: Diotic and dichotic detection using multiplied-noise maskers. J. Acoust. Soc. Am. **103**, 2100–2110 (1998)
75. van de Par, S., Kohlrausch, A.: The role of intrinsic masker fluctuations on the spectral spread of masking. In *Forum Acusticum 2005, Budapest*, pp. 1635–1640, European Acoustics Association (2005)

76. van der Heijden, M., Kohlrausch, A.: The role of envelope fluctuations in spectral masking. J. Acoust. Soc. Am. **97**, 1800–1807 (1995)
77. von Klitzing, R., Kohlrausch, A.: Effect of masker level on overshoot in running- and frozen-noise maskers. J. Acoust. Soc. Am. **95**, 2182–2201 (1994)

***

Biography

Armin Kohlrausch studied physics at the University of Göttingen (Germany) and specialized in acoustics. He received his master degree in 1980 and his Ph.D. in 1984, both on perceptual aspects of sound. From 1985 until 1990 he worked at the Drittes Physikalisches Institut at the University of Göttingen, being responsible for research and teaching in the fields psychoacoustics and room acoustics. In 1991 he joined the Philips Research Laboratories in Eindhoven and worked in the speech and hearing group of the Institute for Perception Research (IPO), a joint venture between Philips and the Eindhoven University of Technology (TU/e). Since 1998, he combines his work at Philips Research Laboratories with a Professor position for auditory and multisensory perception at the TU/e. In 2004 he was appointed research fellow of Philips Research. He is a member of a great number of scientific societies, both in Europe and in the USA. Since 1998 he is a Fellow of the Acoustical Society of America and has served as Associate Editor for the Journal of the Acoustical Society of America, for the Journal of the Association for Research in Otolaryngology and for Acta Acustica United with Acustica. His main scientific interest is the experimental study and modeling of auditory and multisensory perception in humans and the transfer of this knowledge to industrial applications.

Steven van de Par studied physics at the Eindhoven University of Technology, Eindhoven, The Netherlands, and received the Ph.D. degree in 1998 from the Eindhoven University of Technology, on a topic related to binaural hearing. As a Postdoctoral Researcher at the Eindhoven University of Technology, he studied auditory–visual interaction and was a Guest Researcher at the University of Connecticut Health Center. In early 2000, he joined Philips Research, Eindhoven, to do applied research in auditory and multisensory perception, low-bit-rate audio coding and music information retrieval. Since April 2010 he holds a professor position in acoustics at the University of Oldenburg, Germany, with a research focus on the fundamentals of auditory perception and its application to virtual acoustics, vehicle acoustics, and digital signal processing. He has published various papers on binaural auditory perception, auditory–visual synchrony perception, audio coding, and computational auditory scene analysis.

Chapter 8
Neurally Based Acoustic and Visual Design

Yoichi Ando and Peter Cariani

Abstract Manfred Schroeder's ideas concerning acoustics, auditory percepts and preferences, and acoustic design had profound influences on the development of Yoichi Ando's theory of architectural acoustics. Building on Schroeder's theoretical frameworks, over subsequent decades Ando formulated a systematic theory of architectural acoustics design that incorporates acoustics (the structure of sound and how it propagates through an enclosed space), psychoacoustical models of auditory percepts and listener preferences (what we hear, which perceptual attributes are most important for the design of enclosed spaces, what we like to hear, which percepts are most important to our overall satisfaction, how individualized are these preferences), and strategies for optimal design (how a design process can harness psychoacoustical knowledge in order to optimize listener preferences).

In the last two decades, Ando's theory has taken "a neural turn" in which monaural and binaural percepts are grounded respectively in putative central auditory autocorrelation and cross-correlation representations. Experimentally, Ando and co-workers have identified some observable neural correlates of relevant auditory percepts and preferences (e.g., EEG and MEG response latency patterns, spatial extent and temporal persistence of alpha rhythms, hemispheric lateralizations). In theory, the identification of neurophysiological signs of listener satisfaction permits neurally driven design processes that optimize acoustics such that neural processes responsible for listener satisfaction are fulfilled. Finally, the same battery of psychophysical methods, correlation-based representations, and

Note from Peter Cariani: I stood in for Yoichi Ando at the Manfred R. Schroeder Memorial Session at the 116th ASA Meeting, Seattle, USA, May 2011 because, due to illness, he had to be hospitalized in Japan a few days before. I also served as Guest Editor of the book, *Auditory and Visual Sensations*, Springer-Verlag, NY, 2009, and I am honored to contribute here in a similar capacity. In reading this chapter, unless otherwise noted, all first-person perspectives, singular and plural, refer to Ando and his colleagues and not to me.

P. Cariani (✉)
Hearing Research Center, Boston University, 44 Cummington Mall, Boston, MA 02215, USA
e-mail: cariani@bu.edu

N. Xiang and G.M. Sessler (eds.), *Acoustics, Information, and Communication*, Modern Acoustics and Signal Processing, DOI 10.1007/978-3-319-05660-9_8,

neurophysiological experiments has been applied to problems of visual percepts and preferences. The visual results suggest deep similarities with auditory percepts. These many considerations lead to an integrative theory of spatial and temporal design.

8.1 Introduction

Manfred Schroeder had an early and profound influence on the development of my theory of architectural acoustics, and we have maintained our friendship and scientific exchange of ideas throughout our lives.

Our first approach to acoustic design was based on acoustics, psychophysics, and the psychology of preferences, and was inspired in part by Manfred's 1966 assessment of the field of architectural acoustics [1]. This early research program drew from sources outside the listener's nervous system, external acoustical measurements, cochlear biophysics, and overt psychoacoustical judgments, to model acoustics, percepts, and preferences.

Rational design of sensory experience is possible, given a theory of the stimulus, the percepts it produces, and listener preferences for particular stimulus parameters. In acoustics, adaptive methods can be used to optimize opera house and concert hall sound fields, musical instruments, performances, and real time speech recognitions. We employed genetic algorithms (GAs) to design concert halls by using acoustical–psychophysical models to predict what listeners would hear at each seat location, and then optimizing the shape of the hall to best satisfy listener preferences for what they would hear.

In more recent years, our theory has moved inside the listener, grounding acoustic design in terms of brain processes that subserve auditory percepts and preferences. Neural response correlates associated with auditory qualities related to central autocorrelation functions (ACFs) (temporal sensations, such as pitch, timbre, loudness) and those related to interaural cross-correlation functions (IACFs) (spatial sensations) were found to be dominant in one or another cerebral hemisphere. Features of these two central correlation functions describe primary auditory qualities well. The observable neural correlates of subjective preferences and annoyances involve the temporal persistence and spatial extent of electroencephalograms (EEG) and magnetoencephalograms (MEG) alpha rhythms in cortical regions.

Most recently the theory has been applied to visual sensations. Analogous percepts, preferences, and correlation-based representations have been explored for a number of visual patterns (flickers, oscillating objects, and textures). Subjective scale values of listener preferences for sounds, sound fields, and visual patterns all show the same cortical alpha rhythm persistence and extent.

Thus, the research program that was originally outlined by Manfred Schroeder has been greatly advanced, and even expanded in its scope to encompass brain processes and aspects of vision.

This chapter discusses in turn

- People (friendship and scientific collaborations of Manfred and Yoichi).
- Acoustics (how sound propagates through an enclosed space).
- Percepts (modeling of listener percepts, what we hear).
- Preferences (modeling of listener preferences for different percepts, what we like to hear).
- Design (combining acoustics, auditory signal-processing models, psychophysics, psychology of auditory preferences, and architectural constraints to formulate a rational theory of architectural acoustics design, what we need to know to design a building for the purpose of listening to sounds).
- Auditory signal processing (the modeling of signal processing in the auditory system that gives rise to percepts, how we hear, or how acoustics produces what we hear).
- Neural observables (the behavior of neural observables that are associated with percepts and preferences, how we can venture inside the listener's brain to bypass overt judgments to directly see signs of the underlying neural processes involved in the percepts and preferences).
- Vision (the application of the general theory and its methods to the visual modality, what we see and what we like to see).

By necessity, only the essentials of Ando's acoustical theory can be presented here. For the full methodological and empirical details, as well as the application of the theory to visual attributes and preferences, interested readers should consult his most recent book, *Auditory and Visual Sensations* [2] and the original papers in which they were reported.

8.2 People: Manfred R. Schroeder and Yoichi Ando

In his overview of the problem of architectural acoustics, Manfred Schroeder [1] cogently argued that persistent uncertainties in the acoustical design of concert halls resulted in highly uneven results, with both outstanding successes, such as La Grande Salle in Montreal and the Music Pavilion in Los Angeles, and deficiencies, such as London's Royal Festival Hall (1951), New York's Philharmonic Hall (1962), and Berlin's new Philharmonie (1963). Schroeder attributed this inconsistency in the acoustical quality of concert halls, especially large ones, to "an insufficient understanding of the important factors that make for good concert hall acoustics:" the physical (how sound is propagated to the listener), the psychoacoustical (what we hear), and aesthetic (what we prefer to hear).

These three sets of factors that describe sound, percepts, and preferences are the fundamental elements of a unified and comprehensive theory of architectural acoustics. Together with theories of materials and building construction, these form a rational basis for the design of the acoustics-related aspects of buildings.

The immediate problems at that time involved developing better acoustical and psychophysical measures associated with sound reverberation and diffusion.

I first visited Manfred Schroeder for one month in summer of 1971 at the Third Physics Institute, Göttingen, with the intent of investigating aspects of spatial hearing most relevant to the design of concert halls. There, Peter Damaske [3] had investigated subjective diffuseness (SD), and was working to explain the degree of SD of sound fields. He and I worked together to develop methods for measuring degree of diffusion by calculating the magnitude of the interaural cross-correlation function (IACC) [4]. Schroeder and co-workers [5] measured various physical, psychoacoustic, and aesthetic variables at a number of existing concert halls in Europe, and found the IACC to be most effective and consensus factor for sound field preference.

I returned again and again to Göttingen to work out different aspects of the whole theory. During the period of 1975–1977, as an Alexander von Humboldt Fellow [6], I worked with Schroeder's group in Göttingen on modeling subjective preference. In August 1979, we investigated effects of multiple early reflections of the sound field on subjective preference [7]. In 1983 on another visit, we developed a method to quantitatively estimate scaled subjective preference values for each seat of a prospective concert hall design [8]. Soon after, my first book, Concert Hall Acoustics (1985), with Manfred Schroeder's preface, was published.

During the period of 1979–2003, my wife Keiko and I stayed several times at Manfred and Anny Schroeder's house to prepare manuscripts of books [2, 9]. They visited us in Kobe for dinner together after he delivered his invited lectures at the Acoustical Society of Japan, Nagano, and Kobe University (Fig. 8.1). In designing Uhara Hall in Kobe [10], we applied his quadratic residue sequence to the design of the front part of the ceiling to decrease IACC in order to improve listener satisfaction. After the hall was constructed in 1992, Manfred gave a lecture on fractals and chaos in the hall's form and their consequences for the resulting acoustics.

In summer 1995, Manfred was invited to give the keynote lecture at the International Symposium on Music and Concert Hall Acoustics (MCHA), which was held at Kirishima International Music Hall (Miyama Conceru) [59]. Manfred and Anny Schroeder, Peter and Christiane Damaske and Keiko and I all climbed up the strato-volcano Ohachi, near Kirishma. The climb was made memorable by strong sudden winds we encountered on top, which meant that we had to help each other get down the mountain. Our friendship and collaboration continued again at the Third Institute of Physics Institute at Gottingen, and Manfred's many influences have stayed with me even up to the present time.

8.3 Acoustics, Percepts, and Preferences

For a theory of architectural acoustics, one needs to understand how sound is propagated within enclosures in order to predict the sound that will arrive at the listener's ears. Next, one needs to know what the listener hears given the sound presented to the ears, and also what the listener prefers to hear given all possible

Fig. 8.1 Manfred and Anny Schroeder at Zemtsuji-Temple in Nagano, Japan, 1979

choices of listening conditions. Thus theories of acoustics, models of auditory perception, and the psychology of listener preferences are all essential components for a rational theory of architectural acoustics. To these one must also add a theory of architectural design and construction.

8.3.1 Major Acoustical Factors for Concert Hall Design

There are four major acoustical factors, with their corresponding auditory perceptual attributes, that are most important in listener satisfaction in concert hall settings. These are the binaural listening level (LL), the delay time of the first sound reflections (Δt_1), the delay associated with later reverberations (T_{sub}), and the magnitude of the IACC function. The first three factors were appreciated relatively early on in architectual acoustics [11, 12], but the last, binaural factor was a more recent discovery [4, 5, 13–16].

These acoustical factors and perceptual attributes correspond to features in two general temporal correlation representations (Fig. 8.2): the monaural ACF and the binaural IACF. Many more, perhaps all, auditory percepts can be characterized in terms of these two functions, be they descriptions of the acoustical waveform or of central neural representations that subserve auditory percepts. The features and their corresponding auditory attributes are discussed more fully in Sect. 8.5.

In order to systematically characterize these acoustical factors and their perception, we used a sound simulation system in an anechoic enclosure with seven speakers arranged around the listener. Four separate, independent delays could be fed to subsets of the speakers [2, pp. 23–24, 1, 5, 17]. With such a system, we could

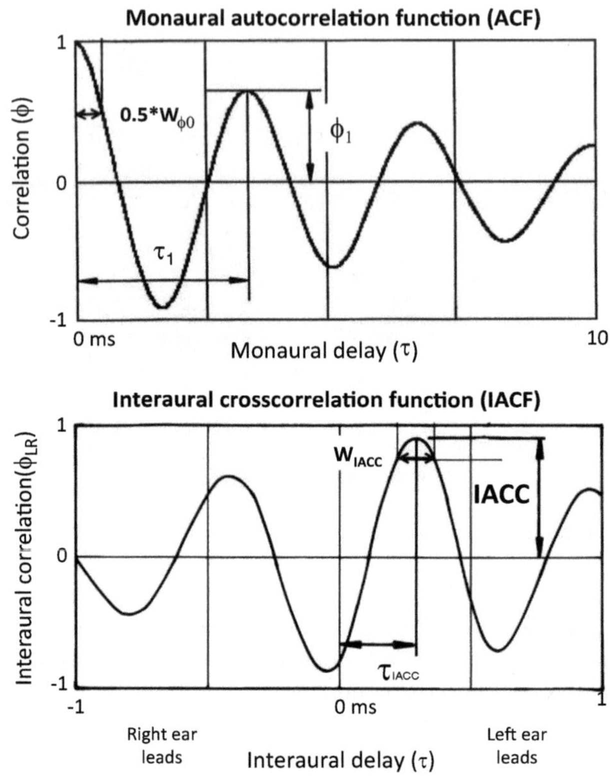

Fig. 8.2 Autocorrelation and cross-correlation functions and their respective features. *Top.* Normalized monaural short-time autocorrelation function (ACF), showing features (respective perceptual correlates in parentheses): time lag of ACF maximum τ_1 (pitch), height of ACF maximum ϕ_1 (pitch strength, regularity), and half-width of ACF zero-lag peak $W_{\phi 0}$ (spectral tilt). *Bottom.* Normalized interaural cross-correlation function (IACF), with respective binaural features: time lag (interaural time difference) of IACF maximum τ_{IACC} (ITD, azimuth), magnitude of the interaural correlation IACC (subjective diffuseness), and width of the IACC peak W_{IACC} (apparent source width ASW)

independently control the delays in the sound field that correspond to apparent right-left position and source width, first reflection time, and later reflection time. Listener preferences were assessed using paired comparison tests.

The four orthogonal and perceptually distinct factors of the sound field most important for concert hall design are:

1. *Binaural listening level* (LL). The LL is the averaged sound pressure level (SPL) at the two ear entrances, in dBA. The corresponding perceptual attribute is loudness. Most listeners prefer listening levels in the 70–90 dBA range, with about 60 % of listener preferences in the 80–85 dBA range [2].

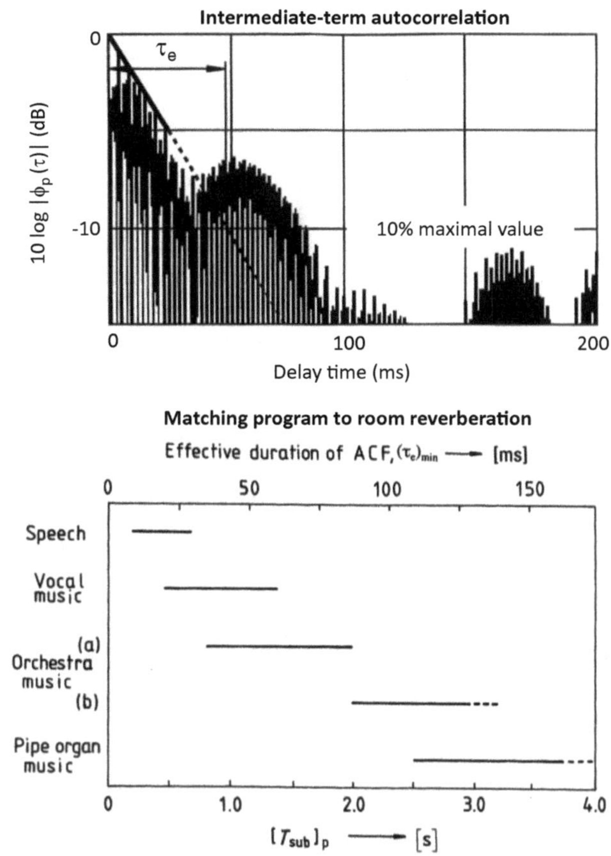

Fig. 8.3 Effective duration, τ_e. *Top*. The computation of effective duration from the extrapolated 90 % decay of the envelope of the intermediate-term ACF of a signal. *Bottom*. Ranges of minimum effective duration $(\tau_e)_{min}$, and the matched, optimal subsequent reverberation times, T_{sub}, for different types of program material

2. *Delay time of the first reflection* after the direct sound (Δt_1). Acoustically, this is the delay (in s) between the arrival of the direct sound and its arrival via the first reflection path at the listener's ears. The first reflection is the earliest and strongest reflection of the generated sound, usually from the stage and its immediate surrounds.

The preferred first reflection time varies with the effective duration of the program material. Effective duration (τ_e) is a measure of the temporal coherence of a sound, the persistence of its regularity, or how fast the sound pattern changes with time. Effective duration can be measured by how fast the ACF decays with increasing time lag (Fig. 8.3, top). Here, effective duration is defined as the delay at which the ACF of the signal has decayed to 10 % of its initial, maximal value.

For a piece of music or speech, the effective duration of the whole is determined by the rate of change of its fastest portion [minimum effective duration $(\tau_e)_{min}$, Fig. 8.3, bottom]. Speech changes its pattern rapidly and has short effective durations (typically 10–15 ms), whereas slow music with sustained notes typically has much longer durations (typically 120 ms). Fast music lies between these two extremes. Listeners prefer the first reflection time to be roughly equal to the effective duration of the program material presented, when the total amplitude of reflections $A = 1.0$ [2, pp. 26–29, 6, 7]. When first reflection times are long compared to this effective duration, echoes may be perceived and the received sounds suffer in clarity. As a consequence, speech in halls with long first reflection times suffers in intelligibility. When first reflection times are short, the delays are perceived as a change in timbre (tone coloration), and when they are very short, they can result in spatial directional effects.

Thus, ideally one wants to match the program material to the specific concert hall in which it is to be played, and vice versa. In an architectural acoustics utopia, each concert hall would have an acoustician on call who could recommend which sound materials would sound best in that hall. How one performs a piece can also be adapted to the concert hall in which it is performed. We have proposed a method of controlling minimum effective duration $(\tau_e)_{min}$ in performance of vocal music that blends the sound source and a given concert hall [9]. If vibrato is introduced during singing, for example, it decreases this value, blending the sound field with a short reverberation time [18].

One can also design different types of concert halls with different kinds of music and speech in mind. Some concert halls are best suited for faster music, such as Mozart's Symphony No. 41, Stravinsky's Le Sacre du Printemps, and Arnold's Sinfonietta, whereas other halls with longer reverberation times would be better suited for slower, more resonant music, such as Brahms' Symphony No. 4, Buckner's Symphony No. 7, and Bach's works for pipe organ.

3. *Subsequent reverberation time* after the first reflection (T_{sub}). Subsequent reverberation time is the decay rate of later reverberations. Sabine's formula [19] estimated this time as a function of room dimensions (volume and surface area) and the average absorption coefficient of its interior surfaces. Late reverberations contribute to the perceived sense of envelopment and the liveliness of the hall. Late reverberations can also be quantified using the effective duration of signals recorded in the sound field. The preferred reverberation time, which is equivalent to that defined by Sabine, is expressed approximately by $[T_{sub}]_p \approx 23(\tau_e)_{min}$. For "Water Music" by Handel, which has a minimal effective duration $\tau_e \sim 63$ ms, listeners preferred subsequent reverberation times between 0 and 4 s, with 45 % preferring delays between 1 and 2 s [2]. A flat frequency response for reverberations is also preferred.
4. *Magnitude of IACC*. This is the magnitude of normalized IACF $\phi_{l,r}(\tau)$, and it indicates the degree of directional dispersion (correlation) of the sound field across the two ears. Highly directional compact sound sources produce maximal, unity-valued IACCs, whereas highly diffuse, spatially distributed sound sources produce minimal, zero-valued IACCs. Perceptually, the IACC covaries with

apparent source width (ASW) and contributes to the sense of envelopment. Listeners in concert hall contexts uniformly prefer minimal values for IACC. This contrasts with the other three parameters, in which there can be considerable variation in individual listener preferences, which follow a U-shaped function with a maximal, optimal value.

For a rational, well-specified theory of acoustics design, one needs to be able to predict listener percepts and preferences given the sound stimulus (program material) and its modification by a particular enclosure (sound field). In the spirit of Manfred Schroeder's research program [1, 20] we developed correlation-based psychoacoustic models for these and other auditory percepts that serve this purpose [2]. Each auditory percept is related to a specific feature of either an internal monaural autocorrelation or binaural cross-correlation representation. This systematic theory of auditory percepts, along with their presumed neural signal-processing operations and their observable neurophysiological correlates, is outlined below in Sect. 8.4. The observable neural correlates of preferences are discussed in Sect. 8.5.

8.3.2 *The Centrality of Listener Preferences for Theories of Design*

There were several reasons that we initially got into the investigation of subjective preference. Preferences are evaluative judgments made by the listener that reflect particular situations or conditions that the listener chooses over other alternatives. A listener's preferences thus guide his or her actions, and serve as indicators of the degree of satisfaction. Preferences are estimated quantitatively using systematic, paired comparison tests that, through the law of comparative judgment, permit numerical, metrical scale values to be obtained.

Subjective preferences are the most fundamental aspects of an acoustic theory of design because they embody the desires of the listener. In biological terms, preferences are the goal states of the organism, the most primitive response that steers behavior. In the nervous systems of every higher organism are such embedded goals and drives. Preferences guide the organism in the direction of maintaining life, toward taking actions that enhance self-preservation and long-term survival. As such, in every brain there are structures and neural processes that realize preferences by steering the behavior of the organism such that preferred conditions are more reliably achieved. Being an essential, fundamental part of the functional organization of the nervous system, it is possible for us to observe brain activities that are related directly to preferences and their influences on behavior.

Preferences are deeply related to an individual's aesthetic sense. More than any other aspect of ourselves, the set of all our preferences, what we prefer most, is what gives each of us our individual unique personalities. Because we are always striving to realize our preferences, sometimes in new ways, our preferences provide a constant motivational source of new creations.

8.3.3 Theory of Multidimensional Listener Preferences

We have listed above the four major factors that matter most for concert hall listening (Figs. 8.3 and 8.4). Each is a separate perceptual dimension with its associated preference curves [5]. In order to optimize the design of an enclosure for listening, we need an overall preference measure that takes the individual preference dimensions and tradeoffs between them into account (see Manfred's discussion of the design problem in [21, pp. 74–79]). Such an integrative model of preference clearly leads to comprehensive criteria for achieving the optimal design of concert halls [8–10, 22].

The integrated preference model takes the form of weighted linear addition of the four normalized preference scale values (Fig. 8.4). The scale values were obtained using paired comparison tests rather than absolute preference magnitude estimation. Paired comparison tests have the advantage that they are simpler and allow relative judgments to be made by relatively inexperienced subjects, thus widening both the pool of people who can usefully participate and the range of acoustic parameters that can be studied.

We found that the four different acoustic factors are in fact independent and are not affected by changes in each other. In a given individual listener, there appear to be no natural groupings or interactions of preferences across these four dimensions [2]. Also, results of such a number of subjective preference tests indicate that the units of scale values (preferences for individual parameters S_i) are almost constant, so that we may add scale values to obtain the total scale value (overall preference S) for a sound field [8]. Thus $S = S_1 + S_2 + S_3 + S_4$.

One can transform and normalize the four preference scale values so that each is an inverted U-shaped curve with a zero value at the most preferred condition (designated by subscripted p). Data plots and curve fits can be found in [2, Chap. 3]. The individual normalized preference factors x_i are given in Table 8.1.

Table 8.1 Preference factors of individual auditory attributes and their relative weights in overall preference

Perceptual attribute on which the preference factor x_i is based	Normalized preference factor x_i[a]	Weighting[b]	
		α_i ($x_i < 0$)	α_i ($x_i > 0$)
Binaural listening level, LL	$x_1 = 20 \log P - 20 \log[P]_p$	0.07	0.04
First reflection time, Δt_1	$x_2 = \log(\Delta t_1/[\Delta t_1]_p)$	1.42	1.11
Subsequent reverberation time, T_{sub}	$x_3 = \log(T_{sub}/[T_{sub}]_p)$	$0.45 + 0.75A$	$2.36 - 0.42A$
Interaural correlation, IACC	$x_4 = \text{IACC}$	1.45	–

[a] P is sound pressure in dBA, p subscripts indicate listener-preferred values
[b] See text below

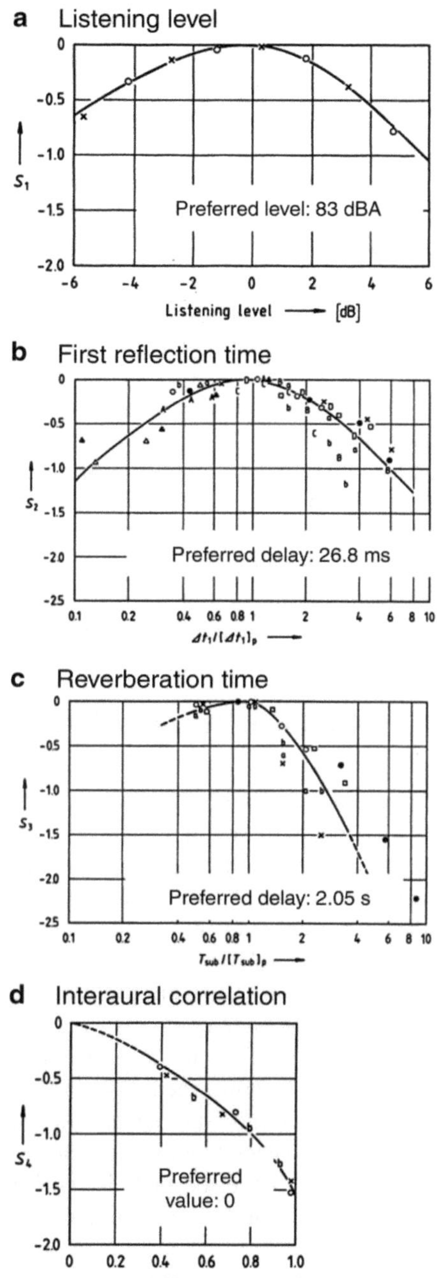

Fig. 8.4 Measured scale preference values of the four major acoustical factors. (**a**) Listening level (LL). (**b**) First reflection time (Δt_1). (**c**) Subsequent reverberation time (T_{sub}). (**d**) Interaural correlation (IACC). Estimated average preferred values were 83 dBA (LL), 26.8 ms (Δt_1), 2.05 s (T_{sub}), and 0 (IACC)

Overall preference S is then the weighted sum of preferences, x_1 through x_4, for the individual dimensions S_i (principle of superposition of preferences).

$$S = S_1 + S_2 + S_3 + S_4 = -\alpha_1|x_1|^{3/2} - \alpha_2|x_2|^{3/2} - \alpha_3|x_3|^{3/2} - \alpha_4|x_4|^{3/2}$$

where x_i are the normalized individual preference factors and α_i are the weighting coefficients for each factor x_i, below and above its optimal value (0), that indicate their relative importance in determining overall preference. The weighting of T_{sub}, α_3, depends in part on the total pressure amplitude of reflections in the room, parameter A in Table 8.1.

Thus, the scale values of preference have been formulated approximately in terms of the 3/2 powers of the normalized objective parameters, expressed as logarithms of parameters, x_1, x_2, and x_3. It is remarkable that the spatial binaural parameter of IACC, $x_4 = \text{IACC}$, is expressed in terms of the 3/2 power of its real value rather than its logarithm, which indicates a much greater relative contribution to overall preference than for the other parameters.

The scale values are not greatly changed in the neighborhood of the most preferred conditions, but decrease rapidly outside of this range. Because the experiments were conducted to find the optimal conditions, this theory holds in the range near the preferred conditions tested for the four factors.

8.4 Rational Theory of Acoustic Design

Once one has developed models for the acoustics of enclosed spaces, the percepts associated with sound fields, and listener preferences for particular values and combinations of those percepts, then it becomes possible to rationalize and automate the design of concert halls. Given particular hall dimensions and geometry, it then is possible to compute the sound fields present for every seat in the house, and subsequently, to compute the values for the four relevant acoustic parameters, and then given a set of listener preferences, to estimate the level of listener satisfaction that will be obtained.

Conversely, if one has a given concert hall, and one knows one's own preferences in quantitative terms, then it is possible to find the seat in the house that will provide the best listening experience, and one could even sell tickets on this basis. We introduced a special facility for seat selection at the Kirishima International Concert Hall in 1994 that tested each listener's own subjective preference and matched these with hall locations in order to make individualized seating recommendations.

8.4.1 Acoustic Design Using GAs

Even if one has fully developed acoustical, psychophysical, and psychological preference models, the optimal choice of concert hall size, shape, geometry, and composition is a complex problem. We used GAs to design two concert halls, the

Kirishima International Music Hall and the Tsuyama Music Cultural Hall, both in Japan. The GA allows optimization of listener preferences for the largest numbers of seats [2, pp. 148–158].

In the case of the Kirishima hall, we began with a shoebox shape and the GA iteratively evolved the shape of the hall, by independently moving the walls, to best satisfy listener preferences at 49 spatial locations distributed within the audience and performer areas [23, 24]. The front and rear walls were vertically bisected to obtain two faces, and each long wall along the side of the seating area was divided into four faces. The coordinates of the two bottom vertices of each surface were encoded on the chromosomes for the GA. Walls were kept vertical to examine the plan of the hall solely in terms of maximizing the average scale value of subjective preference obtained at the greatest number of seating positions, which meant reducing the IACC to its lowest possible value. We focused on this parameter first, because it is the most important in terms of overall preference and because everyone prefers the smallest value that can be attained (IACC $< \varepsilon$) [25]. This contrasts with individual listener preferences for other three parameters, such as LL, where there can be a wide range of preferred values.

The full GA strategy takes into account the listening preferences for both audience and performers (Fig. 8.5, top). Each wall was moved independently of the other walls to optimize the overall shape of the room (left, middle panel). In the acoustical simulation using image method, openings between walls were assumed not to reflect the sound. To minimize IACC, the rear wall of the stage and the rear wall of the audience area took on convex shapes because this avoids reflections from near the median plane.

The resulting shape of optimizing for minimal IACC is similar to the final leaf shape of the Kirishima International Music Hall, Japan (floor plans on the right). Contour lines of equal estimated IACC values are shown in the bottom left panel. To our knowledge, the Kirishima hall is the largest structure that has been designed thus far by GA optimization.

8.4.2 *Rules of Thumb for Optimizing Hall Acoustics for Performances*

From a historical viewpoint, for blending human life and the built environment, architects have been much more concerned with spatial criteria and visual perspective than temporal criteria and sound environments. On the other hand, from Sabine [19] onward, architectural acousticians have been mainly concerned with temporal criteria, represented primarily by reverberation time. Through most of the history of our practice, there has existed no comprehensive theory of design that included the spatial criterion represented by the IACC, so that discussions between acousticians and architects were rarely on the same subject. What we need is a general theory of physical environments in which deeply interrelated temporal and spatial, auditory and visual sensations are taken into account in a combined practice of acoustic and architectural design.

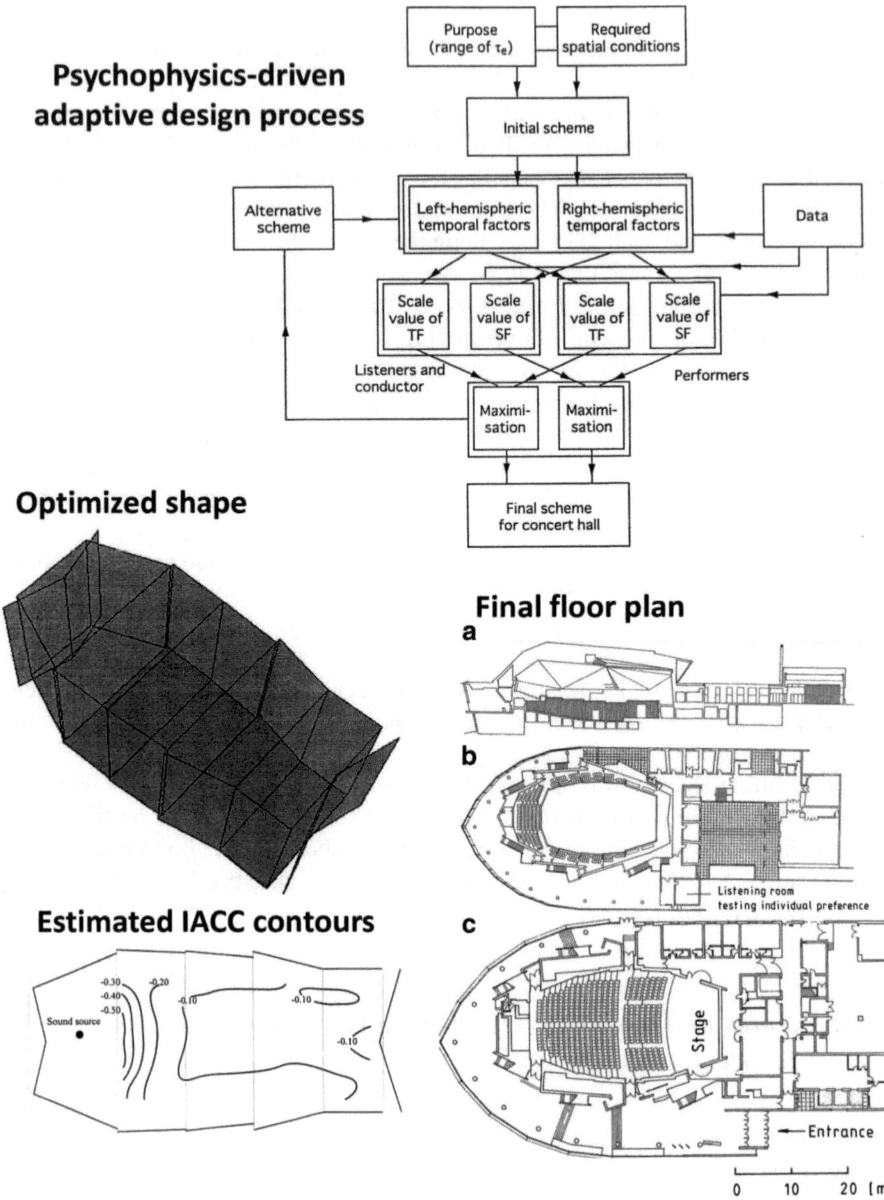

Fig. 8.5 Design of a concert hall using GAs for optimizing listener experience on the basis of acoustic theory, perceptual psychophysics, and the psychology of preferences. *Top*. Schematic of the GA-based iterative, evolutionary design process. *Lower right*. Final design of the Kirishima International Concert Hall. *Left, middle*. Final, optimized shape of the hall. *Lower left*. Predicted listener preference contours for IACC, which was the deciding spatial factor that effectively determined the hall's shape

In designing a concert hall, one must first have some idea of the type of program material that will be performed, so that the dimensions, shape, and composition of the hall can be optimized accordingly for those sounds. The nature of the music or speech material, specifically its minimal effective durations $(\tau_e)_{min}$, will determine the optimal first reflection and subsequent reverberation (T_{sub}) times for the hall. For pipe organ music, for example, values of τ_e about 200 ms or $T_{sub} \sim 4.6$ s are most appropriate, whereas for chamber music, these values would be more like 65 ms and 1.5 s, respectively.

For a given hall, the conductor or the sound coordinator should select music material with effective durations that best blend the music and the sound field within that space. Using the same methods as we used for individualized seat selection, one can also determine the best positions for performers to take by finding those locations that maximize listener preferences over the widest range of seats. The performance can be further optimized by locating the various instruments and performers such that their effective durations are matched to the first reflection times associated with their stage positions. For instance, the values of $(\tau_e)_{min}$ for violins is usually shorter than that of contrabasses that play mainly in lower frequency ranges, so it is advantageous to shift the position of violins closer to a stage sidewall (usually the left wall) and to position the contrabasses more to the center, increasing their distance to the sidewalls, thereby increasing their first reflection time.

In order to provide vertical reflections from places near the median plane of each of the performers on the stage, the back wall on the stage is carefully tilted [9, 26]. The tilted back wall consists of six sub-walls with angles adjusted to provide appropriate reflections within the median plane of the performers. It is worth noticing that tilted sidewalls onstage provide good reflections to the audience sitting close to the stage that reduce IACC. Also, the sidewall can provide useful reflections (arriving from a performer's back) for a piano soloist, for example.

To suppress normal-mode vibration of the stage floor and anomalous sound radiation from the stage floor during performances, arrange joists in triangular configurations, without any neighboring parallel structure. The floor should be thin, e.g., 27 mm, in order to radiate sound effectively through the coupled vibrations of instruments such as cellos and contrabasses. During rehearsal, music performers can control radiation power to some extent by adjusting their position or by inserting a rubber pad under their instrument.

IACC should be kept as small as possible for the audience by suppressing a strong reflection from the ceiling and by appropriate reflections from the left and right sidewall. If the source signal, like most music, contains mainly mid-frequency components, the reflections from the sidewalls should be optimized by orienting the walls at roughly 55° relative to each listener's front [9].

In order to obtain low IACCs for most listeners, we design ceilings using a number of triangular plates with adjusted angles, and the sidewalls are given a 10 % tilt with respect to the main-audience floor. In addition, diffusing elements are added to sidewalls to avoid image shifts of onstage sound sources caused by the strong high frequency (>2 kHz) reflections. These diffusers on the sidewalls, for example, are designed by a deformation of the Schroeder diffuser [27] that eliminates thin partitions that separate the wells.

Under actual listening conditions in an existing hall, the perceived IACC depends on whether or not the amplitudes of reflection exceed hearing thresholds. Thus, a more diffuse sound field may be perceived with increasing power of the sound source. For example, Keet [13] first reported that ASW increases with increased LLs. When the source is weak enough, just above threshold, one hears only the direct sound, and one gets the opposite of a diffuse sound impression. In general, small values of IACC can be realized only by early strong reflections. If the sound source is located in the center line on the stage, then coherent signals arrive at the same time from both sidewalls to produce a high IACC. From this point of view, acoustical-asymmetric properties in shaping the hall may create further advantages.

For music performers on stage, temporal factors, such as pitch, timbre, and sound qualities related to reverberations, are much more critical than the spatial factors, such as LL and sound diffuseness. Since musicians perform over time durations, first reflections with delays commensurate with those of the sounds they produce are of particular importance. We recommend that the sound field for the conductor should be designed as that of a "listener" with appropriate reflections of sidewalls on the stage [28]. Some design iterations may be required to optimize acoustic conditions for performers, conductors, and listeners, before the final scheme of the concert hall is reached.

8.5 Correlation-Based Model of Auditory Percepts

In order to design structures for listening, an adequate theory of auditory function is indispensable. Manfred Schroeder in his 1975 review of the known structure and function of the auditory system concluded that the sense of hearing is a very active field that is fascinating and ultimately still mysterious [20]. In many respects, the situation is not so different almost four decades later.

Over many years, we developed a comprehensive theory of major auditory percepts that is based on two internal representations: the monaural ACF and the binaural IACF. Features of the unnormalized monaural autocorrelation representation are associated with auditory qualities we have called "temporal sensations," namely loudness, pitch, and timbre. Features of the binaural cross-correlation representation are associated with "spatial sensations," namely binaural LL, ASW, and SD. At the cortical level, we found observable neural response correlates associated with temporal sensations predominantly in the left hemisphere, whereas those associated with spatial sensations were found predominantly in the right hemisphere. The perceptual characteristics and neural correlates of all of these sensations are discussed in much greater depth elsewhere [2].

The various dependencies of the percepts on features of these two representations are shown in Fig. 8.6 and Table 8.2. The computation of these features from autocorrelation and cross-correlation functions is shown schematically in Fig. 8.2. We describe the various temporal and spatial perceptual attributes in turn.

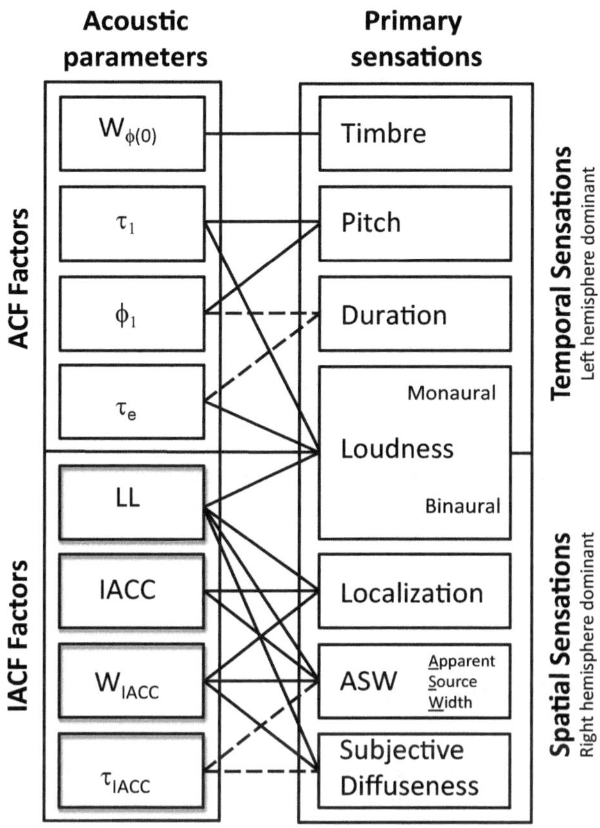

Fig. 8.6 Auditory acoustic features and related percepts. Relations between auditory percepts and features of monaural ACF and IACF. Temporal sensations are related to the temporal forms of signals, whereas spatial sensations are related to the perception of spatial aspects of sounds and the enclosing space. *Solid lines* indicate stronger relations, *dashed lines*, weaker determinants. Observable neural correlates of temporal and spatial sensations and their associated preferences are predominantly seen respectively in either left or right hemisphere

Table 8.2 Hemispheric specializations of temporal and spatial factors observed by analyses of SVR, EEG, and MEG in response to stimuli with changing acoustic parameters

Acoustic factors changed	SVR $A(P_1-N_1)$ latency difference	MEG N1m latency	EEG wave persistence or spatial extent*	MEG wave persistence
Temporal				
Δt_1	**L** > **R** (speech)		**L** > **R** (music)	**L** > **R** (speech)
T_{sub}			**L** > **R** (music)	
Spatial				
LL	**R** > **L** (speech)			
IACC	**R** > **L** (vowel /*a*/)	**R** > **L** (band noise)[a]	**R** > **L** (music)*	
	R > **L** (band noise)			
τ_{IACC}		**R** > **L** (band noise)[a]		
Head-related transfer function[b]		**R** > **L** (vowels)		

[a]Soeta and Nakagawa [29]
[b]Palomaki et al. [30]
*The asterisk indicates that spatial extent was the parameter, not persistence

8.5.1 Loudness

Loudness is a function of total SPL (dB SPL), periodicity (frequency, period) τ_1, normalized amplitude ϕ_1 (regularity), effective duration τ_e (bandwidth), and duration D.

In paired comparison tests of loudness, we found that loudness is both a function of frequency and effective duration. Loudness of a pure tone is greater than that of filtered noise centered on the pure tone, and loudness increased with increasing effective duration τ_e within the critical band [23, 24]. For example, we found that loudness of band pass noise produced by an extremely sharp filter (2068 dB/octave) with identical SPL is not constant within the critical band, quite a different result from previous studies [31]. Practically, this means that if environmental noise, such as aircraft landing noise, contains a pure-tone component or has long effective duration τ_e, then loudness increases and accordingly annoyance increases [32]. This is a typical example, why only measurements of SPL from a sound level meter fail to adequately predict subjective attributes. Just adding +10 dB to a noise with a coherent component is too simple, as loudness and annoyance increases with the value of the effective duration τ_e.

8.5.2 Duration

We studied the perception of duration for short complex tone and pure tones. Perceptually, the duration sensation depends most strongly on the physical signal duration, D, but it can also be influenced slightly by periodicity τ_1, which corresponds to perceived periodicity pitch. The apparent stimulus duration depends on signal periodicity τ_1, i.e., pure tone frequency (f_1) or the fundamental frequency (F_0) of a complex tone, rather than spectral center of gravity. For example, a 3,000 Hz pure tone of duration 150 ms has the same perceived duration as 140 ms duration pure and complex tones that evoke pitches at 500 Hz (a 500 Hz pure tone and a pure tone dyad of 3,000 and 3,500 Hz, $F_0 = 500$ Hz). Thus, when the sound signal evokes a lower periodicity pitch, perceived duration is slightly (~7 %) longer. Duration thus appears to be weakly linked with periodicity, which evokes sensations associated with periodicity pitch, rather than spectral center of gravity, which evokes sensations associated with pitch height.

8.5.3 Pitch

The periodicity pitch of pure and complex tones is associated with the time lag τ_1 and its multiples in the autocorrelation representation. We conducted several pitch-matching studies using stimuli with missing fundamentals and found the pitch matches to be independent of phase (in-phase vs. random phase). The highest missing fundamental matched was ~1,200 Hz [33], which means that the highest frequency components in such a missing fundamental stimulus need to be less than about 5 kHz, a value that is comparable to the frequency limit of significant phase locking in auditory nerve fibers, see Sect. 8.6.1.

Pitch strength or salience, depends on the regularity of the signal, which is indicated by the amplitude ϕ_1 of the corresponding periodicity in the normalized ACF (see also [34, 35]). Maximum pitch strength is observed at $\phi_1 = 1.0$, and no pitch is perceived when $\phi_1 < \varepsilon$, ε being a small positive number associated with the threshold for pitch strength.

8.5.4 Timbre

Timbre includes all those aspects of sound quality that are independent of loudness, duration, pitch, and its spatial attributes, i.e., it encompasses those qualities of sound texture that distinguish two notes of equal pitch, loudness, duration, location, and apparent size that are played by different musical instruments. One of the dimensions of timbre is the relative brightness of a sound, which is often characterized in terms of spectral tilt. The corresponding autocorrelation feature is the factor $W_{\phi(0)}$, defined by width of the central (zero-lag) peak as measured by the delay time at the first 0.5 crossing of the normalized ACF.

8.5.5 Binaural LL

The remaining perceptual attributes are spatial sensations that are associated with features in the IACF. In the binaural listening condition, the LL is the mean amplitude given by $\mathrm{LL} = 10 \log[\Phi_{ll}(0)\Phi_{rr}(0)]^{1/2}/\Phi_{\mathrm{reference}}(0)$ [dBA], with $\Phi_{\mathrm{reference}}(0)$ being the reference amplitude.

8.5.6 Sound Direction in the Horizontal Plane

Apparent sound direction in the horizontal plane depends partially on all of the spatial factors, some of which are features of the IACF. The most important factors are dominant interaural time delay (ITD), τ_{IACC}, and interaural intensity difference (IID), which is a function of the sound energies at the two ears, $\Phi_{ll}(0)$ and $\Phi_{rr}(0)$. Direction can be influenced more weakly by the broadness of the ITD peak W_{IACC} and the degree of IACC.

8.5.7 Apparent Source Width

ASW is the perceived spatial breadth of a sound, how much of the horizontal plane it seems to occupy. We developed a model for estimating the scale value for ASW as a function of IACC, ITD breadth W_{IACC}, and LL. ASW $\approx \alpha(\mathrm{IACC})^{3/2} + \beta(W_{\mathrm{IACC}})^{1/2} + \gamma(\mathrm{LL})^{3/2}$, where $\alpha \approx -1.64$, $\beta \approx 2.42$, $\gamma \approx 0.005$. Thus we found that apparent width increases with decreasing IACC and increasing W_{IACC} as well as LL [36].

8.5.8 Subjective Diffuseness

SD is related to the sense of envelopment of sound. We found the scale values for this quality observed in paired comparison tests to be inversely proportional to the IACC, so that SD $\approx -\alpha(\mathrm{IACC})^{3/2}$, where $\alpha = 2.9$ [37].

8.6 Observable Neural Correlates of Auditory Percepts

The foregoing correlation-based model of auditory perception is based on the notion that the auditory system uses an autocorrelation-like representation for analyzing the internal structure of signals (periodicity, spectrum, duration, and various transients and modulations), and an interaural cross-correlation representation for orienting sounds in space. Like Manfred Schroeder, we have all along had deep interests in how the auditory system might realize these representations and operations, i.e., how the central auditory system works. We subsequently developed an auditory signal-processing model with these two modes of representation in mind, as shown in Fig. 8.7. It can be noted that early on both Licklider [38] and Cherry [39] also proposed such dual auto- and cross-correlation architectures for auditory neurocomputation.

8.6.1 Auditory Signal-Processing Model

This functional, signal-processing model embodies our assumptions about where in the auditory system these two representations might reside. A sound source produces a signal that propagates through space, interacting with surroundings, to reach the listener's two ears. Vibrations in the middle ear and cochlea are mechanically filtered, and in some cases amplified by cochlear hair cells. Mechanical vibrations are transduced into ionic currents that then produce pulsatile action

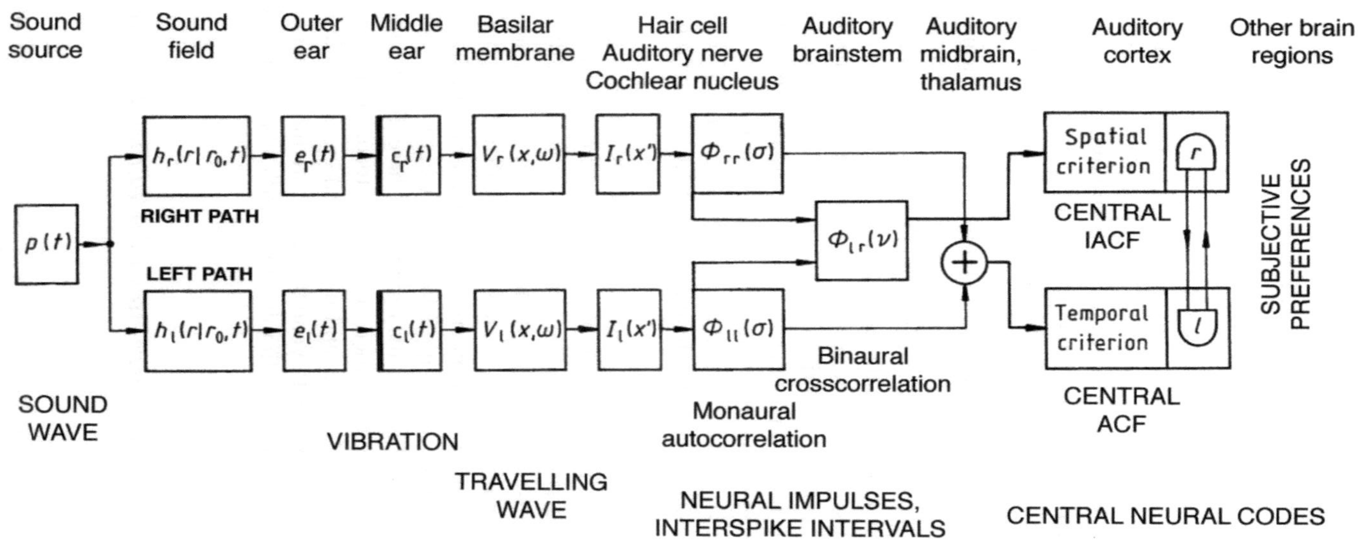

Fig. 8.7 Auditory signal processing model and central correlation representations. The neuroanatomical locations of central autocorrelation (IACF) and binaural cross-correlation (IACF) representations are hypothetical

potentials (spikes) in auditory nerve fibers that are phase locked to stimulus periodicities up to about 5 kHz.

The most direct correlate of the monaural ACF is found in patterns of spike timings in the auditory nerve [2, 34, 35, 40, Chap. 5]. The times between both consecutive and non-consecutive pairs of spikes, called all-order interspike intervals, form an autocorrelation-like, population-based neural representation of the stimulus in the auditory nerve. The representation, which combines the interspike intervals from the 30,000 axonal fibers that make up the human auditory nerve, is robust, precise, and effective up to the ~5 kHz limit of phase locking. The ensuing distribution resembles the positive portion of the stimulus ACF, such that there are direct correspondences between the monaural, temporal ACF parameters outlined in the preceding section, and features of this early temporal, neuronal representation.

The auditory nerve fibers synapse on neurons in the cochlear nucleus, and projections from these neurons in turn synapse on neurons elsewhere in the auditory brainstem and midbrain. Major pathways through the cochlear nucleus preserve spike fine timing and on their way to the midbrain send axon collaterals that synapse on neurons in binaural stations in the brain stem. The bipolar, binaural neurons in the auditory brainstem (medial superior olivary nucleus, MSO) act as coincidence detectors. Because the incoming spike trains are correlated, by virtue of phase locking, with the filtered stimulus at each ear, the population of neurons in this nucleus in effect carries out an interaural cross-correlation (IACF) operation. Thus there are direct neural correlates of the features of the IACF in the binaural stations in the auditory brainstem.

Unfortunately, despite considerable advances over the last several decades, the precise means by which these two temporal, correlation-based early auditory representations might be analyzed by the rest of the system are still poorly understood. As we outline below, we observed the central neural correlates of percepts associated with the monaural ACF, what we have called "temporal sensations" primarily in the left cerebral hemisphere. The observed correlates of percepts associated with the interaural correlation function (IACF), what we have called "spatial sensations," were all found in the right hemisphere. It is for this reason that in the schematic of Fig. 8.7 the combined monaural autocorrelation representations are connected to the left hemisphere, and the binaural cross-correlation representation goes to the right hemisphere. Beyond this, there are further neural-processing stages, cortical and subcortical (e.g., limbic), that mediate auditory preferences.

8.6.2 Observable Neural Correlates of Auditory Percepts and Preferences

The neuronal correlates of the monaural ACF can be observed in single unit responses at the level of the auditory nerve and cochlear nucleus. These also can be seen in averaged gross electrical potentials, such as local field potentials and the frequency-following response (FFR). The neural correlates of the interaural correlation function can be found in single-unit spike train responses, local electrical field potentials, and averaged scalp responses (ABR) from sources in the auditory brainstem.

In a large series of neurophysiological experiments we have examined the neural correlates of auditory percepts and preferences at the level of the auditory brainstem, midbrain, and cortex. Whereas in the auditory nerve and cochlea, the fine timing patterns of neuronal response can be observed through a variety of means, as one ascends the pathway, overt signs of these progressively and dramatically decrease, and the observable neural correlates of auditory percepts then come mainly in the form of changes in neural response latency.

At the level of the auditory brainstem and midbrain, we used the auditory brainstem response (ABR, 0–10 ms latencies) to probe the nature of central auditory representations [41]; [2], pp. 40–48. Here systematic changes can be observed in the relative latencies of corresponding ABR peaks I–V from right and left pathways that covary with stimulus intensity (loudness) [42], sound direction (azimuth [41]), and IACC. These latencies of neuronal responses reflect the electrical effects of large numbers of spikes that are produced roughly synchronously in various neuronal nuclei as volleys of spikes make their way up the auditory pathway.

At the level of the cerebral cortex, we used slow vertex (SV) averaged late evoked electrical responses (SVR, 10–500 ms latencies), EEG, and MEG to probe observable neural correlates of both percepts and preferences [2, 9, 22]. These manifested themselves as changes in response latencies in the slow vertex responses and in the persistence and spatial extent of alpha waves in EEG and MEG recordings.

Slow vertex responses. Slow vertex responses reflect cortical neural responses with longer latencies than ABR, in the range of 170–320 ms. Using transitions between more-preferred to less-preferred stimulus conditions, we were able to observe changes in response latency that covaried with listener preferences. By changing sensation level (SL), first reflection time Δt_1 and IACC we consistently found that the latency increases with increasing preference (Fig. 8.8).

Alpha rhythms. Alpha rhythms can be observed both in running scalp electrical potentials (EEG) and in magnetic signals (MEG) that are generated by cortical populations. In both EEG and MEG, we consistently found that preferred conditions are associated with increased temporal persistence and spatial extent of alpha rhythms. Annoying stimuli, on the other hand, decreased the persistence and extent of the rhythms. Remarkably, these neural correlates of auditory preferences also were found to hold for visual preferences, as we will discuss in the next section.

The temporal persistence of the alpha rhythm is measured by computing the effective duration of the alpha-band EEG or MEG signal. The longer the effective duration, the more steady the rhythm over longer periods of time [43].

The spatial extent of the alpha rhythm was assessed in terms of how correlated were the alpha-band neural response signals across different electrode or sensor

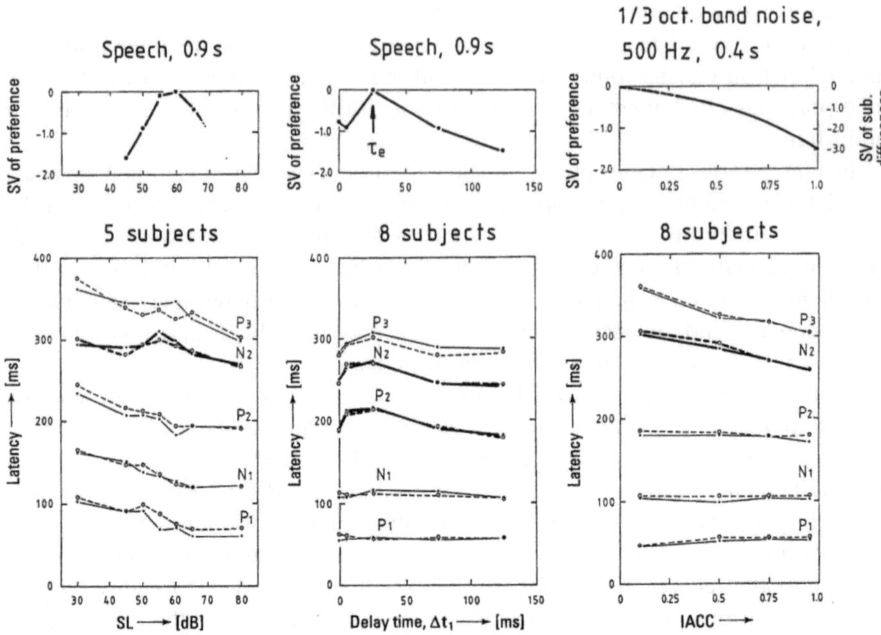

Fig. 8.8 Observable neural correlates of three acoustical factors in latencies of neural responses to two stimuli, speech and noise, as seen in averaged evoked auditory potentials (slow vertex responses). *Top plots*. Preference curves for the three factors, sound level (SL), first reflection time (Δt_1), and IACC. *Bottom plots*. Latencies of successive peaks and troughs in the slow vertex responses from left (*solid lines*) and right side (*dashed lines*) scalp recording sites, averaged over multiple subjects. Note correspondences between preferred values and longer latencies and interhemispheric relative latency differences [2, pp. 52–55]

sites. The higher the average correlation, the more similar are the alpha-band signals across different cortical locations, and the greater the spatial extent of the rhythm over the whole cortical surface. Interestingly, we found that the signal power in the alpha-band did not correlate with preference.

Hemispheric lateralization. In a series of experiments using SVR, EEG, and MEG in response to transitions between more-preferred and less-preferred stimulus conditions [44–47], we observed clear and consistent differences between neural responses associated with each of the two cerebral hemispheres (Table 8.2). In all cases, temporal sensations associated with first reflection time Δt_1 and subsequent reflection time T_{sub}, which correspond to features in the monaural ACF, were predominant at recording locations over the left hemisphere. Conversely, neural responses for spatial sensations that were related to changes in LL, IACC, predominant interaural delay (ITD), and head-related transfer function (HRTF) were observed in recording locations over the right hemisphere.

8.7 Visual Sensations

We believe that there are many deep parallels between auditory and visual perception because the two modalities may share common underlying neurally based correlational information-processing mechanisms. In the last decade we have applied our correlation theory and psychophysical methods to the study of visual sensations. Here we very briefly summarize some of the specific visual percepts and preferences. Details of the psychophysical and neurophysiological investigations are available in the primary published reports and are also presented in depth in book form [2].

As with audition, many visual percepts can be explained in terms of correlation-based representations. Temporal aspects of visual sensation, such as the perception of flickering lights, including a "missing fundamental" perception of flicker periodicity [48], can be described in terms of temporal ACFs (TACFs), whereas spatial aspects of visual sensation, such as the perception of texture can be described in terms of spatial ACFs (SACFs) [49].

Different basic aspects of visual perception and their corresponding features in temporal and spatial autocorrelation representations are illustrated in Fig. 8.9. For temporally modulated visual stimuli, the visual percepts of brightness, duration, flicker periodicity, and flicker frequency spectrum, correspond to auditory percepts of loudness, duration, pitch, and spectral aspects of timbre. Perceived visual contrast, coarseness, and regularity across spatial positions correspond roughly to auditory perceptual dimensions of loudness, low vs. high pitch, and low vs. high regularity (pitch salience, noise vs. tonal quality).

We also studied the preferences for these visual percepts as well as the observable neurophysiological signs that are associated with them [58]. There appear to be universal preferences for the rates and regularities of visual stimuli, such as flickering lights (e.g., twinkling stars, fires, water flows, sunlight playing on water), oscillating forms (leaves fluttering in the breeze), and visual textures (clouds, cloth patterns). One notices the pleasing correlated movements and the sound that leaves of bamboo trees make in the wind. As the classical Japanese poet Yakamochi Otomo (c. 715–785) observed "... A small bamboo clump of my house, we might listen to bracing sound moving in the wind of this evening" (poem *Manyoshu*, Y. Ando, translation).

As with auditory stimuli, there are inverted U-shaped preference curves for individual visual preferences—we like visual patterns that are neither too ordered nor too disordered. These also can be combined in a multifactor model to predict overall viewer satisfaction. In addition, we also found that preferred visual perceptual states produced coherent EEG and MEG alpha rhythm activity that persisted longer, i.e., with longer effective duration, and was more spatially widespread over the brain [43, 50–52].

Remarkably, as with auditory percepts and preferences, we also observed hemispheric lateralizations associated with visual sensations. The two cerebral hemispheres appear to play somewhat different roles, with the left cerebral hemisphere concerned more with linear, sequential modes of thinking, such as speech and

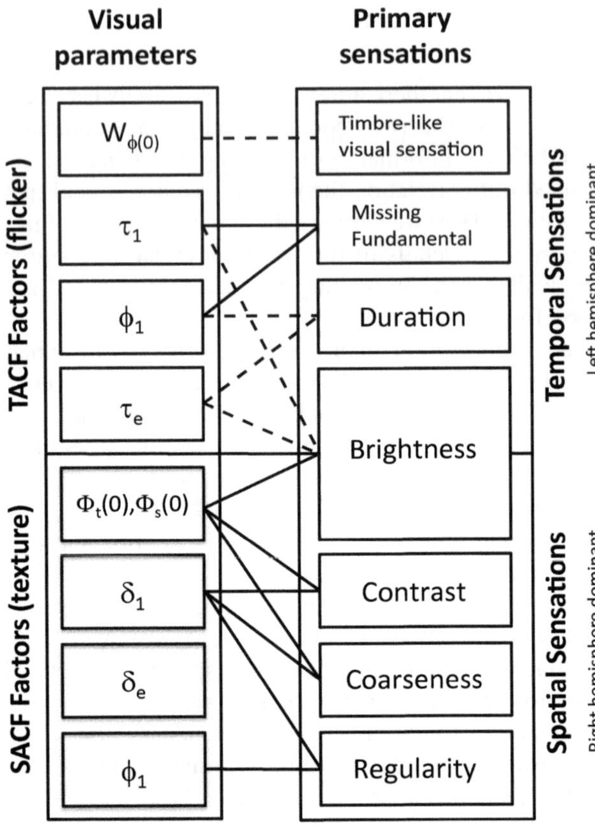

Fig. 8.9 Relations between visual percepts and features of visual temporal and spatial ACFs (TACF, SACF). Temporal sensations are related to the temporal forms of visual signals, such as flicker rate and duration, whereas spatial sensations are related to spatial vision, such as textural coarseness and regularity. Like their auditory counterparts, cf. Fig. 8.6, observable neural correlates of temporal and spatial visual sensations and their associated preferences are predominantly seen respectively in either left or right hemisphere

calculation, and with the right hemisphere more engaged with visual space in multidimensional and non-temporal terms [53–56].

In our investigations we found that the observable neural correlates of temporal sensations in both auditory and visual modalities were almost always found predominantly in the left hemisphere, whereas spatial sensations in both modalities were found predominantly in the right hemisphere [2]. Thus, there may exist a high degree of independence between the temporal and spatial factors for any subjective attribute, and this in turn suggests a basis for a theory of environmental design. Whenever possible, we should strive to design for both hemispheres in order to facilitate their complementary capabilities (Fig. 8.10).

8.8 Theory of Temporal and Spatial Environmental Design

We seek to develop a comprehensive human-centered theory of design that takes into account both the spatial and temporal dimensions of things. In architectural acoustics we need theories of sound and its propagation, models of human percepts

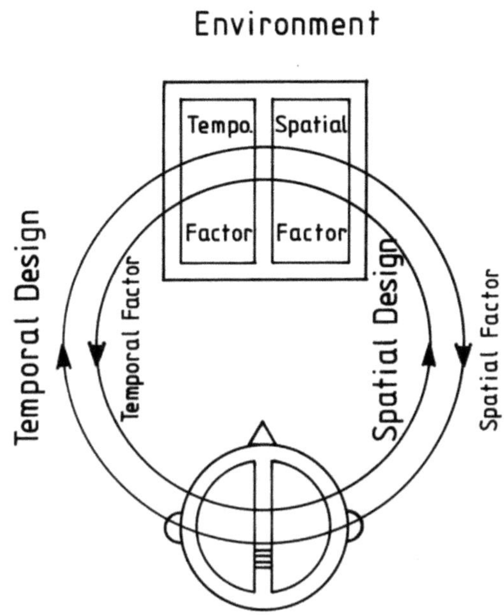

Fig. 8.10 Toward an integrated theory of temporal and spatial environmental design. The two cerebral hemispheres cooperate with each other to respond to and create spatial and temporal patterns in their environments

and preferences, techniques for optimizing the shapes and compositions of structures, and the means for building them. Our methods are rational to the extent that we attempt to make as clear as possible the basis of our design process, what human preferences we are attempting to optimize, and how we will go about that process.

In the general spirit of Manfred Schroeder's [1] influential *Science* article on architectural acoustics, a general theory of environmental design should include models of the physical world, models of how human perceive and interact with that world, as well as models of what human beings want [2, 57]. Such a theory needs to embrace and accommodate both individual tastes and group preferences. In taking the neural turn, in venturing inside the brain, we acquire more insights into the workings of those percepts and preferences, and consequently into the means by which both individual and group preferences can be better satisfied. We envision creative design practices that explicitly take into account temporal and spatial factors to best suit and utilize the respective needs and capabilities of the two cerebral hemispheres.

Philosophically, we hold great importance for the creative role of the individual. Personality is uniquely formed by both genetics and environmental influences on development. Temporal design is that part of an integrated theory of design that deals with the rhythms and temporal flows of life that occur over different time spans, from momentary awareness to daily, weekly, monthly, and yearly cycles, to even longer spans of phases of our lives (Fig. 8.11, top). Our theories of design should on a very fundamental level self-consciously take these different temporal regimes, each with its own special properties and possibilities, into account. Ultimately, we want to create design theories and practices that feed back upon

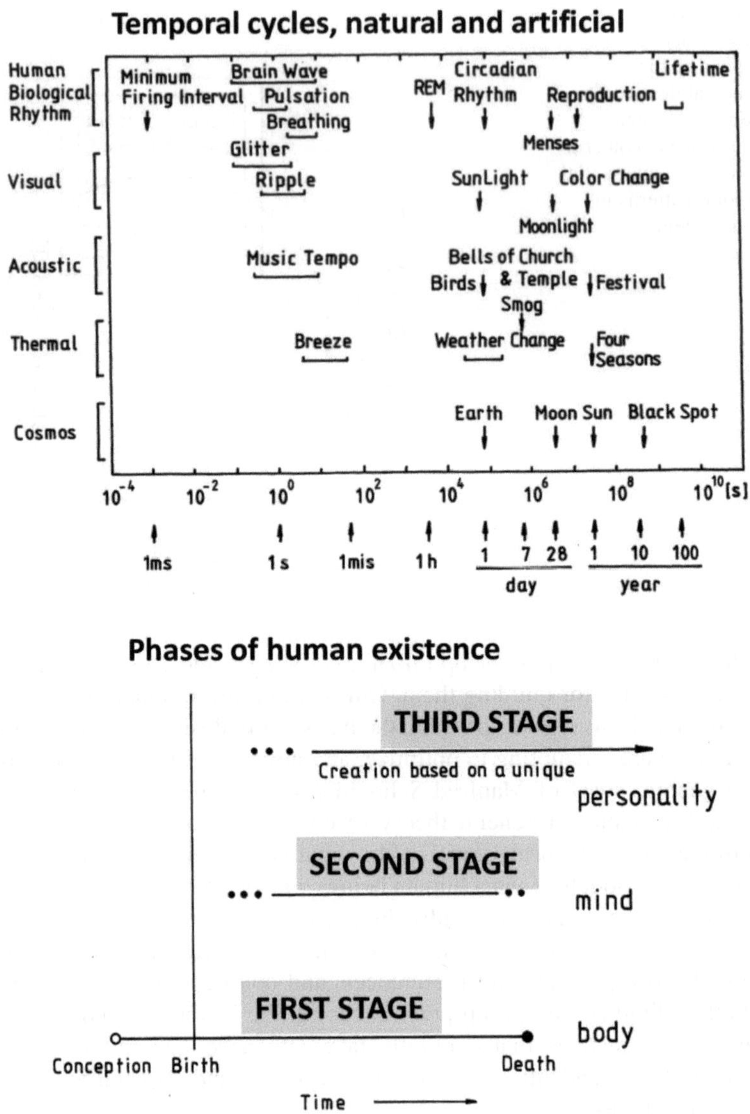

Fig. 8.11 Timescales of natural and artificial cycles and existential phases of human life. *Top*. The timescales of our world. *Bottom*. Phases of human life. The first stage is the lifespan of the mortal body. The second stage is the life of the individual mind. The third stage is the continuing effect of an individual's personality and ideas on the rest of humanity. While the first two stages of an individual are limited in time, the third stage can persist almost indefinitely into the future through the enduring memories of others

ourselves to enhance our own physical and mental development. When we close this final self-modifying loop, we enable an open-ended process of improved designs that further realize human potentials.

Over the arc of our lives, each of us inevitably passes through several overlapping phases (Fig. 8.11, bottom). The first stage is the life of the body. Our mortal bodies are conceived, born, mature, and die. The second stage is the life of the mind in which our minds develop, grow, realize their possibilities, and for some, eventually senesce. We share these first two phases with animals, which have bodies and minds that are basically like our own, the differences being only those of degrees. In some cases the capabilities of other animals exceed those of our own bodies and minds. The third stage of life is the life of personalities and ideas that are shared among human beings. Because of our social natures and the continual regeneration of our individual memories and collective cultures, personalities and ideas can reverberate long after the first two stages have ended.

Manfred Schroeder in the first and second stages of his life had many very positive and formative influences on my life and ideas. Manfred Schroeder is now continuing on in his third stage of life, and we believe and hope that his personality and ideas will continue to influence us and many others far into the future.

References

1. Schroeder, M.R.: Architectural acoustics. Science **151**, 1355–1359 (1966)
2. Ando, Y., Cariani, P.: Auditory and Visual Sensations. Springer, New York (2009)
3. Damaske, P.: Subjektive Untersuchungen von Schallfeldern. Acustica **19**, 199–213 (1967)
4. Damaske, P., Ando, Y.: Interaural cross-correlation for multichannel loudspeaker reproduction. Acustica **27**, 232–238 (1972)
5. Schroeder, M.R., Gottlob, D., Siebrasse, K.F.: Comparative study of European concert halls: correlation of subjective preference with geometric and acoustic parameters. J. Acoust. Soc. Am. **56**, 1195–1201 (1974)
6. Ando, Y.: Subjective preference in relation to objective parameters of music sound fields with a single echo. J. Acoust. Soc. Am. **62**, 1436–1441 (1977)
7. Ando, Y., Gottlob, D.: Effects of early multiple reflections on subjective preference judgments on music sound fields. J. Acoust. Soc. Am. **65**, 524–527 (1979)
8. Ando, Y.: Calculation of subjective preference at each seat in a concert hall. J. Acoust. Soc. Am. **74**, 873–887 (1983)
9. Ando, Y.: Architectural Acoustics, Blending Sound Sources, Sound Fields, and Listeners. AIP Press/Springer-Verlag, New York (1998)
10. Ando, Y.: Concert hall acoustics based on subjective preference theory. In: Rossing, T.D. (ed.) Springer Handbook of Acoustics. Springer, New York (2007)
11. Sabine, W.C.: Architectural Acoustics (J. Franklin Institute, January, 1915). In: Sabine, W.C. (ed.) Collected Papers on Acoustics, pp. 219–236. Dover reprint, New York (1915) (original edition Harvard, 1922)
12. Beranek, L.L.: Music, Acoustics and Architecture. Wiley, New York (1962)
13. Keet, M.V.: The influence of early lateral reflections on the spatial impression. In: Proceedings of the 6th International Congress of Acoustics, Tokyo, Paper E-2-4 (1968)
14. Barron, M., Marshall, A.H.: Spatial impression due to early lateral reflections in concert halls: the derivation of a physical measure. J. Sound Vib. **77**(2), 211–232 (1981)
15. Cremer, L., Müller, H.A., Schultz, T.J.: Principles and Applications of Room Acoustics, vols. 1 & 2. Applied Science Publishers, New York (1982)

16. Schroeder, M.: Concert halls: from magic to number theory. In: Deutsch, D. (ed.) The Psychology of Music, 2nd edn, pp. 25–46. Academic, Amsterdam (1999)
17. Schroeder, M.R.: Acoustics in human communications: room acoustics, music, and speech. J. Acoust. Soc. Am. **68**(1), 22–28 (1980)
18. Kato, K., Fujii, K., Hirawa, T., Kawai, K., Yano, T., Ando, Y.: Investigation of the relation between minimum effective duration of running autocorrelation function and operatic singing with different interpretation styles. Acta Acust. United Acust. **93**, 421–434 (2007)
19. Sabine, W.C.: Reverberation (The American Architect and the Engineering Record, 1900). In: Sabine, W.C. (ed.) Collected Papers on Acoustics, pp. 3–68. Dover reprint, New York (1900) (original edition Harvard, 1922)
20. Schroeder, M.R.: Models of hearing. Proc. IEEE **63**, 1332–1352 (1975)
21. Schroeder, M.R.: Fractals, Chaos, Power Laws: Minutes from an Infinite Paradise. Freeman, New York (1991)
22. Ando, Y.: Concert Hall Acoustics. Springer, Heidelberg (1985)
23. Sato, S., Otori, K., Takizawa, A., Sakai, H., Ando, Y., Kawamura, H.: Applying genetic algorithms to the optimum design of a concert hall. J. Sound Vib. **258**, 517–526 (2002)
24. Sato, S., Kitamura, T., Ando, Y.: Loudness of sharply (2068 dB/Octave) filtered noises in relation to the factors extracted from the autocorrelation function. J. Sound Vib. **250**, 47–52 (2002)
25. Singh, P.K., Ando, Y., Kurihara, Y.: Individual subjective diffuseness responses of filtered noise sound fields. Acustica **80**, 471–477 (1994)
26. Nakayama, I.: Preferred time delay of a single reflection for performers. Acustica **54**, 217–221 (1984)
27. Schroeder, M.R.: Binaural dissimilarity and optimum ceilings for concert halls: more lateral sound diffusion. J. Acoust. Soc. Am. **65**, 58–963 (1979)
28. Meyer, J.: Influence of communication on stage on musical quality. In: Proceedings of the 15th International Congress on Acoustics, Trondheim, pp. 573–576. (1995)
29. Soeta, Y., Nakagawa, S.: Auditory evoked magnetic fields in relation to interaural time delay and interaural crosscorrelation. Hear. Res. **220**, 106–115 (2006)
30. Palomaki, K., Tiitinen, H., Makinen, V., May, P., Alku, P.: Cortical processing of speech sounds and their analogues in a spatial auditory environment. Cogn. Brain Res. **14**, 294–299 (2002)
31. Zwicker, E., Flottorp, G., Stevens, S.S.: Critical band width in loudness summation. J. Acoust. Soc. Am. **29**, 548–557 (1957)
32. Soeta, Y., Maruo, T., Ando, Y.: Annoyance of bandpass filtered noises in relation to the factor extracted from autocorrelation function. J. Acoust. Soc. Am **116**, 3275–3278 (2004)
33. Inoue, M., Ando, Y., Taguti, T.: The frequency range applicable to pitch identification based upon the auto-correlation function model. J. Sound Vib. **241**, 105–116 (2001)
34. Cariani, P.A., Delgutte, B.: Neural correlates of the pitch of complex tones. I. Pitch and pitch salience. II. Pitch shift, pitch ambiguity, phase-invariance, pitch circularity, and the dominance region for pitch. J. Neurophysiol. **76**, 1698–1734 (1996)
35. Cariani, P.: Temporal coding of periodicity pitch in the auditory system: an overview. Neural Plast. **6**, 147–172 (1999)
36. Sato, S., Ando, Y.: On the apparent source width (ASW) for bandpass noises related to the IACC and the width of the interaural cross-correlation function (WIACC). J. Acoust. Soc. Am. **105**, 1234 (1999)
37. Ando, Y., Kurihara, Y.: Nonlinear response in evaluating the subjective diffuseness of sound field. J. Acoust. Soc. Am. **80**, 833–836 (1986)
38. Licklider, J.C.R.: Three auditory theories. In: Koch, S. (ed.) Psychology: A Study of a Science. Study I. Conceptual and Systematic. vol. I, pp. 41–144. McGraw-Hill, New York (1959)
39. Cherry, C.: Two ears – but one world. In: Rosenblith, W.A. (ed.) Sensory Communication, pp. 99–117. MIT Press/John Wiley, New York (1961)

40. Secker-Walker, H.E., Searle, C.L.: Time domain analysis of auditory-nerve-fiber firing rates. J. Acoust. Soc. Am. **88**, 1427–1436 (1990)
41. Ando, Y., Yamamoto, K., Nagamastu, H., Kang, S.H.: Auditory brainstem response (ABR) in relation to the horizontal angle of sound incidence. Acoust. Lett. **15**, 57–64 (1991)
42. Hecox, K., Galambos, R.: Brain stem auditory evoked responses in human infants and adults. Arch. Otolaryngol. **99**, 30–33 (1974)
43. Soeta, Y., Okamoto, Y., Nakagawa, S., Tonoike, M., Ando, Y.: Autocorrelation analyses of MEG alpha waves in relation to subjective preference of a flickering light. Neuro Rep. **13**, 527–533 (2002)
44. Ando, Y., Kang, S.H., Morita, K.: On the relationship between auditory-evoked potential and subjective preference for sound field. J. Acoust. Soc. Jpn. (E) **8**, 197–204 (1987)
45. Ando, Y., Kang, S.H., Nagamatsu, H.: On the auditory-evoked potentials in relation to the IACC of sound field. J. Acoust. Soc. Jpn. (E) **8**, 183–190 (1987)
46. Ando, Y., Chen, C.: On the analysis of the autocorrelation function of α-waves on the left and right cerebral hemispheres in relation to the delay time of single sound reflection. J. Archit. Plan. Environ. Eng., AIJ **488**, 67–73 (1996)
47. Chen, C., Ando, Y.: On the relationship between the autocorrelation function of the α-waves on the left and right cerebral hemispheres and subjective preference for the reverberation time of music sound field. J. Archit. Plan. Environ. Eng., AIJ **489**, 73–80 (1996)
48. Fujii, K., Kita, S., Matsushima, T., Ando, Y.: The missing fundamental phenomenon in temporal vision. Psychol. Res. **64**, 149–154 (2000)
49. Fujii, K., Sugi, S., Ando, Y.: Textural properties corresponding to visual perception based on the correlation mechanism in the visual system. Psychol. Res. **67**, 197–208 (2003)
50. Soeta, Y., Nakagawa, S., Tonoike, M., Ando, Y.: Magnetoencephalographic responses corresponding to individual subjective preference of sound fields. J. Sound Vib. **258**, 419–428 (2002)
51. Soeta, Y., Uetani, S., Ando, Y.: Relationship between subjective preference and alpha wave activity in relation to temporal frequency and mean luminance of a flickering light. J. Opt. Soc. Am. A **19**, 289–294 (2002)
52. Soeta, Y., Uetani, S., Ando, Y.: Propagation of repetitive alpha waves over the scalp in relation to subjective preferences for a flickering light. Int. J. Psychophysiol. **46**, 41–52 (2002)
53. Sperry, R.W.: Lateral specialization in the surgically separated hemispheres. In: Schmitt, F.O., Worden, F.C. (eds.) The Neurosciences: Third Study Program. MIT Press, Cambridge (1974) (Chapter 1)
54. Davis, A.E., Wada, J.A.: Hemispheric asymmetry: frequency analysis of visual and auditory evoked responses to non-verbal stimuli. Electroencephalogr. Clin. Neurophysiol. **37**, 1–9 (1974)
55. Galin, D., Ellis, R.R.: Asymmetry in evoked potentials as an index of lateralized cognitive processes: relation to EEG alpha asymmetry. Neuropsychologia **13**, 45–50 (1975)
56. Levy, J., Trevarthen, C.: Metacontrol of hemispheric function in human split-brain patient. J. Exp. Psychol. Hum. Percept. Perform. **2**, 299–312 (1976)
57. Ando, Y.: Theory of temporal and spatial environmental design. In: McGraw-Hill Yearbook of Science & Technology 2009, pp. 384–389. McGraw-Hill, New York (2009)
58. Soeta, Y., Mizuma, K., Okamoto, Y., Ando, Y.: Effects of the degree of fluctuation on subjective preference for a 1 Hz flickering light. Perception **34**, 587–593 (2005)
59. Ando, Y., Noson, D.: Music and Concert Hall Acoustics. Academic, London (1997)

***

Biography

Yoichi Ando received his Ph.D. degree from Waseda University, 1975 (adviser Takeshi Itoh). Early in his career he was an Alexander-von-Humboldt Fellow at the Third Physics Institute, University of Goettingen, where he worked with Manfred Schroeder. He has held a number of positions at Kobe University (Associate Professor, Professor, and Associate Dean) and is currently Professor Emeritus there. He has authored over 130 papers on acoustics and has written several books on architectural acoustics (Concert Hall Acoustics. Springer-Verlag, Heidelberg, 1985; Architectural Acoustics, Blending Sound Sources, Sound Fields, and Listeners. AIP Press/Springer-Verlag, New York, 1998; Auditory and Visual Sensations, Springer-Verlag, New York, 2009). His honors include a "Laurea Honoris Causa" from the University of Ferrara and a member of the Academy of Sciences of the Institute of Bologna.

Peter Cariani's training and work has involved theoretical biology, biological cybernetics, and auditory neuroscience (B.S. 1978, MIT, biology; M.S. 1983, Ph.D. 1989, Binghamton University, systems science). His doctoral work explored epistemic implications of self-constructing adaptive systems (evolutionary robotics). Subsequent work at Eaton-Peabody Laboratory for Auditory Physiology investigated autocorrelation-like neural time codes for pitch, timbre, and consonance and proposed artificial neural timing nets for sound separation. In 2009 he guest-edited Yoichi Ando's book, *Auditory and Visual Sensations.* He is presently a Senior Research Scientist at the Hearing Research Center at Boston University and a Clinical Instructor in Otology and Laryngology at Harvard Medical School. He currently teaches courses at Harvard and MIT related to psychology of music.

Chapter 9
Manfred R. Schroeder: A Personal Memoir, Optimizing the Reflection Phase Grating

Peter D'Antonio

I did not have the privilege of working directly with Manfred Schroeder, but rather I was motivated and inspired by the development of the reflection phase grating in the late 1970s. This research was the spark that ignited my passion for acoustics and converted my avocation for music and recording into a second vocation and the birth of RPG Diffusor Systems, Inc.

This story begins in 1980, in the conference room of the Laboratory for the Structure of Matter at the Naval Research Laboratory (NRL) in Washington, DC, Fig. 9.1, where I was employed as a diffraction physicist. Knowing my interest in music, a colleague handed me the latest issue of Physics Today, Fig. 9.2, with a cover photo of Manfred Schroeder seated in an anechoic chamber. This was my first virtual meeting with Manfred Schroeder. The article suggested using number theoretic diffusors in concert halls to provide lateral reflections. While my interest at the time was not in concert halls, and in fact at this time my only link to the field of acoustics was a love of composing, recording and performing music, I became fascinated with the thought of using these diffusors in a renovation of Underground Sound, Fig. 9.3, a private recording studio I originally built in 1972 with Jerry Ressler, a colleague and fellow musician. The acoustic renovation utilized a new concept called Live End Dead End proposed by Don and Carolyn Davis, of Synergetic Audio Concepts ["The LEDE concept for the control of acoustic and psychoacoustic parameters in recording control rooms," J. Audio Eng. Soc., 28, 585–95 (1980)].

At NRL, I was examining the three-dimensional structure of matter in various phases, using electron, X-ray, and neutron diffraction techniques, Fig. 9.4. I shared the article with John Konnert, a colleague in my group, and it became apparent that the "reflection phase gratings" suggested by Schroeder were in effect,

P. D'Antonio (✉)
RPG Diffusor Systems, Inc., Upper Marlboro, MD 20774, USA
e-mail: pdantonio@rpginc.com

N. Xiang and G.M. Sessler (eds.), *Acoustics, Information, and Communication*, Modern Acoustics and Signal Processing, DOI 10.1007/978-3-319-05660-9_9,

Fig. 9.1 U.S. Naval Research Laboratory, Washington, DC

two-dimensional sonic crystals, which scatter sound in the same way that three-dimensional crystal lattices scatter electromagnetic waves, Fig. 9.4. Since the diffraction theory employed in X-ray crystallographic studies were applicable to reflection phase gratings, it was straightforward for us to model and design the reflection phase gratings. The central theoretical connection was the coherent diffraction equation developed by Sir Lawrence Bragg, shown in Fig. 9.4, where the product of two times the periodicity repeat, d, and the sin of the diffraction direction, θ, equaled the wavelength, λ, times and integer n.

Having scientific backgrounds, John Konnert and I approached acoustics as we did the field of diffraction physics and began researching and publishing findings in the scientific literature. The Audio Engineering Society and Syn-Aud-Con offered a unique forum and community for discussing the research. In October 1983, at the 74th AES Convention in New York, I presented our research on the Schroeder diffusor in the Studio Design Session C, shown in Fig. 9.5, with a bit of intimidation, because Manfred Schroeder was the lead-off invited speaker. As part of this presentation, I described an Apple II program, shown in Fig. 9.6, which allowed acousticians to design these phase gratings, as well as plot their diffraction patterns. Following the session, Manfred and I had our first personal meeting where he enlightened me on the use of the Chinese Remainder Theorem, which enabled the creation of a two-dimensional primitive root sequence from a longer one-dimensional sequence, maintaining the beneficial Fourier property of a flat power spectrum. In Fig. 9.7, I show his handwritten notes in the 74th Technical Meeting & Professional Exhibits AES Program Oct. 8–12, 1983, illustrating the diagonal filling process, making use of periodicity, for several primes, N, for which $N-1$ could be factored into two relative coprimes, i.e., it cannot be used for $N=5$.

Fig. 9.2 Cover of Physics Today, October 1980, where I first was introduced to Manfred Schroeder

Two periods of a reflection phase grating, with $N = 17$ divided wells and well width W, are shown in Fig. 9.8. The angles of incidence and diffraction are α_i and α_d, respectively. There are three aspects of the phase grating that are important, namely the number of periods, the number of wells, and their relative depth with respect to a reference surface plane. Any periodic surface, a diffraction grating, a crystal or a reflection phase grating, scatters sound coherently when a certain condition is satisfied, as described in Fig. 9.9. Incident ray AB is reflected as BD. Incident ray EG is reflected as ray GH. When the difference in path length BC-FG is equal to an integral number of wavelengths, mλ, coherent scattering occurs in the diffraction direction α_d, according to Eq. 1, where N is an odd prime and W is the width of a well.

$$\sin \alpha_\mathrm{d} = \frac{\mathrm{m}\lambda}{\mathrm{NW}} - \sin \alpha_\mathrm{i} \qquad (9.1)$$

In far field theory, sin α_i is assumed to be 0. The second and third aspects pertaining to the number of wells and their relative depths, is where Schroeder's

Fig. 9.3 Underground Sound Recording Studios, where my interest in diffusion was born

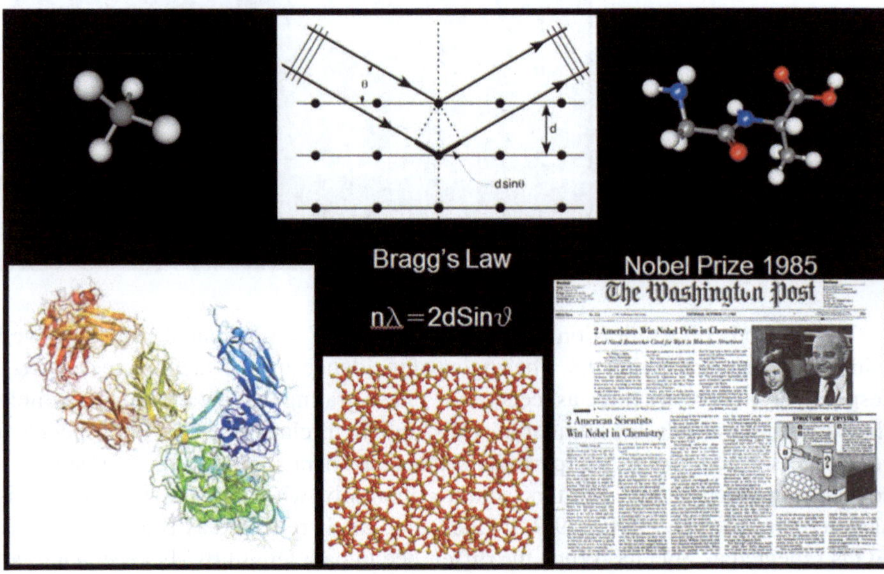

Fig. 9.4 A collage of matter in gas, crystalline, and amorphous phases that I was studying in my work as a diffraction physicist at NRL, in a group headed by Nobel Laureate Dr. Jerome Karle. Bragg's Law was directly transferrable from crystallography to the reflection phase grating

insight and genius came into play. Being a proficient mathematician, he was very aware of the power and "magic" of prime numbers. He often marveled at the unreasonable effectiveness of number theory in science and communication and described numerous applications in his books. Much of this early research was done by Carl Friedrich Gauss in the eighteenth century.

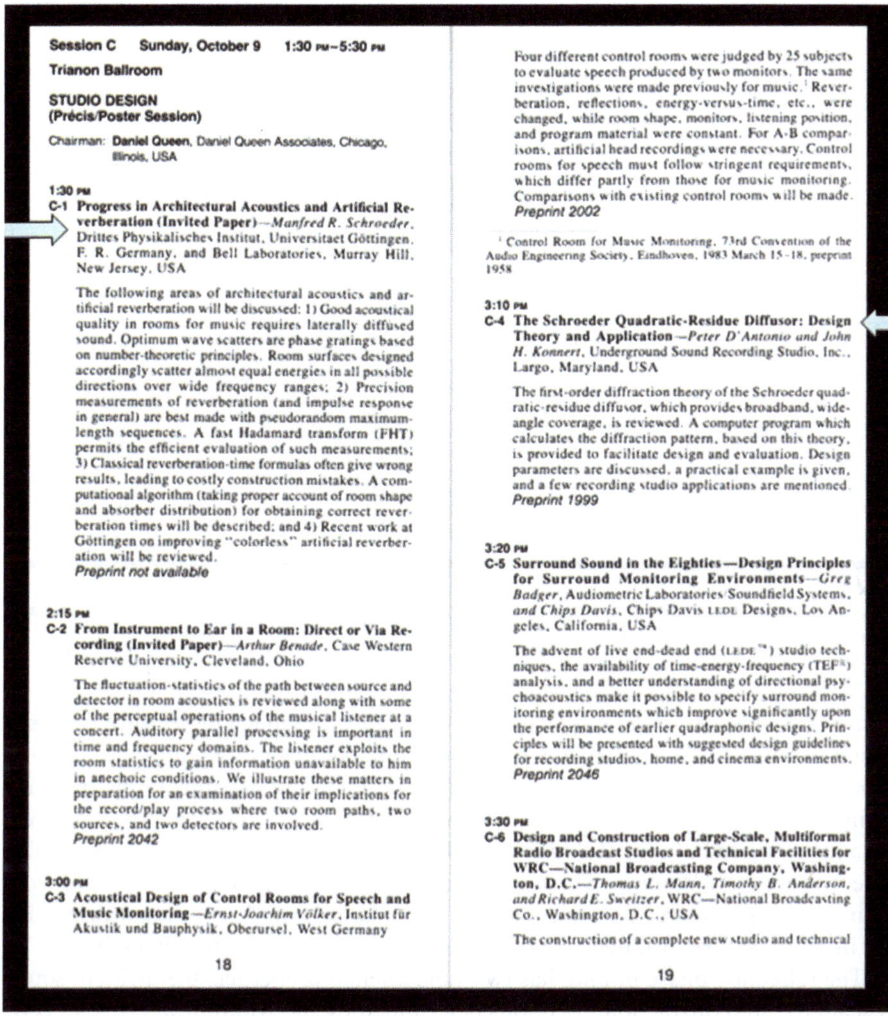

Session C Sunday, October 9 1:30 PM–5:30 PM
Trianon Ballroom

STUDIO DESIGN
(Précis/Poster Session)

Chairman: **Daniel Queen**, Daniel Queen Associates, Chicago, Illinois, USA

1:30 PM
C-1 Progress in Architectural Acoustics and Artificial Reverberation (Invited Paper)—*Manfred R. Schroeder*, Drittes Physikalisches Institut, Universitaet Göttingen, F. R. Germany, and Bell Laboratories, Murray Hill, New Jersey, USA

The following areas of architectural acoustics and artificial reverberation will be discussed: 1) Good acoustical quality in rooms for music requires laterally diffused sound. Optimum wave scatters are phase gratings based on number-theoretic principles. Room surfaces designed accordingly scatter almost equal energies in all possible directions over wide frequency ranges; 2) Precision measurements of reverberation (and impulse response in general) are best made with pseudorandom maximum-length sequences. A fast Hadamard transform (FHT) permits the efficient evaluation of such measurements; 3) Classical reverberation-time formulas often give wrong results, leading to costly construction mistakes. A computational algorithm (taking proper account of room shape and absorber distribution) for obtaining correct reverberation times will be described; and 4) Recent work at Göttingen on improving "colorless" artificial reverberation will be reviewed.
Preprint not available

2:15 PM
C-2 From Instrument to Ear in a Room: Direct or Via Recording (Invited Paper)—*Arthur Benade*, Case Western Reserve University, Cleveland, Ohio

The fluctuation-statistics of the path between source and detector in room acoustics is reviewed along with some of the perceptual operations of the musical listener at a concert. Auditory parallel processing is important in time and frequency domains. The listener exploits the room statistics to gain information unavailable to him in anechoic conditions. We illustrate these matters in preparation for an examination of their implications for the record/play process where two room paths, two sources, and two detectors are involved.
Preprint 2042

3:00 PM
C-3 Acoustical Design of Control Rooms for Speech and Music Monitoring—*Ernst-Joachim Völker*, Institut für Akustik und Bauphysik, Oberursel, West Germany

Four different control rooms were judged by 25 subjects to evaluate speech produced by two monitors. The same investigations were made previously for music.[1] Reverberation, reflections, energy-versus-time, etc., were changed, while room shape, monitors, listening position, and program material were constant. For A-B comparisons, artificial head recordings were necessary. Control rooms for speech must follow stringent requirements, which differ partly from those for music monitoring. Comparisons with existing control rooms will be made.
Preprint 2002

[1] Control Room for Music Monitoring, 73rd Convention of the Audio Engineering Society, Eindhoven, 1983 March 15–18, preprint 1958

3:10 PM
C-4 The Schroeder Quadratic-Residue Diffusor: Design Theory and Application—*Peter D'Antonio and John H. Konnert*, Underground Sound Recording Studio, Inc., Largo, Maryland, USA

The first-order diffraction theory of the Schroeder quadratic-residue diffusor, which provides broadband, wide-angle coverage, is reviewed. A computer program which calculates the diffraction pattern, based on this theory, is provided to facilitate design and evaluation. Design parameters are discussed, a practical example is given, and a few recording studio applications are mentioned.
Preprint 1999

3:20 PM
C-5 Surround Sound in the Eighties—Design Principles for Surround Monitoring Environments—*Greg Badger*, Audiometric Laboratories/Soundfield Systems, *and Chips Davis*, Chips Davis LEDE Designs, Los Angeles, California, USA

The advent of live end-dead end (LEDE™) studio techniques, the availability of time-energy-frequency (TEF®) analysis, and a better understanding of directional psychoacoustics make it possible to specify surround monitoring environments which improve significantly upon the performance of earlier quadraphonic designs. Principles will be presented with suggested design guidelines for recording studios, home, and cinema environments.
Preprint 2046

3:30 PM
C-6 Design and Construction of Large-Scale, Multiformat Radio Broadcast Studios and Technical Facilities for WRC—National Broadcasting Company, Washington, D.C.—*Thomas L. Mann, Timothy B. Anderson, and Richard E. Sweitzer*, WRC—National Broadcasting Co., Washington, D.C., USA

The construction of a complete new studio and technical

18 19

Fig. 9.5 Program for the Studio Design Session C, at the 74th AES Convention in New York, where I first met Manfred Schroeder

I would like to digress a bit to present a statement made by the New York Times writer John Tierney in which he stated that "No matter how its practitioners of mathematics try to deliberately ignore the physical world, they consistently produce the best tools for understanding it." A few examples help to emphasize this idea. The Greeks decide to study a strange curve called an ellipse and 2000 years later astronomers discover that it describes the orbits of the planets. In 1854 Bernhard Riemann conjectured that it's not possible to draw two parallel lines ad infinitum and described curved space, which 60 years later Einstein announced as the shape of the universe. In the eighteenth century in Gottingen, Carl Friedrich Gauss

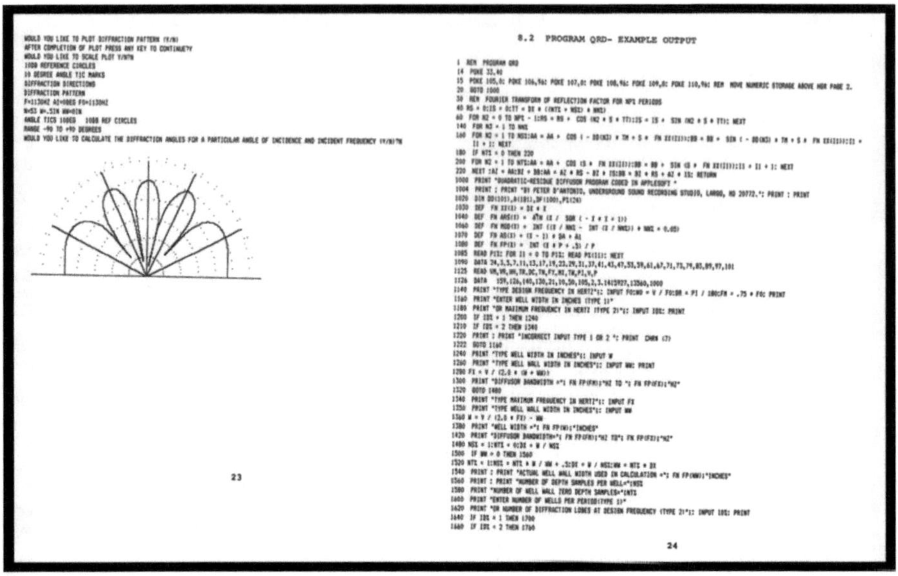

Fig. 9.6 Apple II program presented at the Studio Design Session C, which calculated the well depths and the diffraction pattern of a QRD

discovered quadratic residues, quadratic reciprocity and much more, with no application in mind. In 1975 Schroeder introduced number theory into the world of room acoustics from simple binary m-sequences to multivariate sequences, one of which being the quadratic residue sequence, with good autocorrelation properties and broader bandwidth. In his 1987 Rayleigh Lecture, the topic was "The Unreasonable Effectiveness of Number Theory in Science and Communication." Schroeder pointed out that in wave interference it is not the path differences that determine the interference pattern, but the residues after dividing by the wavelength.

The reflection phase grating has two fundamental properties, the incident sound is scattered into diffraction directions determined by the width of the period, NW, and the energy in the diffraction directions is equal, because the exponentiated well depths have a flat power spectrum. The first property of grating lobes from periodic surfaces is well established in optics, i.e., diffraction gratings. However, the second property, namely the uniformity of the energy in the diffraction lobes, is where Schroeder made one of his brilliant realizations. Flat surfaces reflect energy preferentially in one direction, the specular direction. Schroeder realized that one way to scatter sound uniformly into all of the diffraction lobes was to create a periodic scattering surface consisting of divided wells whose depths were based on the number theory sequences that Gauss developed, e.g., the quadratic residue sequence. These reflection phase gratings, based on quadratic residue sequences shown in Fig. 9.10, have the unique property that the energy in the seven diffraction

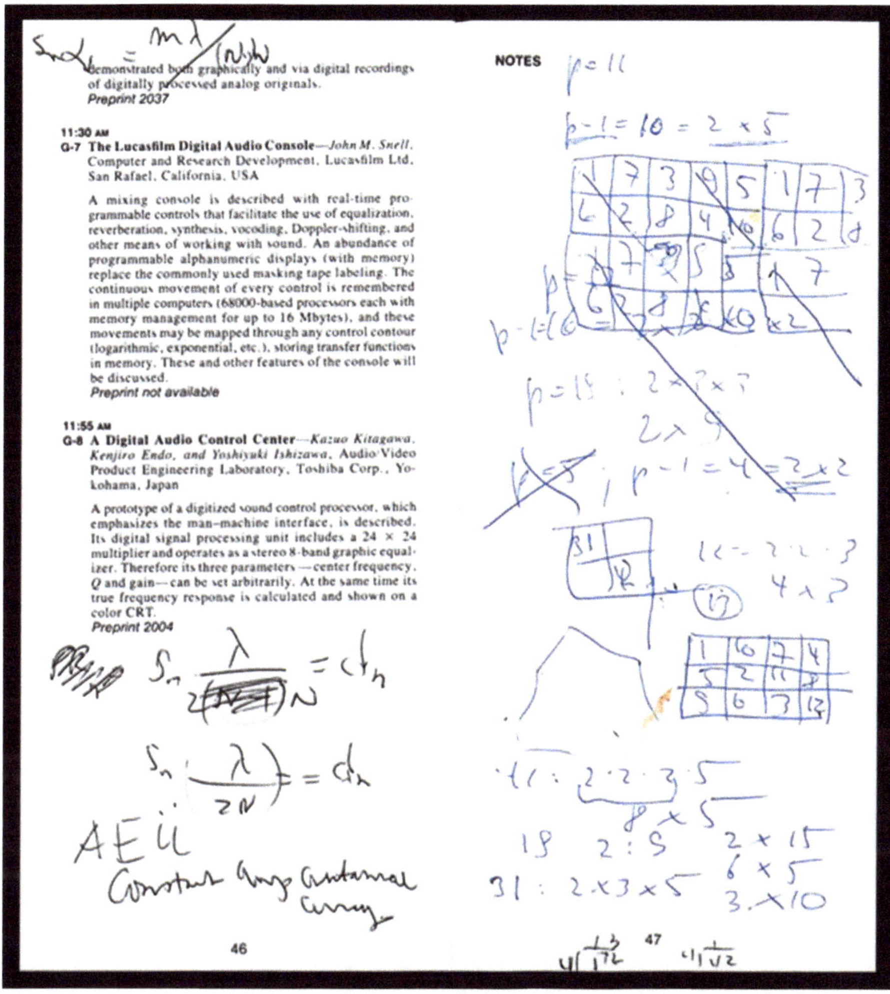
demonstrated both graphically and via digital recordings of digitally processed analog originals.
Preprint 2037

11:30 AM
G-7 The Lucasfilm Digital Audio Console—*John M. Snell*, Computer and Research Development, Lucasfilm Ltd, San Rafael, California, USA

A mixing console is described with real-time programmable controls that facilitate the use of equalization, reverberation, synthesis, vocoding, Doppler-shifting, and other means of working with sound. An abundance of programmable alphanumeric displays (with memory) replace the commonly used masking tape labeling. The continuous movement of every control is remembered in multiple computers (68000-based processors each with memory management for up to 16 Mbytes), and these movements may be mapped through any control contour (logarithmic, exponential, etc.), storing transfer functions in memory. These and other features of the console will be discussed.
Preprint not available

11:55 AM
G-8 A Digital Audio Control Center—*Kazuo Kitagawa, Kenjiro Endo, and Yoshiyuki Ishizawa*, Audio/Video Product Engineering Laboratory, Toshiba Corp., Yokohama, Japan

A prototype of a digitized sound control processor, which emphasizes the man-machine interface, is described. Its digital signal processing unit includes a 24 × 24 multiplier and operates as a stereo 8-band graphic equalizer. Therefore its three parameters—center frequency, Q and gain—can be set arbitrarily. At the same time its true frequency response is calculated and shown on a color CRT.
Preprint 2004

46

NOTES

47

Fig. 9.7 Schroeder's hand written notes in the back of the 74th Technical Meeting & Professional Exhibits Program describing the Chinese Remainder Theorem to me

directions shown is equal. In Fig. 9.11, d_n are the well depths based on the sequence values S_n, the wavelength λ and the prime N. k is the wavenumber, α_i and α_d are the angles of incidence and diffraction, $R(x)$ is the reflection factor, and the Fourier transform of the reflection factor yields a constant energy, $|p(k)|^2$ equal to $1/N$ for all of the diffraction orders. Another way to think of this is that the autocorrelation of the sequence values is zero except for the zero shift modulo N. And it is well known that the spectrum of a two valued autocorrelation is flat. In this case, the frequencies are spatial frequencies or directions.

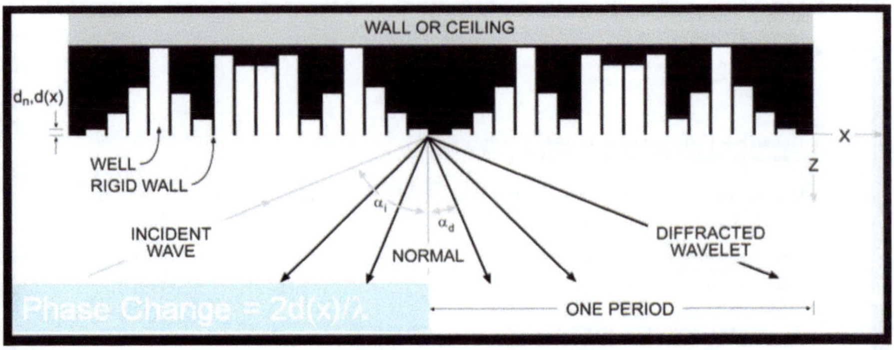

Fig. 9.8 Two periods of a QRD $N = 17$ diffusor showing incident and diffracted waves

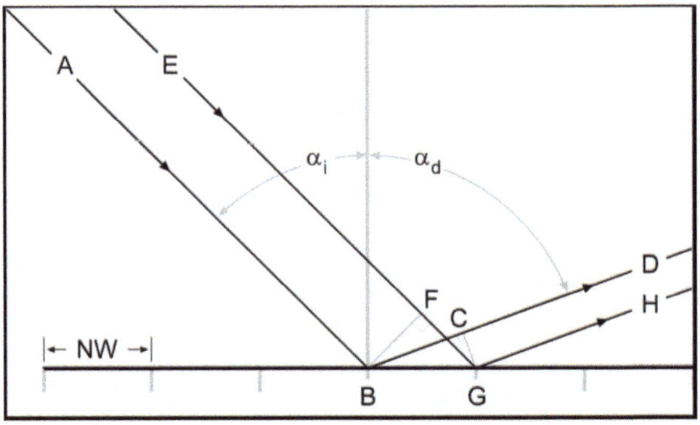

Fig. 9.9 Construction illustrating constructive interference condition in which BC-FG = mλ

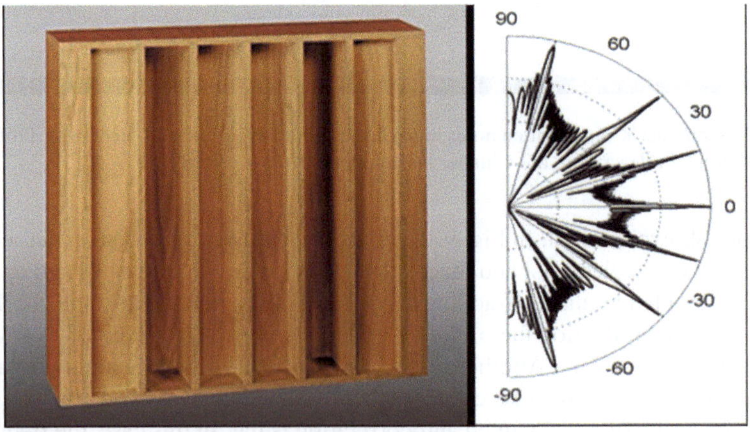

Fig. 9.10 *Left*: commercial QRD $N = 7$. *Right*: diffraction pattern at 3,000 Hz for 50 periods where the energy is concentrated in the diffraction directions (Taken from "*Acoustic Absorbers and Diffusers: Theory, Design and Application*", T.J. Cox and P. D'Antonio, Taylor & Francis, 2nd Edition (2009))

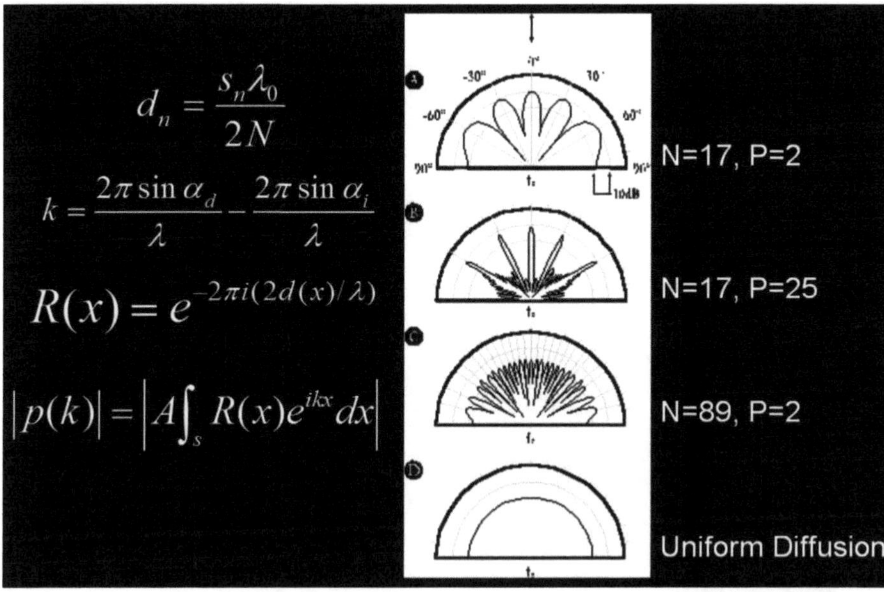

Fig. 9.11 Equations describing the well depths, d_n, wavenumber, k, reflection factor, $R(x)$, and scattered pressure, $p(k)$, along with the diffraction patterns for several prime numbers, N, and repeats, P

Following my presentation and meeting with Schroeder, I met Bob Todrank at an evening reception. Bob was designing a new studio for the Oak Ridge Boys in Hendersonville, TN and was interested in utilizing these new acoustical surfaces. The Oak Ridge Boy's Acorn Sound Recorders project, Fig. 9.12, was celebrated with a Syn-Aud-Con control room design workshop in 1984. This project was a resounding success and turned out to be a harbinger of many exciting things to come. It also led to many other projects and collaborations with a growing community of new studio designers. Use in recording studios soon led to broadcast studios, high end listening rooms, worship spaces, and eventually to performance spaces and schools.

In 1983, I carried out the first measurements of quadratic residue and primitive root diffusors with a TEF 10 analyzer at a Syn-Aud-Con seminar in Dallas, Texas, with the assistance of Don Eger of Techron, shown in Fig. 9.13. In 1984, an intensive measurement program was carried out using Richard Heyser's time delay spectrometry. Farrell Becker was very helpful in the initial evaluation of these exciting new surfaces. Not having access to an anechoic chamber, a boundary measurement technique was developed. These measurements were initially carried out at full scale in large spaces, like open fields and parking lots, eventually moving indoors to a sports arena, a motion picture sound stage, and a local high school gymnasium. The measurements enabled the theories to be validated.

Fig. 9.12 First application of commercial diffusors at Acorn Sound Recorders, Hendersonville, TN in 1984

It was clear that to properly evaluate these surfaces, a standard needed to be created. This turned out to be a 28-year process! The diffusion coefficient is now standardized as ISO 17497-2. The goal was to measure the scattered polar responses and extract from these data a diffusion coefficient, which was a measure of how uniformly these surfaces scattered sound versus frequency, as a complement to the absorption coefficient. In the early 1980s polar response measurements were made by measuring impulse responses one at a time, as shown in the left panel of Fig. 9.14, from a loudspeaker at a given angle of incidence to 37 microphones separated by 5°. This was an incredibly laborious process, but yielded polar responses that allowed evaluation of these early surfaces. As RPG began to grow from a cottage industry, it became necessary to make these measurements routinely, so a 1:5 scale boundary plane measurement goniometer was built, using a microphone switcher, shown in the right panel of Fig. 9.14. Under computer control, the TEF analyzer emitted 37 sequential MLS test signals and the switcher automatically switched to adjacent microphones. This was a great time savings and eliminated the need to constantly find large open spaces to make full scale measurements. As computer hardware evolved, it became possible in 2011 to measure all of the observation positions, for a given angle of incidence, simultaneously with one MLS test signal. This setup is shown in Fig. 9.15, using

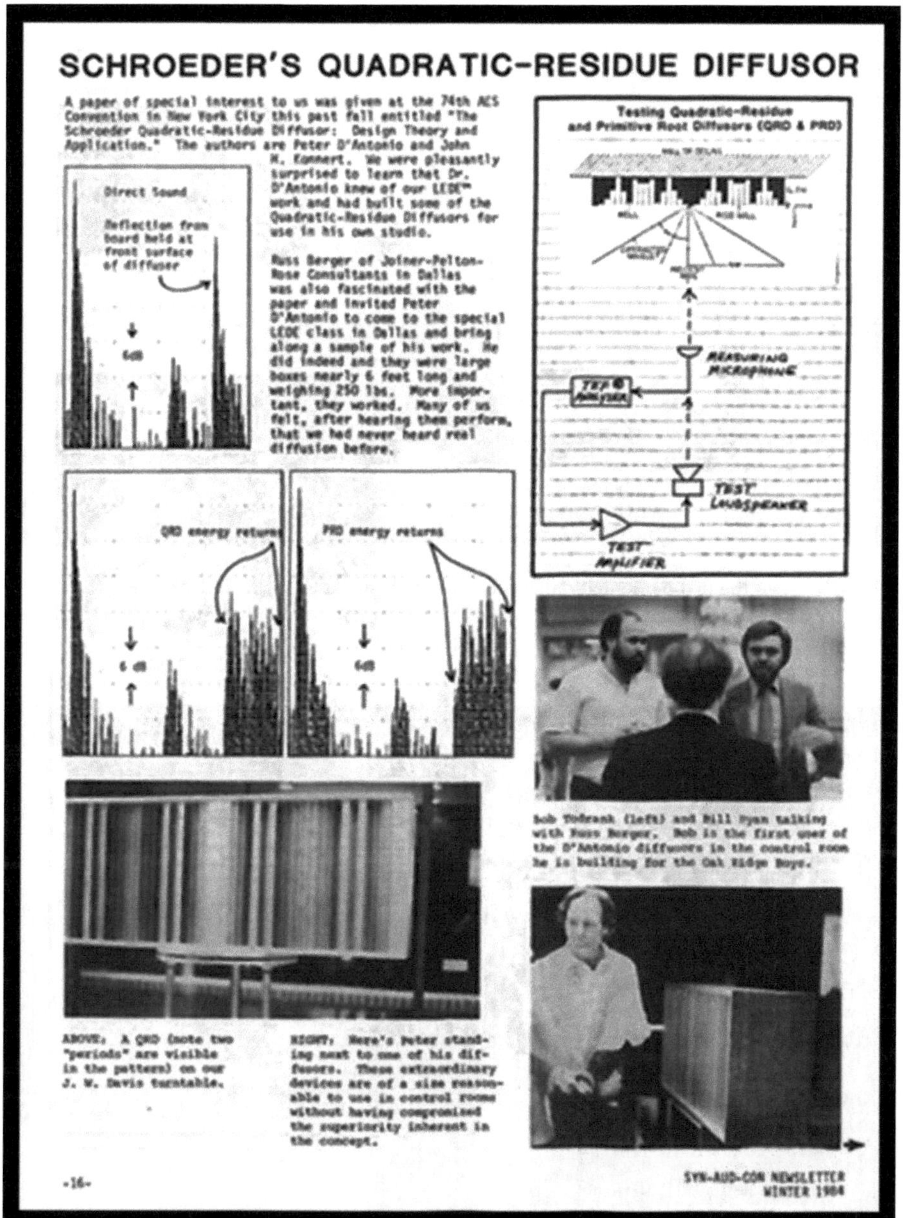

SCHROEDER'S QUADRATIC-RESIDUE DIFFUSOR

A paper of special interest to us was given at the 74th AES Convention in New York City this past fall entitled "The Schroeder Quadratic-Residue Diffusor: Design Theory and Application." The authors are Peter D'Antonio and John H. Konnert. We were pleasantly surprised to learn that Dr. D'Antonio knew of our LEDE™ work and had built some of the Quadratic-Residue Diffusors for use in his own studio.

Russ Berger of Joiner-Pelton-Rose Consultants in Dallas was also fascinated with the paper and invited Peter D'Antonio to come to the special LEDE class in Dallas and bring along a sample of his work. He did indeed and they were large boxes nearly 6 feet long and weighing 250 lbs. More important, they worked. Many of us felt, after hearing them perform, that we had never heard real diffusion before.

Bob Todrank (left) and Bill Ryan talking with Russ Berger. Bob is the first user of the D'Antonio diffusors in the control room he is building for the Oak Ridge Boys.

ABOVE: A QRD (note two "periods" are visible in the pattern) on our J. W. Davis turntable.

RIGHT: Here's Peter standing next to one of his diffusors. These extraordinary devices are of a size reasonable to use in control rooms without having compromised the superiority inherent in the concept.

-16-

SYN-AUD-CON NEWSLETTER
WINTER 1984

Fig. 9.13 First measurements of QRD and PRD commercial diffusors, using the TEF Analyzer in the Winter 1984 issue of the Syn-Aud-Con Newsletter

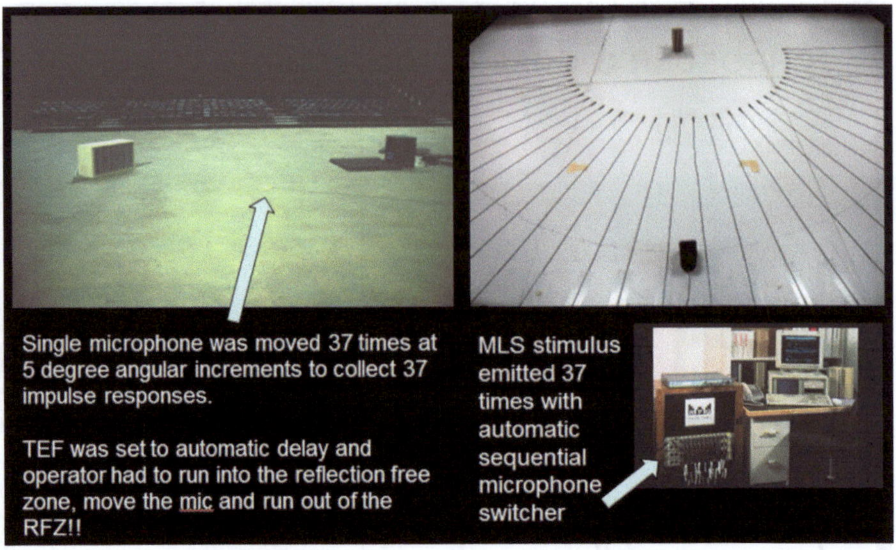

Fig. 9.14 *Left*: the first goniometer measurements were made full scale one at a time with a TEF analyzer with a microphone radius of 5 m and a speaker radius of 10 m. *Right*: in 1994 a 1:5 scale goniometer was built with a 1 m mic radius and a 2 m speaker radius

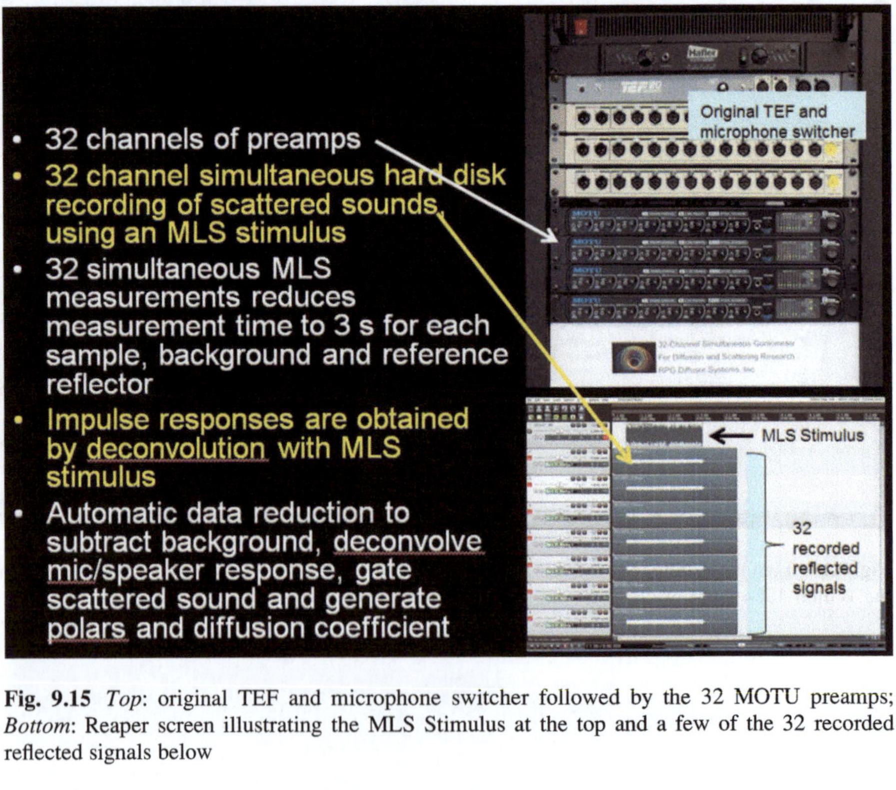

Fig. 9.15 *Top*: original TEF and microphone switcher followed by the 32 MOTU preamps; *Bottom*: Reaper screen illustrating the MLS Stimulus at the top and a few of the 32 recorded reflected signals below

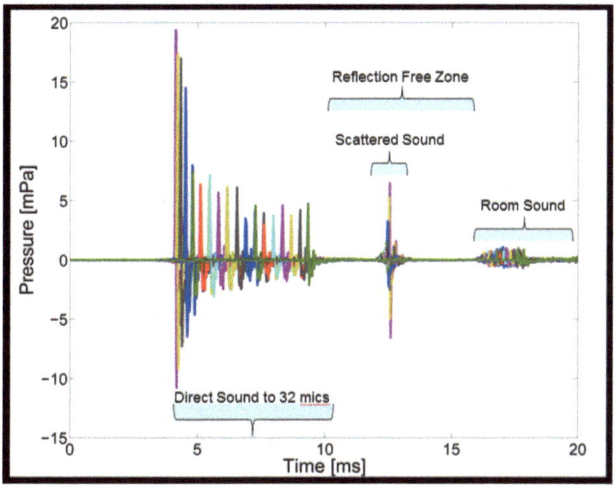

Fig. 9.16 Impulse responses for the 32 direct sounds and scattered reflections, along with the room sound interference

32 microphones. Each microphone was connected to a MOTU preamp and the Firewire output was sent to a computer, which recorded the scattered MLS signals on hard disk. The scattered signals were deconvolved, using the MLS test stimulus, to obtain the impulse responses. The impulse responses for the 32 direct sounds and scattered reflections are shown in Fig. 9.16, along with the room sound interference. The scattered sound was extracted via a multistep process illustrated in the left panel of Fig. 9.17, in which a background response, $h2(t)$, with no sample present, is subtracted from the full impulse response, $h1(t)$, to minimize the direct sound and interfering room reflections. $h2(t) - h1(t)$ is then deconvolved with the loudspeaker/microphone response, $h3(t)$, to yield $h4(t)$ and windowed to isolate the scattered impulse response. In the right panel of Fig. 9.17, we show the entire process leading to the diffusion coefficient. (A) the goniometer with a speaker at 150 degrees, (B) the total impulse response at one microphone position, with the scattered sound outlined, (C) the isolated impulse responses at all microphone positions, (D) 5 selected Fourier transforms of the scattered impulse responses and three selected 1/3rd octave polar responses, (E) the diffusion coefficient obtained from the circular autocorrelation of these polar responses, without and with normalization. To remove edge diffraction, the diffusion coefficient of the sample is normalized by the diffusion coefficient of the reference reflector. In Fig. 9.18, we show a photo of a test sample, in this case, three hemicylinders, the diffusion coefficient for the sample and reference reflector and the normalized diffusion coefficient for normal incidence. Below we show the 1/3rd octave polar responses for the sample (red) and the reference reflector (blue).

While the QRD was revolutionary, there were three aspects that we investigated to improve performance, shown in Fig. 9.19. These included extending the bandwidth, minimizing the effect of grating lobes, i.e., making the response uniform, and lastly

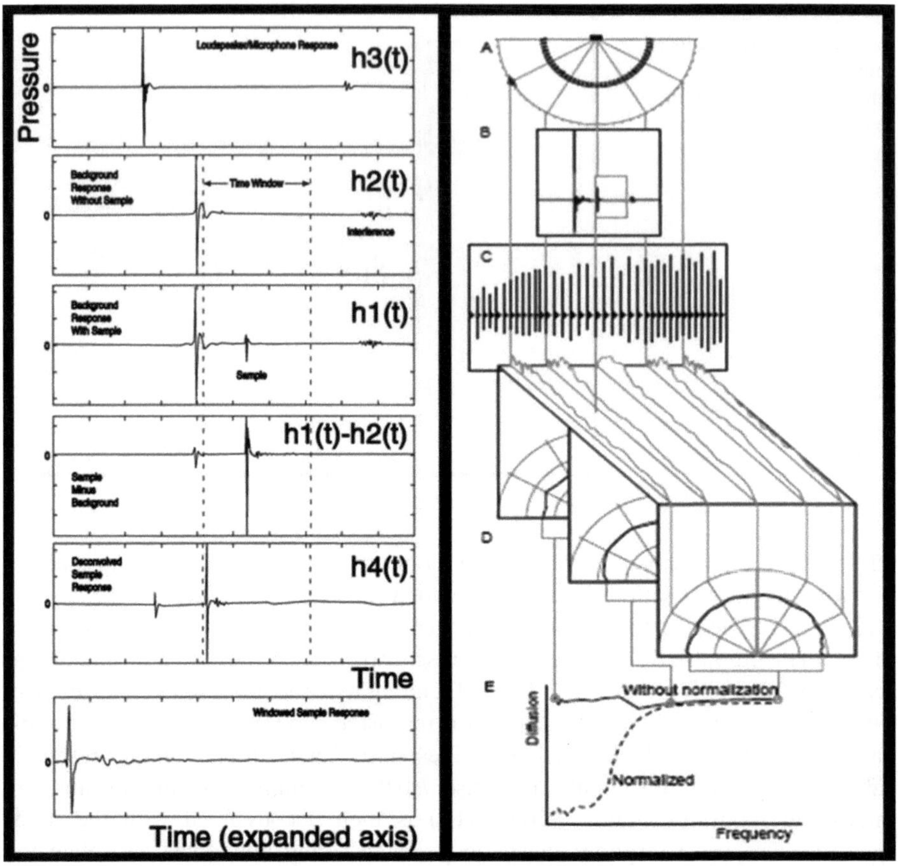

Fig. 9.17 *Left*: data reduction procedure to extract scattered sound at a given microphone position from a given angle of incidence; *Right*: complete process to determine the diffusion coefficient from the scattered impulse responses (Taken from "*Acoustic Absorbers and Diffusers: Theory, Design and Application*", T.J. Cox and P. D'Antonio, Taylor & Francis, 2nd Edition (2009))

eliminating the quantized well depth effect, which results in a specular reflection at the frequency where all wells scatter in phase. For a QRD, these frequencies occur at integer multiples of the prime multiplied by the design frequency.

When considering how to expand the bandwidth, I was intrigued by the idea of the self-similarity of fractals and proposed nesting, scaled versions of the QRD forming a self-similar design, in which each generation of nesting would cover different frequency ranges [P. D'Antonio, "A new 1 or 2-dimensional fractal sound diffusor," J. Acoust. Soc. Am., Suppl. 1, 87, S10]. On my way to an Acoustical Society meeting at Penn State, I accidentally met Manfred Schroeder in the Philadelphia airport and we flew together to the meeting. During the flight we discuss many things, including how his diffusors were being accepted, the success of RPG and the diffusing fractal, now called a Diffractal, which fascinated him. He then contacted Freeman, the publisher of his forthcoming book, *Fractals, Chaos,*

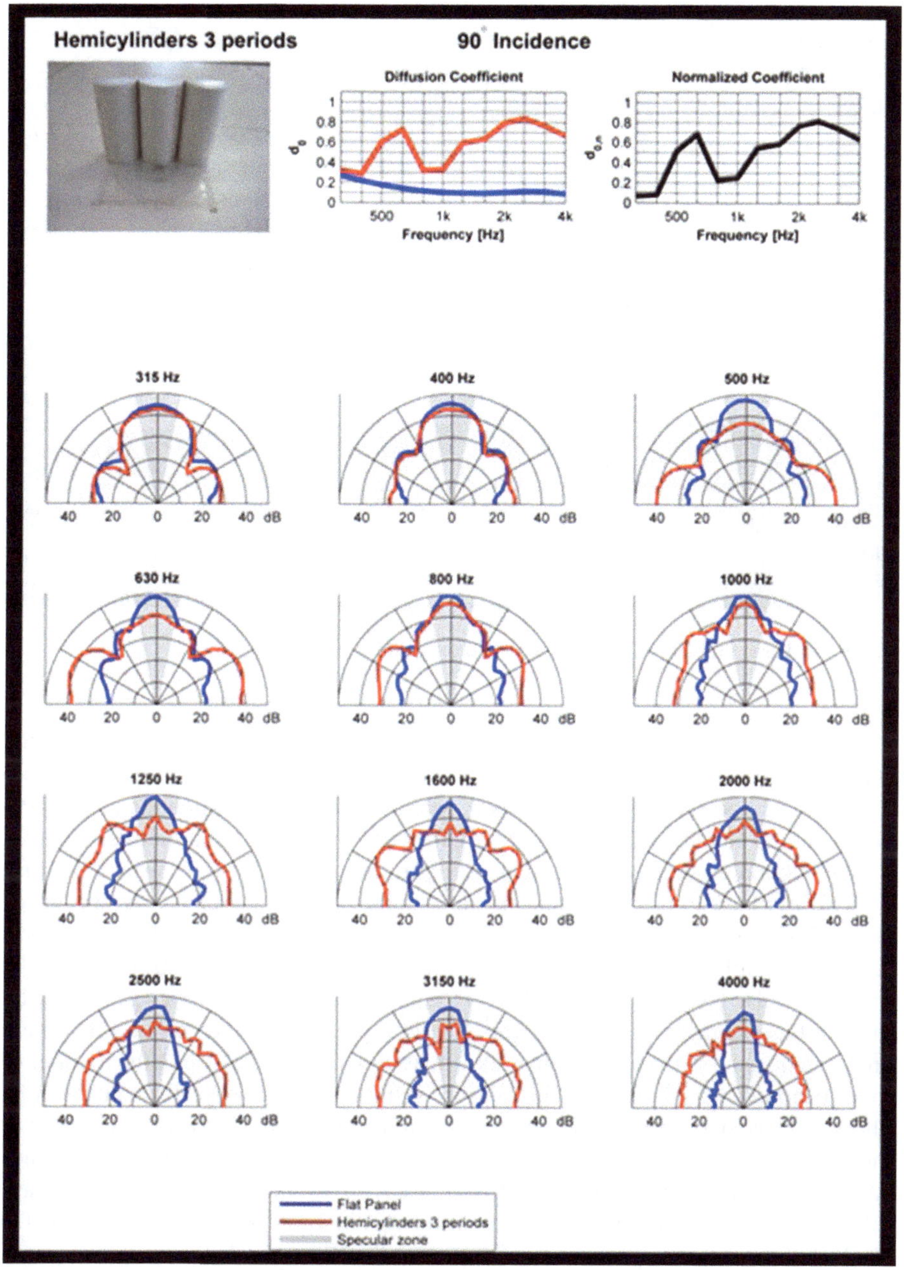

Fig. 9.18 *Top*: photo of 3 hemicylinders, diffusion coefficient of the sample (*red*) and reference reflector (*blue*), normalized diffusion coefficient of sample (*black*). *Bottom*: third octave polar responses of the sample (*red*) and reference reflector (*blue*)

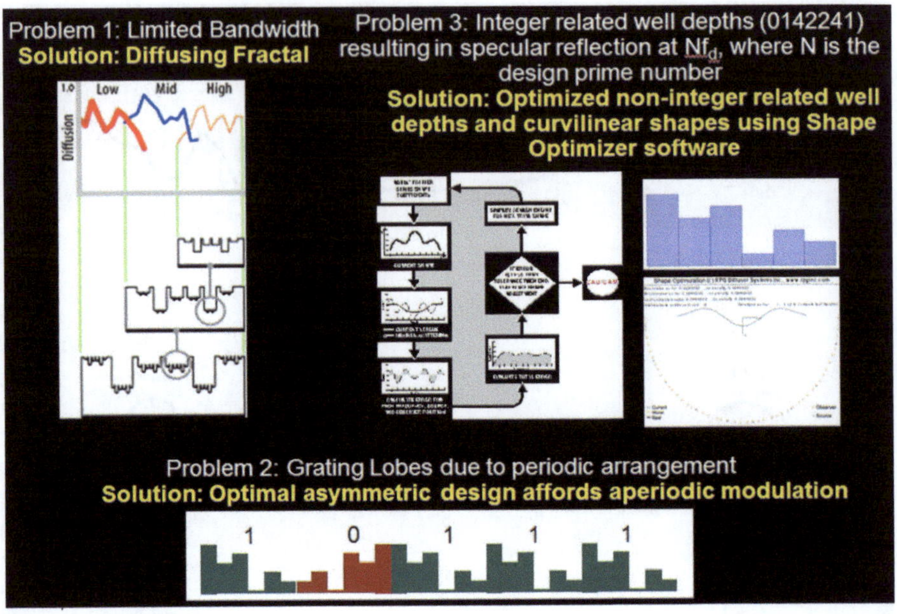

Fig. 9.19 Three problems that were mitigated to optimize the phase grating

Power Laws: Minutes from an Infinite Paradise, and asked them to include the statement at the bottom of Fig. 9.20 prior to publication. During a subsequent Audio Engineering Society Convention in NY, Manfred Schroeder visited me at the RPG booth and he can be seen pointing to a Diffractal with product literature in hand.

The second problem is associated with grating lobes and is very ironic, because the QRD is based on the concept of periodicity, using number theory sequences which insure equal energy in the diffraction directions. Yet to achieve uniform scattering in all directions, a way had to be found to minimize grating lobes! The scattered polar responses in Fig. 9.18 are dominated by grating lobes generated by the fact that the diffusors are periodic. The lobe energy may be constant, but there are large minima between the lobes, except at high frequencies when the number of lobes becomes very large. For this reason, significantly better performance can be obtained if the periodicity lobes can be removed by making the diffusor aperiodic or increasing the repeat distance. It seemed the QRD was cursed by periodicity. James Angus came up [J.A.S. Angus, "Large area diffusors using modulated phase reflection gratings," Proc. 98th Convention Audio Eng. Soc., Preprint 3954, D4 (1995)] with a solution in a series of papers outlining methods for using two phase grating base shapes in a modulation scheme to minimize periodicity. Another approach Trevor Cox and I developed is to form an asymmetric QRD sequence and instead of repeating it periodically, one would follow the prescription of an optimal binary sequence whose aperiodic Fourier transform is as flat as possible.

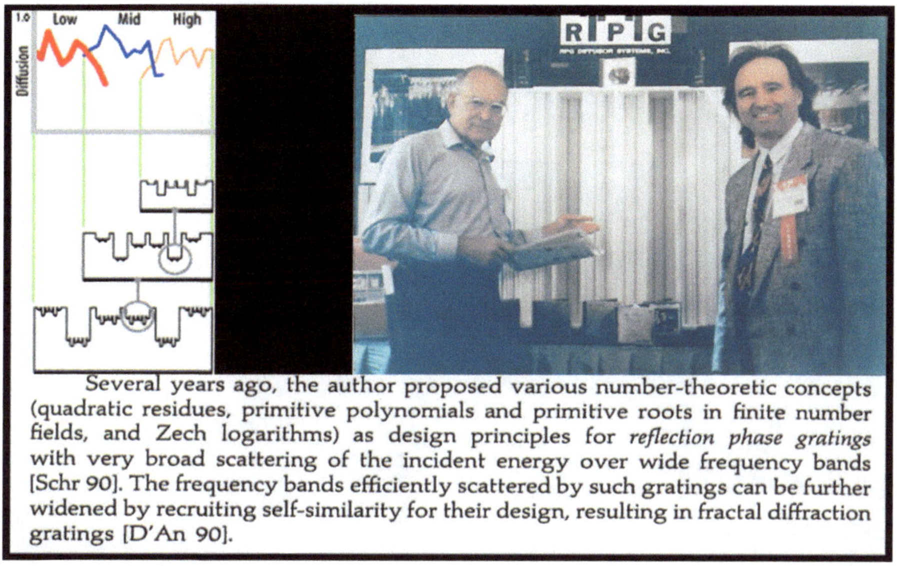

Fig. 9.20 Manfred Schroeder and Peter D'Antonio in the RPG booth at AES in New York in front of a Diffractal along with a mention in his new book with reference [D'An 90]

That is, if the binary sequence is a zero, the base shape is used, if the sequence value is one, the asymmetric QRD is flipped. In this way an aperiodic modulation is formed and grating lobes are minimized.

The last problem deals with specular scattering at a frequency where all of the wells scatter in phase. For the QRD, this frequency is equal to integer multiples of the prime times the design frequency. This occurs because the well depths are integer multiples of one another. To minimize these flat plate frequencies an optimization program was created, in collaboration with Trevor Cox, that combined boundary element prediction, multidimensional optimization techniques, and the diffusion coefficient. It is an iterative program which cycles until the shape produces a desired diffusion coefficient. When used for divided wells or nondivided steps, the goal is to find non-integer related wells or steps, thus avoiding the flat plate frequencies. It can also be used to define a wide range of curvilinear shapes which can complement contemporary architecture, shown in Fig. 9.21.

Thus by utilizing a variety of techniques including, boundary element prediction, multidimensional minimization, defining a diffusion coefficient, fractal geometry, optimal aperiodic modulation, etc. we have been able to optimize Schroeder's seminal idea of an acoustical reflection phase grating. This research has yielded essentially three types of diffusive surfaces, optimized and modulated reflection

Fig. 9.21 A wide variety of curvilinear diffusive shapes obtained with the Shape Optimizer software

phase gratings, planar binary absorption-reflection amplitude gratings, and optimized curvilinear shapes, seen in Fig. 9.22.

It was a great pleasure to meet Manfred Schroeder a few additional times. One was in Rome at the ICA, where Michael Vorlander convened a special session on diffusors. It was a great privilege for me to present a paper outlining the progress we had made in optimizing his Schroeder diffusor. Unfortunately he was recovering from a stroke, but was still in good spirits. Our last meeting was at an ASA meeting in Paris, where we shared some drinks in the hotel lobby and met again briefly in the Louvre.

As a small way to thank Manfred Schroeder for inspiring us and launching our careers in acoustics, Trevor Cox and I dedicated our book *Acoustic Absorbers and Diffusers: Theory, design and application* to him. His thank you letter and a photo of the Second Edition are seen in Fig. 9.23.

Manfred Schroeder had many hobbies and cycling was one of them. For comic relief, I compiled a collage of Trevor Cox, James Angus, Manfred Schroeder, and myself entitled Diffusor Docs, in Fig. 9.24. As a fitting tribute to Manfred Schroeder, the RPG Diffusor was inducted into the music industry's Technology Hall of Fame, in 2013. My closing sentiments are expressed in Fig. 9.25.

Fig. 9.22 Three types of diffusors that evolved from Schroeder's seminal suggestion. *Left* to *right*: a family of 1D and 2D reflection phase grating diffusors; a family of 2D binary amplitude diffusor-absorbers (diffsorbers); and a family of wood and glass reinforced gypsum curvilinear optimized diffusors

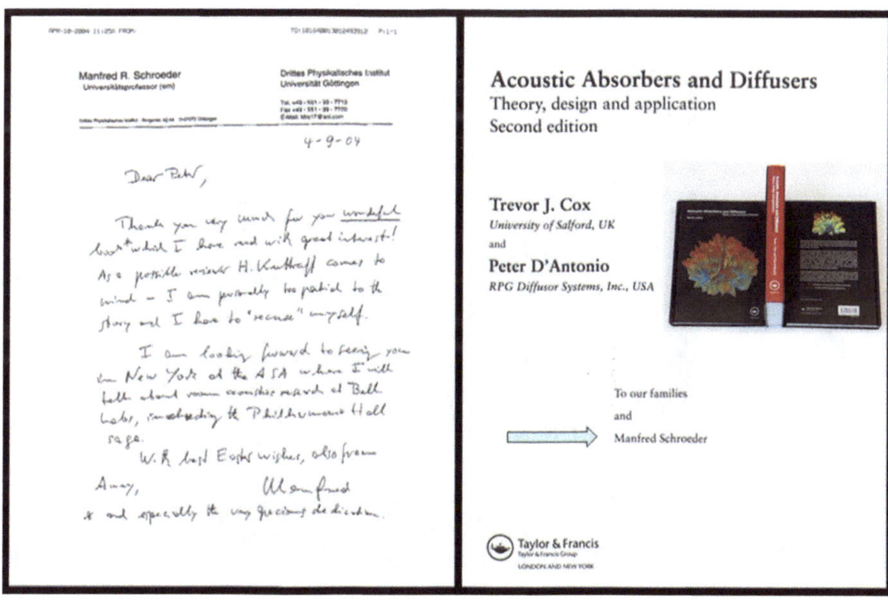

Manfred R. Schroeder
Universitätsprofessor (em)

Drittes Physikalisches Institut
Universität Göttingen

4-9-04

Dear Peter,

Thank you very much for your wonderful book* which I have read with great interest! As a possible reviewer H. Kuttruff comes to mind – I am personally too partial to the story and I have to "recuse" myself.

I am looking forward to seeing you in New York at the ASA where I will talk about room acoustics research at Bell Labs, including the Philharmonic Hall saga.

With best Easter wishes, also from Anny,

Manfred

* and especially the very gracious dedication.

Acoustic Absorbers and Diffusers
Theory, design and application
Second edition

Trevor J. Cox
University of Salford, UK
and
Peter D'Antonio
RPG Diffusor Systems, Inc., USA

To our families
and
Manfred Schroeder

Taylor & Francis
Taylor & Francis Group
LONDON AND NEW YORK

Fig. 9.23 A photo of a book written by Trevor J. Cox and Peter D'Antonio, with a dedication to Manfred Schroeder, along with his thank you letter

Fig. 9.24 Diffusor Docs: *Top left* clockwise, Manfred Schroeder, Trevor Cox, James Angus, and Peter D'Antonio

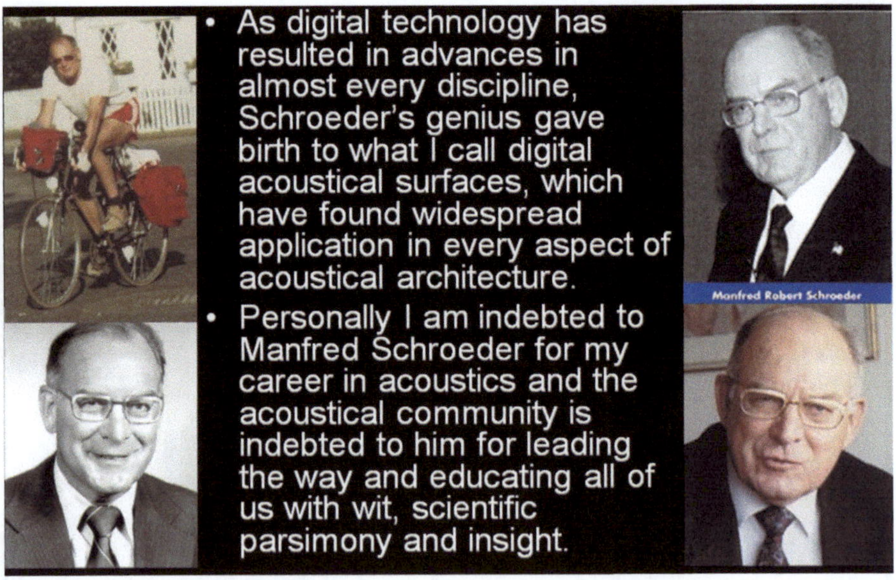

Fig. 9.25 Vielen Dank und Abschied

***

Biography

Dr. **Peter D'Antonio** was born in Brooklyn, NY in 1941 and is Founder/Chairman of RPG Diffusor Systems, established in 1983. He has co-authored *Acoustic Absorbers and Diffusers: Theory, Design and Application,* 2nd Edition, Taylor & Francis 2009 and contributed several chapters to the *Master Handbook of Acoustics,* 5th Edition, McGraw Hill 2009. Dr. D'Antonio served as Chairman of the AES SC-04-02 and is a member of ISO 17497-1 and ISO 17497-2. He is an adjunct professor of acoustics at the Cleveland Institute of Music, since 1991 and a Fellow of the Acoustical Society of America and the Audio Engineering Society.

Chapter 10
Manfred R. Schroeder: Challenge the Present and There Is a Better Way

Bishnu S. Atal

Abstract The chapter author worked with Manfred R. Schroeder for about 25 years at Bell Telephone Laboratories starting in May 1961. Manfred always inspired this author to look for things one has never seen before. Digital computers were emerging in the 1960s with great promise and we used computers to do many new things: simulation of acoustics of concert halls, studying sound decay in rooms using ray tracing, and producing high-quality speech at low bit rates. This chapter will highlight a few examples. Manfred had been deep in the area of vocoder research at that time, but we started on a new track leading us ultimately to code-excited linear prediction. This set the spark for expanding the use of cell phones worldwide. As a manager, Manfred Schroeder was uncompromising in pushing for scientific excellence resulting in success; Bell Labs had the reputation of being the best.

10.1 Introduction

I worked with Manfred Schroeder for about 25 years at Bell Telephone Laboratories starting in May 1961. Manfred always inspired me to look for things one has never seen before. Digital computers were emerging in the 1960s with great promise and we used computers to do many new things: simulation of acoustics of concert halls, studying sound decay in rooms using ray tracing, and producing high-quality speech at low bit rates. This chapter will highlight a few examples. Manfred had been deep in the area of vocoder research at that time, but we started on a new track leading us ultimately to code-excited linear prediction (CELP). This

B.S. Atal (✉)
Department of Electrical Engineering, University of Washington, Seattle, WA, USA
e-mail: bsatal@bishnu.net

N. Xiang and G.M. Sessler (eds.), *Acoustics, Information, and Communication*, Modern Acoustics and Signal Processing, DOI 10.1007/978-3-319-05660-9_10,

set the spark for expanding the use of cell phones worldwide. As a manager, Manfred Schroeder was uncompromising in pushing for scientific excellence resulting in success; Bell Labs had the reputation of being the best.

10.2 Computer Simulation of Acoustics of Concert Halls

In early 1960s, digital computers were primarily used for solving numerical problems. However, it was becoming increasingly obvious that computer simulation was a powerful tool for understanding complicated processes. Manfred Schroeder was a pioneer in using such tools. In one of the early applications of computer simulation, a digital computer (IBM 7099) was programmed to "simulate" sound transmission in rooms [1]. The idea was that one could evaluate the "acoustics" of a new concert hall even before the hall was built. The digital simulation could help to "preaudit" architectural designs before construction and to investigate subjective correlates of a wide variety of reverberation processes. If a design was found to be unsatisfactory, as a result of subjective evaluations of the simulation, modifications could be made in the architectural plans.

Thus, one could start with reverberation-free signals and program a digital computer to add echoes and reverberation with specified delays, spectral content, and decay characteristics. Several output signals could be generated on the computer; these signals when radiated from loudspeakers in an anechoic chamber could produce sound pressure waves at listener's ears resembling those in real halls. Since we have only two ears to listen with, it should be possible to accomplish the result by radiating from only two loudspeakers. If done properly, the acoustical cross-talk between two loudspeakers at the two ears can be eliminated.

Figure 10.1 illustrates the sequence of pulses to be emitted from the two loudspeakers to achieve the desired response from a source S on the right. The details of these operations are contained in [1]. Sound arrivals from about ten different directions were created on the computer and reproduced in the anechoic chamber at Bell Laboratories by means of two loudspeakers located several meters in front of the listener. Although the computations were based on an essentially fixed head position, the experiments showed that head movements up to $\pm 10°$ did not affect the percept; "externalization" (perceiving the sound as originating outside one's head) was perfect. In fact, the synthesis of lateral arrival, including arrivals from $\pm 90°$ was so perfect that listeners often turned their heads to see the nonexistent sound source.

10.3 Study of Sound Decay in Rooms Using Ray Tracing on Computers

Shortcomings of classical reverberation time formulas, named after Sabine, Eyring, and Millington, to predict reverberation time correctly or to calculate absorption coefficients accurately have been known for a long time. Occasionally, one

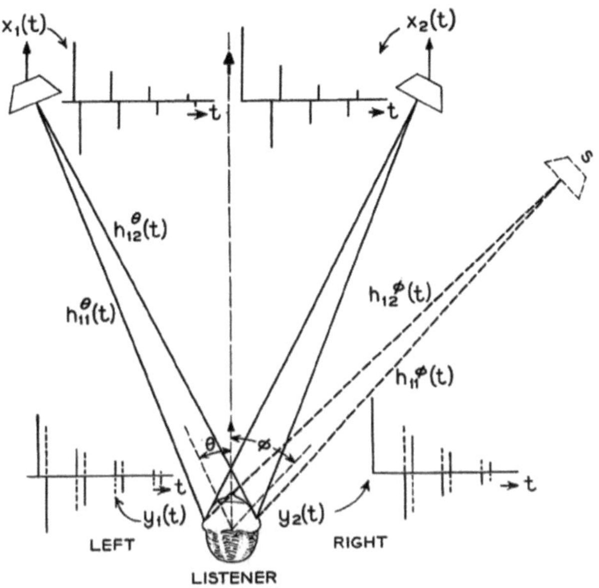

Fig. 10.1 Creation of virtual sound image at an arbitrary lateral direction by means of two loudspeakers in front of the listener. The signals $x_1(t)$, $x_2(t)$ from the two loudspeakers combine at the listener's ears to resemble those from a real lateral sound source S

obtained absorption coefficients greater than 1, a physically impossible result! Such problems were often ascribed to "lack of diffusion" in the reverberation chambers. Manfred and I felt that the basic assumptions that go into these formulas were incorrect.

Another argument raised by many people then was that the existing reverberation theories were based on ray approximation. Since, sound does not travel like rays, no wonder, theory and experimental results do not agree. It turned out that the existing reverberation theories were not correct even under the ray approximation [2, 3].

In order to develop an accurate theory of reverberation based on ray acoustics, I computed sound decays in two-dimensional "rooms" of different shapes and sizes by simulating the propagation of rays on a digital computer. The method of ray tracing on a digital computer is shown in the Fig. 10.2. It shows a two-dimensional enclosure with a patch of absorbing material on one of its walls and an omnidirectional sound source emitting 300 equal-energy rays spaced 1.2° apart (although any other directional characteristic could have been used). The computer program tracked the path of each of the 300 rays following the laws of geometrical acoustics and the energy of each of these rays as they traveled in the room. Each time a ray hit the wall, its energy was diminished by a factor $(1-\alpha)$, where α is the absorption coefficient. The computer program could follow a ray over any desired length of time. The total sound energy after the introduction of a short pulse of sound in the enclosure was obtained by adding the energies of all the component rays.

An example of the sound decay in a two-dimensional non-rectangular enclosure with one of the walls covered with an absorbing material with $\alpha = 0.8$ is shown in

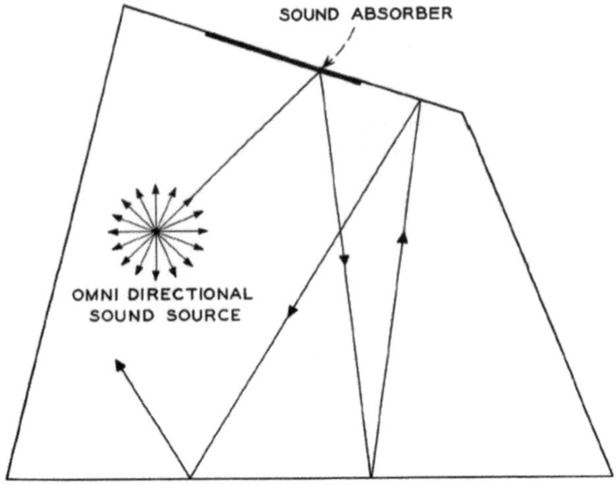

Fig. 10.2 The computer calculates the path of 300 rays from an omnidirectional source. Every time a ray hits an absorbing material, its energy is reduced depending on the absorption coefficient of the material. The computer keeps a running account of the remaining energy

Fig. 10.3. The figure shows a comparison between the decay obtained on the computer (reverberation time $T = 0.38$ s) and the decays predicted by the two-dimensional forms of the Sabine ($T = 0.63$ s), Eyring ($T = 0.56$ s), and Millington formulas ($T = 0.31$ s), respectively. The reverberation time predicted by the Sabine formula is 65 % too large. The Millington value is 17 % too small. The Eyring formula provides an intermediate value with an error of +45 %.

I made a modest beginning in developing a new theory of sound decay in rooms. I will give here a brief outline of steps that provide a more accurate representation of how sound decays in rooms. For simplicity, the case of absorbing material on a single wall is considered. The theory can easily be generalized to case when the absorbing material is spread over more than one wall. Let $\varepsilon(\alpha, t)$ represent the total sound energy in the room at a time t sec after the sound source is stopped. It is easily seen that $\varepsilon(\alpha, t)$ is written as

$$\varepsilon(\alpha, t) = \sum_{k=0}^{\infty} (1-\alpha)^k P(k, t) \tag{10.1}$$

where $P(k, t)$ is the probability of a ray hitting the sound absorber k times in t sec. The logarithmic decay $\ln \varepsilon(\alpha, t)$ can be expressed in terms of cumulants $q_n(t)$ of the probability distribution $P(k, t)$. Thus,

$$\ln \varepsilon(\alpha, t) = \overline{n}(t)\ln(1-\alpha) + \sum_{k=2}^{\infty} \frac{q_k(t)}{k!}[\ln(1-\alpha)]^k \tag{10.2}$$

where $\overline{n}(t) = \sum_{n=0}^{\infty} nP(n, t)$. All of the classical reverberation time formulas can be derived from Eq. (10.2) by substituting the cumulants appropriate to the probability distribution implicit in the formulas. For example, Sabine's formula is

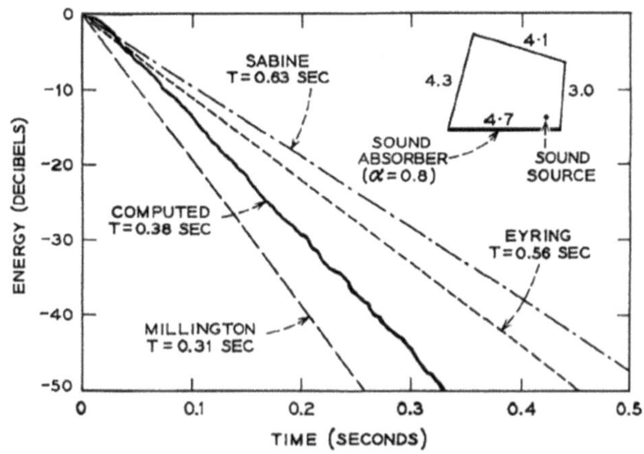

Fig. 10.3 Comparison of sound decay curves found by computer ray-tracing with decay rates predicted by the Sabine, Eyring, and Millington formulas

obtained by setting all the cumulants equal to $\overline{n}(t)$. The above equations can be easily generalized when more walls are covered with absorbing material. The probability distribution and cumulants are replaced by the joint distribution and joint cumulants.

Further analysis of Sabine and Eyring formulas has revealed that the underlying probability distributions implicit in these formulas are based on the assumption that the probability that a ray hits a given wall is independent of the wall the ray came from. This cannot be correct, since a ray cannot hit the same wall twice in succession. The probability of hitting a wall at any reflection depends upon the entire past history of the ray. One can thus regard the Eyring formula as equivalent to a zero-order Markov process. A natural way of improving the Eyring formula will be to consider the effect of taking the first-order Markov dependence into account in developing the theory.

10.3.1 Reverberation Theory Based on First-Order Markov Dependence

Here is a brief description of the new theory. Consider a room with K walls. Let α_i be the absorption coefficient of the absorbing material on wall i. Let p_{ij} be the probability that a ray starting from the wall i hits the wall j at the next reflection. It is reasonably straightforward to compute these probabilities from the geometry of the room. Consider now a matrix **P** with its term in the row i and column j given by p_{ij} and a diagonal matrix **A** with its i-th diagonal term given by $1 - \alpha_i$. Then, each element of the matrix **PA** is nonnegative and the matrix **PA** has an eigenvalue $\lambda_1(\alpha_1, \ldots, \alpha_K)$ which is greater than the absolute value of every other eigenvalue of the matrix **PA** [4]. The reverberation time is given by

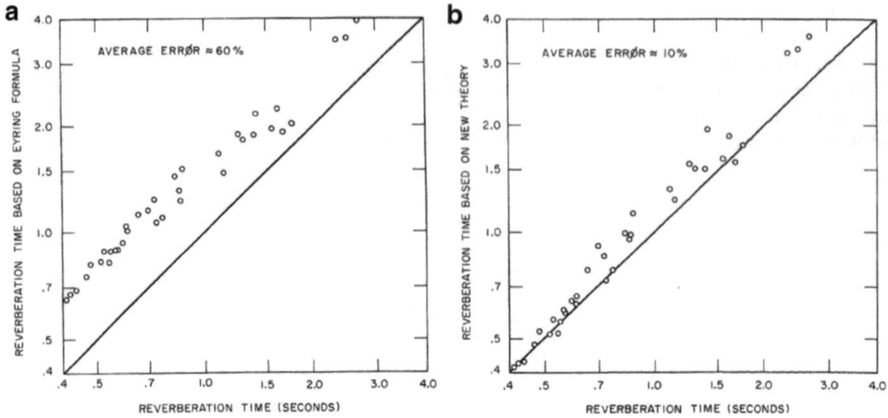

Fig. 10.4 (**a**) Comparison of reverberation times, obtained by computer ray-tracing (abscissa), with reverberation times computed by Eyring formula for 36 different shapes (ordinate) and (**b**) comparison of reverberation times, obtained by computer ray-tracing, with reverberation times computed using the new theory based on first-order Markov dependence for the same 36 shapes used in (**a**)

$$T(\alpha_1, \ldots, \alpha_K) = \frac{-13.82}{\overline{n}\ln[\lambda_1(\alpha_1, \ldots, \alpha_K)]} \tag{10.3}$$

where $\overline{n}$ is the mean collision frequency of the rays. Equation (10.3) reduces to Eyring formula if p_{ij} is independent of i (Fig. 10.4).

10.4 Early Predictive Coding Research

The research in speech coding (or speech bandwidth compression) originated about 75 years ago. At that time, the demand for telephones was growing rapidly. Telephone service was available coast-to-coast, with close to ten million telephones in service. The first under-the-ocean transatlantic telephone cable was still years away. Western Electric asked its research engineers at Bell Telephone Laboratories if voice signals could be transmitted over existing telegraph cables.

The bandwidth required for voice transmission is approximately 3,000 Hz, but the transatlantic telegraph cables had a bandwidth of only a few hundred Hz. Homer Dudley, who had joined Western Electric (parent of Bell Laboratories) in 1921, argued that the real information in speech was carried by telegraph-like low-frequency modulation signals corresponding to slow motion of the vocal organs. Dudley implemented his ideas in a device named "Vocoder," but the signal processing technology available then was primitive and it was very difficult to realize the full potential of his ideas. Although vocoders did not produce speech of quality good enough for telephone application, they were used during World War II by President

Roosevelt and Prime Minister Churchill for secure voice communication. Vocoders remained the central theme of speech coding research for about 35 years.

During the time speech scientists were occupied with vocoders, several major developments of fundamental importance were taking place outside speech research. A major contribution to the mathematical theory of communication of signals was made by Claude Shanon, represented in his published work in 1948 [5]. Shanon's work established a mathematical framework for coding and transmission of signals. In the 1940s, Norbert Wiener developed a mathematical theory for calculating the best filters and predictors for detecting signals hidden in noise.

Wiener worked during the Second World War on the problem of aiming anti-aircraft guns to shoot down an airplane. Since the plane moves, one must predict its position at the time the shell will reach the plane. Wiener's work appeared in his famous monograph [6], "Extrapolation, Interpolation and Smoothing of Time Series," published in 1949.

Linear Prediction is now a major tool in signal processing. Another development was Pulse Code Modulation (PCM), a method for sampling a continuous signal and quantizing each sample into binary digits. It enabled coding of speech with telephone bandwidth at a bit rate of 64 kb/s with negligible loss of quality. PCM marked the beginning of the digital age.

Following the work of Shanon and Wiener, Peter Elias published two papers in 1955 [7] on predictive coding. In predictive coding, both the transmitter and the receiver store the past values of the transmitted signal and from them predict the current value of the signal. The transmitter transmits, not the signal, but the prediction error, that is the quantized difference between the signal and its predicted value. At the receiver, this transmitted prediction error is added to the predicted value to reproduce the signal. For efficient coding, the successive terms of the prediction error should be uncorrelated and the entropy of its distribution should be as small as possible. In predictive coding, the transmitted prediction error is made as white as possible (Fig. 10.5) to achieve coding efficiency. Predictive coding is a remarkably simple concept.

In 1965, as part of my Ph.D. course work at the Polytechnic Institute of Brooklyn, New York, I had to review two papers of my choice and present the material in the class. One of them was Peter Elias's paper on predictive coding, published in 1955. Just a few months later, in 1966, I was one day in Manfred Schroeder's office at Bell Labs when John Pierce brought a tape showing a new speech time compression system. Schroeder was not impressed. After listening to the tape, he said that *there had to be a better way* of compressing speech and mentioned the work in image coding by Chape Cutler at Bell Labs based on differential pulse code modulation (DPCM) technique, which was a simplified version of predictive coding. Our discussions that afternoon kept me thinking. Schroeder's remarks at our meeting made a deep impression. Waiting at the subway station for a train to Brooklyn, I convinced myself that I should do some exploratory investigation to determine if predictive coding could work for speech signals. A first step was to find out if the first-order entropy of the distribution of prediction error signal is significantly smaller than the corresponding entropy of the speech signal; indeed, smaller entropy of the prediction error could produce a lower bit

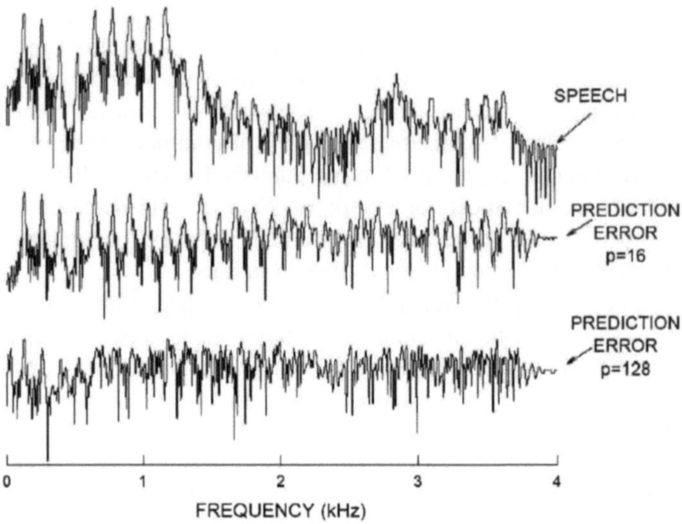

Fig. 10.5 Spectrum of the speech signal (first waveform), the spectrum of the prediction error after prediction with a linear predictor of order 16 (second waveform), and the spectrum of the prediction error after prediction with a linear predictor of order 128 (third waveform). The speech signal was sampled at 8 kHz. The 128th-order linear predictor was able to remove most of the pitch-related fine structure of the signal

rate. The results were encouraging. For speech sampled at 6.67 kHz, the first-order entropy of prediction error turned out to be 1.3 bits/sample as compared to 3.3 bits/sample for the speech signal.

Of course, for a signal that changes rapidly from one speech sound to the next the prediction has to be *adaptive*; hence the name adaptive predictive coding (APC). We demonstrated the first APC system and its speech quality at the IEEE International Conference on Speech Communication held in Boston in 1967 [8], using two-level encoding of the prediction error. The predictive coder produced natural-sounding speech and speech quality was good [9], except for the presence of a low-level crackling noise that could be heard with careful listening over headphones.

10.5 Exploiting Masking for Noise Reduction in Speech Coders

Digital coders invariably introduce errors in the coded speech signal. Our next advance came in 1978 when we started to introduce *perceptual error* criteria. In other words, instead of minimizing a mathematical root-mean-square error, we minimized the *subjective loudness* of the quantizing noise as perceived by the human ear in the presence of the speech signal [10, 11]. This required measurements of loudness and masking was performed in collaboration with J. L. Hall [10],

BISHNU ATAL, right, a member of the Speech and Communications Research Department, Murray Hill, and Manfred Schroeder, head, Hearing and Speech Synthesis Research Department, Murray Hill, have won the Senior Award of the IEEE Acoustics, Speech, and Signal Processing Society for their paper entitled, "Predictive Coding of Speech Signals and Subjective Error Criteria." The paper appeared in the June 1979 issue of the Transactions in Acoustics, Speech and Signal Processing *and describes their recent work on encoding speech efficiently for digital transmission. The award was presented April 10 at the International Conference on Acoustics, Speech, and Signal Processing in Denver.*

Fig. 10.6 Manfred and I were working hard to reduce noise in speech coders to compress speech at low bit rates

to educate us as to the precise causes of audible distortion in a speech signal coded at low bit rates. Our new directions were recognized by IEEE with an award in 1980 (Fig. 10.6). The importance of adjusting the spectrum of the error so as to minimize the subjective distortion is now well established.

10.6 Developments in Speech Coders Leading to CELP

The optimum noise spectrum for minimum subjective distortion is in general non-flat. Early predictive coders designed for minimum subjective distortion used instantaneous quantizers; but instantaneous quantizers are suboptimal. We therefore decided to investigate *non*-instantaneous quantizing. Ideally in speech coding, one is interested in finding a sequence of binary digits which after decoding produces speech with minimal audible distortion. We therefore considered the use of block codes using tree search procedures to determine the optimal code. We first

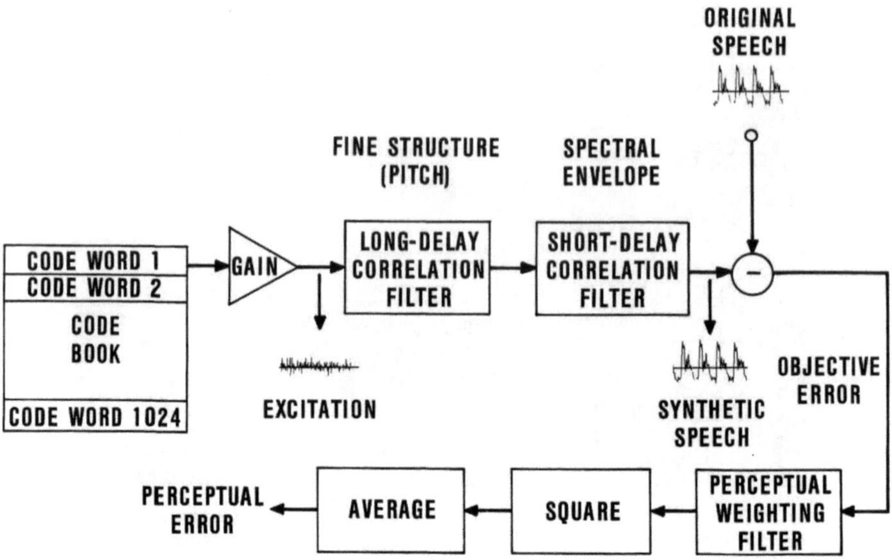

Fig. 10.7 Illustration of the search procedure in a code-excited linear prediction (CELP) coder

employed a binary tree search procedure for coding the excitation at 1 bit/sample which produced speech of high quality.

The use of block codes is necessary when only a fraction of a bit is available for coding each sample of the excitation. Very low bit rates for the excitation are necessary to bring the total bit rate for coding the speech signal down to 8 kb/s or even lower. In code-book coding, a set of possible sequences (may be random) is stored in a code book. For a given speech segment, the optimum sequence is selected by exhaustive search of the code book and an index specifying the optimum sequence is transmitted to the receiver. Code-book coding is often impractical due to the large size of code books. At very low bit rates, however, exhaustive search of the code book became practical, when the processing speed of digital processing chips was increasing three times every 2 years. The search procedure was still computationally very expensive; it took 125 s of Cray-1 CPU time to process 1 s of the speech signal. Here is a summary of various coding methods that we investigated before converging to code-excited linear prediction (CELP) (Fig. 10.7) [12].

[1978] Subjective error criteria
[1979] Exploiting masking properties of human ear
[1981] Tree coding based on rate distortion theory
[1982] Tree coding using block codes
[1982] Multipulse LPC
[1983] Stochastic coding of speech
[1984] Code-excited linear prediction

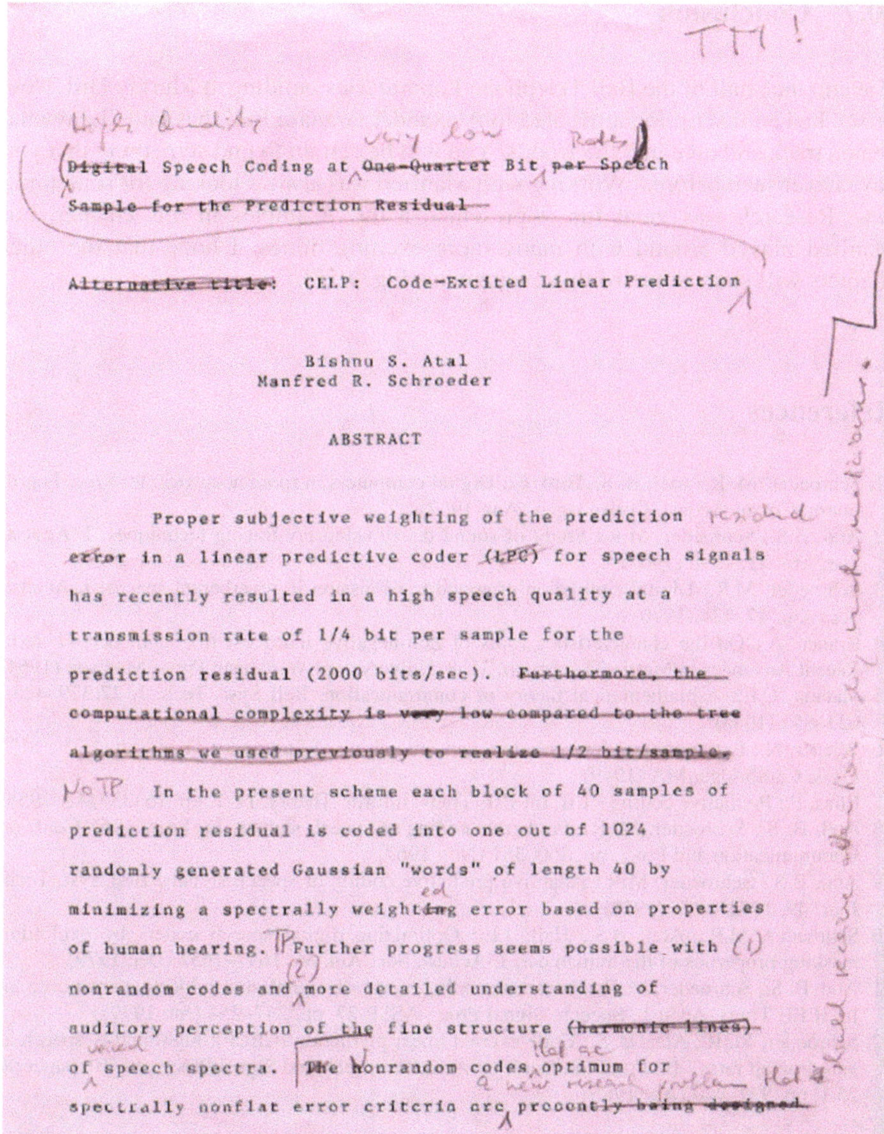
Digital Speech Coding at One-Quarter Bit per Speech Sample for the Prediction Residual

Alternative title: CELP: Code-Excited Linear Prediction

Bishnu S. Atal
Manfred R. Schroeder

ABSTRACT

Proper subjective weighting of the prediction error in a linear predictive coder (LPC) for speech signals has recently resulted in a high speech quality at a transmission rate of 1/4 bit per sample for the prediction residual (2000 bits/sec). Furthermore, the computational complexity is very low compared to the tree algorithms we used previously to realize 1/2 bit/sample.

In the present scheme each block of 40 samples of prediction residual is coded into one out of 1024 randomly generated Gaussian "words" of length 40 by minimizing a spectrally weighting error based on properties of human hearing. Further progress seems possible with nonrandom codes and more detailed understanding of auditory perception of the fine structure (harmonic lines) of speech spectra. The nonrandom codes optimum for spectrally nonflat error criteria are presently being designed.

Fig. 10.8 Corrections from Manfred on the first manuscript of the code-excited linear prediction (CELP) coder

10.7 Conclusion

The entrance hall of the Bell Telephone Laboratories building at Murray Hill, New Jersey had an inscription attributed to Alexander Graham Bell that said "Leave the beaten track and dive into the woods. You will be certain to find something that you have never seen before." Working with Manfred was always looking for something new. Research was great fun. I presented a few examples in this chapter, but Manfred played around with many more exciting things. I hope that the other chapters will share some of this excitement (Fig. 10.8).

References

1. Schroeder, M. R., Atal, B. S., Bird, C.: Digital computers in room acoustics. In: Proc. Fourth Intern. Congr. Acoust, Copenhagen (Aug 1962)
2. Atal, B.S., Schroeder, M.R.: Study of sound decay using ray-tracing techniques. J. Acoust. Soc. Am. **41**, 1598 (1967)
3. Schroeder, M.R.: Digital simulation of sound transmission in reverberant spaces. J. Acoust. Soc. Am. **47**, 424 (1970)
4. Braner, A.: On the characteristics roots of non-negative matrices. In: Schnieder, H. (ed.) Recent Advances in Matrix Theory, pp. 3–38. University of Wisconsin Press, Madison (1964)
5. Shanon, C.E.: A mathematical theory of communication. Bell Syst. Tech. J. **27**(379–423), 623–656 (1948)
6. Wiener, N.: Extrapolation, Interpolation, and Smoothing of Stationary Time Series. The MIT Press, Cambridge, MA (1949)
7. Elias, P.: Predictive coding - I II. In: IRE Trans. Inform. Theory, IT-1, pp. 16–33 (Mar 1955)
8. Atal, B. S., Schroeder, M. R.: Predictive coding of speech signals. In: Proc. 1967 Conf. on Communication and Proc., pp. 360–361 (Nov 1967)
9. Atal, B.S., Schroeder, M.R.: Adaptive predictive coding of speech signals. Bell Syst. Tech. Jour. **49**, 1973–1986 (1970)
10. Schroeder, M.R., Atal, B.S., Hall, J.L.: Optimizing digital speech coders by exploiting masking properties of the human ear. J. Acoust. Soc. Am. **66**, 1647–1652 (Dec 1979)
11. Atal, B. S., Schroeder, M. R.: Predictive coding of speech signals and subjective error criteria. In: IEEE Trans. Acoust., Speech, Signal Proc., ASSP-27, pp. 247–254 (Jun 1979)
12. Schroeder, M. R., Atal, B. S.: Code-excited linear prediction (CELP): high-quality speech at very low bit rates.' In: Proc. Int. Conf. on Acoust., Speech and Signal Proc., vol. 10, paper no 25.1, pp. 937–940 (Apr 1985)

***

Biography

Bishnu S. Atal is an Affiliate Professor in the Electrical Engineering Department at the University of Washington, Seattle, WA. He retired in March 2002 after working for more than 40 years at Lucent Bell Labs, and AT&T Labs. He was made a Bell Laboratories Fellow in 1994 and an AT&T Fellow in 1997. He is internationally recognized for his many contributions to speech analysis, synthesis, and coding. His pioneering work in linear predictive coding of speech established linear prediction as one of the most important speech analysis technique leading to many applications in coding, recognition, and synthesis of speech. He was elected to the National Academy of Engineering in 1987, and to the National Academy of Sciences in 1993. He is a Fellow of the Acoustical Society of America and the IEEE. He is the recipient of the IEEE Centennial Medal in 1984 and the IEEE Morris N. Liebmann Memorial Field Award in 1986. He received the Thomas Edison Patent Award from the Research & Development Council of New Jersey in 1994 and the New Jersey Inventors Hall of Fame Inventor of the Year Award in 2000. He was awarded the Benjamin Franklin Medal in Electrical Engineering in 2003 and the IEEE Jack S. Kilby Signal Processing Medal in 2013.

Chapter 11
Transducer Research at Bell Laboratories Under Manfred Schroeder

Gerhard M. Sessler

Abstract In this chapter, the work on electroacoustic transducers and the corresponding materials research in Manfred Schroeder's Department at Bell Laboratories is reviewed, together with ensuing work at other laboratories. The emphasis is on studies of electret-type transducers and materials. This includes the description of the first polymer electret microphone, the first use of fluorocarbon electrets of high thermal stability soon thereafter, and the introduction of directional electret microphones. Other electret transducers designed during this period, such as microphone arrays, headphones, and touch actuators, are also discussed. Next, early materials research on polymer electrets at Bell Labs, utilizing isothermal and thermally stimulated methods as well as electron radiation experiments, is reviewed. Finally, the later studies in a number of other laboratories, based on the early work under Schroeder, are outlined. This includes, among others, research on digital electret headphones, two-dimensional arrays, silicon (MEMS) condenser microphones, and, extending into the present time, transducers based on piezoelectric cellular polymers.

11.1 Introduction

Bell Laboratories and its parent organization, the Bell Telephone Company, had a long and rich history of contributions to electroacoustic transducer research. Foremost, and right at the beginning of Bell Telephone, was the work by Alexander Graham Bell on transducers needed to operate a telephone. His first transducers were moving armature microphones and receivers, described in his 1876 patent on

---

G.M. Sessler (✉)
Department of Communications Technology, Technische Universität Darmstadt,
Merckstrasse 25, 64283 Darmstadt, Germany
e-mail: g.sessler@nt.tu-darmstadt.de

N. Xiang and G.M. Sessler (eds.), *Acoustics, Information, and Communication*,
Modern Acoustics and Signal Processing, DOI 10.1007/978-3-319-05660-9_11,

telephony [1]. Later achievements at Bell Telephone were the invention of the condenser microphone by E. C. Wente in 1917 [2], the use of the thermophone as a precision source of sound energy by Arnold and Crandall, also in 1917 [3], and in 1942 the development and commercialization of a condenser microphone which was the standard precision microphone for several decades, the legendary 640 AA [4] by Western Electric, the manufacturing branch of Bell.

Manfred Schroeder came to Bell Laboratories in 1954 and was appointed Head of the Acoustics Research Department in 1958 and Director of the Acoustics and Speech Research Laboratory in 1962, a position which he held until his departure from Bell Labs in 1969. One of the research topics in Manfred's area was the field of electroacoustic transducers, particularly transducers for the audio frequency range, suitable for use in telephony. The liberal scientific atmosphere at Bell Labs in those days made it possible to extend this work into the fields of infrasonic and ultrasonic transducers and to also include research work on materials of interest in transducer design. Work in these areas continued into the 1980s, long after Manfred's departure from Bell Labs.

In the chapter, the work on electroacoustic transducers and the corresponding materials research in Manfred Schroeder's Laboratory in the 1960s is reviewed. The emphasis is on the work on electret-type transducers and materials. Although this was not one of Manfred's own main activities, he did contribute many interesting ideas, particularly on directional transducers and he always had a vivid interest in the progress of this work and gave it the managerial support necessary for its successful execution.

11.2 Electret Microphones

The work on electret microphones in Schroeder's laboratory began around 1960. As opposed to much older Japanese work on electret microphones, in which the bias voltage in a condenser microphone was supplied by a thick wax electret with a dipolar polarization [5], the new studies were based on the use of thin polymer films carrying surface or space charges. Initial work showed that condenser microphones utilizing such "foil electrets" are of simple design and have high sensitivity, good frequency response, low distortion, and relatively high capacitance, with the latter being advantageous for the processing of the output signal [6]. The possible usefulness of these transducers in telephony was therefore soon recognized and the work on electret microphones intensified over the following years [7–9].

The first such microphone, built with a polyethyleneterephthalate electret (PETP, Trade name Mylar) film, is shown in Fig. 11.1. The essential parts of this transducer are the electret film, metalized on its outer surface, and a metal backplate with circular ridges which establish and control an air gap between electret and plate. If the membrane is deflected by an impinging sound wave, a voltage proportional to the sound pressure is generated between the metal layer of the film and the backplate. The air cavity behind the backplate is connected by holes to the

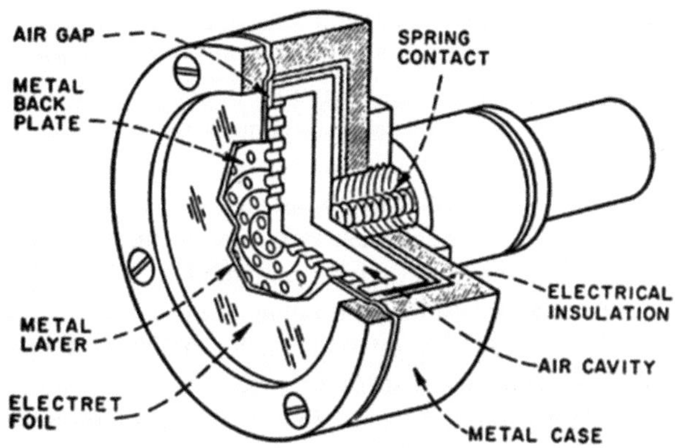

Fig. 11.1 First electret microphone with polymer film electret [6, 7]

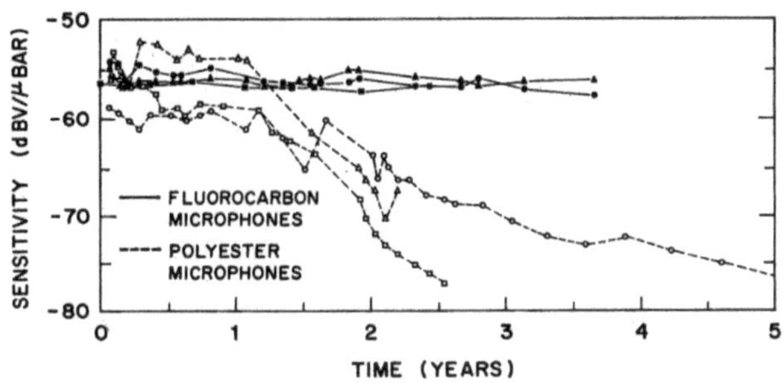

Fig. 11.2 Decay of sensitivity of early electret microphones [11]

air gap to soften the restoring force on the film and thus ensures larger vibration amplitudes and thus a larger sensitivity of the microphone.

It was soon recognized that the sensitivity of microphones with PETP electret under normal environmental conditions was constant for about a year but decreased thereafter. Typical results are shown in Fig. 11.2. The decrease was, of course, due to the decay of the electret charge. This made it necessary to look for electrets with improved charge stability. Experiments with a number of polymers eventually showed that electrets made of fluoroethylenepropylene (FEP, Trade name Teflon) had significantly longer time constants of the charge decay [10, 11]. It appeared that FEP and its homopolymer polytetrafluoroethylenepropylene (PTFE) were the best polymer electret materials available. Actually they were, and still are, the polymers of choice in almost all electret microphones.

The operation of electret microphones was analyzed by assuming a charge layer on the free surface of the electret film and induction charges on the film and back electrodes. The sensitivity of the device then follows from the density of the charge and the geometric dimensions of the system [12]. Sensitivities obtained from this analysis are, for the typical charge densities used on the electrets, of the order of 20 mV/Pa and thus in good agreement with experimental values. The calculations also showed that the electret charge corresponds to a bias voltage of about 100–200 V in conventional electrostatic microphones. Later numerical and experimental work at Bell Labs was devoted to an analysis of the static deflection of an electret membrane by the electrostatic field of the charges and the resulting stability problem, and also to the dynamic deflections of the membrane, due to the sound pressure, for particular air gap geometries [13–15].

The development of prototypes of electret microphones, suitable for use in telephony, started in various laboratories around the world in the mid-1960s. At Bell Labs, the Development Department came up with several prototypes designed for use in telephony and denominated as EL-1 to EL-4. These transducers had initially external dimensions equal to those of the carbon microphones in use at that time, but were later on built with smaller dimensions. The microphones were equipped with a transistor preamplifier integrated directly into the microphone case. At Northern Electric in Canada a microphone prototype, also for use in telephony, was designed with a polycarbonate electret [16] and was later evaluated by field tests. In 1968 Sony in Japan brought the first commercial electret microphone developed for use in HiFi on the market. Soon other companies followed, production of electret microphones increased at a rapid pace and has now reached about three billion annually.

11.3 Directional Electret Microphones

It was recognized early that electret microphones, due to their simplicity in mechanical design, would be very suitable sensors for directional transducers. Particularly microphones based on the gradient principle could be realized easily with electret microphones since these are small and permit ready access to the rear cavity behind the membrane.

In a discussion about the application of the gradient principle to electret microphones, Manfred Schroeder suggested to build second-order microphones with toroidal directional characteristics by utilizing the gradient combination shown in Fig. 11.3. Toroidal microphones were of particular interest in conference telephony, where participants seated around a table have to be covered with equal sensitivity by a directional microphone. To implement such a microphone, the sound field has to be sampled in eight points as indicated in Fig. 11.3, left [17]. This sampling may be performed by the opening of tubes which guide the sound waves into two cylindrical cavities, as shown in the center part of the figure. Using two cavities, positioned on either side of the membrane of an electret microphone and each collecting the sound of four tubes, one obtains the opposite sensitivities designated by "plus" and "minus" in the left part of the figure. The tubes together with the cavities form a Helmholtz resonator while the tubes by themselves have a length resonance. The Helmholtz and length resonances are placed at the lower and upper corner frequencies of the telephone band, respectively. With such a design, a flat frequency response is obtained within this frequency band since the ω^2-dependence of the second-order gradient is compensated by the $1/\omega^2$-dependence of the Helmholtz resonator above its resonance.

Fig. 11.3 Basic design and directional characteristics of the second-order toroidal microphone [17]. *Left*: schematic. *Center*: implementation. *Right*: directivity pattern

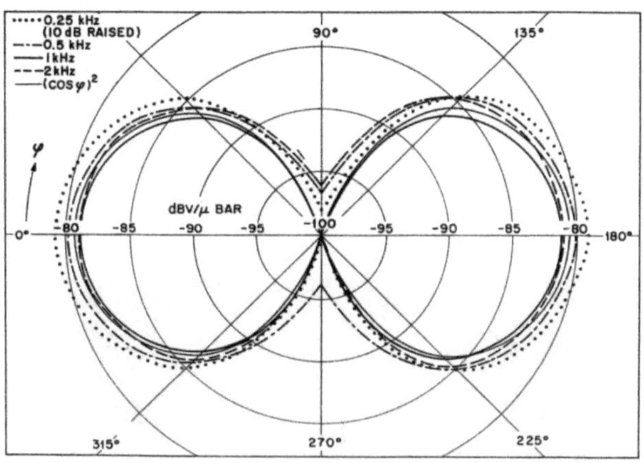

Fig. 11.4 Polar pattern of a toroidal microphone designed according to Fig. 11.3 [17]

The directivity pattern of a toroidal microphone of this kind is illustrated in the right-hand part of Fig. 11.3. The calculated and measured polar patterns are depicted in Fig. 11.4 [17]. Expected are cosine-square dependencies which are independent of frequency. The figure shows the relatively good agreement between the calculated directivity and the measured patterns for three frequencies. The directivity index of these microphones is 3 dB.

Other directional gradient-type microphones realized in Schroeder's laboratory were unidirectional and close-talking transducers. The unidirectional microphones were again second-order transducers. They were similarly designed with two tube-cavity systems, each containing two tubes, and thus had also constant sensitivity in the telephone band. The calculated and measured polar patterns are depicted in Fig. 11.5 [18, 19] and compare favorably in the direction of the main lobe but show some deviations in the other directions where the sensitivity is lower. The directivity index of such systems is about 9 dB. These unidirectional transducers, just as the toroidal transducers, are relatively simple systems. Nowadays they can be easily realized with spherical microphone arrays, as shown in Fig. 11.6, which have proper signal processing capabilities [20].

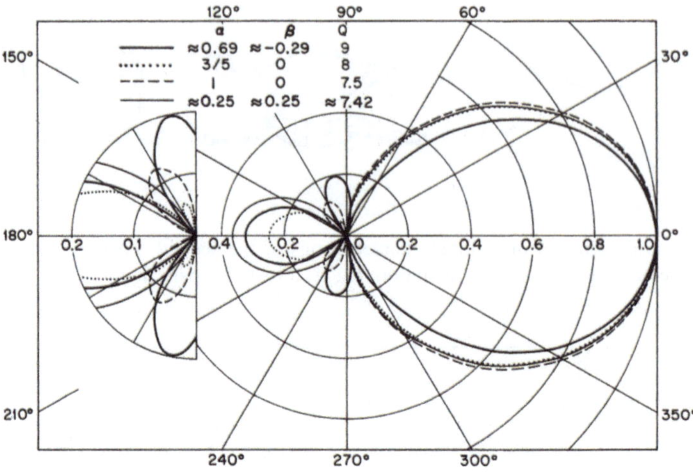

Fig. 11.5 Polar pattern of unidirectional microphone [18, 19]

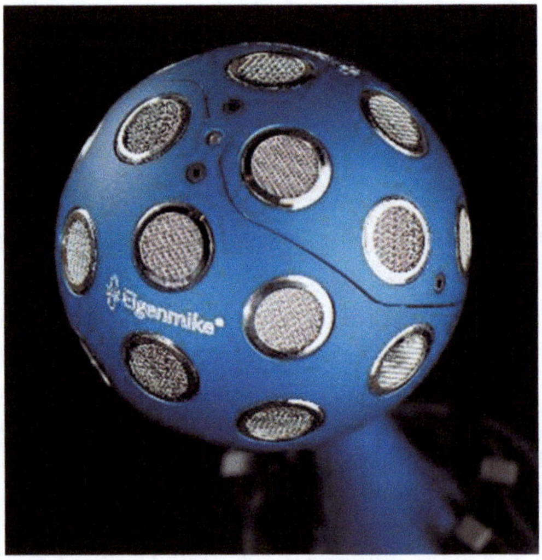

Fig. 11.6 "Eigenmike" with adjustable directional characteristics (Meyer and Elko [20])

Close-talking microphones were designed with the first-order gradient arrangement in which the membrane of an electret microphone is accessible to the sound waves from both sides [21]. These microphones have, without further compensation, a flat frequency response for a close sound source but a response decreasing toward low frequencies for far sound sources. Thus, low-frequency noise generated at a large distance from the microphone is suppressed by up to 30 dB compared to speech or other sounds originating nearby.

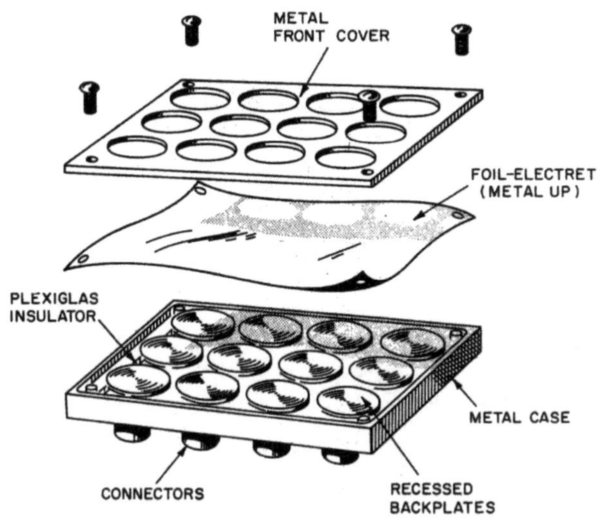

Fig. 11.7 Exploded view of foil-electret touch dial [23]

11.4 Other Electret Transducers

Apart from electret microphones, other electret transducers were realized in the 1960s and early 1970s in the Acoustics Research Laboratory in Murray Hill. Among these were headphones, touch-sensitive transducers, ultrasonic transducers, and holographic arrays.

Electret headphones were actually introduced in 1961 [22], prior to the electret microphones. The first transducers of this kind comprised a 6 μm thick PETP membrane which carried an electrostatic charge and therefore did not require an external DC-bias. Because of their high resonance frequency, such headphones had an excellent impulse response, a flat frequency response in the audio range and also low distortion. Electret headphones were brought on the market shortly after electret microphones and have since been in commercial use.

Touch-sensitive electret transducers were of interest in the 1960s for many applications, in particular as telephone dials. An exploded view of such a dial is shown in Fig. 11.7 [23]. It consists of a metalized foil electret and 12 backplates which are recessed in a dielectric plate. By touching the foil through one of the holes of the front cover, the film is deflected and an electric signal is generated. Such dials were attractive in those days because of their simplicity and reliability due to the fact that no metallic contacts are required.

Ultrasonic two-dimensional arrays allow one to sample sound fields in a large number of points, as required in acoustic holography. A possible approach to such a sampling device is a subdivided, large-area electret microphone [24, 25]. The design of a transducer of this kind is shown in Fig. 11.8, left. It consists of a film electret with n metal strips and a backplate with an equal number of metal strips. If film and backplate are arranged such that the strip patterns are perpendicular to each other, as shown in Fig. 11.8, right, a square array with n^2 microphone elements is

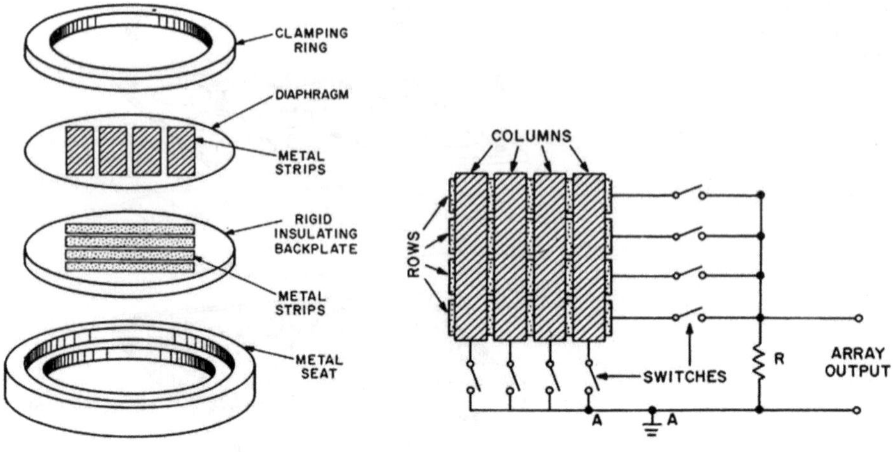

Fig. 11.8 Two-dimensional electret microphone array for acoustic holography [24, 25]. *Left part*: exploded view of the array. *Right part*: electrical connections

obtained. With switches connected to each film and backplate strip, the output of any particular microphone element is obtained by closing the corresponding column and row switches. The advantage of this array is that n^2 microphones are realized with only $2n$ components and that the gating requires only $2n$ gates. This is particularly important if the number of elements is large.

11.5 Research on Electret Materials for Use in Electroacoustic Transducers

Research on electret materials was carried out in the Acoustics Research Department in support of the transducer work. The electret research also started in the early 1960s and was initially directed toward a search for polymers with superior charge-storage capabilities. The electret material used initially was the above-mentioned PETP which was widely employed in condenser transducers with solid dielectric, the so-called Sell-transducers. It had been noticed before that PETP assumes a quasi-permanent charge if exposed to an electric field [26]. Systematic experiments revealed, however, that charge in this material is not sufficiently permanent for use in electret microphones (see Sect. 11.2). After some search and extensive testing, FEP was identified as an extremely stable electret material and then used in electret microphones [10, 11].

Among the experimental methods for the investigation of charge retention in polymer films were, for example, the study of isothermal charge decay at room temperature and at elevated temperatures, the measurement of thermally stimulated currents at linearly increasing temperatures, the analysis of charge deposition

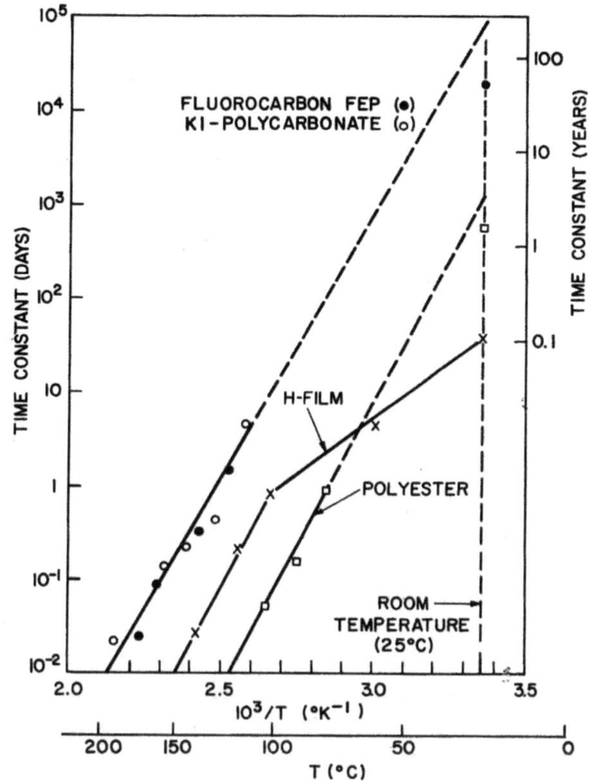

Fig. 11.9 Time constants of surface-charge decay of various 25 μm polymer films as function of inverse temperature (Arrhenius plots) [27]

by electron beams and of the accompanying radiation-induced effects, and the evaluation and extrapolation of time constants of the charge decay obtained at elevated temperatures. The latter method was particularly useful in view of the long time constants of the charge decay at room temperature expected for some of the polymers.

Results for the time constants τ of the charge decay for four different polymers at elevated temperatures and at room temperature are shown in Fig. 11.9 [27]. The log τ vs. $1/T$- plots usually yield straight lines (the so-called Arrhenius plots) which indicate that the activation energy in the temperature range considered is constant. The data show that the time constants evaluated directly at room temperature are in order-of-magnitude agreement but somewhat smaller than the constants obtained by the straight-line extrapolations of the high-temperature data. The deviations are not surprising since the isothermal decay is always non-exponential in the sense that it slows down with time, such that longer time constants than those obtained by extrapolation of relatively short-term RT-data have to be expected. The longest time constants are those for FEP and polycarbonate. It was found, however, that FEP is much better in humid atmospheres, which made it the material of choice for electret microphones (see Sect. 11.2). The polyester (PETP) electrets exhibit the

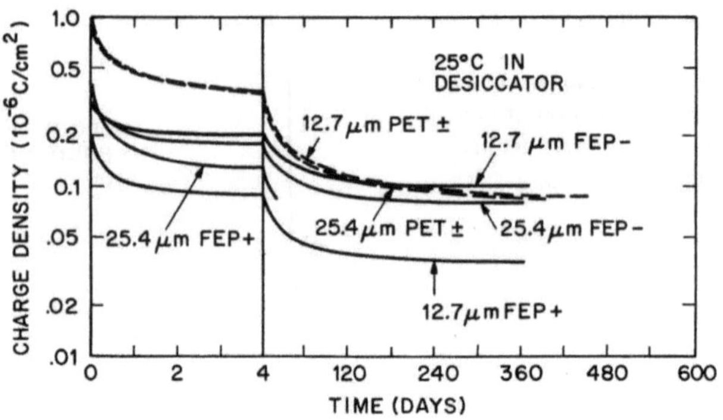

Fig. 11.10 Charge decay for high charge densities on various polymer films [29]

relatively short lifetimes that had already been found in the experimental results for the first electret microphones, as shown in Fig. 11.2. As opposed to these results, the polyimide (H-film) data have to be approximated by two straight lines, indicating a change in activation energy of the charge-activation process at about 100 °C.

In the late 1960s, first experiments with electron-beam charging were carried out [28]. The beam energy was chosen such that the electrons did not penetrate the entire thickness of the film but were trapped within the film volume. It was recognized that this method allows one to obtain stable electrets and that the charges deposited in the films had a very uniform lateral density distribution. These experiments paved the way for the later cooperation with Professor Bernhard Gross, who visited the Acoustics Research Department regularly in the 1970s and with whom many experiments on charge storage and charge dynamics in electrets irradiated with low-energy electrons were performed (see Sect. 11.6).

Around 1970, interest focused also on the potential of electrets for other applications. In this context, the maximum charge densities that can be achieved in various polymers were investigated. The limits for stored charge are determined by the dielectric strength of the materials and are as high as 10^{-2} C/m^2 for PETP and 0.5×10^{-2} C/m^2 for FEP, as shown in Fig. 11.10 [29]. Charge decays were always found to be non-exponential in the sense that time constant increase with time, as the charge density gets lower. For example, for charge densities below about 10^{-3} C/m^2, the charge decays slow down considerably and time constants reach about 10 years for FEP. This indicates the presence of traps of different energetic depth with the shallower traps discharging more quickly than the deeper traps. As Fig. 11.10 also shows, the superior charge retention properties of FEP are only observed for negative charging. Positive charge dissipates much more quickly. Later on it was found that the stability of positive charges in FEP could be improved significantly by corona-charging at elevated temperatures [30].

11.6 A Few Results of Later Work

The work on electroacoustic transducers and on electro-active materials, based on the results obtained in the 1960s, continued after Manfred's departure. The studies were carried out at Bell Labs and later on also at a number of other places where new research laboratories devoted to this kind of work were set up by people departing from the Acoustics Research Department in Murray Hill. Just a few examples, mostly related to the authors' own activities, will be discussed in the following.

Transducer work at Bell Labs in the 1970s was still concerned with directional microphones, particularly second-order gradients [19], but also with other electret devices. One such gadget was an earphone that converts digital PCM signals directly into an analog acoustic output signal by using a backplate subdivided such that areas scaled by powers of 2 are obtained [31]. By covering these areas with an electret film, one obtains an array of individual sound radiators whose backplate segments are addressed with the proper digital bit. Thus, an acoustic output corresponding to the digital input signal is obtained.—A few years later, in the early 1980s, transducer work intensified again. Topics were once more electret earphones and directional microphones, but major activities were also in signal processing for microphone arrays and in silicon or MEMS microphones.

The work on electret earphones was directed toward a better understanding of edge-supported systems. While a theoretical model described the static and dynamic deflections of the membrane, predicted the nonlinear response, and allowed one to maximize the earphone sensitivity, experimental results confirmed the predictions and offered additional insight into the importance of transducer parameters, such as ridge height and air loading [32]. Later on, these studies were extended to earphones with radially varying parameters, such as charge density or air gap thickness [33].

A directional microphone in the form of an acoustic antenna was realized with a long strip of electret foil placed over a metalized backplate [34]. By locally adjusting parameters such as the charge density on the electret along the length of the strip, a shading of the sensitivity may be achieved which results in a desired directivity pattern. Compared to a linear microphone array with the same directionality, the antenna offers some advantages such as easier construction and higher signal-to-noise ratio.

Two-dimensional arrays consisting of hundreds of electret microphones for capturing speech signals of persons seated in large auditoria were also designed in the 1980s. Signal processing was needed to discriminate speech from interfering noise and to automatically steer the arrays in the speaker direction by a microprocessor system [35]. Similar arrays, used for teleconferencing, included hands-free voice control [36]. Another design feature of such arrays was the control of the pattern of the beam, including a constant beam width, within a broad frequency range [37].

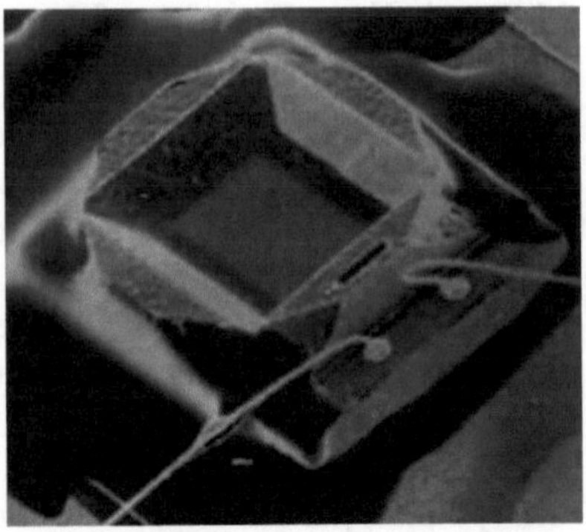

Fig. 11.11 First silicon condenser (MEMS) microphone [41, 42]

The work on silicon microphones was initiated independently at Bell Labs and at the Darmstadt University of Technology, where the author established a new laboratory in 1975. In both places, silicon *condenser* microphones were studied, although in the Darmstadt laboratory later on also other transduction principles, such as piezoelectric, FET-modulating, and opto-acoustic systems [38], were investigated. Several design concepts for the condenser microphones were published or patented between 1983 and 1985 [see [39, 40] and references therein]. They were mostly based on sacrificial-layer etching technology to obtain a single-chip transducer with the required air gap. As far as biasing is concerned, either an external voltage in the 10–28 V range or electret biasing by charging a silicon dioxide layer was suggested. Implementations of externally biased microphones, as shown in Fig. 11.11 and consisting of a backplate chip and a membrane chip, came in 1986 [41, 42]. These were actually the first operational silicon condenser microphones worldwide. Improved transducers, mostly consisting of a single chip, followed. These were designed in several other laboratories, including Darmstadt, in the 1990s [38] but it took until 2002 until commercial MEMS microphones were built.

Electret work in the 1970s shifted to a powerful diagnostic tool already developed before, namely electron-beam irradiation (see above) [28]. Starting in 1972, Bernhard Gross was invited annually to Bell Labs to cooperate on experiments with electron beams. He introduced the method of the split-Faraday cup to investigate charge deposition, charge distribution, and radiation-induced conductivity (RIC) in polymers [43]. From such measurements conclusions were drawn about charge dynamics, i.e., transport, recombination, and trapping of deposited charges in polymers used as electret materials. At Bell Labs, this work was supported by computer simulations of charge dynamics in electron-irradiated polymer films [44]. The calculations were performed by solving the charge-transport equations for open- and short-circuit conditions. Other electret work in the 1980s at Bell Labs concerned the experimental discrimination between surface and bulk traps in polymers [45] and the thermal stabilization of positive charges in FEP [30]. The collaboration with Bernhard Gross was also extended to the Darmstadt laboratory and continued into the 1990s [46].

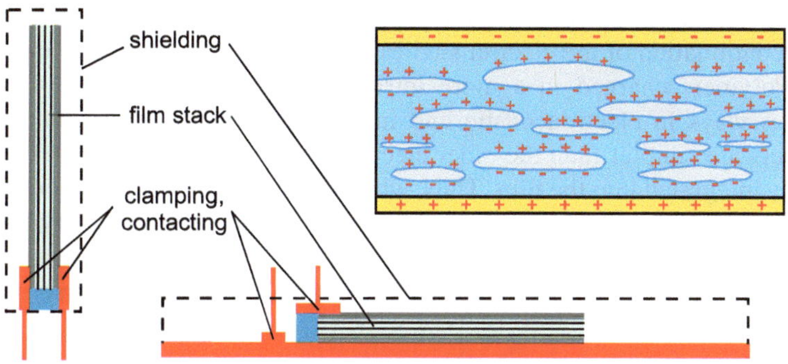

Fig. 11.12 Schematic cross section of piezoelectret (*right top*) and design of piezoelectret microphone (*left bottom*) [52, 53]

In the late 1990s, a new type of piezoelectric material became available [47, 48]. This material consists of a cellular polymer, e.g., cellular polypropylene, which is charged such that all the voids in the material carry positive charges on their upper surface and negative charges on their lower surface, see Fig. 11.12 [49]. Such films show a strong longitudinal piezoelectric effect with d_{33}-coefficients above 1,000 pC/N at 0.001 Hz and 500 pC/N in the audio frequency range [50]. These coefficients are more than an order of magnitude larger than those of poly(vinylidene fluoride), the best piezoelectric polymer known before. Further improvements up to 1,500 pC/N in the audio frequency range were obtained by applying an additional DC-bias to the films [51].

This large piezoelectric activity suggested the use of such films as microphones: A sound wave impinging on a film of this type will periodically modulate the film thickness and thus cause an electrical output signal proportional to the sound pressure as long as the film dimensions are small compared to the wavelength. The only additional components in such a microphone are suitable contacts, output leads, and electric shielding, as illustrated in the lower part of Fig. 11.12. First microphones of this type showed relatively flat frequency responses extending from the infrasonic into the ultrasonic region and sensitivities of about 0.5 mV/Pa [52]. Later transducers using stacked film designs and improved cellular electrets exhibited sensitivities of up to 15 mV/Pa [53].

The transduction mechanism of these microphones can, of course, be described either as a piezoelectric effect or as electret-biasing of a capacitive microphone. The former kind of view has been largely adopted nowadays since the electro-activity of the piezoelectrets is phenomenologically identical to that of a classical piezoelectric medium, including the hysteresis effects exhibited by the old and the new materials [54]. However, looking at the intrinsic causes of the piezoelectricity, the charges trapped at the walls of the voids in the material are responsible for the observed piezoelectric activity and thus the transducers are also of the electret type. Therefore, the term "piezoelectret" has been coined for these materials and the corresponding transducers.

This new transducer work and the supporting materials research are thus, in a sense, a continuation of the electret studies started under Manfred Schroeder in the 1960s: Polymers charged in a more intricate way, the piezoelectrets, are used in various ways in electroacoustic and in electromechanical transducers. These transducers are simpler than the older electret devices in the sense that the air gap is directly integrated into the film. Thus, while the electret replaces the external bias by a built-in charge, the piezoelectret additionally replaces the air gap by built-in air cavities. With the recent activities in the field of piezoelectret devices, we are witnessing another step in the evolution of electroacoustic transducers.

References

1. Bell, A. G.: U. S. Patent 174465, Improvement in Telegraphy (1876)
2. Wente, E.C.: A condenser transmitter as a uniformly sensitive instrument for the absolute measurement of sound intensity. Phys. Rev. Ser. II **10**, 39–63 (1917)
3. Arnold, H.D., Crandall, I.B.: The thermophone as a precision source of sound. Phys. Rev. **10**, 22–38 (1917)
4. Western Electric Improved 640 AA condenser microphone and RA-1095 Amplifier. Copyright: Western Electric Company, Incorporated (1946)
5. Nishikawa, S., Nukiyama, D.: A use of the electret in a condenser transmitter. Proc. Imp. Acad. Tokyo **4**, 290–291 (1928)
6. Sessler, G.M., West, J.E.: Self-biased condenser microphone with high capacitance. J. Acoust. Soc. Am. **34**, 1787–1788 (1962)
7. Sessler, G.M., West, J.E.: Electret transducers: a review. J. Acoust. Soc. Am. **53**, 1589–1600 (1973)
8. Elko, G.W., Harney, K.P.: A history of consumer microphones: the electret condenser microphone meets micro-electro-mechanical-systems. Acoustics Today **5**(2), 4–13 (2009)
9. Flanagan, J.L.: A singular advance in conversion of acoustic signals to electrical form: the electret microphone. IEEE Signal Process. Mag. **27**(5), 102–116 (2010)
10. Sessler, G.M., West, J.E.: Electrostatic microphones with foil electret. J. Acoust. Soc. Am. **37**, 1209–1210 (1965)
11. Sessler, G.M., West, J.E.: Foil-electret microphones. J. Acoust. Soc. Am. **40**, 1433–1440 (1966)
12. Sessler, G.M.: Electrostatic microphones with electret foil. J. Acoust. Soc. Am. **35**, 1354–1357 (1963)
13. Warren, J.E., Brzezinski, A.M., Hamilton, J.F.: Capacitance-microphone static membrane deflections. J. Acoust. Soc. Am. **52**, 711–719 (1972)
14. Warren, J.E.: Capacitance microphone static membrane deflections: comments and further results. J. Acoust. Soc. Am. **58**, 733–740 (1975)
15. Warren, J.E., Brzezinski, A.M., Hamilton, J.F.: Capacitance microphone dynamic membrane deflections. J. Acoust. Soc. Am. **54**, 1201–1213 (1973)
16. Reedyk, C.W.: An electret transmitter for the telephone set. Electrochem. Technol. **6**, 6–10 (1968)
17. Sessler, G.M., West, J.E., Schroeder, M.R.: Toroidal microphones. J. Acoust. Soc. Am. **46**, 28–36 (1969)
18. Sessler, G.M., West, J.E.: Directional transducers. IEEE Trans. Audio Electroacoust. **AU-19**, 19–23 (1971)
19. Sessler, G.M., West, J.E.: Second-order gradient unidirectional microphones utilizing electret transducers. J. Acoust. Soc. Am. **58**, 273–278 (1975)

20. Meyer, J., Elko, G. W.: Realization of a spherical microphone array. In: Proc. 19th Internat. Congr. Acoust., Paper ELE-02-007, Madrid (2007)
21. Sessler, G.M., West, J.E.: First-order gradient microphone based on the foil-electret principle: discrimination against air-borne and solid-borne noises. J. Acoust. Soc. Am. **46**, 1081–1086 (1969)
22. Sessler, G.M., West, J.E.: Condenser earphone with solid dielectric, Audio Engineering Society Preprint No. 187, (Audio Engineering Society, 1961). J. Audio Eng. Soc. **10**, 212–215 (1962)
23. Sessler, G.M., West, J.E., Wallace Jr., R.L.: New touch actuator based on the foil-electret principle. IEEE Trans. Commun. **COM-21**, 61–65 (1973)
24. Nigam, A.K., Taylor, K.J., Sessler, G.M.: Foil-electret transducer arrays for real-time acoustical holography. In: Wade, G. (ed.) Acoustical Holography, vol. 4, pp. 173–194. Plenum Publishing Corp, New York (1972)
25. Nigam, A.K., Sessler, G.M.: Condenser–transducer array for acoustical holography. Appl. Phys. Lett. **21**, 229–231 (1972)
26. Kuhl, W., Schodder, G.R., Schröder, F.-K.: Condenser transmitters and microphones with solid dielectric for airborne ultrasonics. Acustica **4**, 519–532 (1954)
27. Sessler, G.M., West, J.E.: Foil electrets and their use in condenser microphones. J. Electrochem. Soc. **115**, 836–841 (1968)
28. Sessler, G.M., West, J.E.: Charging of polymer foils with monoenergetic low-energy electron beams. Appl. Phys. Lett. **17**, 507–509 (1970)
29. Sessler, G.M., West, J.E.: Production of high quasipermanent charge densities on polymer foils by application of breakdown fields. J. Appl. Phys. **43**, 922–926 (1972)
30. von Seggern, H., West, J.E.: Stabilization of positive charge in fluorinated ethylene propylene copolymer. J. Appl. Phys. **55**, 2754–2757 (1984)
31. Flanagan, J.L.: Direct digital-to-analog conversion of acoustic signals. Bell Syst. Tech. J. **59**, 1693–1719 (1979)
32. Busch-Vishniac, I.J.: Response of an edge-supported circular membrane electret earphone. Parts I and II. J. Acoust. Soc. Am. **75**, 977–989 (1984), 990–995
33. Busch-Vishniac, I.J.: Response of an edge-supported, circular, foil-electret earphone with radially varying parameters. J. Acoust. Soc. Am. **83**, 796–804 (1988)
34. Busch-Vishniac, I.J., West, J.E., Wallace Jr., R.L.: A new approach to transducer design applied to a foil electret acoustic antenna. J. Acoust. Soc. Am. **76**, 1609–1616 (1984)
35. Flanagan, J.L., Johnston, J.D., Zahn, R., Elko, G.W.: Computer-steered microphone arrays for sound transduction in large rooms. J. Acoust. Soc. Am. **78**, 1508–1518 (1985)
36. Flanagan, J.L., Berkley, D.A., Elko, G.W., West, J.E., Sondhi, M.M.: Autodirective microphone systems. Acustica **73**, 58–71 (1991)
37. Elko, G.W.: Microphone array systems for hands-free telecommunication. Speech Commun. **20**, 229–240 (1996)
38. Sessler, G.M.: Silicon microphones. J. Audio Eng. Soc. **44**, 16–22 (1996)
39. Hohm, D., Sessler, G. M.: An integrated silicon-electret-condenser microphone. In: Proc. of the 11th Int. Congress on Acoustics, vol. 6, pp. 29–32. Paris (1983)
40. Poteat, T. L., Bush-Vishniac, I. J.: Silicon micromechanics: new technology for sensors. In: Proc. of the 12th Int. Congress on Acoustics, vol. 3, paper L 1–2. Toronto (1986)
41. Hohm, D.: Silicon sensor for airborne sound. In: Proc. of the 12th Int. Congress on Acoustics, vol. 3, paper L 3–3. Toronto (1986)
42. Hohm, D., Hess, G.: A subminiature condenser microphone with silicon nitride membrane and silicon backplate. J. Acoust. Soc. Am. **85**, 476–480 (1989)
43. Gross, B., Sessler, G.M., West, J.E.: Charge dynamics for electron-irradiated polymer-foil electrets. J. Appl. Phys. **45**, 2841–2851 (1974)
44. Berkley, D.A.: Computer simulation of charge dynamics in electron-irradiated polymer foils. J. Appl. Phys. **50**, 3447–3453 (1979)
45. von Seggern, H.: Detection of surface and bulk traps. J. Appl. Phys. **52**, 4086–4089 (1981)
46. Sessler, G.M.: Bernhard Gross and electret research: his contributions, our collaboration, and what followed. IEEE Trans. Dielectr. Electr. Insul. **13**, 942–952 (2006)

47. Lekkala, J., Poramo, R., Nyholm, K., Kaikkonen, T.: EMF force sensor—a flexible and sensitive electret film for physiological applications. Med. Biol. Eng. Comput. **34**, 67–69 (1996)
48. Sessler, G.M., Hillenbrand, J.: Electromechanical response of cellular electret films. Appl. Phys. Lett. **75**, 3405–3407 (1999)
49. Hillenbrand, J., Sessler, G.M.: High-sensitivity piezoelectric microphones based on stacked cellular polymer films. J. Acoust. Soc. Am. **116**, 3267–3270 (2004)
50. Zhang, X., Hillenbrand, J., Sessler, G.M.: Improvement of piezoelectric activity of cellular polymers using a double-expansion process. J. Phys. D. Appl. Phys. **37**, 2146–2150 (2004)
51. Hillenbrand, J., Sessler, G.M.: DC-biased ferroelectrets with large piezoelectric d_{33}-coefficients. J. Appl. Phys. **103**, 074103-1–074103-7 (2008)
52. Kressmann, R.: New piezoelectric polymer for air-borne and water-borne sound transducers. J. Acoust. Soc. Am. **109**, 1412–1416 (2001)
53. Hillenbrand, J., Sessler, G.M.: Stacked piezoelectret microphones of simple design and high sensitivity. IEEE Trans. Dielectr. Electr. Insul. **13**, 973–978 (2006)
54. Bauer, S., Gerhard-Multhaupt, R., Sessler, G.M.: Ferroelectrets: soft electroactive foams for transducers. Phys. Today **57**(2), 37–43 (2004)

***

Biography

Gerhard M. Sessler studied physics at the University of Goettingen in Germany and joined Bell Laboratories in 1959 as a member of the Acoustics Research Department in Murray Hill, N.J. He had responsibilities for the work on electroacoustic transducers, concert hall acoustics, and charge storage phenomena in solids. Together with James E. West he invented the first polymer electret microphone which was commercialized later on and still is the predominant microphone type. In 1975, he was appointed a professor of electroacoustics at the University of Technology in Darmstadt, Germany. He has since been involved in various fields of acoustics and solid-state physics, in particular in acoustic silicon transducers and acoustic signal processing. In 1983, he and his co-worker Dietmar Hohm designed the first MEMS condenser microphone based on silicon micromachining. Presently, he is an Emeritus Professor and he and his group are working on new piezoelectric polymers and their applications.

Chapter 12
Manfred Schroeder and Acoustical Impedance

Roland Kruse and Volker Mellert

Abstract Manfred Schroeder was a man with many interests. Between the years 1966 and 1967, he invented devices for measuring the surface impedance of materials as well as the vocal tract impedance. At the time, he would only consider normal incidence; however, he already anticipated that the behavior of materials at grazing incidence would have relevance to room acoustics, a field of research he is well known for.

To achieve this aim, an experimental set-up to determine the characteristic impedance of absorbers at grazing incidence has been developed. The sample is placed on the bottom of a rectangular impedance tube and the horizontal wave number above the sample is calculated from the pressure transfer function between two microphones. From this wave number, the absorber properties can be deduced. While the method works reasonably well for highly absorbing samples, the non-ideal sound field in the tube—as confirmed by finite element simulations—limits its usefulness in case of less absorbent materials. Improvements of the method are suggested.

12.1 Introduction

Manfred Schroeder was a man with many interests. Between the years 1966 and 1967, he invented devices for measuring the surface impedance of materials as well as the vocal tract impedance. At the time, he would only consider normal incidence; however, he already anticipated that the behavior of materials at grazing incidence would have relevance to room acoustics, a field of research he is well known for.

R. Kruse (✉)
Acoustics Group, Institute of Physics, Oldenburg University, 26111 Oldenburg, Germany
e-mail: r.kruse@tu-bs.de

N. Xiang and G.M. Sessler (eds.), *Acoustics, Information, and Communication*, Modern Acoustics and Signal Processing, DOI 10.1007/978-3-319-05660-9_12,

To achieve this aim, an experimental set-up to determine the characteristic impedance of absorbers at grazing incidence has been developed. The sample is placed on the bottom of a rectangular impedance tube and the horizontal wave number above the sample is calculated from the pressure transfer function between two microphones. From this wave number, the absorber properties can be deduced. While the method works reasonably well for highly absorbing samples, the non-ideal sound field in the tube—as confirmed by finite element simulations—limits its usefulness in case of less absorbent materials. Improvements of the method are suggested.

12.2 System for Determining Acoustic Reflection Coefficients

The first sign of Manfred Schroeder's research on measuring impedances is his patent number 3,346,067 [10], granted in October 1967 but filed already in March 1966, which deals with a "system for determining acoustic reflection coefficients" which equals, as we have plane wave propagation, a device for impedance determination. Before having a detailed look at this invention, let us have a look at how reflection/absorption coefficients of materials (at normal incidence) were measured at that time: using the Kundt's tube. This meant that for each frequency of interest, one had to find—by moving the microphone along the tube—the minimum and maximum of the sound pressure. In addition, if one was interested in the complex reflection coefficient, one had to locate the position of the first minimum with respect to the sample's surface. Obviously, that was a time-consuming task, and for materials with high absorption, like porous absorbers at high frequencies, it was difficult to locate the extrema of the sound pressure.

Manfred Schroeder sought to overcome this problem by using broadband signals and two microphones, relying on the definition of the reflection coefficient: "The reflection coefficient, an important parameter in determining the acoustic characteristics of a material, is defined as the ratio of the amplitude of a reflected pressure wave to the amplitude of an incident pressure wave, both amplitudes being measured at the face of the material" [10]. In other words, we try to separate incident and reflected wave and to predict the sound pressure at the sample's surface from measurements done at some distance to it.

Figure 12.1 shows a sketch of the system. It shows what would nowadays be called a two-microphone impedance tube, the only difference being that the sound source is not located opposite the sample but at the side of the tube.

From the measured pressure amplitudes at the two locations, first the amplitudes of the incident and reflected wave were derived with Eq. (12.1), then the (complex) reflection coefficient: Eq. (12.2). Analog electronics (delays, multipliers, etc.) were used to perform these operations, as this was the established solution for such a

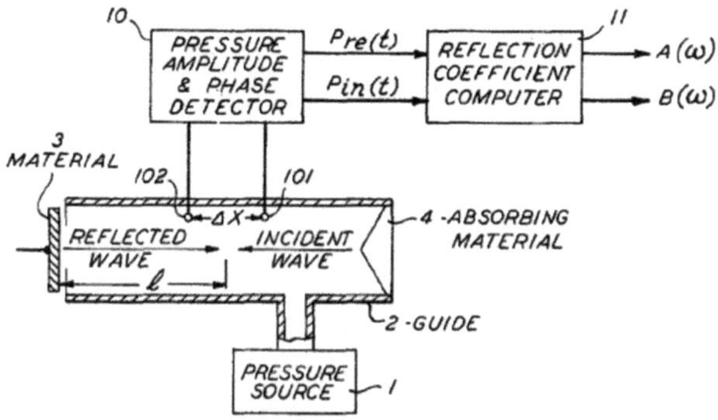

Fig. 12.1 Sketch of the impedance measurement system invented by M.Schroeder in 1966 [10]

problem at that time (the term "reflection coefficient computer" is rather misleading from today's point of view on digital signal processing). Broadband processing was achieved by using a bandpass filter bank and having multiple "computers" evaluating Eq. (12.2) simultaneously. Hence, one could obtain the frequency-dependent, complex reflection coefficient almost instantaneously.

$$\begin{aligned} P_{\text{in}}(\omega) &= \frac{1}{e^{i\omega\Delta t} - e^{-i\omega\Delta t}}\left[P_1(\omega)e^{i\omega\Delta t/2} - P_2(\omega)e^{-i\omega\Delta t/2}\right] \\ P_{\text{re}}(\omega) &= \frac{1}{e^{i\omega\Delta t} - e^{-i\omega\Delta t}}\left[P_2(\omega)e^{i\omega\Delta t/2} - P_1(\omega)e^{-i\omega\Delta t/2}\right] \end{aligned} \tag{12.1}$$

$$\Re(R) = \frac{\overline{P_{\text{re}}(t)P_{\text{in}}(\omega_j, t)}}{\overline{P_{\text{in}}^2(\omega_j, t)}} \quad \Im(R) = \frac{\overline{P_{\text{re}}(t)\left(P_{\text{in}}(\omega_j, t)e^{i\pi/2}\right)}}{\overline{P_{\text{in}}^2(\omega_j, t)}} \tag{12.2}$$

Thinking about the work of Manfred Schroeder one wonders how this invention fits into his line of work, as it seems to be neither related to room acoustics (surprisingly, the patent does not refer to the use of reflection/absorption coefficients in this area) nor to speech processing or perception. A closer look reveals that it is most likely a spin-off from his research on speech (re-) production.

12.3 Determination of the Geometry of the Human Vocal Tract

In 1967, Manfred Schroeder submitted a manuscript [11] to JASA, published the next year, in which he describes a procedure involving, first, the determination of the frequency-dependent impedance of the vocal tract and, second, the derivation of the vocal tract geometry using an appropriate model. Reading the motivation behind this study, it becomes clear why Schroeder was interested in such a topic: "Knowledge of vocal-tract configuration (tongue and jaw position, lip rounding, etc.) is basic to a better understanding of the physical and physiological processes involved in the human speech process—one of man's important means of communication. Present lack of knowledge in this area is epitomized by the variable quality and limited vocabulary exhibited by present speaking machines...." No, Manfred Schroeder was not interested in a mere measurement technique but had a much more visionary goal in mind, something that was also part of the zeitgeist with science fiction movies featuring talking robots and "computers".

There're several factors making this publication particularly interesting: on the one hand, the setup used to measure the impedance is very similar to the one in the patent, with the material sample replaced by the subject's mouth, and on the other hand, because of the extensive use of digital processing: the two microphone signals are digitized right away, the impedance is calculated, poles and zeros of the impedance curves (representing resonances in the vocal tract) are located and from them the geometry (cross-sectional area versus distance) is estimated by numerical inversion of Webster's horn equation, a process involving an iterative procedure which digital computers are well suited for. We may, consequently, guess that the system filed for patent is a scaled-down version of the system presented in this chapter, making use of established components.

This in not the first occasion demonstrating Schroeder's interest in digital computers, he also used them for measuring room-acoustical parameters [12] and for simulating the effect of such parameters, e.g., the frequency-dependent reverberation time, on sound perceived in a "virtual" room [13]. And all this despite the limited processing power available in the Sixties and the enormous price of digital computers. Obviously, someone believed in the future of these machines.

Now where do we stand today? The two-microphone impedance tube (with digital processing) is a standardized method (ISO 10534-2) for measuring the normal-incidence reflection coefficient. In addition, in situ methods are used to determine the impedance at arbitrary angles of incidence [14, 15]. What one may miss is a laboratory method for determining the impedance at grazing incidence.

Manfred Schroeder gives us a hint on what this can be useful for: in his paper on measurements in the Philharmonic Hall [16], he and his colleagues made an interesting observation, the seat effect: "This lack of low frequencies in the first overhead reflection revealed another low-frequency deficiency that ad hitherto gone unnoticed: A progressive attenuation of low frequencies in the direct sound as it grazes across the rows of seats. (This 'seat effect' must exist in many other halls, but it is usually masked by the presence of low-frequency components in the early overhead reflections)."

The acoustical properties of materials or structures can vary with the angle of incidence (non-local reaction), and for a correct simulation of the room acoustics data for other than normal incidence is needed.

Therefore, we shall now outline how such data can be obtained using a standing-wave tube.

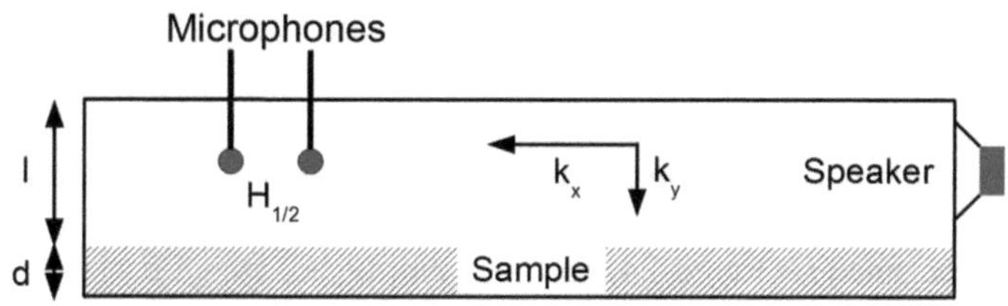

Fig. 12.2 A standing-wave tube with grazing sound incidence

12.4 Standing-Wave Tube at Grazing Incidence

Figure 12.2 shows an overview of the proposed setup. It is similar to a two-microphone impedance tube, but the sample is now placed at the bottom of the tube, which has a square cross-section of $10 \times 20\ \mathrm{cm}^2$ and a length of 91 cm. It is made from 20 mm thick aluminum to guarantee a sufficient bending stiffness. The upper cover is height adjustable and allows the channel height to be increase by up to 8 cm. The loudspeaker at the right radiates broadband noise, which is recorded by two microphones having the same height above the sample's surface. The quantity to be measured is the transfer function $H_{1/2}$ between the microphones. To allow for differences between the sensitivities and phase responses of the microphones, the switch-microphones calibration technique [17] has been used.

This technique is an advancement of a technique described by [18] who used a Kundt's tube-like approach with a movable microphone with an otherwise identical setup.

The general effect of the (absorbing) sample at the bottom of the channel is that waves traveling above it are attenuated and do not remain plane waves. This can be exploited to find the acoustical properties of the material under test.

12.4.1 Theoretical Background

The sound field inside the tube is assumed to be two-dimensional as the depth of the channel is only 10 cm. The field can then be described by a horizontal wave number k_x and a vertical wave number k_y. In the first step, k_x is deduced from the measured transfer function $H_{1/2}$ and the distances x_1, x_2 of the microphones with respect to the hard wall at the end of the tube. We shall assume that this wall is perfectly reflecting. Equation (12.3) can then be used, keeping in mind that the microphones are located at the same height y above the sample; it can be solved numerically for k_x.

$$H_{1/2} = \frac{p_1}{p_2} = \frac{e^{-ik_x x_1} + e^{ik_x x_1}}{e^{-ik_x x_2} + e^{ik_x x_2}} = \frac{\cosh(ik_x x_1)}{\cosh(ik_x x_2)} \tag{12.3}$$

The wave number k_x is linked to the wave number k_a and characteristic impedance Z_a of the material and the geometry of the problem. The theory of Scott [19] as outlined by Eq. (12.4) has been employed, which basically states that the vertical wave components are to be continuous at the sample's surface. It is important to note that this theory (and the similar one by Mechel [20]) assumes an infinitely long channel; especially, it does not take any additional effects, i.e. the boundary condition of a source into account.

$$\sqrt{k^2 - k_x^2} \tan\left(\sqrt{k^2 - k_x^2} l\right) = \frac{1}{Z_a' k_a'} \sqrt{k_a^2 - k_x^2} \tan\left(\sqrt{k_a^2 - k_x^2} d\right) \tag{12.4}$$

Equation (12.4), too, can be solved numerically for the absorber properties. Because this would require deriving two complex quantities from a single one, the solution may be non-unique. In addition, just like in the case of in situ impedance deduction, pretests indicate that it is prone to small measurement errors [21]. Therefore, we will assume that the absorber's impedance and wave number can be described by, e.g., the Delany–Bazley model [22], so that only the single parameter "flow resistivity" has to be deduced. Other absorber models would serve as well.

12.4.2 Results and Discussion

To verify the functionality of setup and data processing, measurements were first done with an empty tube, so that the expected value for k_x is $(1 + 0i)$. Figure 12.3 shows the actual results. At low frequencies, the wave number deviates from the expected value, likely due to the (relative) proximity of the loudspeaker and hence the initially non-plane wave field. Unexpectedly, the wave number is also incorrect around 800 Hz (and at other, higher frequency ranges). In addition, within these frequency ranges, the coherency between the microphone signals (not shown) was reduced due to the sound pressure at one of the microphones being very low. Assuming plane waves, the lowest frequency a pressure minimum is expected to occur at a microphone 10 cm in front of a hard wall is 850 Hz but the minimum actually occurred at the microphone 2 cm in front of the wall.

A finite element simulation of the sound propagation inside a rigid channel with known geometry and position of the sound source was done to better understand how the sound field looks like. In Fig. 12.4, the pressure distribution inside the empty tube is shown for four frequencies. At 500 and 700 Hz, one sees the non-plane field close to the source, whereas at the left side (where the microphones are located), the wave fronts are mostly flat. On the other hand, at 850 Hz, there's a

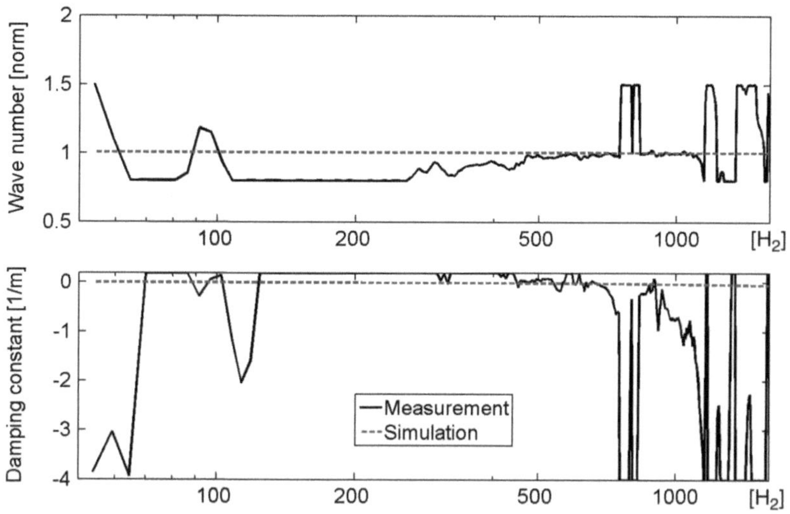

Fig. 12.3 Measured horizontal wave number in an empty tube with the microphones 2 and 10 cm in front of the hard backing. Expected value is $k_x = (1 + 0i)$

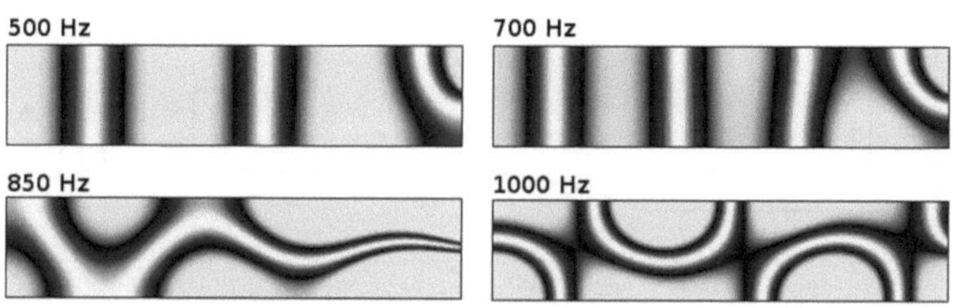

Fig. 12.4 Sound pressure distribution inside a channel of $20 \times 91\ \text{cm}^2$ with rigid walls and a 8 cm loudspeaker at the *right side. White indicates* low, black high pressure

pressure minimum in the top left corner leading to the low coherency of the microphone signals. Though this is unexpected from a room acoustics point of view, where one would always expect pressure maxima in the corners, this is simply the result of the source not being small compared to the channel dimensions. By use of a probe microphone, the course of the pressure minimum in this corner has been verified to agree well with the FE simulation.

Further simulations show that this unwanted "source effect" can be reduced by placing a second loudspeaker at the right wall, thus making the excitation more symmetric, and by reducing the height of the channel above the sample.

With the given setup, another option was to place the microphones further away from the reflecting wall, as then the propagation loss due to the absorbing material would reduce the interference between incident and reflected wave, hence increasing the pressure in the mimima. One must be aware that this places the microphones closer to the source and might reduce the accuracy of the method at low frequencies.

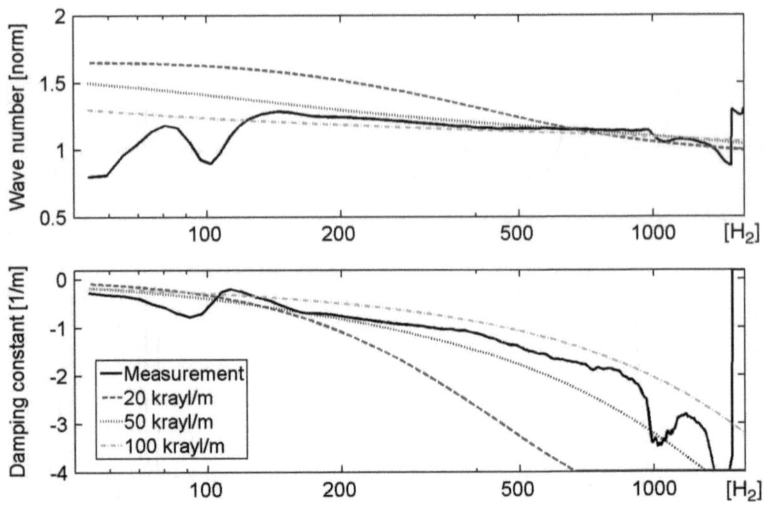

Fig. 12.5 Measured wave number above 9 cm polyurethane foam in comparison with the wave number predicted for a Delany–Bazley type absorber with three different flow resistivities

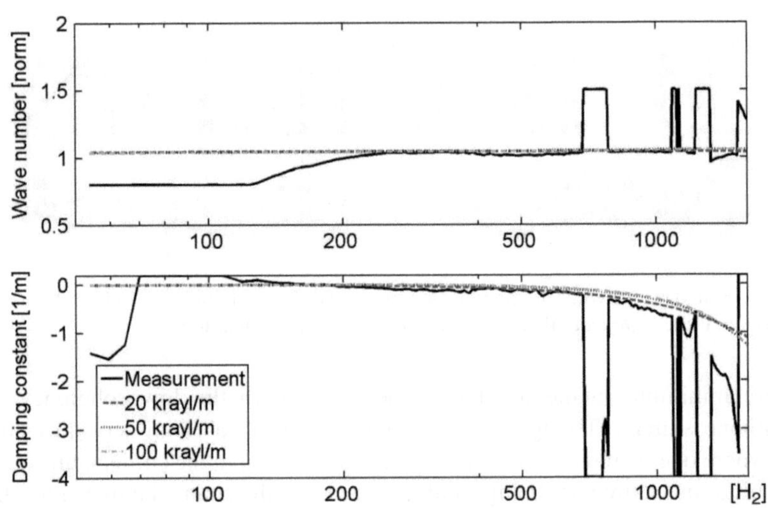

Fig. 12.6 Measured wave number above 1 cm polyurethane foam in comparison with the wave number predicted for a Delany–Bazley type absorber with three different flow resistivities

Figures 12.5 and 12.6 show the measured wave number in comparison with the wave number expected for a Delany–Bazley type absorber of same thickness. The microphones were now located 22 resp. 31.5 cm in front of the left wall. For the 9 cm thick foam, the (average) course of the wave number above 150 Hz agrees well with the one expected for a porous absorber with a flow resistivity between

50 and 100 krayl/m. Due to the high propagation loss, the sharp peaks as observed with an empty tube have disappeared.

The results for the otherwise comparable 1 cm thick foam do not prove satisfactory. On the one hand, due to the reduced absorption, the undesired extrema have reappeared. On the other hand, even the predicted courses of the wave number do hardly vary for flow resistivities of 20–100 krayl/m because a 1 cm thick layer simply does not produce a significant propagation loss within the investigated frequency range.

12.4.3 Summary and Conclusion

In 1966, Manfred Schroeder demonstrated his interest in measurement of the acoustic impedance through a publication in JASA, dealing with the impedance of the human vocal tract, and through a patent describing a two-microphone impedance tube. The knowledge of the (surface) impedance of materials and structures is required for room-acoustical simulations. The standard procedure ISO 10534-2 measures the impedance for normal incidence, though observations by Schroeder in relation to the seat effect indicate that a measurement at grazing incidence can be desirable.

In our study, we have advanced an existing but hardly known and inconvenient technique to gain a procedure which is similar to the ISO standard. The tested material is placed on the bottom of a rectangular impedance tube and the wave number in the air space above it is deduced from the sound pressure transfer function between two microphones. By use of a model for the sound propagation in such a situation, the absorber characteristics can be deduced. Measurements show that the new technique does still have its limitations: the sound field deviates from what the model predicts because the existence of the source of finite size is not taken into account. As a result, at low frequencies the incident field is not reasonably flat; additionally, pressure minima leading to poor coherency between the microphone signals occur at other locations than expected. Finite element simulations verify these findings. By careful choice of the microphone positions, credible results could be achieved for absorbent material samples. But for samples with low absorption, like a 1 cm thick open cell foam, the method is not sensitive enough to distinguish between materials with largely different flow resistivities, at least with the current measurement setup and geometry.

All in all, though improvements are still required, the new method has demonstrated its ability to determine the impedance at grazing incidence in a well-defined laboratory setting allowing for the comparison between different materials.

References

1. Schroeder, M.R.: Die statistischen Parameter der Frequenzkurve von großen Räumen. Acustica **4**(2), 594–600 (1954)
2. Schroeder, M.R.: New method of measuring reverberation time. J. Acoust. Soc. Am. **37**(3), 409–412 (1965)
3. Schroeder, M.R.: Diffuse sound reflection by maximum-length sequences. J. Acoust. Soc. Am. **57**(1), 149–150 (1975)
4. Schroeder, M.R.: An artificial stereophonic effect obtained from a single audio signal. J. Audio Eng. Soc. **6**(2), 74–79 (1958)
5. Harvey, F.K., Schroeder, M.R.: Subjective evaluation of factors affecting two-channel stereophony. J. Audio Eng. Soc. **9**(1), 19–28 (1961)
6. Schroeder, M.R.: Computer Speech: Recognition, Compression, Synthesis. Springer, New York (1999)
7. Schroeder, M.R.: Improvement of feedback stability of public address systems by frequency shifting. J. Audio Eng. Soc. **10**(2), 108–109 (1962)
8. Schroeder, M.R., Logan, B.F.: Colorless artificial reverberation. J. Audio Eng. Soc. **9**(3), 192 (1961)
9. Schroeder, M.R.: Natural sounding artificial reverberation. J. Audio Eng. Soc. **10**(3), 219–223 (1962)
10. United States Patent 3,346,067: System for determining acoustic reflection coefficients. Manfred R.Schroeder, Bell Telephone Laboratories (1967)
11. Schroeder, M.R.: Determination of the geometry of the human vocal tract by acoustic measurement. J. Acoust. Soc. Am. **41**(4), 1002–1010 (1967)
12. Atal, B.S., Schroeder, M.R., Sessler, G.M., West, J.E.: Evaluation of acoustic properties of enclosures by means of digital computers. J. Acoust. Soc. Am. **40**(2), 428–433 (1966)
13. Schroeder, M.R., Atal, B.S.: Computer simulation of sound transmission in rooms. Proc. IEEE **51**(3), 536–537 (1963)
14. Mommertz, E.: Angle-dependent in situ measurement of reflection coefficients using a subtraction technique. Appl. Acoust. **46**, 251–263 (1995)
15. ANSI/ASA S1.18: American National Standard Method for Determining the Acoustic Impedance of Ground Surfaces (2010)
16. Schroeder, M.R., Atal, B.S., Sessler, G.M., West, J.E.: Acoustical measurements in Philharmonic Hall (New York). J. Acoust. Soc. Am. **40**(2), 434–440 (1966)
17. Chung, J.Y., Blaser, D.A.: Transfer function method of measuring in-duct acoustic properties. I. Theory. J. Acoust. Soc. Am. **63**(3), 907–913 (1980)
18. Müller, W.: Messung von Absorptionseitenschaften von porösen Materialien bei streifendem Einfall (measurement of absorption properties of porous material at grazing incidence), Diplomarbeit, University of Oldenburg (in German) (1990)
19. Scott, R.A.: The propagation of sound between walls of porous material. Proc. Phys. Soc. **58**, 358–368 (1946)
20. Mechel, F.P.: Schallabsorber, vol. I, chapter 11.1: "Ebene Wellenleiter". Hirzel Verlag, Stuttgart (in German) (1989)
21. Kruse, R., Mellert, V.: Effect and minimization of errors in in situ ground impedance measurements. Appl. Acoust. **69**(10), 884–890 (2008)
22. Delany, M.E., Bazley, E.N.: Acoustical properties of fibrous absorbent materials. Appl. Acoust. **3**, 105–116 (1970)

***

Biography

Dr. **Roland Kruse** received his Ph.D. degree in 2009 from the University of Oldenburg, under the supervision of Prof. Volker Mellert, for his work on the in situ measurement of the acoustical ground impedance. From 2009 to 2010 he continued his research at the Acoustics Group, The Open University, Milton Keynes studying the influence of the environmental conditions on source localization outdoors. After returning to the Acoustics Group at Oldenburg University, he became a lecturer at the Technical University of Braunschweig in 2011. His current research is aimed at finite element technologies and material parameter identification for biological and other composite materials.

Prof. **Volker Mellert** was head of the Acoustics Group at the Institute for Physics of Oldenburg University and retired in 2009. He received his Ph.D. in physics in Göttingen 1972 and became a Professor for Applied Physics in Oldenburg in 1976. He was President of the German Association for Acoustics (DEGA) and of the European Association for Acoustics (EAA). He served as board member of the International Commission of Acoustics and as Editor-in-Chief for ACUSTICA. He received the German Medal of Merit for his contributions in acoustics and the Helmholtz-Medal of the DEGA. He is Fellow of the Acoustical Society of America.

Chapter 13
Research on Underwater Acoustics in Göttingen

Dieter Guicking

Abstract An overview is given of the research work on hydroacoustics at the "Drittes Physikalisches Institut" of the University of Göttingen (Germany) from 1947 till 2002, under the leadership of Erwin Meyer, Manfred Schroeder, and Werner Lauterborn. Details are presented on absorbers of underwater sound, on the Institute's experimental facilities (most of them newly developed), on the statistical sound field theory, on the measurement of sound absorption and radiated power in a reverberation tank, and on natural vibrations of and sound radiation from flat plates and cylindrical shells submerged in water. Brief summaries are given on the investigation of viscometers, on self-imaging in dispersive hydroacoustic wave-guides, on Schroeder diffusers for waterborne sound, on resonance scattering as applied to localize metallic objects at the sea floor, and on the construction of adjustable absorbers of underwater sound using electrorheological fluids. Much of the material has previously been published in German only. The bibliography comprises 64 entries.

13.1 Background

When Manfred Schroeder in 1968 returned to Göttingen as Director of the "Drittes Physikalisches Institut" in succession of the Institute founder Erwin Meyer, he took over a research contract with the British Navy by which investigations on certain aspects of underwater sound were sponsored. During WWII, Meyer and his team—then at the Heinrich-Hertz Institute for Vibration Research in Berlin—were forced by the Authorities to work on sound propagation and sound absorption in water, a consequence of the emerging sonar technology by which most of the German

D. Guicking (✉)
Drittes Physikalisches Institut der Universität Göttingen, Friedrich-Hund-Platz 1, 37077 Göttingen, Germany
e-mail: Dieter.Guicking@phys.uni-goettingen.de

N. Xiang and G.M. Sessler (eds.), *Acoustics, Information, and Communication*, Modern Acoustics and Signal Processing, DOI 10.1007/978-3-319-05660-9_13,

submarines were detected and destroyed. Meyer, who disliked any war and other conflicts, was convinced that the Nazi regime would not persist very long and saved secretly copies of his classified reports on the hydroacoustics research (ignoring the mortal danger if discovered) and offered them after the war to the British and American occupying forces with the intention to get the material published for international access. The resulting extensive report [1] is still today estimated by researchers on hydroacoustics all over the world as a treasury of important fundamentals and detailed experimental results.

In 1947, Meyer was appointed Director of the newly founded third Physics Institute at the University of Göttingen, and in 1948 he succeeded in placing the research contract with the British Navy, initially for 1 year, but the contract was extended year after year for a record breaking period of 30 years until 1978—long beyond Meyer's death (March 1972). Main topics of research were the further development of absorbers of underwater sound, sound absorption in sea water and related liquids, development and improvement of experimental techniques, later on sound radiation from submerged objects, and, finally, (with another sponsor) resonance classification of metallic objects on the sea bottom. Manfred Schroeder did not involve himself in underwater acoustics very much, but he supported the hydroacoustics group and observed their achievements with interest. A close cooperation with him on statistical sound fields resulted from investigations on a "reverberation tank" (Sect. 13.3.7) and a "shallow water basin" (Sect. 13.3.6). Much of this material has been published in German only (except for the reports delivered to the Navy, which were—of course—not available to the public). A selection of the research projects is outlined here in English for the first time unless otherwise referenced.

Not included are two major hydroacoustic research fields of the Institute which have been documented elsewhere. Continuing the earlier work of Meyer's team in Berlin [1], the sound absorption and dispersion in aqueous solutions and other liquids were studied, now with emphasis on the molecular mechanisms. The extensive investigations with acoustic and microwave technologies covered a very wide frequency range. The team was headed by Reinhard Pottel and Udo Kaatze ([2], 13th and 14th Chapter, pp. 333–404). Also continuing the work described in [1], acoustic cavitation and sonoluminescence were studied in Göttingen from the Institute's founding until today, since 1968 by Werner Lauterborn and his team ([2], 6th and 7th Chapter, pp. 139–198).

13.2 Resonance Absorbers of Underwater Sound

During the war time, Meyer's team developed a thin-layer sound absorbing rubber coating for possible application to submarines. The absorber carried the cover name "Alberich" after a Germanic mythical dwarf who became invisible by a magic hat. The absorber was very effective, but to Meyer's knowledge not applied in combat until to the end of the war. However, in 1998 the wreck of the German submarine U-480 was detected by chance in the sediment of the British Channel. Divers saw

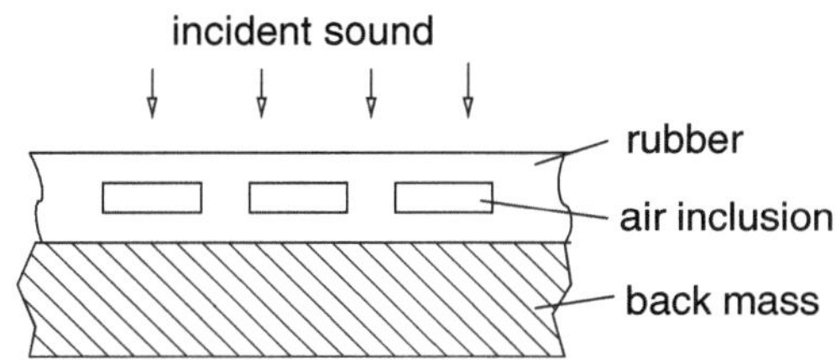

Fig. 13.1 "Alberich" absorber

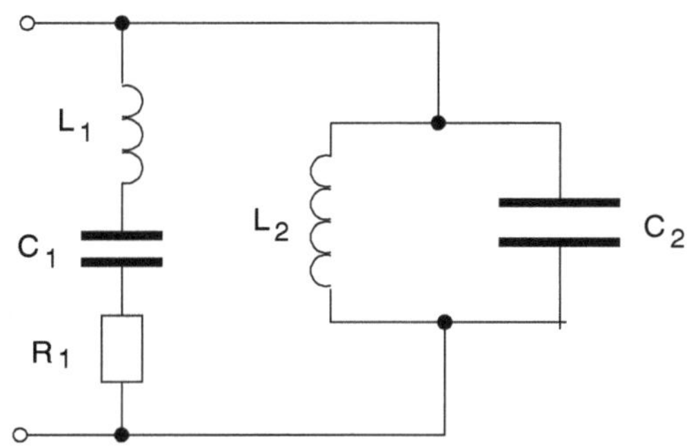

Fig. 13.2 Electrical equivalent of a mechanical two-circuit resonance absorber

the characteristic Alberich coating and realized that they had found the first "Stealth U-boat" ever built. The TV transmitter *arte* has produced an emotional film (in German) about U-480 and its history (successful operation in the Channel in August 1944, but destroyed by a mine on 24 February 1945). The film was broadcast on February 18, 2008 [3].

Figure 13.1 shows the principle of the absorber. The air inclusions in the rubber layer act as mechanical parallel resonance circuits. The sound pressure incident on the rubber surface causes compressions and depressions of the air, the vibrating rubber layer above the air cushions acts as a membrane with specific mass M_1, its bending stiffness and the air hole compression give the spring part F_1, and the loss factor of the bending vibrations determines the resistance W_1. Applying the force/voltage analogy, the electrical equivalent of a mechanical parallel circuit is a series circuit, $M_1 \Rightarrow L_1$, $F_1 \Rightarrow C_1$, $W_1 \Rightarrow R_1$. The back mass $M_2 \Rightarrow L_2$ (e.g., a submarine hull) and the resilient rubber layer with spring constant $F_2 \Rightarrow C_2$ form a mechanical series resonance circuit, i.e., an analog electrical parallel circuit, see Fig. 13.2. If both circuits are tuned to the same resonance frequency ω_0, the input impedance of the system varies rather slowly in the vicinity of ω_0, and the bandwidth of the absorber increases with the loss factor of the rubber. In the war-time development of "Alberich," the loss factor of the synthetic rubber ("Buna S") was enhanced by about 40 % carbon black filler ([1], p. 296).

After the war, more and more plastic materials with a wide variety of elastomeric properties became available, and Meyer hoped to improve the resonance absorbers by a suitable polymer instead of the rubber. It was no problem to find materials with very high loss factors, but it turned out soon that high loss is inevitably related to a strong temperature (and frequency) dispersion of the elastic modulus

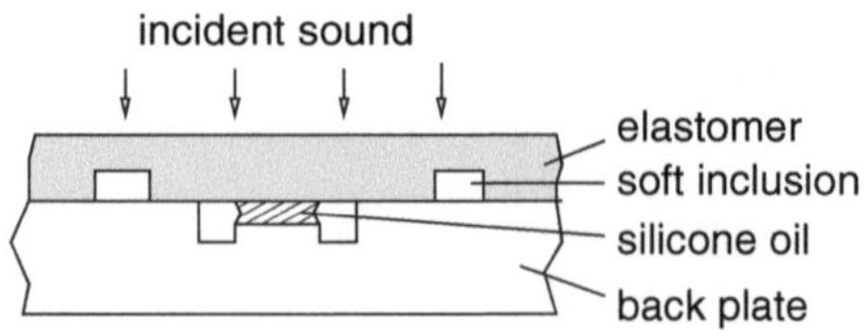

Fig. 13.3 Squeeze film absorber

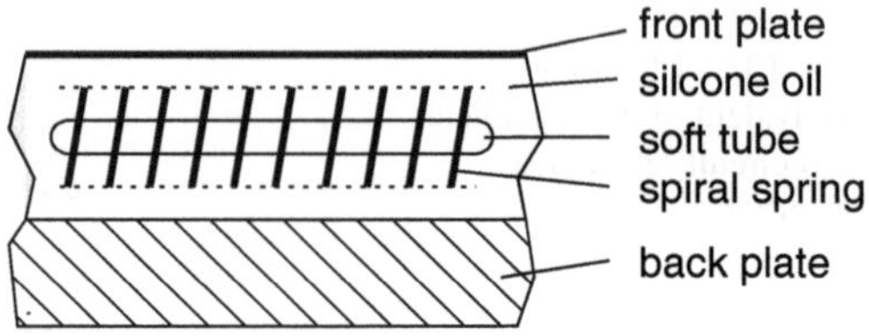

Fig. 13.4 Coil spring absorber

(quantitatively described by the Kramers–Kronig relations) so that highly effective absorbers of this type are useful only in a narrow temperature interval which prohibits practical application.

Searching for an alternative loss mechanism with only moderate temperature dependence, viscous flow of thin fluid layers was chosen. Figure 13.3 shows an early design where between the massive back plate and an elastic cover layer a thin film (0.25–0.5 mm) of viscous silicone oil is placed which serves as absorbing element. The incident sound wave deforms the elastic cover layer which acts like a piston squeezing the fluid into the ring slot and back, thereby causing viscous loss (the slots are filled with compressible material, e.g., closed-cell sponge rubber). The cover layer is a nearly lossless castable polymer (e.g., silicone rubber) which is adjusted to the specific impedance of water by soft inclusions of sponge rubber. Again, this mechanism forms a damped mechanical parallel circuit, and the cover layer with the back mass a rather lossless series circuit so that the electrical equivalent circuit of Fig. 13.2 is valid for this absorber type, too. Test samples of 5 cm diameter with a few such "flow elements" were tested in a hydroacoustic impedance tube ("Kundt's tube") and showed a frequency bandwidth (reflection coefficient below 20 %) of about 20 kHz in a temperature interval of more than 20 °C, which is better than achieved with pure relaxation loss [4–7].

This absorber construction was not suited for applications at variable depth of water because static pressure reduces the thickness of the silicone oil layer and thereby alters the flow resistance. Less sensitive to ambient pressure was the construction sketched in Fig. 13.4. Central part is a slightly stretched coil spring so that the surrounding silicone oil can oscillate through the slits between the coil turns. Inside the spring is an air-filled soft plastic tube to enable compression. Again, the setup is a two-circuit resonance absorber. With a slit width of 0.1 mm, a wire diameter of 0.4 mm, viscosity of the silicone oil of 1,000 cSt, and a back plate of 9 mm brass, the reflection coefficient stayed below 10 % from 4.5 through 13 kHz. With slightly modified dimensions, a really broadband absorber of 13.5 cm total thickness was developed, consisting of a 10.5 cm wedge-type

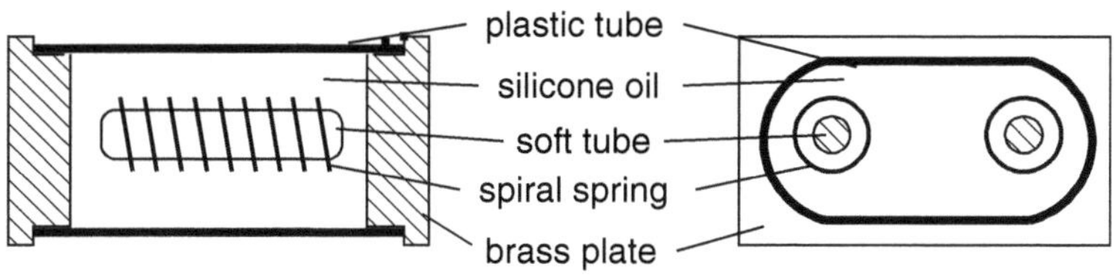

Fig. 13.5 Absorber element

absorber backed by a coil spring absorber. Its bandwidth extended from 5 kHz up to very high frequencies [8].

As a step towards commercial production, compact absorber elements were developed, consisting of an oval plastic tube with two coil spring sections inside (Fig. 13.5). The idea was to produce many such elements, fasten them on a structure to be coated, and then to embed the elements by a casting resin. The absorption coefficient of an optimized absorber of this type fell below 12 % in the frequency interval from 3.5 through 10 kHz.

The research contract with the British Navy ended in 1978 because they would no longer support unclassified research. Since also the respective German authorities insisted on classification of research results sponsored by them, Manfred Schroeder and myself decided not to continue work on absorbers of underwater sound because it was important for us to do research work within the scope of diploma and doctoral dissertations which have to be published. For a later resumption see Sect. 13.11.5.

13.3 Experimental Facilities

The hydroacoustics laboratory was equipped with conventional devices and several newly developed special appliances.

13.3.1 Deep Water Tank

A large anechoic water tank, 7 m long, 4 m wide, and 4 m deep, was installed beneath the basement of the institute [10]. The tank was lined with rib-type absorbers [11], the reflectivity of which is lower than 10 % between 7 and 70 kHz. Figure 13.6 shows a view into the empty tank, here with an installation for the measurement of sound scattering from sound-soft spheres [12].

Fig. 13.6 Deep water tank

13.3.2 Impedance Tube

Several thick-walled steel tubes (Kundt's tubes) with 5 cm inner diameter and up to 3 m length facilitated impedance measurements of absorbers by the pulse-echo method ([1], Chap. IX). A PZT transducer at the lower end of the upright standing tube transmits a sine-wave pulse of typically six periods and receives a sequence of echo pulses, reflected between the test object at the upper tube end and the transmitter. The first echo is compensated electronically. Amplitude and phase of the compensation signal are compared to the corresponding values obtained with a sound-soft reference reflector (or the free water surface) and allow the complex impedance of the test object to be calculated. Measurements were possible in the frequency range from about 3 kHz through 15 kHz. At lower frequencies the primary pulse and the echo would overlap, and at higher frequencies tube modes would be excited, violating the assumption of plane waves.

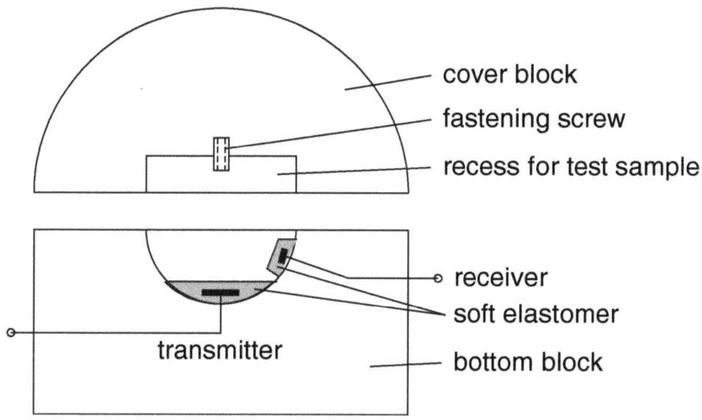

Fig. 13.7 Pressure chamber for impedance measurements

13.3.3 Pressure Chamber

In an attempt to extend impedance measurements to lower frequencies than possible with the impedance tube, a pressure chamber was developed for the frequency range from about 200 Hz to 5 kHz. After preliminary experiments the construction as shown in Fig. 13.7 turned out to be optimal. The bottom block (brass cylinder, diameter 140 mm, height 70 mm) contained a semispherical cavity (diameter 52 mm) with a transmitter (PZT or $BaTiO_4$ disk, diameter 20 mm) at the bottom and a receiver (diameter 5 mm), mounted about 3 cm apart. In order to avoid mechanical cross talk between the transducers via the chamber body, they were embedded in a very soft casting resin, the surface of which was silver coated to reduce electrical cross talk. The semispherical brass cover block contained a 15 mm deep recess of 52 mm diameter in which the test sample could be fastened by a headless screw. The two brass blocks were bolted together by six strong screws (not shown). In order to avoid air bubbles in the chamber, the two parts were assembled in a water bath. The cross talk attenuation exceeded 40 dB, which was sufficient for absorber measurements.

Assuming that the inner chamber dimensions are small compared to the wavelength, an a.c. voltage u_t applied to the transmitter creates a sound pressure p which is constant over the chamber volume. The complex compliance $\underline{C}_s$ of a sample introduced into the chamber is

$$\underline{C}_s = \frac{1}{C_0}\left[\left(\frac{u_t}{\underline{u}_r}\right)_s - \left(\frac{u_t}{\underline{u}_r}\right)_0\right] + \kappa V \tag{13.1}$$

where C_0 is the calibration constant of the chamber, to be determined with a sample of known compliance; $\left(u_t/\underline{u}_r\right)_s$ is the complex ratio of input and output voltage with the test sample in the chamber; $\left(u_t/\underline{u}_r\right)_0$ is the complex ratio of input and output voltage if the chamber is filled with water only; κ is the compressibility of the chamber fluid (here: degassed water); and V is the volume of the test sample.

Then the acoustic impedance $\underline{Z}_s$ of the test sample is

$$\underline{Z}_s = \frac{1}{j\omega \underline{C}_s}. \tag{13.2}$$

More details, including electrical equivalent circuits, calibration results, and test measurements with different fluids and resonance absorbers, have been published in [13].

13.3.4 Resiliently Lined Water Duct

The necessity of small test samples in impedance tubes or a pressure chamber is in general favorable; it can, however, be disadvantageous in case of inhomogeneous absorbers where a small sample does not represent the acoustic properties, or if it cannot be produced from constructional reasons. This is especially problematic at higher frequencies: the width of an impedance tube for measurements up to 80 kHz would be as small as 1 cm.

In order to enable hydroacoustic impedance measurements of "large area" specimens, a resiliently lined water duct of 2 m length and 34 cm width with open water surface at a maximum height of 34 cm was constructed (it is the acoustic analogue of a square waveguide in microwave technology). In the basic experimental design, a test sample of size 34 × 34 cm filled the cross-section at one end of the duct, an equally sized capacitive transmitter was placed at the other end. The measurements were made with a procedure known from the Kundt's tubes ([1], Chap. IX): a hydrophone is moved lengthwise through the duct, evaluated is the standing wave ratio and the shift of a pressure minimum in comparison to a sound-soft reference reflector (the modern two-microphone technique with standing-wave separation would also be applicable).

In contrast to hard-walled waveguides (which can hardly be realized for water-borne sound), the sound-soft edge conditions (acoustic "short circuit") cause dispersion and lead to a higher phase velocity in the duct than in the free field. If f is the frequency of a sinusoidal wave, the free-field wavelength is $\lambda_0 = c_0 f$ with the free-field sound velocity c_0 (1,484 m/s in water at room temperature). Sound propagation between sound-soft walls can be understood as zigzag reflection ([27], Sect. 2.1.3). The oblique lines in Fig. 13.8 indicate wave fronts of a wave propagating with free-field sound velocity c_0, incident on the test object at an angle $\varphi > 0°$. Zero sound pressure at $z = 0$ and $z = h$ requires $h = n\lambda_z/2 = n\lambda_0/2 \sin\varphi$, $n = 1, 2, \ldots$ The phase velocity in x-direction is $c_x = \lambda_x f = c_0/\cos\varphi$. For a water height h and square cross-section, a unique wave mode (the 1,1-mode, $n = 1$) exists if λ_z lies in the range $h < \lambda_z < 2h$. If the hydrophone slides exactly along the center line, the next duct mode ($h = \lambda_z$) is suppressed because it has a pressure node at the center line so that the usable frequency range is extended to $2h/3 < \lambda_z < 2h$. The angle of incidence is

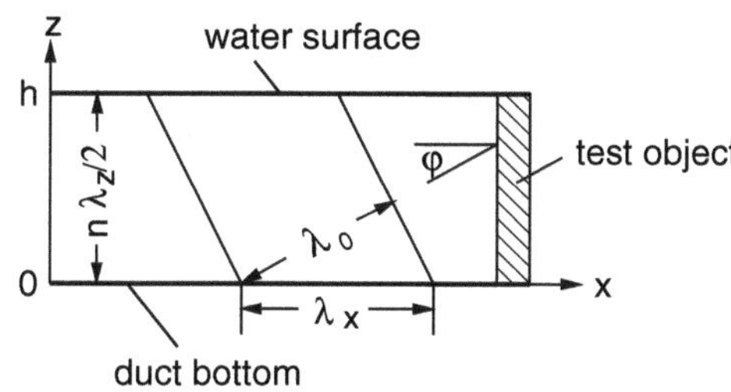

Fig. 13.8 Resiliently lined water duct

$$\varphi = \arccos\frac{c_0}{c_x} = \arccos\left(\sqrt{1-(\lambda_0/2h)^2}\right), \tag{13.3}$$

which means that it is frequency dependent at constant water height (and width in the square duct). To avoid this, measurements at different frequencies can be made with correspondingly adjusted duct cross-section, realized by a movable length-wall, and adjusted height of water. At 6.8 kHz the full cross-section of 34 × 34 cm is appropriate, and at 15 kHz: 15.5 × 15.5 cm. The angle of incidence in both cases is $\varphi \approx 18°$. Experimental results for various absorbers have been published in [8, 9].

13.3.5 Wide Impedance Tube

Since the aforementioned water duct applies to comparatively low frequencies only, we developed a device for impedance measurements in water up to about 200 kHz, employing a tube with diameter much larger than the wavelength. The problem of generating a plane wave over the whole tube cross-section was solved with a sound-soft paraboloid reflector, machined from polystyrene foam ("Styrofoam"), as sketched in Fig. 13.9. Two miniature PZT transducers were placed at the focal point. Their separation was much smaller than a wavelength at the highest frequency of interest. The usual pulse-echo method was applied, demanding for a minimum pulse length of six periods. Thus, in order to avoid overlap of signal and echo, the tube length must exceed three wavelengths at the lowest measuring frequency. Since, on the other hand, the tube width should be much larger than a wavelength, the tube had—in contrast to Kundt's tubes—not to be much longer than wide. The "screen" above the transducers (made from an elastomer filled with cork particles) served as a wedge-type absorber for the spherical direct sound wave which would interfere with the reflected pulse. The screen was fixed by three thin, stiff wires which did not disturb the sound field. The tube wall above the paraboloid was also lined resiliently in order to weaken the discontinuity at the bend which is source of a weak diffraction wave. An advantage of the parabolic reflector is that disturbance waves are not focused to the receiver and thus do not significantly contribute to the output signal. The PVC stop at the top served to fix the foam against buoyancy, and it was the support for the test samples resp. the reference

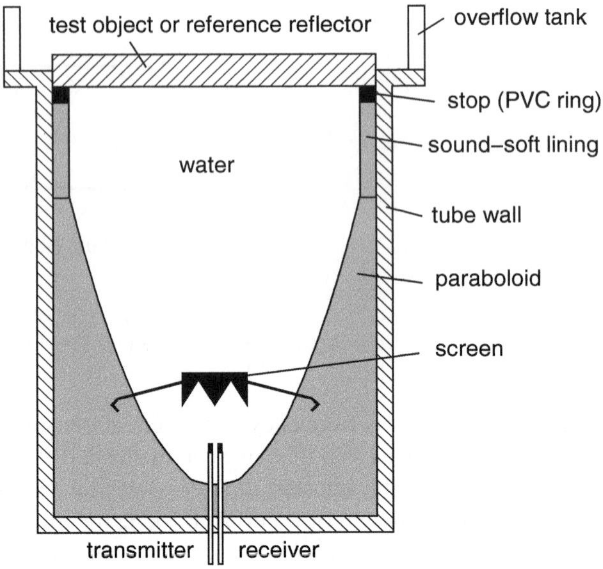

Fig. 13.9 Wide impedance tube (not to scale)

reflector. The "overflow tank" allowed to raise the water level so that the test objects could be inserted obliquely to remove air bubbles (which would strongly impair the experimental results).

Two tubes were built: Tube I with inner diameter 29 cm and tube length 60 cm, for frequencies above 10 kHz, and tube II with inner diameter 12.5 cm and length 30 cm, for frequencies above 30 kHz. With the sound-soft reference reflector, sequences of typically four echoes were observed. The impedance measurements were performed as with the standard impedance tube by compensating the first echo pulse and comparing amplitude and phase of the compensation signal with the corresponding values for the reference reflector.

More details about the experimental technique and measurements with water layers of varying height as well as various absorber samples have been published in [14].

13.3.6 Shallow Water Basin

A two-dimensional sound field is suited for certain experimental investigations, e.g., sound scattering, radiation, and absorption measurements, and also to check theoretical predictions for two-dimensional statistical sound fields, see Sect. 13.4.2.

To this end, a shallow water basin (4 m long and 3 m wide) with horizontally leveled chipboard bottom and 30 cm high sidewalls was constructed. Sheets of 3 mm thick polystyrene foam were laid on the chipboard and served as sound-soft

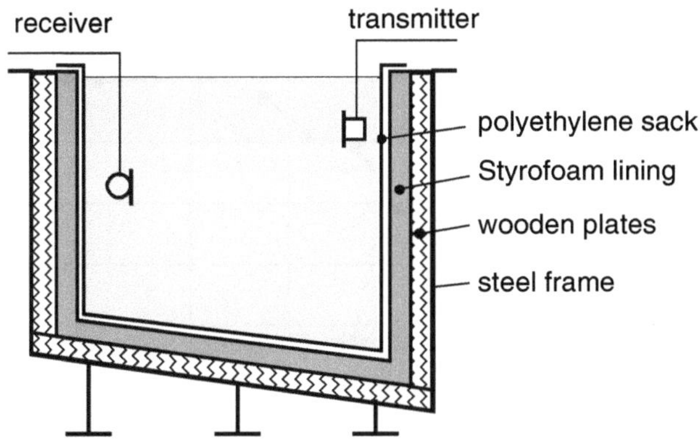

Fig. 13.10 Reverberation tank

underside. A watertight plastic foil covered the foam and was pulled up along the sides. The sidewalls could be lined with sound-soft reflector plates to create a two-dimensional reverberation basin, or by rib-type absorbers. The sound-field in the basin was scanned by a small hydrophone (Celesco LC-5) with motor-driven support. It could be moved either in two directions through the whole basin area with constant speed and at constant depth, or by a rotating cantilever arm along a semicircle to record sound radiation patterns of objects placed near the center of one long side. The speed of the hydrophone motion was slow enough to avoid generation of surface waves and eddies in the water.

13.3.7 Reverberation Tank

In analogy to the reverberation rooms in airborne acoustics for the measurement of sound absorption and sound power radiation, a reverberation tank for similar hydroacoustic measurements was constructed as sketched in Fig. 13.10. The tank had an irregular shape as common for reverberation rooms. The walls consisted of wooden boards which were supported by a steel frame. In contrast to former reverberation tanks with steel walls [15–17], this tank was lined with polystyrene foam ("Styrofoam" of Dow Chemical), glued to the wooden plates. The mean tank dimensions were $1.60 \times 1.43 \times 1.03$ m. The sound-soft lining covered an area of 8.7 m^2, the free water surface was 2.3 m^2, and the water volume 2.4 m^3. The water was softened to prevent lime precipitation, and it was degassed to avoid air bubbles which are known to absorb sound ([1], Chap. II).

The reverberation time T_{60} of the tank in this state varied between 80 ms and 160 ms in the frequency range from 4 kHz through 200 kHz. It was found in preliminary experiments that the reflectivity of the polystyrene foam could be enhanced drastically by coating it with a thin plastic foil (polyethylene, thickness

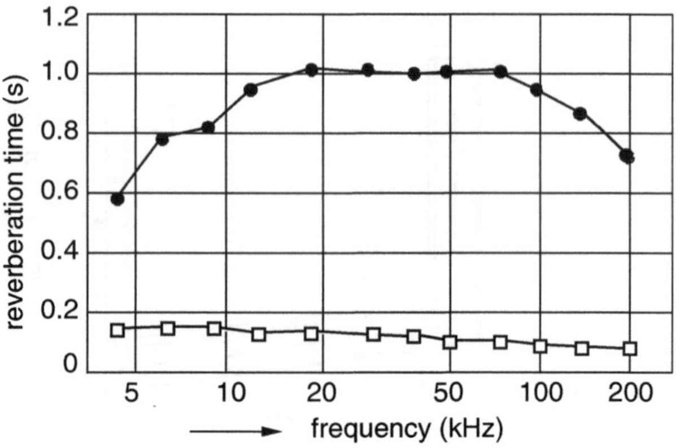

Fig. 13.11 Reverberation time of the water tank. *Lower curve*: without plastic foil, *upper curve*: with foil

150 μm): T_{60} exceeds 500 ms in the whole frequency range with a maximum of 1 s between 18 and 70 kHz [18], see Fig. 13.11. The reason for this effect is that the water is prevented from penetrating into the narrow surface pores of the foam where it forms tiny flow absorbers. The absorption coefficient of the tank walls was about 0.02, the Q-factor reached 1,400 at 5 kHz, 23,000 at 50 kHz, and 62,000 at 200 kHz.

13.4 Statistical Sound Field Theory

13.4.1 Three-Dimensional Sound Field

Manfred Schroeder, as a young student at Erwin Meyer's Institute, has developed the statistical theory of reverberant sound fields for airborne sound [20–23], but the formalism can be applied to water with slight modifications. Experiments with the reverberation tank were performed in the scope of a Ph.D. Thesis [19] and have not been published somewhere else. Therefore, they shall be described here in some detail. The measurement of absorption coefficients requires reliable values of the reverberation time. Sufficient accuracy can be achieved only by averaging over many independent measurements. This is also true for determining the total acoustic power radiated from vibrating objects. Strategies for getting independent measurements can be deduced from Schroeder's theory.

A reverberation room is a resonator, operated in the frequency range of overlapping modes. Manfred Schroeder has stated in [22] that the average spacing df of the resonance frequencies must be smaller than one-third of the halfwidth Δf of a resonance peak, which is related to the reverberation time T_{60} by

$$\Delta f = \frac{3 \ln 10}{\pi T_{60}} \approx \frac{2.2}{T_{60}}. \tag{13.4}$$

The value of df can be obtained in the following way. The number N of modes in a room with volume V below a frequency limit f_g is

$$N = \frac{4\pi V}{3c^3} f_g^3 \tag{13.5}$$

where c is the sound velocity. (The usually added correction terms for tangential modes in reverberation rooms do not apply here because such modes do not exist on sound-soft walls.) The average interval df between two successive eigenfrequencies, as calculated from (13.5), is

$$\mathrm{d}f = \frac{c^3}{4\pi V f^2}, \tag{13.6}$$

provided $\mathrm{d}f \ll f$. The condition of $\Delta f/\mathrm{d}f > 3$ was fulfilled in the reverberation tank above 4 kHz.

A stationary sound field in the reverberation tank can be understood as the superposition of resonant modes which are excited by the transmitter signal (nearly 200 modes in the lowest third-octave band about 4.5 kHz; the number increases in proportion to f^3 due to (13.5)). An essential change of the sound field, i.e., a new mode configuration, is achieved by a frequency shift greater than

$$\delta f = \frac{\Delta f}{2} \approx \frac{1.1}{T_{60}} \tag{13.7}$$

if the modes overlap. An equivalent variation can be reached by a volume change δV, in the reverberation tank easily done by draining some water. According to (13.5), δV is related to δf by

$$\frac{\delta V}{V} = 3\frac{\delta f}{f}. \tag{13.8}$$

Thus, the volume change δV is equivalent to δf of (13.7) if

$$\delta V = 3\frac{V}{f} \cdot \frac{\Delta f}{2} \approx \frac{3.3V}{f T_{60}}. \tag{13.9}$$

Another way to alter the sound field is to change the room shape, e.g., by submerging an object in the tank. Obviously, the sound field at the hydrophone location is changed significantly if the object volume is of the order of δV as given by (13.9).

The sound field modification by changing the tank shape is also equivalent to changing the hydrophone position. For two points half a wavelength apart, the spatial cross-correlation coefficient of the sound energy vanishes, and it stays below 0.05 for larger distances. This was proved for excitation with a single frequency or narrow-band noise [24]; for excitation by third-octave noise the spatial decay of the cross-correlation function will be still faster. Thus, the sound fields at positions half an average wavelength apart (or more) are statistically independent.

Summarizing, there are four possibilities to obtain statistically independent measures of the sound field in the reverberation tank:

(a) Change of the excitation signal by frequency shift.
(b) Change of water volume.
(c) Submerging objects.
(d) Different hydrophone positions.

Assuming Gaussian processes, the four methods are theoretically equivalent, and their effects on measurable quantities (e.g., stationary sound pressure level and reverberation time) can be predicted theoretically. According to [23, 24], the standard deviation σ_L of the stationary sound pressure level L depends on the reverberation time T_{60}, the bandwidth B of the excitation signal, and the number n of independent measurements:

$$\sigma_L = \frac{5.57}{\sqrt{1 + 0.238 T_{60} B} \cdot \sqrt{n}} \text{ dB}, \tag{13.10}$$

assuming interference of at least ten excited modes.

From experimental reasons, the time interval $T_{15} = T_{60}/4$ for 15 dB decay was measured instead of T_{60}, in practice as the time interval between two instants at which the reverberation curve crossed for the first time two pre-adjusted levels $\Delta L = 15$ dB apart. The scatter of the switching instants is $\sigma_t = \sigma_L \cdot \Delta t / \Delta L$ where $\Delta t / \Delta L = T_{60}/60$ dB is the slope of the reverberation curve. The dependence of T_{15} on two such measurements leads to $\sigma_{T_{15}} = \sqrt{2}\sigma_t$ and $\sigma_{T_{60}} = 4\sigma_{T_{15}}$, thus with (13.10)

$$\frac{\sigma_{T_{60}}}{T_{60}} = \frac{\sqrt{2}}{15}\sigma_L = \frac{5.57 \cdot \sqrt{2}}{15\sqrt{1 + 0.238 T_{60} B}\sqrt{n}}. \tag{13.11}$$

The bandwidth $B = 0.2316f$ of third-octave band noise yields

$$\frac{\sigma_{T_{60}}}{T_{60}} = \frac{0.525}{\sqrt{1 + 0.0551 T_{60} f} \cdot \sqrt{n}} \approx \frac{2.24}{\sqrt{T_{60} f n}} \tag{13.12}$$

if the term "1" under the root is neglected.

The scatter of reverberation time measurements was investigated experimentally. In a first series, the scatter was recorded for excitation with pulses of third-octave band random noise about 6.4 kHz. Without averaging ($n = 1$) and $T_{60} \approx 0.75$ s, (13.12) yields a value of 0.032 or 3.2 %. The finite 3 % resolution of the level recorder used in the experiments has to be considered, too: $\sqrt{3.2^2+3^2}=4.4\,\%$, which is a bit smaller than the measured standard deviation of 5.6 %. Averaging over 25 repetitive recordings should reduce the scatter to $1/\sqrt{n}=1/5$ of the former value, i.e., 0.9 %, and this was found experimentally.

While in this test the sound field was kept constant, in subsequent experiments the sound field was modified by the above-mentioned methods (b) and (d). In one series, the tank was excited by third-octave band random noise about 6.4 kHz. The water volume was stepwise reduced 25 times by decrements of $\Delta V=\delta V/2=1.65V/(fT_{60})=1{,}400\ \text{cm}^3$. For each of these 26 volumes, 25 repetitive decay processes were recorded. Single values for different tank volumes jumped statistically with a standard deviation of 7 %. Averaging over the 25 repetitions reduced this value to 2 %. Similar values were obtained from measurements with third-octave band noise around 12.7, 25.5, 50.9, and 101.9 kHz.

In a further series of scatter measurements, the water volume and the transmitter position were kept constant, but the receiver, a small hydrophone of the type Celesco LC-5, was shifted 25 times by $\Delta x=\lambda/4$ of the mean frequency, taking again 25 repetitive measurements for each hydrophone position. The displacement of the hydrophone support (a thin brass tube) modified the mode pattern a little, thereby further reducing the correlation of subsequent data. As excitation noise pulses, the same frequency bands were applied as before. The acoustically active PZT element of the hydrophone was a cylinder of 3.5 mm length and 2 mm diameter which is much smaller than the shortest wavelength used so that spatial averaging by the finite hydrophone size is insignificant. The measured scatter was almost the same as in the series with volume change, confirming the theoretical predictions.

13.4.2 Two-Dimensional Sound Field

A two-dimensional sound field can also be treated with statistical methods for higher frequencies where sufficiently many modes overlap. As in the resiliently lined water duct (Sect. 13.3.4), the basin wavelength λ_b depends on frequency $f=c_0/\lambda_0$ and water depth h:

$$\lambda_b=\frac{\lambda_0}{\sqrt{1-\left(\frac{\lambda_0}{2h}\right)^2}} \tag{13.13}$$

(c_0 and λ_0 are the free-field sound velocity and wavelength). According to Cremer [25], the number of eigenfrequencies below a limit f_g is

$$N=\frac{\pi S f_g^2}{c_b^2} \tag{13.14}$$

where S is the surface area of the basin and $c_b = \lambda_b f$ the phase velocity of the waves in the basin. Other than in the 3D field, N increases with f_g^2 instead of f_g^3 (13.5). From (13.14) we get the mean spacing between adjacent eigenfrequencies in the vicinity of f_g:

$$\mathrm{d}f = \frac{c_b^2}{2\pi S f_g} \tag{13.15}$$

(instead of (13.6) for the 3D field). A unique sound pressure distribution in the vertical direction occurs for water depths h if

$$\frac{\lambda_0}{2} < h < \lambda_0. \tag{13.16}$$

For $h = \lambda_0$, (13.13) yields $\lambda_b = 1.155\lambda_0$ as lower limit of λ_b. With the condition of Schroeder's large room frequency f_{cr}: $\mathrm{d}f/\Delta f \geq 3$, Δf from (13.4), we get for the shallow water basin

$$f_{cr} = \frac{c_b^2}{4S} T_{60}. \tag{13.17}$$

Applying this to our basin (3.9 × 2.9 m within the resilient lining) with $T_{60} \approx 0.5$ s, $S = 11.4$ m^2, and a two-dimensional phase velocity of $c_b = 1{,}700$ m/s, we calculate $f_{cr} = 32$ kHz, or a maximum water depth of 5 cm for the statistical theory to be applicable.

Theoretical probability density functions of the complex sound pressure amplitudes and phases, of the sound pressure level, of spatial frequencies in an isotropic diffuse sound field and in a semi-diffuse sound field, statistical parameters of the spatial transfer functions, energy flux density and phase rotation and derivatives have been derived and published in [26], together with experimental results from single-frequency wave fields in the shallow water basin. The experimental data confirmed the theoretical predictions, except for excitation frequencies below f_{cr} if one eigenmode dominates. The semi-diffuse sound field was realized by lining part of the basin with wedge-type absorbers, and with maximum-length diffusers elsewhere. It turned out that the weak reflections from the absorber cause a small maximum in the power spectrum at negative spatial frequencies and modulate the spatial pattern of energy density so that the mean height of maxima is drastically reduced.

13.5 Sound Absorption Measurement

The reverberation tank was used for the measurement of sound absorption. The test object was a "large" (34 × 34 cm) coil spring absorber (of the type sketched in Fig. 13.4). The absorber was suspended closely below the water surface. The absorption coefficient α can be evaluated from the reverberation times T_0 and T of the tank without and with absorber, respectively. Assuming Eyring's theory (e.g., [27], p.73) the sound energy E decays from an initial value E_0, after switching off the sound source, according to

$$E = E_0(1 - \overline{\alpha})^{(Act/4V)} \tag{13.18}$$

where V is the tank volume, A its total surface area, $c = 1{,}484$ m/s the sound velocity, and $\overline{\alpha} = \left(\sum_n \alpha_n A_n\right)/A$ the average absorption coefficient of all boundary surfaces. With $E/E_0 = 10^{-6}$ for the reverberation time, $t = T$, (13.18) yields

$$1 - \overline{\alpha} = \mathrm{e}^{-\frac{kV}{AT}} \tag{13.19}$$

where $k = (24 \ln 10)/c = 0.03724$ s/m. Let α_0 and T_0 be the quantities for the tank without absorber, $\overline{\alpha}$ and T with absorber, A_s the absorber surface, and α its absorption coefficient. $\overline{\alpha}$ is defined as

$$\overline{\alpha} = (\alpha_0(A - A_s) + \alpha A_s)/A, \tag{13.20}$$

and with

$$\alpha_0 = 1 - \mathrm{e}^{-\frac{kV}{AT_0}} \tag{13.21}$$

we get

$$\alpha = 1 - \mathrm{e}^{-\frac{kV}{AT_0}} + \frac{A}{A_s}\left(\mathrm{e}^{-\frac{kV}{AT_0}} - \mathrm{e}^{-\frac{kV}{AT}}\right). \tag{13.22}$$

Assuming small exponents and $A_s \ll A$, (13.22) is approximated by

$$\alpha \approx \frac{kV}{A_s}\left(\frac{1}{T} - \frac{1}{T_0}\right). \tag{13.23}$$

The influence of the scatter of the experimental reverberation times T_0 and T on the accuracy of α can be estimated. Let σ_T and σ_0 be the standard deviations of T and T_0, respectively. Then the standard deviation of α is

$$\sigma_\alpha = \sqrt{\left(\frac{\partial \alpha}{\partial T}\right)^2 \sigma_T^2 + \left(\frac{\partial \alpha}{\partial T_0}\right)^2 \sigma_0^2} = \frac{kV}{A_s}\sqrt{\frac{\sigma_T^2}{T^4} + \frac{\sigma_0^2}{T_0^4}} \tag{13.24}$$

and the relative error

$$\frac{\sigma_\alpha}{\alpha} = \sqrt{\frac{T_0^2\left(\frac{\sigma_T}{T}\right)^2 + T^2\left(\frac{\sigma_0}{T_0}\right)^2}{T_0 - T}} \tag{13.25}$$

or, assuming $\sigma_T/T \approx \sigma_0/T_0$,

$$\frac{\sigma_\alpha}{\alpha} = \frac{\sigma_T}{T}\frac{\sqrt{1 + T^2/T_0^2}}{1 - T/T_0}. \tag{13.26}$$

If, as in the measurements described in the following, $T = 0.7T_0$, we conclude from (13.26) that the error of α is about four times greater than that of T and T_0.

Measurements with the "large area" coil spring absorber were made with third-octave band random noise between 4.5 and 72 kHz and showed a rather wide scatter of single measurements, typically $\alpha = 0.7 \pm 0.3$ (about 40 %) at 4.5–15 kHz, shrinking to about 4 % at 72 kHz. Averaging over 13 hydrophone positions and either 10 repetitions reduces the scatter to 20 % at low and 2 % at high frequencies. Obviously, reliable values of thc absorption coefficient at low frequencies demand for averaging over many more independent measurements.

An effect which has not been considered in this context is the occurrence of (cylindrical) diffraction waves originating from the edges of the test absorber. They propagate across the absorbing surface and cause excess absorption. This "edge effect" or "area effect" influences the experimental results. We showed in a later theoretical study that the edge effect is much stronger for absorbers on pressure-release boundaries than on hard walls [28]. This was confirmed by experiments [29].

13.6 Measurement of Power Radiation

The acoustical power P radiated by a vibrating object into a reverberation room is related to the average steady state energy density E_0 of the stationary sound field by

$$P = \frac{E_0\overline{\alpha}Ac}{4} \tag{13.27}$$

where $\overline{\alpha}$ is the average absorption coefficient and A the total surface area of the reverberation room [30]. According to (13.19) $\overline{\alpha}$ is given by the reverberation time T:

$$\overline{\alpha} = 1 - \mathrm{e}^{-\frac{kV}{AT}} \approx \frac{kV}{AT} = 24\ \ln 10\frac{V}{AcT} = 55.262\frac{V}{AcT}, \tag{13.28}$$

and the stationary energy density

$$E_0 = \frac{\overline{p_{\text{eff}}^2}}{\rho c^2} \tag{13.29}$$

by the r.m.s. sound pressure p_{eff} (ρ is the mass density of the medium, and the bar denotes spatial averaging). Hence, the sound power P can be determined by measuring the sound pressure p and the reverberation time T:

$$P = 13.82 \frac{\overline{p_{\text{eff}}^2} V}{\rho c^2 T}. \tag{13.30}$$

The reverberation room technique for the measurement of power radiation is particularly useful for objects with pronounced directional characteristics since the room integrates over all angles of radiation. The sound pressure measurements have to be taken in the diffuse field, i.e., at sufficient distance r from the radiator. According to [22], the following condition has to be met:

$$r > \frac{1}{4}\sqrt{\frac{Q\overline{\alpha}A}{1-\overline{\alpha}}} \approx 0.048\sqrt{\frac{QV}{T}} \tag{13.31}$$

where Q is the directivity factor of the source (in the direction of the receiver). The approximation is obtained with $\overline{\alpha}$ from (13.28), $c = 1{,}484$ m/s, and $\overline{\alpha} \ll 1$.

When investigating vibrating plates or similar objects one often is not so much interested in the absolute power radiated but in the radiation efficiency s which is defined as the ratio of the actual power P radiated by the object to the power which would be radiated by an equally large area of an infinite piston vibrating with the same r.m.s. particle velocity v_{eff} [31]:

$$s = \frac{P}{\rho c S \overline{v_{\text{eff}}}^2} \tag{13.32}$$

where ρc is the characteristic impedance of the surrounding medium, S the radiating surface, and $\overline{v_{\text{eff}}}$ the spatial average of the r.m.s. velocity of the radiating surface. According to (13.30) and (13.32), the radiation efficiency can be determined in a reverberation room by measuring the radiated power P and the reverberation time T, or with (13.30)

$$s = \frac{13.82\overline{p_{\text{eff}}^2} V}{\rho^2 c^3 T S \overline{v_{\text{eff}}}^2}. \tag{13.33}$$

13.7 Diffuseness in the Reverberation Tank

A sound field is diffuse in a point if the acoustic energy flux density through this point is equally distributed over all solid angles [32]. Diffuseness is assumed in the reverberation theories and has therefore to be guaranteed, e.g., for the measurement of sound absorption and power radiation in reverberation rooms. Besides measurements with directional microphones [33] or deviations from the spatial $1/r^2$ decay of the energy density with distance r from a sound source [34], Manfred Schroeder has developed methods for the measurement of diffuseness based on the statistical theory of reverberant sound fields. We have supplemented them as outlined in the following [35].

13.7.1 Angular Distribution

A point source at the origin of a coordinate system may radiate a sinusoidal signal $\underline{a}_1(t) = \hat{a}_1 e^{j\omega t}$ and create a stationary sound field in a room. A microphone or hydrophone displaced in x-direction may record the output signal $\underline{a}_2(t) = \hat{\underline{a}}_2 e^{j\omega t}$. The ratio

$$\underline{H}(x) = \frac{\underline{a}_2(t)}{\underline{a}_1(t)} = R(x) + jX(x) = |\underline{H}(x)| e^{j\varphi(x)} \tag{13.34}$$

is the complex transfer function between transmitter and receiver. A plane wave propagating at an angle α against the x-direction has the wavenumber component $k_x = 2\pi(\cos\alpha)/\lambda$. The sound field $\underline{a}_2(t)$ and the function $\underline{H}(x)$ at the receiver position are composed of many plane waves incident under angles α_m:

$$\underline{H}(x) = \sum_m \hat{\underline{a}}_m e^{-2\pi j \frac{\cos\alpha_m}{\lambda} x}, \tag{13.35}$$

and equivalently for the other directions: $\underline{H}(y)$ with $\hat{\underline{b}}_m$ and β_m, $\underline{H}(z)$ with $\hat{\underline{c}}_m$ and γ_m. Angular equal distribution of the energy flux means

$$\overline{\cos\alpha} = \left(\int_0^\pi |\hat{\underline{a}}(\alpha)|^2 \cos\alpha \mathrm{d}\alpha \right) \Big/ \left(\int_0^\pi |\hat{\underline{a}}(\alpha)|^2 \mathrm{d}\alpha \right) = 0 \tag{13.36}$$

and correspondingly $\overline{\cos\beta} = \overline{\cos\gamma} = 0$. Since generally

$$\overline{\cos^2\alpha} + \overline{\cos^2\beta} + \overline{\cos^2\gamma} = 1, \tag{13.37}$$

in a diffuse sound field $\overline{\cos^2\alpha} = \overline{\cos^2\beta} = \overline{\cos^2\gamma} = 1/3$. $\overline{\cos\alpha}$ measures the mean energy flux, $\overline{\cos^2\alpha}$ the width of the energy flux distribution. Experimental data of $\overline{\cos\alpha}$, $\overline{\cos\beta}$, $\overline{\cos\gamma}$ and, e.g., $\overline{\cos^2\alpha}$, $\overline{\cos^2\beta}$ ($\overline{\cos^2\gamma}$ follows from (13.37)) provide information on the directional energy flux distribution. Omitting detailed derivations, the following comparatively simple experimental procedures can be deduced.

13.7.2 Spatial Distance of Pressure Maxima

Assuming that the first (y') and second (y'') spatial derivatives of $y = |\underline{H}(x)|^2$ are Gaussian distributed, S. O. Rice [36] has shown that the mean spatial distance of sound pressure maxima is given by

$$\overline{\Delta x_{\max}} = 2\pi\sqrt{\frac{\overline{y'^2}}{\overline{y''^2}}} \approx \lambda\sqrt{\frac{\overline{\cos^2\alpha} - \overline{\cos\alpha}^2}{\overline{\cos^4\alpha} - 4\,\overline{\cos^3\alpha}\cdot\overline{\cos\alpha} + 3\,\overline{\cos^2\alpha}^2}}. \tag{13.38}$$

For a diffuse three-dimensional sound field follows $\overline{\Delta x_{max}} \approx 0.79\lambda$. In a "semi-diffuse" sound field, e.g., in a reverberation room with one totally absorbing wall (perpendicular to the x-direction) is $\overline{\cos\alpha} = 1/2$ and further $\overline{\Delta x_{\max}} \approx 1.5\lambda$. One concludes that the mean distance of maxima in a monofrequent stationary sound field depends on the angular distributions.

13.7.3 Zeros of R(x)

The average spatial distance of the zeros of $R(x)$ (see (13.34)) is

$$\overline{\Delta x_0} = \frac{\lambda}{2\sqrt{\overline{\cos^2\alpha}}} \tag{13.39}$$

and correspondingly for the other directions. From (13.37) follows

$$\frac{1}{\overline{\Delta x_0}^2} + \frac{1}{\overline{\Delta y_0}^2} + \frac{1}{\overline{\Delta z_0}^2} = \frac{4}{\lambda^2}, \tag{13.40}$$

and for the three-dimensional diffuse sound field

$$\overline{\Delta x_0} = \overline{\Delta y_0} = \overline{\Delta z_0} = \lambda\frac{\sqrt{3}}{2}. \tag{13.41}$$

If deviations from this equality are found, they indicate lacking diffuseness but do not give information on the energy flux direction since only the quadratic moments are considered.

13.7.4 Spatial Phase Rotation

The average over x of the spatial phase rotation $\mathrm{d}\varphi/\mathrm{d}x$ (see (13.34)) must vanish. Generally holds

$$\overline{\left(\frac{\mathrm{d}\varphi}{\mathrm{d}x}\right)} = -\frac{2\pi}{\lambda}\ \overline{\cos\alpha}, \tag{13.42}$$

i.e., this quantity measures the mean energy transport.

13.7.5 Spatial Correlation Functions

The measurement of diffuseness by spatial correlation functions is known since long [37, 38]. A few supplements are given in the following. Let

$$c_{RR}(\Delta x) = \left(\int R(x)R(x+\Delta x)\mathrm{d}x\right) / \left(\int R^2(x)\mathrm{d}x\right) \tag{13.43}$$

and

$$c_{RX}(\Delta x) = \left(\int R(x)X(x+\Delta x)\mathrm{d}x\right) / \sqrt{\int R^2(x)\mathrm{d}x \int X^2\mathrm{d}x}. \tag{13.44}$$

If Δx_1 is the first zero of $c_{RR}(\Delta x)$, a good approximation is

$$\overline{\cos^2\alpha} \approx \left(\frac{\lambda}{3.6\Delta x_1}\right)^2 \tag{13.45}$$

(correspondingly for the y- and z-direction). The measurement can also be done with octave-band filtered noise; let λ be the geometric mean of the wavelengths at the band limits, then it follows for the diffuse sound field $\Delta x_1 = 0.49\lambda$ instead of 0.50λ for sinusoidal excitation. Well beyond the first zero, $c_{RR}(\Delta x)$ is almost equal for both cases.

While $c_{RR}(\Delta x)$ depends only on the width of the angular distribution, $c_{RX}(\Delta x)$ informs about the energy flux; assuming $\Delta x \ll \lambda/2\pi$, we find

$$c_{RX}(\Delta x) \approx \frac{4\pi}{\lambda}\Delta x\,\overline{\cos\alpha}. \tag{13.46}$$

If measurements with frequency mixtures are made, the higher frequency components are overestimated due to the factor $1/\lambda$. This can be avoided by temporal integration of one component. Let $p_1(t)$ and $p_2(t)$ be the receiver signals at two points $\Delta x \ll \lambda/2\pi$ apart. Then the mean energy flux is

$$I = \frac{1}{\rho\Delta x}\overline{\langle p_1\rangle p_2}^{\,t} \tag{13.47}$$

where ρ is the density of the medium and $\langle p_1\rangle = \int p_1(t)\mathrm{d}t$.

13.7.6 Height of Maxima in the Frequency Response Curve

Let the level difference h between a sound pressure maximum and the next minimum be called height of a maximum. Its mean value $\overline{h}$ in a diffuse sound field is 12.5 dB for spatial averaging at a fixed frequency, and 7.8 dB for spectral averaging at a fixed position (because of the different "spectra"). For the latter case the influence of direct sound can be estimated as

$$\overline{h} \approx \frac{6.3\,\mathrm{dB}}{\left(0.4 + d^4\right)^{1/4}} \tag{13.48}$$

where d is the amplitude ratio of direct and diffuse sound.

13.7.7 Experiments

Measurements in the resiliently lined reverberation tank revealed that for sinusoidal excitation the average distance of maxima in front of a 20 cm wide absorber strip is larger, and the mean height of maxima smaller than in a diffuse sound field. But at 60 cm distance from the absorber the sound field is apparently diffuse. Further experiments were planned, but abandoned because of the end of the research contract with the British Navy.

13.8 Vibration and Sound Radiation of Flat Plates

A representative of the British Navy visited us every summer to discuss our past and future work within the research contract. By the end of the 1960s, we were asked to do some fundamental investigations about "how does the sound get out of the ship into the water." Since a ship hull is a rather thick curved steel plate, we decided to study the vibrational behavior and sound radiation from thick-walled cylindrical shells. We realized soon that it would be helpful to study beforehand the vibration and sound radiation of flat plates in air and water. Both are dominated by natural frequencies of the flexural resonances which can be calculated from the wave equation and the boundary conditions.

The particle velocity v of *thin* plate flexural vibrations obeys the fourth order partial differential equation

$$\frac{D}{\rho h}\Delta^2 v + \frac{\partial^2 v}{\partial t^2} = 0 \tag{13.49}$$

where $\Delta = \partial^2/\partial x^2 + \partial^2/\partial y^2$ is the Laplacian operator, $D = Eh^3/(12(1-\mu^2))$ the flexural rigidity, E Young's modulus, μ Poisson's ratio, h the plate thickness, and

ρ the mass density. For *thick* plates additional terms account for rotary inertia and shear deformation:

$$\frac{D}{\rho h}\Delta^2 v + \frac{\partial^2 v}{\partial t^2} - \left(\frac{h^2}{12} + \frac{D\gamma}{hG}\right)\frac{\partial^2}{\partial t^2}\Delta v + \frac{\rho h^2 \gamma}{12G}\frac{\partial^4 v}{\partial t^4} = 0 \tag{13.50}$$

where G is the shear modulus and $\gamma \approx 1.2$ the shear distribution constant [39, 40]. Assuming harmonic vibration, $v(x, y, t) = V(x, y)\cos \omega t$, (13.49) yields for the mode shapes $V(x, y)$ of thin plates

$$\frac{D}{\rho h}\Delta^2 V - \omega^2 V = 0, \tag{13.51}$$

and (13.50) for thick plates:

$$\frac{D}{\rho h}\Delta^2 V - \omega^2 V + \left(\frac{h^2}{12} + \frac{D\gamma}{hG}\right)\omega^2 \Delta V + \frac{\rho h^2 \gamma}{12G}\omega^4 V = 0. \tag{13.52}$$

Equation (13.52) can be solved rigorously by a superposition of sin, cos, sinh and cosh terms in x and y, the weights being determined by the boundary conditions. In order to calculate the natural frequencies of a plate with free edges and dimensions a and b in x- and y-direction, it is convenient to assume

$$V(x, y) \approx \hat{V}\left(\cos k_x x\right)\left(\cos k_y y\right) \tag{13.53}$$

with

$$k_x = \left(m - \frac{1}{3}\right)\frac{\pi}{a}, \qquad k_y = \left(n - \frac{1}{3}\right)\frac{\pi}{b} \tag{13.54}$$

where m and n are the number of nodes in x- and y-direction. The terms $-1/3$ account for the fact that the exponential near-fields at the edges shift the outer node lines to $\lambda/6$ distance from the edges if λ is the flexural wavelength in the respective direction. For $m = 0$ holds $k_x = 0$, and for $n = 0$: $k_y = 0$. $m = 1$ or $n = 1$ would mean rotations without restoring forces so that for $a \geq b$ the lowest vibrational mode is $(m, n) = (2, 0)$. Substituting (13.53) in (13.51) yields the frequency equation for thin plate modes

$$\omega_{mn}^2 = \frac{D}{\rho h}\left(k_x^4 + k_y^4\right), \tag{13.55}$$

and in (13.52) for the thick plate modes:

$$\omega_{mn}^4 - \omega_{mn}^2 \frac{12G}{\rho h^2 \gamma}\left(1 + \left(\frac{h^2}{12} + \frac{D\gamma}{hG}\right)\left(k_x^2 + k_y^2\right)\right) + \frac{12GD}{\rho^2 h^3 \gamma}\left(k_x^4 + k_y^4\right) = 0. \quad (13.56)$$

The natural frequency of the thick plate flexural vibration is the lower one of the real-valued solutions of (13.56).

Our test object was a flat steel plate with dimensions $a = 36$ cm, $b = 12$ cm and $h = 2$ cm, excited at the center by a longitudinally vibrating piezoelectric tube (10 mm diameter) with a damped countermass—a sandwich of lead plates and plastic layers so that its resonances were sufficiently damped and a constant driving voltage excited the plate with nearly constant force in the frequency range of interest (up to 35 kHz). To realize free edge conditions on all sides, the plate and the countermass were suspended from thin plastic threads to hang freely in the water. With sliding sine wave excitation, the frequency response curves of the plate in air and water were measured with a small accelerometer, mounted by a magnetic clamp at the plate center (opposite to the exciter), and recorded on a level recorder. The response curve for the plate in air showed sharp resonances up to 35 kHz, but in water only up to about 12 kHz. Due to the mass load and greater radiation loss, the resonance frequencies are shifted to lower values as compared to the results in air, and at higher frequencies the damping by sound radiation into water is so strong that no resonance peaks occur.

The resonance shift caused by the added mass $m_w = \rho_w h_w$ of water per unit area can be estimated by modifying (13.55):

$$\omega_{mn}^2 = \frac{D}{\rho h + m_w}\left(k_x^4 + k_y^4\right). \quad (13.57)$$

According to [41], m_w for an infinite plate with one-sided load at low frequencies is the mass of a water layer with thickness $h_w \approx 1/6$ of the flexural wavelength of the plate. Our measurements of the resonance frequencies in air and water showed, however, that—on the average over many modes—the mass load amounts to a water layer of only 1/10 of the flexural wavelength on both sides of the submerged plate, presumably because of the acoustic short circuit around the free edges of the plate. Calculated flexural wavelengths for the $(m, 0)$ modes agreed exactly to our measured values in air and water.

Vibration patterns of the plate were scanned by the accelerometer which could be moved motor-driven along the x-direction sliding on a grease film at any height (the y-coordinate). The vibration patterns of some modes were also visualized by holographic interferometry. Both methods showed that the node lines never intersect ("node lines avoid each other").

The far-field sound radiation pattern was recorded in the shallow water basin (Sect. 13.3.6). A hydrophone could be moved along a semicircle of radius 1.5 m with the aid of a motor-driven turning arm. The far-field radiation pattern $p(\Theta, \phi)$ and the vibration pattern $V(x, y)$ are a pair of Fourier transforms. With Θ the azimuth and ϕ the elevation angle, $\alpha = k \sin \Theta$ and $\beta = k \sin \phi$ (k the wavenumber in the medium) holds

$$p(\alpha, \beta) = \frac{j\omega\rho}{2\pi R} e^{-jkR} \int_{-a/2}^{+a/2} \int_{-b/2}^{+b/2} V(x, y) e^{j\alpha x} e^{j\beta y} dx dy. \qquad (13.58)$$

$\Theta = \phi = 0$ is the direction of the plate normal. Confining to the horizontal plane, we find

$$p(\Theta) \propto \int_{-a/2}^{+a/2} V(x) e^{jk \sin \Theta} dx. \qquad (13.59)$$

With the approximation $V(x) \propto \cos k_x x$ follows for the sound pressure magnitude in the far-field

$$|p(\Theta)| \propto \left| \frac{\sin[(a/2)(k_x - k \sin \Theta)]}{k_x - k \sin \Theta} + \frac{\sin[a/2(k_x + k \sin \Theta)]}{k_x + k \sin \Theta} \right| \qquad (13.60)$$

which compared well with our experimental results, except for $\Theta \approx \pm 90°$.

The radiation efficiency s of flexurally vibrating plates (see (13.32), (13.33)) is strongly frequency dependent because of the dispersion of bending waves. The flexural wave velocity is proportional to $\sqrt{\omega}$ so that at low frequencies the acoustic short circuit along the plate surface reduces the power radiation. At the critical frequency f_c, where the flexural wave velocity equals the sound velocity in the surrounding medium, s runs through a maximum $s(f_c) > 1$, and at higher frequencies s approaches unity (see also the next Section). Many of our experimental results on sound radiation from plates and their comparison with theoretical predictions have been published in [42, 43], including the influence of rib-stiffeners.

13.9 Critical Frequency of Plates in Dense Media

Ignoring the influence of the surrounding medium on the plate bending wave velocity, the critical frequency for thin plates is

$$\omega_{c0} = c_0^2 \sqrt{\frac{\rho h}{D}} \qquad (13.61)$$

where c_0 is the sound velocity in the medium, ρ, h, and D as in (13.49). But the high mass and radiation load at f_c leads to a splitting of the dispersion curve, the frequency dependence of the plate bending wave velocity $c_b(\omega)$ [44] so that c_b

never equals c_0. The separation is small for plates vibrating in air, but it becomes important in water. A stability criterion allows to determine f_c. We could show that for steel plates in water $\omega_c = 1.34\omega_{c0}$ and for aluminum plates $\omega_c = 1.52\omega_{c0}$. The deduction and illustrations have been published in [45].

13.10 Vibration of Cylindrical Shells

While flexural vibrations of flat plates are governed by fourth order partial differential equations (Sect. 13.8) thick-walled cylindrical shells obey tenth order equations ([46], a), and even the neglection of rotary inertia and shear deformation (as for thin-walled shells) leads to still eighth order equations ([46], b). If the added mass of the surrounding medium, e.g., water, is not negligible, the rigorous treatment demands for solving integral equations [47]. In order to reduce the computational effort we tried to develop simplified equations for the natural frequencies of flexural vibrations of thick-walled, finite-length cylindrical shells.

The natural modes are described by (m, n) where m is the number of node circles and n the number of node lines parallel to the cylinder axis along half the circumference. Let l be the cylinder length, r_a the outer radius, and h the thickness of wall. Assuming cosine-shaped particle velocity distribution, the circumferential wavenumber is

$$k_u = \frac{2\pi}{\lambda_u} = \frac{n}{r_a} \tag{13.62}$$

(λ_u circumferential wavelength). The axial wavenumber can be approximated by

$$k_a = \left(m - \frac{1}{3}\right)\frac{\pi}{l} \tag{13.63}$$

for the free-free cylinder (see also (13.54) for the plate with free edges). It is suitable to normalize the calculated natural frequencies $\omega_{mn} = 2\pi f_{mn}$ with ω_r', the frequency of the infinitely long cylinder in the mode $(0, 0)$, a uniform radial vibration ("breathing cylinder"):

$$\omega_r' = 2\pi f_r' = \frac{c_L}{r_0} \tag{13.64}$$

where $r_0 = r_a - h/2$ is the mean cylinder radius and $c_L = \sqrt{E/\rho(1-\mu^2)}$ the longitudinal wave velocity, E Young's modulus, ρ mass density, and μ Poisson's ratio. f_r' is the frequency at which the longitudinal wavelength equals the cylinder perimeter. Slightly lower than ω_r' is the "ring frequency" $\omega_r = 2\pi f_r = c_D/r_0$ with $c_D = \sqrt{E/\rho}$. f_r is important for sound radiation since in its vicinity resonances with high radiation efficiencies are clustered. In the following the normalized frequency

$$\Omega_{mn} = \frac{\omega_{mn}}{\omega_r'} \tag{13.65}$$

will be considered.

The theory of elasticity leads to different eighth order partial differential equations for thin-walled cylindrical shells, depending on the way of deduction ([46], b). Assuming

$$v(\varphi, z, t) = \hat{v}\,(\cos n\varphi)(\cos k_a z)(\cos \omega t) \tag{13.66}$$

for the particle velocity (φ in circumferential, z in axial direction) yields cubic algebraic equations for Ω_{mn}^2 ([46], a). The three solutions correspond to radial, axial, and circumferential vibrations, of which the radial vibration is relevant for sound radiation. With the assumption $\Omega \ll k_a r_0$ the cubic equations reduce to quadratic equations, as M. Heckl [48] has shown for the "simplified Kennard equation," ending up with

$$\Omega^2 = \left(1-\mu^2\right)\frac{x^4}{\left(x^2+n^2\right)^2} + \delta A \tag{13.67}$$

where $x = k_a r_0$, $\delta = h^2/12r_0^2$, and $A = (x^2+n^2)^2 - [n^2(4-\mu) - 2 - \mu]/(2-2\mu)$. Other cubic equations for Ω^2 were deduced by Flügge and Arnold/Warburton:

$$\Omega^6 + p\Omega^4 + q\Omega^2 + r = 0 \tag{13.68}$$

with (for the Flügge equation):

$$p_F = -\left[\frac{3-\mu}{2}\left(x^2+n^2\right) + 1 + \delta\left(x^2+n^2\right)^2\right], \tag{13.69}$$

$$q_F = \frac{1-\mu}{2}\left[\left(x^2+n^2\right)^2 + (3-2\mu)x^2 + n^2\right] + \delta\frac{3-\mu}{2}\left(x^2+n^2\right)^3,$$

$$r_F = -\frac{1-\mu}{2}\left[\left(1-\mu^2\right)x^4 + \delta A_F\right],$$

$$A_F = \left(x^2+n^2\right)^4 - 2\left[\mu x^6 + 3x^4n^2 + (4-\mu)x^2n^4 - (2-\mu)x^2n^2 + n^6\right] + n^4,$$

and for the Arnold/Warburton equation:

$$p_{AW} = p_F - \delta\left[n^2 + 2(1-\mu)x^2\right], \tag{13.70}$$

$$q_{AW} = q_F + \delta\left[2x^2(1-\mu)\left(x^2+1\right) - \left(2-\mu^2\right)x^2n^2 - \frac{3+\mu}{2}n^4 + n^2\right],$$

$$r_{\mathrm{AW}} = r_F - (1-\mu)x^2\delta\left[\mu\left(x^4 - n^4 + n^2\right) + x^2\left(1-\mu^2\right)\left(2-n^2\right)\right].$$

Applying Heckl's approximation $\Omega \ll k_a r_0$ to the Flügge equation, one obtains

$$\Omega^2 = \frac{(1-\mu^2)x^4 + \delta A_F}{\left(x^2+n^2\right)^2 + n^2.} \tag{13.71}$$

In an attempt to account for the influence of rotary inertia and shear deformation, we had the idea to apply the relation of thick and thin plates. Imagine the cylinder cut open along one axial node line and developed into a plane, then you get the "equivalent plate" of the cylinder. Its edge conditions are "simply supported" at the new edges, and free at the former cylinder top and bottom. Let Ω_{C0} be the frequency of the thin-walled cylinder (calculated from one of the aforementioned theories), Ω_{P0} that of the thin plate, and $\Omega_{\mathrm{P}h}$ that of the thick plate, then we proposed to calculate the frequency $\Omega_{\mathrm{C}h}$ of the thick-walled cylinder by

$$\Omega_{\mathrm{C}h} = \Omega_{\mathrm{C0}}\frac{\Omega_{\mathrm{P}h}}{\Omega_{\mathrm{P0}}} \tag{13.72}$$

where Ω_{P0} and $\Omega_{\mathrm{P}h}$ are obtained from (13.55) and (13.56).

In order to check this suggestion and to compare the different theoretical approaches for the cylinder eigenfrequencies, we performed careful measurements in air and in the deep water tank with two cylinder shells: Cylinder I with $l = 40$ cm, $r_0 = 20$ cm, $h = 1.7$ cm, and Cylinder II with $l = 40$ cm, $r_0 = 11$ cm, $h = 2.4$ cm. Both cylinders had free edges. Cyl. II was excited by a point force (piezo transducer with countermass) at the upper edge from inside, Cyl. I at half height to suppress modes with odd m (otherwise the resonances would overlap too much). The resonance frequencies could be read from the maxima of the admittance frequency response, the mode shapes were determined by scanning the particle velocity along the outer surface with a motor-driven accelerometer. For the measurements in water, the end faces of the cylinders were sealed with plastic foils, so that the cylinders were air filled, and water loaded only at the outside.

We could identify in air 26 modes of Cyl. I and 15 of Cyl. II between 800 and 8,300 Hz. The reference frequencies f_r were 4.3 kHz for Cyl. I and 7.8 kHz for Cyl. II. The measured resonance frequencies were compared to those calculated from the cubic equations without and with Heckl's approximation, and without and with the thickness correction (13.72). The best results were obtained with Heckl's approximation of the Flügge equation and the thickness correction ((13.71) and (13.72)): mean error 0.1 %, standard deviation ±1.9 %. This is better than could be expected. Probably some of the approximation errors cancel each other. The complete list of results, estimation of the added mass in water, diagrams of the natural frequencies as functions of m and n have been published in [49]. Theory and experiments on sound radiation from the two cylinders are described in [50], and for the cylinders with ring stiffeners in [51].

13.11 Miscellaneous

13.11.1 Viscometer

During the development of flow absorbers (Figs. 13.3, 13.4, and 13.5) it turned out to be advantageous to measure the viscosity of the fluids and the impedance of thin fluid layers. Our student Joachim Richter developed experimental facilities and the associated theory. A fluid volume of less than 10 mm^3 was brought between two circular PZT (or barium titanate) transducers of 8 mm diameter, the distance of which could be adjusted by a precision thread to an accuracy of about 1 μm. The apparatus was predominantly suited for Newtonian fluids (without elasticity), but also for visco-elastic fluids. The theory, construction and experimental results have been published in [52] for the viscometer and in [53] for the impedance measurements.

13.11.2 Self-Imaging

Self-imaging is a phenomenon occurring in dispersive waveguides, e.g., in optical fibers [54] and the deep sea SOFAR channel [55]. Because each mode in the waveguide propagates with its own phase velocity, the modes soon become dephased and an object in the entrance plane cannot be reconstructed, unless after a certain distance the accumulated phase differences between any two modes are approximately multiples of 2π and so produce a "self-image." Since self-imaging occurs at different distances for different frequencies, it is possible with monofrequent sound fields only.

Our student Gero Timann proved theoretically that self-imaging exists in any waveguide with discrete phase velocities of the guided modes and confirmed this experimentally in the resiliently lined water duct (Sect. 13.3.4). He proved also that self-imaging is possible over long distances. However, because a waveguide is a spatial low-pass filter, the resolution of the images is smaller than that of the original [56].

13.11.3 Schroeder Diffusers

One of Manfred Schroeder's great contributions to room acoustics was the development of diffusively reflecting wall structures, based on number theoretical principles. I remember the origin. When Manfred Schroeder returned to Göttingen he had to give a four-semester lecture course on the Institute's research topics, in the summer semester 1971 on the fundamentals "Physics of vibration." I was Erwin Meyer's co-author for a book on this subject [57], and Schroeder—as usual during the German university vacation in the USA—asked me to send him the manuscript.

He thanked me writing “this was exactly what the doctor has prescribed.” He found a short chapter on Barker sequences and maximum length sequences which were new to him (both have “thumb tack” autocorrelation functions). Back in Göttingen, he asked me if I had thought about the spatial equivalent of these time sequences, and I had to confess that I did not. But he had immediately realized that such structures could solve the problem of diffuse sound reflection. He engaged students to start experiments, and, typical for him, asked himself how to optimize such sequences. He found the solution in number theory as is well known [58].

The success in room acoustics inspired our hydroacoustics group to apply Schroeder diffusers (or, as he wrote, diffusors) to underwater sound, too. We thought that diffuse reflectors might be of some interest for underwater structures, e.g., to prevent sonar localization. Since rigid reflectors are not readily available for underwater sound, we made experiments with sound-soft structures in the shallow water basin (Sect. 13.3.6), applying quadratic residue sequences. We encountered problems for oblique incidence since separations between adjacent diffuser elements are difficult to realize with sound-soft material, but for rather vertical incidence they showed a wide scatter in the directivity patterns [59].

13.11.4 Resonance Scattering

By the end of 1986, a representative of the German BWB: “Bundesamt für Wehrtechnik und Beschaffung” (Federal Office of Defense Technology and Procurement) offered us a research contract on sonar classification of submerged objects. We learnt that thousands of undiscovered mines are still buried in the North Sea and the Baltic Sea at unknown locations, presenting a potential danger for ships leaving the normal routes. Scanning the sea bottom with sonar yields on the sonar screen many “mine-like objects,” the majority of which are stones. We learnt also that a few very experienced sonar operators listening to the acoustic echoes can discriminate metallic objects from stones with high reliability, but they are unable to tell how they do it. This shows that the information about the nature of a sound scattering object is contained in the echo, but in a not very obvious way. This encouraged us to apply artificial neural networks as discriminators.

In analogy to airborne acoustics, where a metallic object “sounds” different than a stone when being hit, a sonar impulse exciting eigenfrequencies causes a specific ringing of the object. In the same manner as atomic particles can be characterized by the frequencies and halfwidths of their scattering resonances, macroscopic underwater bodies should theoretically be identifiable from the line spectra of their backscattered echoes.

Experiments with the two steel cylinders introduced in Sect. 13.10, the flat steel plate of Sect. 13.8, a high-ρc ceramic rod, and several stones showed resonance backscatter by more or less pronounced ringing in the time domain, and resonance peaks in the spectrograms.

Initial experiments were made in our deep water tank. Test signals were rectangular or Gaussian pulses of a few ms duration at frequencies around 100 kHz. Later on, test facilities of the German Navy on a lake in Northern Germany could be used. The test objects and transducers were suspended from a raft to a depth of 10 m below the water surface (depth of the lake 40 m). It turned out soon that isolated resonances could be identified at low frequencies only, while at higher frequencies the resonances strongly overlap. In order to distinguish four different objects (big cylinder, small cylinder, stone, and "no object," i.e., the direct sound only) we applied an artificial neural network, a conventional perceptron with 16 input channels, one hidden layer with 3 neurons, and two output neurons. After training the network with echoes at 4 aspect angles, the four targets were reliably identified for other angles, and also new objects (cylinder/stone) were discriminated with 80–90 % reliability.

Computer simulations showed that a neural network can much better discriminate between slightly different signals if time series data are offered instead of power spectra. The reason is the loss of information on trans-spectral coherence by the transformation to the power spectrum.

More details on theory and experiments have been published in [60], and a resonance scattering theory for strongly overlapping resonances in [61].

13.11.5 Electrorheological Fluids

In the mid 1990s, the BWB (see Sect. 13.11.4) asked us to do some research on the application of electrorheological fluids (ERF) to absorbers for underwater sound. Suspensions of polarizable particles in insulating fluids show the effect that the application of an electric field changes the viscosity of the mixture from liquid towards solid or gel. The electric field polarizes the individual particles, and their electrostatic attraction forms particle chains between the electrodes, thus enhancing the flow resistance. Standard applications of ERF were clutches, low frequency vibration dampers, engine mounts, and electro-hydraulic devices. Acoustic applications were not yet known.

We started experiments to study the ERF properties at acoustic frequencies with an appropriately constructed impedance tube. Sound velocity and attenuation of ERFs were measured as functions of frequency, field strength, and temperature. Further, a laboratory model of a squeeze film absorber (Fig. 13.3) for underwater sound was built and tested. Within certain limits, the absorber impedance could be controlled by the electric field strength applied to the ERF. In another construction, the field strength and the gap width were adjusted under computer control. A genetic algorithm was applied to adaptively minimize the reflection coefficient. Convergence to $|\underline{r}| \approx 0.13$ was reached after about 20 "generations" of the genetic algorithm. Details and experimental results have been published in [62] and [63], including a theoretical model based on the Biot theory for fluid-saturated marine sediments.

13.12 Résumé

Summarizing, I have experienced a very fruitful period of acoustic research under the Institute's directors Erwin Meyer, Manfred Schroeder, and finally (since 1994) Werner Lauterborn. I enjoyed stimulating discussions with all of them and all other colleagues on topical problems, and I estimated also the freedom to select research projects of my own preference, particularly during the "Schroeder era." Our Institute was highly esteemed during Meyer's leadership, and even more under Manfred Schroeder's guidance. At an ASA meeting in the mid 1990s, the chairman announced my talk saying "Dieter comes from the *legendary* Third Physical Institute in Göttingen." The high reputation of our Institute made it also attractive to excellent students, and we benefitted from their skill in theoretical and computer work, in my research group not only on underwater acoustics, but also in the field of active noise and vibration control (ANVC) which we started upon Manfred Schroeder's suggestion when the British Navy contract ended in 1978. Our research group became known as the first one propagating active impedance control both for sound and vibration. I have given an overview of the whole field of ANVC in ([2], 5th Chapter, pp. 107–138). Readers may also be interested in my biography of Erwin Meyer [64].

(If any reader of this text is interested in an English translation of one of the originally German publications, please contact me at Dieter.Guicking@phys.uni-goettingen.de.)

References

1. Meyer, E. (ed), Kuhl, W., Oberst, H., Skudrzyk, E., Tamm, K.: Sound Absorption and Sound Absorbers in Water (Dynamic Properties of Rubber and Rubberlike Substances in the Acoustic Frequency Region). Translated by Charles E. Mongan Jr., Department of the Navy, Bureau of Ships, Washington, DC, Report NavShips 900,164, Vol. 1. Manuscript delivered 1946, published 1950 (XVIII + 519 pp.) (1950)
2. Kurz, T., Parlitz, U., Kaatze, U. (eds.): Oscillations, Waves, and Interactions. Sixty Years Drittes Physikalisches Institut. A Festschrift. Universitätsverlag Göttingen 2007. ISBN: 978-3-938616-96-3. Vii + 474 pp. http://www.webdoc.sub.gwdg.de/univerlag/2007/oscillations.pdf (2007)
3. TV film "Vom Jäger zum Gejagten. Das erste Stealth-U-Boot der Welt" ("From the hunter to the hunted. The first stealth U-boat of the world") transmitted by *arte* on 18 February 2008, 21:00–21:50. http://de.wikipedia.org/wiki/U_480, http://www.uboat.net/boats/U480.htm (2008)
4. Meyer, E., Brendel, K.: Reflexionsarme Dünnschichtabsorber für Wasserschall. In: Sixth International Congress on Acoustics (ICA), August 21–28, 1968. Proceedings: Paper L-3-1, pp. L-65–L-68. Tokyo, JP (1968)
5. Meyer, E., Brendel, K., Richter, J.: Absorption von Wasserschall durch erzwungene Wechselströmung von Flüssigkeiten. Acustica **19**, 8–20 (1967/1968)
6. Brendel, K., Leuschner, H.-W.: Wasserschall-Resonanzabsorber mit Reibungsverlusten in Flüssigkeitsschichten. Acustica **21**, 260–268 (1969)
7. Leuschner, H.-W.: Ölgedämpfter Resonanzabsorber für Wasserschall. Acustica **25**, 280–288 (1971)

8. Ahlswede, Th., Guicking, D.: Breitband-Wasserschallabsorber geringer Schichtdicke. In: Fortschritte der Akustik, DAGA'75, pp. 663–666. Physik-Verlag, Weinheim (1975)
9. Ahlswede, T.: Breitband-Wasserschallabsorber geringer Schichtdicke. Dissertation (Ph.D. Thesis) Göttingen (1974)
10. Meyer, E., Schilz, W., Tamm, K.: Über den Bau eines reflexionsfreien Wasserschall-Meßbeckens. Acustica **10**(AB 1), 281–287 (1960)
11. Meyer, E., Tamm, K.: Breitbandabsorber für Flüssigkeitsschall. Acustica **1/2**, AB 91–AB 104 (1951/1952)
12. Wille, P.: Experimentelle Untersuchungen zur Schallstreuung an schallweichen Objekten. Acustica **15**, 11–25 (1965)
13. Richter, J., Leuschner, H.-W., Ahlswede, T.: Druckkammerverfahren zur Messung von akustischen Impedanzen in Wasser. Acustica **28**, 90–97 (1973)
14. Guicking, D., Karcher, K.: Hydroacoustic impedance measurement of large-area samples in the frequency range up to 200 kHz. Acustica **54**, 200–208 (1984)
15. Lubman, D.: Reverberation measurements of narrow-band noise power in a water tank. J. Acoust. Soc. Am. **37**, 1209 (1965)
16. Watters, B.C.: A reverberant water tank for underwater measurements. J. Acoust. Soc. Am. **53**, 357 (1972)
17. Blake, W.K., Maga, L.J.: Chamber for reverberant acoustic power measurements in air and in water. J. Acoust. Soc. Am. **57**, 380–384 (1975)
18. Guicking, D., Wetterling, Th.: A Reverberation Tank for Hydroacoustic Measurements. In: Eighth International Congress on Acoustics (ICA). Proceedings: Contributed Papers, vol. II, p. 482. London (1974)
19. Wetterling, Th.: Hydroakustische Messungen in einem schallweich begrenzten Halltank. Dissertation (Ph.D. Thesis) Göttingen (July 1975)
20. Schroeder, M[anfred R.]: Eigenfrequenzstatistik und Anregungsstatistik in Räumen. Modellversuche mit elektrischen Wellen. Acustica 4(AB 1) (1954) 456–468. English translation: Normal Frequency and Excitation Statistics in Rooms: Model Experiments with Electric Waves. J. Audio Eng. Soc. **35**, 307–316 (1987)
21. Schroeder, M[anfred R.]: Die statistischen Parameter der Frequenzkurven von großen Räumen. Acustica 4(AB 2) (1954) 594–600. English translation: Statistical Parameters of the Frequency Response Curves of Large Rooms. J. Audio Eng. Soc. **35**, 299–306 (1987)
22. Schroeder, M.R.: Frequency-correlation functions of frequency responses in rooms. J. Acoust. Soc. Am. **34**, 1819–1823 (1962)
23. Schroeder, M.R.: Effect of frequency and space averaging on the transmission responses of multimode media. J. Acoust. Soc. Am. **46**, 277–283 (1969)
24. Lubman, D.: Spatial averaging in a diffuse sound field. J. Acoust. Soc. Am. **46**, 532–534 (1969)
25. Cremer, L., Müller, H.L.: Die wissenschaftlichen Grundlagen der Raumakustik. Band 2, 2. Aufl. S. Hirzel Verlag, Stuttgart (1976)
26. Rollwage, M., Ebeling, K.J., Guicking, D.: Experimental investigation of two-dimensional diffuse sound fields in a shallow water basin. Acustica **58**, 149–161 (1985)
27. Meyer, E., Neumann, E.-G.: Physical and Applied Acoustics. An Introduction. Academic, New York, London (1972)
28. Guicking, D., Voronovich, A.: Theoretical evaluation of the edge effect of an absorbing strip on a pressure-release boundary. Acustica **70**, 66–75 (1990)
29. Zogalla, P., Guicking, D.: Untersuchungen zum Absorber-Kanteneffekt im schallweich berandeten Halltank. In: Fortschritte der Akustik, DAGA'92, pp. 269–272. DPG GmbH, Bad Honnef (1992)
30. Meyer, E., Just, P.: Messung der Gesamtenergie von Schallquellen. Zeitschrift für technische Physik **10**, 309–316 (1929)
31. Gösele, W.: Schallabstrahlung von Platten, die zu Biegeschwingungen angeregt sind. Acustica **3**, 243–248 (1953)

32. Schroeder, M.R.: Methoden zur Messung der Diffusität in Hallräumen. Acustica **9**(AB 1), 256–264 (1959)
33. Furduev, V.V., Tung, C.: Measurement of diffuseness of the acoustic field in rooms by the directional microphone method. Sov. Phys. Acoust. **6**, 103–111 (1960)
34. Reichardt, W., Sarkov, N.: Definition und Messung von Hall-Diffusitätsgraden. Hochfrequenztechnik und Elektroakustik **80**, 80–85 (1971)
35. Guicking, D., Lutze, R., Schroeder, M.R.: Messungen zur Diffusität im Halltank. In: Fortschritte der Akustik, DAGA'78, pp. 309–312. VDE-Verlag, Berlin (1978)
36. Rice, S.O.: Mathematical analysis of random noise. In: Wax, N. (ed.) Selected Papers on Noise and Stochastic Processes. Dover, New York (1954)
37. Dämmig, P.: Zur Messung der Diffusität von Schallfeldern durch Korrelation. Acustica **7**, 387–398 (1957)
38. Koyasu, M., Yamashita, M.: Evaluation of the degree of diffuseness in reverberation rooms by cross correlation technique. Bull. Kobayasi Inst. Phys. Res. **18**, 1–10 (1976)
39. Mindlin, R.D.: Influence of rotatory inertia and shear on flexural motions of isotropic, elastic plates. J. Appl. Mech. **18**, 31–38 (1951)
40. Leissa, A.W.: Vibration of Plates. NASA SP-160, US Government Printing Office, Washington, DC (1969)
41. Morse, P.M., Ingard, K.U.: Linear acoustic theory. In: Flügge, S. (ed.) Handbuch der Physik, Vol. XI/1, Akustik 1. Springer, Berlin (1961)
42. Guicking, D., Warhonowicz, T., Buddruss, C.: Schallabstrahlung von ebenen, zu Biegeschwingungen angeregten Platten in Wasser. Acustica **23**, 131–141 (1970)
43. Teubner, V.: Wasserschallabstrahlung von Platten mit und ohne Versteifungen. Acustica **31**, 203–215 (1974)
44. Kurtze, G., Bolt, R.H.: On the interaction between plate bending waves and their radiation load. Acustica **9**(AB 1), 238–242 (1959)
45. Guicking, D., Boisch, R.: Zur Grenzfrequenz ebener Platten in dichten Medien. Acustica **44**, 41–45 (1980)
46. Leissa, A.W.: Vibration of Shells. NASA SP-288, US Government Printing Office. a) Chapter 3.5.2, b) Chapter 2.1, c) Chapter 2.3 (1973)
47. Sandman, B.E.: Fluid-loading influence coefficients for a finite cylindrical shell. J. Acoust. Soc. Am. **60**, 1256–1264 (1976)
48. Heckl, M.: Vibrations of point-driven cylindrical shells. J. Acoust. Soc. Am. **34**, 1553–1557 (1962)
49. Guicking, D., Boisch, R.: Vereinfachte Berechnung der Eigenfrequenzen dickwandiger Zylinder in Luft und Wasser. Acustica **42**, 89–97 (1979)
50. Boisch, R., Guicking, D.: Schallabstrahlung von dickwandigen Stahlzylindern in Wasser. Acustica **45**, 322–339 (1980)
51. Boisch, R., Guicking, D.: Schallabstrahlung von ringversteiften Stahlzylindern in Wasser. Acustica **51**, 29–43 (1982)
52. Richter, J.: Viskosimeter für sehr geringe Flüsssigkeitsmengen. Rheol. Acta **9**, 550–553 (1970)
53. Richter, J.: Mechanische Impedanzen dünner Flüssigkeitsschichten. Acustica **22**, 95–106 (1969/1970)
54. Friesem, A.A., Levy, U., Silberberg, Y.: Parallel transmission of images through single optical fibers. Proc. IEEE **71**, 208–221 (1983)
55. Semenov, A.T.: Transmission of acoustic images in natural ocean waveguides. Sov. Phys. Acoust. **27**, 171–175 (1981)
56. Timann, G., Guicking, D.: Acoustic Self-Imaging in Resiliently Lined Waveguides. In: Ermert, H., Harjes, H.-P. (eds.) Nineteenth International Symposium on Acoustical Imaging, Bochum, April 3–5, 1991. Proceedings: Acoustic Imaging, vol. 19, pp. 149–154. Plenum Press, New York (1992)
57. Meyer, E., Guicking, D.: Schwingungslehre. ISBN 3-528-08254-2. Friedr. Vieweg + Sohn, Braunschweig (1974)

58. Schroeder, M.: Number Theory in Science and Communication. With Applications in Cryptography, Physics, Digital Information, Computing, and Self-Similarity, 5th ed. ISBN 978-3-540-85297-1. Springer, Berlin (1984, 1986, 1997, 2006, 2009)
59. Wicker, K., Guicking, D.: Diffus reflektierende Oberflächen im Wasserschall. In: Fortschritte der Akustik, DAGA 94. Teil B, pp. 837–840. DPG GmbH, Bad Honnef (1994)
60. Guicking, D., Görk, K., Peine, H.: Classification of underwater objects using resonance scattering and neural networks. In: Weydert M. (ed.) Proceedings of the European Conference on Underwater Acoustics, September 14–18, pp. 567–570. Luxembourg (1992)
61. Peine, H., Guicking, D.: Acoustical resonance scattering theory for strongly overlapping resonances. Acta Acustica **3**, 233–241 (1995)
62. Guicking, D., Wicker, K., Eberius, C.: Electrorheological fluids as an electrically controllable medium: I. Experimental arrangement and application to an absorber of underwater sound. Acustica **88**, 886–895 (2002)
63. Wicker, K., Guicking, D.: Electrorheological fluids as an electrically controllable medium: II. Medium properties and description by the Biot theory. Acta Acustica **88**, 896–903 (2002)
64. Guicking, D.: Erwin Meyer—A Great German Acoustician. Biographical Notes. http://www.dpi.physik.uni-goettingen.de/guicking/Erwin-Meyer-Eng.pdf

***

Biography

Dieter Guicking, born 1935, studied physics in Göttingen, Germany, and joined the "Drittes Physikalisches Institut" in 1959, starting with experiments on dynamic properties of high polymers. Since 1966 staff member, 1971–1998 in tenured position. Research fields: underwater acoustics, models of biological circadian rhythms, and active control of sound and vibration.

Part II
Manfred Schroeder's Memoirs

Fig. 1 Baby, 1 year old, summer 1927

Fig. 2 1929 with mother Hertha

Fig. 3 As a child with father Karl

Fig. 4 With parents and sisters Ingrid (*left*) and Helga (*right*) at age 16 or 17

Fig. 5 As a student in Göttingen with Erhard Scheibe and Jobst von Behr

Fig. 6 Building a radio as a student

Fig. 7 With parents and Helga and Ingrid (sisters) as student visiting in Hamm

Fig. 8 Immigrating to America on the Andrea Doria, 1954

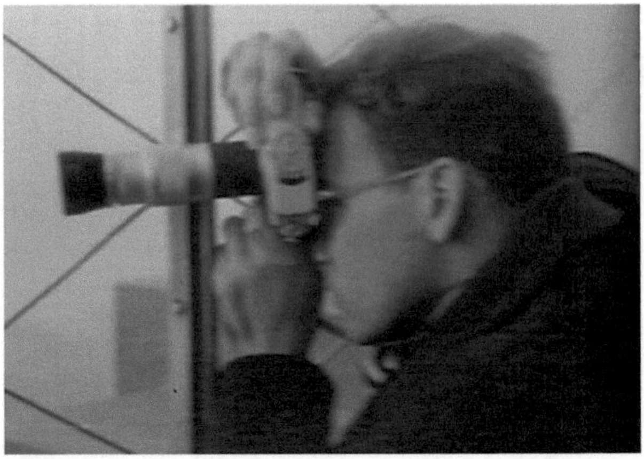

Fig. 9 First good camera, Leica, bought 1954

Fig. 10 First date with Anny, October 3rd 1954

Fig. 11 Wedding, February 25th 1956, Anny and Manfred

Fig. 12 First car 1957, Plymouth, purchased in 1954

Fig. 13 With Anny and children, Marion, Alexander, and Julian, 1964

Fig. 14 Gillette, New Jersey with wife and three kids: Alexander, Julian, and Marion, 1966

Fig. 15 Young Manfred

Fig. 16 Magazine cover, 1960

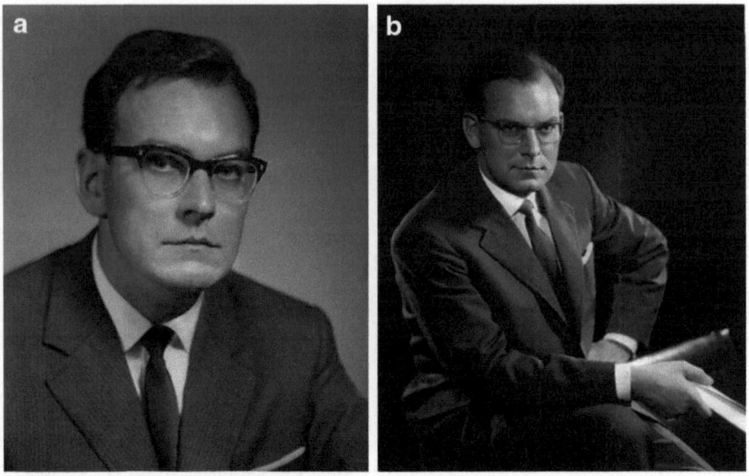

Fig. 17 (**a**, **b**) Manfred in Bell Labs

Fig. 18 Concert hall acoustics research

Fig. 19 Manfred with Ben Logan in Bell Labs (invention of the all-pass reverberator)

Fig. 20 Manfred with McLean in Bell Labs (invention of the anti-feedback system)

Fig. 21 Manfred in Bell Labs anechoic chamber (simulation of concert halls)

Fig. 22 Bicycling with Anny on Nantucket (1981)

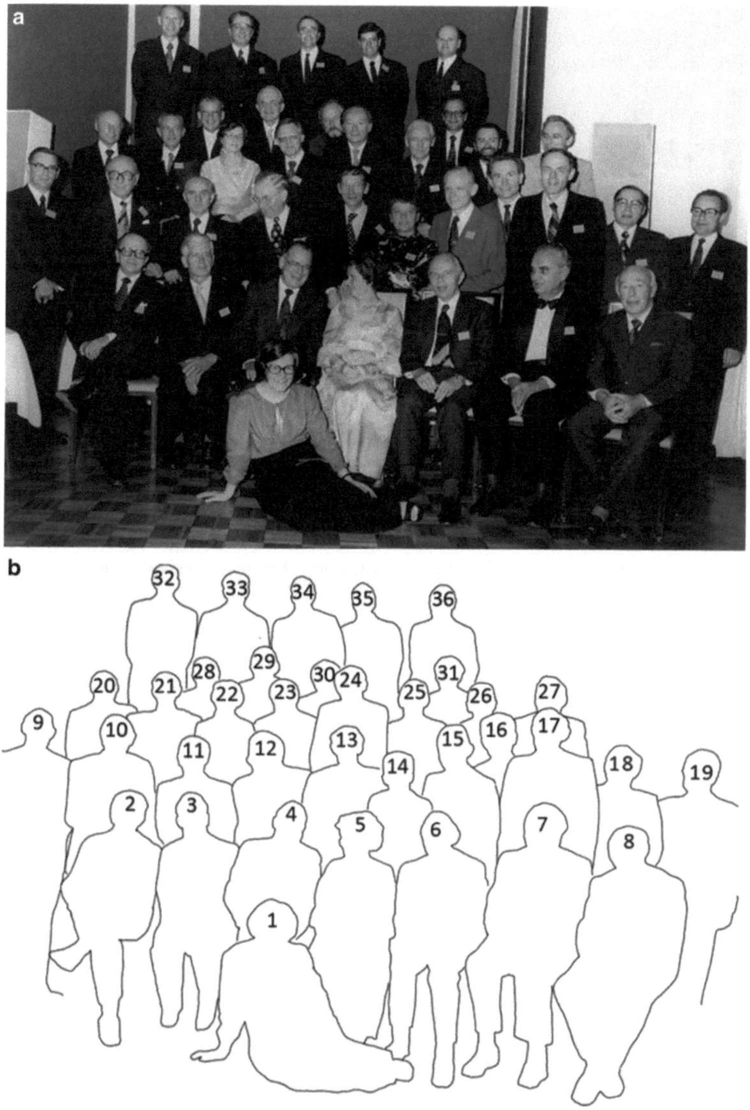

Fig. 23 (**a**) Göttingen 1977: 30th Anniversary of the Third Physics Institute. (**b**) *1*. Inge Schmid, geb. Laskowski, *2*. Horst Haeske, *3*. Horst-Gunther Diestel, *4*. Manfred R. Schroeder, *5*. Edith Kuhfuß, *6*. Günther Kurtze, *7*. Hans Severin, *8*. Konrad Tamm, *9*. Klaus Brendel, *10*. Clemens Starke, *11*. Peter Dämmig, *12*. Rolf Thiele, *13*. Georg-Renatus Schodder, *14*. Lucie Hackenbroich, geb. Bursian, *15*. Wolfgang Westphal, *16*. Helmut Siegert, *17*. Gerhard Sessler, *18*. Heinrich Henze, *19*. Reinhard Pottel, *20*. Karl Werner, *21*. Joachim Malzfeldt, *22*. Marie-Luise Beyer, geb. Exner, *23*. Klaus Helbig, *24*. Hans-Jürgen Schmitt, *25*. Peter-Paul Heusinger, *26*. Eberhard Mundry, *27*. Ernst-August Hampe, *28*. Wolfgang Burgtorf, *29*. Walter Kuhl, *30*. Wilhelm Ebrecht, *31*. Heinrich Kuttruff, *32*. Hans-Wilhelm Helberg, *33*. Wolfgang Eisenmenger, *34*. Walther Junius, *35*. Jörg Schmid, *36*. Frieder Eggers

Fig. 24 With secretary Mrs. Edith Kuhfuß in Göttingen, Drittes Physikalisches Institut

Fig. 25 (**a**) Göttingen 1987: 40th Anniversary of the Third Physics Institute with all people who joined the Institute in the first 10 years. (**b**) *1*. Edith Kuhfuß, *2*. Inge Schmid, geb. Laskowski, *3*. Marie-Luise Beyer, geb. Exner, *4*. Manfred R. Schroeder, *5*. Horst Haeske, *6*. Rolf Thiele, *10*. Georg-Renatus Schodder, *12*. Frieder Eggers, *13*. Reinhard Pottel, *14*. Horst-Gunther Diestel, *16*. Klaus Brendel, *17*. Wolfgang Eisenmenger, *18*. Heinrich Kuttruff, *22*. Jörg Schmid, *25*. Hans-Wilhelm Helberg, *26*. Gerhard Sessler. The following list was not very certain by the editors: *8*. Peter Dämmig, *9*. Clemens Starke, *19*. Walther Junius, *24*. Heinrich Henze *27*. Sanat Kumar Mukherjee, *28*. Eberhard Mundry, *29*. Friedrich Wiekhorst

Fig. 26 Göttingen, October 11th 1997: 50th Anniversary of the Third Physics Institute

Fig. 27 *1991: The frontispiece in page vii is derived from this photo using Eikonal equation algorithm by Wolfgang Moeller*

Fig. 28 *ICA Congress 1983 Paris (from left: Anny, Manfred, Gerhard and Renate Sessler, Mildred and Jim Flanagan)*

Fig. 29 (**a**, **b**) At *Darmstadt University of Technology 1998*

Fig. 30 With Ning Xiang and his daughter Charlotte at the 75th Anniversary of the Acoustical Society of America, New York City 2004

Fig. 31 *At the ASA Meeting Banquet, Honolulu 2006 (from left: Gerhard Sessler, Renate Sessler, Anny Schroeder, Ingrid Kuttruff, Manfred, Heinrich Kuttruff)*

Fig. 32 Family get together December 2007, from *left* to *right* standing: Nicola, Julian, Marion, Uwe, Alexander, Marion, Manfred, and Anny; kneeling: Nora, Lilly, Shadow, and Julia

Fig. 33 Manfred's hobby Photography (on the board: the Old Man from the Mountain, courtesy: A. Kohlrausch)

Fig. 34 With Heinrich Kuttruff

Fig. 35 With Armin Kohlrausch at the Honoring session for Manfred sponsored by the Acoustical Society of America Meeting, 2006 (Board: Freihalten für Anlieferungen)—courtesy by Omoto

Fig. 36 With Hiroya Fujisaki at the Honoring session for Manfred sponsored by the Acoustical Society of America Meeting, 2006—courtesy by Omoto

Fig. 37 Brief remarks at the end of the Honoring session for Manfred: with high appreciation and pleasure, sponsored by the Acoustical Society of America Meeting, 2006—courtesy by Omoto

Fig. 38 Honoring Manfred 2006 at the ASA Meeting Hawaii jointed with the Acoustical Society of Japan, with all invited speakers and chair (from *left*: Ning Xiang (chair), Bishnu Atal, James West, Gerhard Sessler, Volker Mellert, Armin Kohlrausch, Manfred Schroeder, Heinirch Kuttruff, Anny Schroeder, Jean-Dominique Polack, Michael Vorlaender, Fumitada Itakura, James Flanagan, Birger Kollmeier, Hiroya Fujisaki)—courtesy by Omoto

Fig. 39 In Memory of Manfred 2011 at the Acoustical Society of America meeting in Seattle, WA (from *left*: Birger Kollmeier, Joshua Atkins, Julian Schroeder, James West, Peter D'Antonio, Alexander Schroeder, Anny Schroeder, Jean-Dominique Polack, Ning Xiang, Roland Kruse, Gerhard Sessler, Peter Cariani, Bishnu Atal)—courtesy by the ASA

Chapter 14
Right Turns

Manfred R. Schroeder

In the course of my lifetime, I have made many bad decisions and a few good ones; some without further consequences, others with lasting results. Here I want to focus on the good decisions—the Right Turns—that have had major impacts on my life.

14.1 Pulling Little Roots

Even as a child I was never much interested in sports. When the neighborhood boys rang the doorbell for me to join them in *Schlagball* (Soft ball), my mother said "Manfred is in his room 'pulling little roots'" (*Würzelchen ziehen*), her metaphor for solving math problems. To this day, for better or worse, I have never been interested in participating in competitive sports. (But I do like to watch tennis, downhill skiing, skating, and ballroom dancing, activities that I actually practiced when I was younger.)

In my entire life, I rarely did things I didn't like and that, I think, is good advice for a happy life. For example, I abhorred chemistry and never studied it as I was actually required to do as a physics student. Later in life, as an institute director, I had to supervise many theses involving a heavy dose of (physical) chemistry. So I got hold of some text books and read up on the chemistry I needed to help other people in need.

Author was deceased at the time of publication.

M.R. Schroeder (✉)
Drittes Physikalisches Institut, Georg-August-University Göttingen, Friedrich Hund-Platz 1, 37077 Göttingen, Germany

N. Xiang and G.M. Sessler (eds.), *Acoustics, Information, and Communication*, Modern Acoustics and Signal Processing, DOI 10.1007/978-3-319-05660-9_14,

14.2 Air Force to Navy

A potentially life-saving Right Turn I took in September 1944, 2 months after I had turned 18. Just before reporting to my Air Force radar unit in Detmold for boot camp, I had cut my right toe on a glass shard during my last civilian swim. I therefore missed boot camp. The Air Force, not wanting to bother with a half-invalid, packed me off to the Navy. (The Navy and the Air Force had an understanding to exchange technical personnel if and when the need arose.) I was not amused and my father, a lifelong flying enthusiast (and combat pilot in World War I) was incensed. He was about to call the Air Force personnel office in Berlin, where he had some old friends, to have my transfer rescinded. But I thought "Why bother—maybe I'll be better off in the Navy." I prevailed and on or about September 30th 1944, I reported to the corresponding Navy radar unit in Stralsund—which was just concluding its own boot camp. But with my lack of basic training, I was demoted to a (live!) hand grenade practicing target. Luckily the grenades that were lobbed at me were low-yield and I survived the demonstration unhurt (see my Chap. 17).

In early April 1945, when the war in Europe drew to a close, my classmates who had stayed with the Luftwaffe (Air Force) were pulled out of their classrooms at the *Luftnachrichtenschule* (Air force communications school) in Halle near Leipzig in central Germany and sent against the—by then—highly experienced American army. My buddies were of course totally unprepared for trench warfare, equipped only with ancient (Italian) rifles. About a third of my class perished on a *single* day, April 12th 1945, including my Austrian friend, Heinrich Brenner, on the banks of the Unstrut river, west of Halle.

By contrast, at the same time in the last weeks of the war, I was operating coastal radars on the North Sea coast in Holland, where we never saw enemy action after the ill-fated Market Garden offensive (17–25 September 1944) by British and Canadian paratroopers near Arnhem (commemorated in the 1977 film "A Bridge Too Far").

14.3 Göttingen to Murray Hill

My next Right Turn came when I decided to commence my studies of physics in Göttingen (rather than nearby Münster) because of the presence there of the "Father of Uncertainty," Werner Heisenberg. But, as so often in life, that main motivation proved irrelevant. I attended just a single lecture (on the shell model of the atomic nucleus) by the famous physicist and I decided I had enough of Heisenberg: I didn't understand a word ("spin-orbit coupling" and other strange concepts). But my interest continued to lie in theoretical physics; I was married to theory you might say. Yet even so, we were obliged to do a lab course in applied physics. One of the experiments in that lab course was to measure the speed of light (electromagnetic waves) on a long cable coiled up in the basement of the Drittes Physikalisches Institut (DPI). The standard (and expected) method was to send a very short electric impulse into one end of the cable and measure the delay with which it appeared at the other end. Before starting out on this not exactly simple task, I thought it would be much easier to connect the output of the cable to its input via an audio amplifier. Once I cranked up the amplifier gain enough, this feedback set-up would become unstable and the cable would start to "sing." The easily measured frequency of this tone is the reciprocal delay on the cable. This simple method worked like a charm and the lab supervisor (Dr. Hans Severin) alerted the institute director (Prof. Erwin Meyer) who, on the spot, offered me a position as a graduate student at the DPI. Thus, while maintaining my connection with the theory group, I was now working on a thesis in applied physics (involving the statistics of microwave cavity resonances).

And it was precisely this combination of theoretical and experimental physics that propelled me on my trajectory from Germany to the United States. Through my friend the theorist Peter Hasen I learned that William Shockley, Bell Labs physicist and co-inventor of the transistor, was visiting Göttingen looking for good German students for Bell.[1] My thesis advisor, the experimentalist Meyer, was kind enough to write a letter on my behalf to one of Bell's research directors (Winston Kock) that resulted in my employment interview (at the Dorchester in London, no less) and subsequent employment offer (cum visa) to join Bell Labs.

I had initially planned to stay in the States for a year or two, but I soon got married (to Anny Menschik of Radio Free Europe in New York), raised a family (Marion, Julian, Alexander), got promoted (to department head and, later, to director of acoustics and speech research), became a US citizen (in April 1963) and now, 55 years later, I am still living in the States (in Berkeley Heights, N.J.) for 5 months every year.

All of this happened, I believe, as a result of a few Right Turns:

- The decision to study in Göttingen
- To study physics rather than mathematics (my first love)
- To measure the speed of light by an audio feedback loop
- To stay connected with the theory group in Göttingen
- To emigrate to America

14.4 Murray Hill to Göttingen

Ironically, my next Right Turn—after 15 years in America—was to return to Germany. In 1966, Wolfgang Eisenmanger, whom I had invited to spend a summer at Bell Labs, asked me—as charged by the faculty of physics—whether I would be

[1] Shockley's primary interest was in one-dimensional metals on which he had just published himself (*Phys. Rev.* **91**, 228 (1953)).

interested in a call to Göttingen, to succeed Erwin Meyer as a professor of physics and director of the DPI. I was very happy then with my work at Bell and never thought of leaving Murray Hill. But—out of politeness—I answered yes, I would be interested. (I mean, what could I possibly say? "I am not interested"?? That would have been very impolite.)

When the call actually came, in 1968, I was getting a bit bored by Bell and I gladly accepted (after a few details had been negotiated with the state government in Hanover, represented by Herr Ministerial-something Ulrich Hopfe). When Hopfe asked me what salary I expected, I showed him my last salary slip from Bell (for $39,000). Well, Hopfe said, "we can't match *that*," but he would give me all raises and extras until retirement—which amounted to barely half my American income. But, instead of quibbling, I said "fine" and accepted the offer (which included *seven* new permanent positions for the institute: one associate professor, two assistant professors, and four shop people—very generous indeed).

Although taking a 50 % salary cut seemed like a big gamble, but with the Deutschmark's (and the Euro's) unremitting rise against the dollar the income gap was soon bridged and, in any case, it wasn't for the money I had changed jobs. I always thought I would like teaching and, in fact, it turned out I *loved* it.

In addition, and perhaps most importantly, I became a full-fledged physicist again (at Bell my work involved a heavy dose of engineering—speech compression for cell phones and the future Internet). I also widened my horizon far beyond physics and engineering by weekly contacts with colleagues from the humanities (history, religion, philosophy, linguistics, art).

Chapter 15
School Years

Manfred R. Schroeder

15.1 Ahlen

I was born in the house of my grandparents, a large director's villa near the entrance to the coal mine where both my father and grandfather worked. There was a large orchard behind the house (with delicious red and black currents and goose berry bushes and a tall pear tree) and a small kennel for *Hector*, my grandpa's hunting dog.

When I was first born, I was given up for dead and received an emergency baptism (*Nottaufe*) in the kitchen sink because I didn't make any sounds. Later I didn't stop crying for 4 years and my parents had to string woollen blankets across my room to deaden the sound because grandpa was very sensitive to every little noise (*Höllenspektakel*, as he called it).

Not far from the house was a city bus stop where I befriended one of the drivers, Karl Jaxtin, who took me on extended bus tours around the city. I loved it. This was before 1933 but Jaxtin was already a Nazi, an *SA-Mann*, feared by the local communists among the miners for his combativeness. During the war he became a tank driver and perished in the early part of the attack on the Soviet Union.

My first "school" was a Roman Catholic kindergarten, St. Joseph's, in Ahlen in Westphalia (now part of the state of North-Rhine-Westphalia) not far from the coal mine *Gewerkschaft Westfalen* where my father worked as an engineer and my grandfather, Robert Kraemer, was director of below-ground operations. *Westfalen* was the north-eastern most mine of the Ruhr coal fields and they had to go to a depth of 1 km and more to find the black gold.

I remember my nickname from those preschool days: *Kohlenklau* (coal thief). I don't know how I acquired that moniker: I certainly never stole any coal as a child and I wasn't particularly dirty. But most of my friends were miners' children and,

Author was deceased at the time of publication.

M.R. Schroeder (✉)
Drittes Physikalisches Institut, Georg-August-University Göttingen, Friedrich Hund-Platz 1, 37077 Göttingen, Germany

N. Xiang and G.M. Sessler (eds.), *Acoustics, Information, and Communication*, Modern Acoustics and Signal Processing, DOI 10.1007/978-3-319-05660-9_15,

during the depression, *Kohlenklau* was a meaningful concept among them. Sixty years later, one of my Internet passwords was based on this uncommon word. (At another time I chose as a password the non-word name I gave myself as a kid because I couldn't properly pronounce *Manfred*, containing as it does a cluster of three consonants: *nfr*.)

After kindergarten, I attended the parochial *Volksschule* (grade school) for 4 years. One of the few things I remember from those years were the canings (*Stockhiebe*) on the palms of your hand when you had dared to snicker during the singing of the national anthem or forgot a line in a poem. I soon developed calluses and hardly felt the whipping but my friend Eugen and others always howled with pain on such occasions.

It was also at St. Joseph's that I went to Confession and Holy Communion. Since I really didn't know how sin was defined, I confessed some innocent doings simply because they felt good—so, I figured, they must be sinful. And I remember the perennial question in the confessional: *Alleine oder mit Anderen*? (Alone or with others?).

One thing I remember from my early school days was in 1934 (it must have been early July) when I saw police patrols near our house on *Stapelstrasse*: the "Röhm Putsch" was in full swing with many "disloyal" *S.A.*-men being murdered by Hitler and his gang—Goering, Himmler, Heydrich and members of the loyal, elite *Schutzstaffel* (SS)—during and after the "Night of the Long Knives" as it became known. Although barely 8 years old, I sensed that something was very wrong with Germany's leadership.—But we, as children, still got a kick out of the rambling rantings of the *Führer*, sometimes lasting 3 h or more.[1]

On my 10th birthday, in 1936, I got my first bicycle which I rode with abandon, always pretending that I was flying a fighter plane. The bicycle was a gift from my aunt Else Schroeder, a missionary in China, who was on a rare leave from Szechuan where she taught English and Bible classes. As a young woman she had joined the "China Inland Mission" and, after basic training in England, had set out, in 1913, on the 6-week journey, via Berlin, Moscow, Vladivostok, Shanghai, and up the Yangtse river, to her mission near the Tibetan border. She stayed until 1952, 3 years after Mao Tse Dong had taken over and she and the other missionaries were expelled from China—after having "donated" their schools and hospitals to the state, the Christian crosses being replaced by Red Stars. But in those 39 years in China she had become so Chinese that she didn't feel comfortable living among non-Chinese any more. So

[1] Around 1920 Hitler became member # 7 of the *Deutsche Arbeiter Partei* (German Worker's Party) which he soon re-christened *National-Sozialistische Deutsche Arbeiter Partei* (abbreviated as NSDAP)—a clever name, seemingly uniting the opposing political poles: National and Socialist. Hitler also designed the Swastika flag, the swastika derived from the Germanic sun-wheel still rolled down the hillsides at summer solstice in Nordic countries. (The Hindu swastika has a different parentage, I believe.)

she moved to Taiwan to continue her missionary work. She also kept up a lively correspondence with several of her Christian converts from the mainland.[2]

Even before leaving Szechuan, but after the Communist take-over, she was asked by kids in the street where she was from. She answered "Germany." The next day the kids wanted to know "which Germany?" But aunt Else pretended not to understand and stuck to her previous answer: "Germany." Another day later, the kids came back (obviously primed by their parents) asking "who is your president?" And aunt Else answered "Theodor Heuss," whereupon the children scolded her: "*Heuss*—bad Germany. *Pieck* (the East German communist party boss)—good Germany!" The gospel of communism was obviously spreading to some distant corners of the world.

At age 10, I left the *Volksschule* and transferred to the local *Gymnasium* (high school emphasizing the humanities, Latin, etc.). The only thing I remember from that high school is that I was always late. Why couldn't I get to class, like everybody else, before 8 o'clock instead of habitually arriving around 8:05? The teacher was not amused and once turned full circle on his heel in despair. I also remember a few classmates: Hugo Kuhweide, who later perished in Russia, and a very handsome boy, Norbert Plate.

Another thing I remember from my *Gymnasium* days were two students, a boy and a girl, who were always walking to school in the opposite direction. My parents explained that they were Jewish children attending a different school. But one day I didn't see them any more. My parents said that the family had moved away. (Later it became clear to me that, since this was in 1936, the kids and their family had probably avoided a more sinister fate.)

15.2 Rotenburg

In the meantime my father, a World War I combat pilot, had joined the reconstituted German Luftwaffe and, after a brief stay in Göttingen, was assigned to the gigantic *Luftzeugamt* (air force supply center) Rotenburg on the Wümme, halfway between Bremen and Hamburg. So, in early 1937, the family moved to Rotenburg where I started to attend the Middle School. (The nearest high school was 50 km away, in Bremen.) In Rotenburg, the first foreign language was English and I needed private tutoring to catch up. I learned to pronounce the "English" *r* by saying, ever faster over and over again, "Pedinz von Pedeussen," until it became "Prinz von Preussen" (Prince of Prussia). But the result was more Italian than English. I found the whole English experience frustrating and once started crying and the teacher explained:

[2] In the eighteenth century, some Jesuits, who were sent by the Vatican to the Qing imperial court in Beijing to convert the Chinese to Christianity were, instead, themselves "infected" by Confucianism.

"You never had to study very hard; everything was easy for you. But now, for the first time in your life, you have to really exert yourself." I thought the teacher was exactly right and henceforth I started to study more diligently.

It was also at Rotenburg that I first heard the word Gestapo (*Geheime Staatspolizei*). For a time, we were living in a beautiful thatch-roof building that housed the airport architects, assorted secretaries and some officers' families. One day the building burnt down (luckily we had just moved out) and sabotage was immediately suspected. Soon the Gestapo took over, but they couldn't find any culprits. (The cause of the fire was probably one of the tenants, the bohemian kind, smoking in bed and falling asleep—for longer than intended.) The design of the building called *Bauleitung* (construction supervision)—all wood, and as mentioned, a thatch roof—was the brainchild of the commanding officer (Colonel Christensen) who was strong on aesthetic appeal but weak on fire safety. (It's amazing what whims people got away with even after 5 years of Nazi rule.) The Christensens had a beautiful daughter, Elfie, on whom I had cast an early (and distant) eye. But she was very closely watched by the family ever since she had been a victim of statutory rape.

15.3 Jungvolk

In 1937, after I turned 10, I had to join the *Jungvolk*, the junior branch of the *Hitlerjugend.* I had volunteered when I was 7 or 8 because I was envious of the boys marching in the street outside our house. This looked like great fun. I applied and was accepted. But, being the smallest kid, I always marched in the last row with "giant" strides and was still unable to keep up with the older boys. My uniform was of course much too large for me and my belt was sometimes slipping down near my knees. Everyone, my parents included, got a great laugh out of watching little Manfred parade by the house. Later, when I *had* to join, it was no longer such fun. I remember with particular dread the *Kriegsspiele* (war games) where we had to box and wrangle with each other.

A little later, I was appointed to the *Fanfarenzug* (fanfare platoon) but I never learned to play the *Fanfare* properly, and few of the other boy "musicians" did. But when we all played together the tune might have been recognizable—by the law of averages, I presume. To this day I remember some of the campfire songs like the melodious *Flamme empor* (Rise flame).

15.4 Rübeland im Harz

One of the worst experiences while a member of the *Jungvolk* was a summer camp in Rübeland in the Harz mountains in 1937. We had to bicycle through the Lüneburg Heath from Rotenburg to Celle, some 70 km—easy. The next day, we biked 130 km (80 miles) to Bad Harzburg, including an extended tour of the old city of Braunschweig which lay enroute. One of the sights we saw and that I still remember was the *Alte Waage* (Old Weighing Station) near the center of the medieval town which miraculously survived the war (and which I revisited 50 years later). By the time we got to our youth hostel in Bad Harzburg we were really pooped. During an evening stroll I had lost a 2-Mark coin. It was pitch dark except under the street lights and, naturally, that's where I looked, but with no success.

The next day was really bad. It was only 40 km (25 miles) but all the way up, up, up to a place called Rübeland. Our bikes had of course only one gear and even some of the toughest boys from our class of 11-year olds including Hermann "Bully" Sander cried on his bike. If we could have only dismounted and pushed our bikes (burdened by extra clothing and gear). But this was after all Nazi Deutschland and we were being prepared (without our realizing it) for war.

This warlike spirit became even more pervasive in the camp itself. The *Zeltlager* (tent camp), situated in the midst of the woods, had of course no regular toilets and we had to do with a latrine—I for the first (and last!) time in my life. One of the exercises I remember was throwing (wooden) hand grenades. The food was abominable and the rations much too small.

One rainy day was scheduled for a 60-km bike trip to the Brocken, the highest peak in northern Germany (1,142 m, 3,747 ft). The first half was a strenuous pedalling uphill. Then we ditched our bikes and continued climbing on foot. When we finally reached the top, the fog was so dense that we could hardly make out the outlines of the tourist hotel and the new TV-tower, one of the first in Germany, used during the 1936 Berlin Olympics for demonstrating live country-wide television to gaping audiences—German and foreign. But we never stopped at the top and continued marching right down again. Eventually, the group leaders gave us a little rest and we could munch a sandwich in the roadside ditch. Talk about steeling German youths to the deprivations of war!

After 2 or 3 weeks my parents and grandparents came to visit. When they saw me and my classmates (including the formerly *dicke* (fat) *Peinemann*) reduced to thin, ghostlike states, they immediately invented some pretext to rescue me from my camp ordeal. Now liberated, we spent a wonderful week in the Harz, visiting the famous *Roßtrappe*, according to legend the imprint of a horse's hoof before leaping across a wide gorge. And at the top of the Brocken we actually went *into* the restaurant for delicious cakes and *Torten mit Schlagsahne*.

Before we checked into a hotel for the night my mother inspected the beds and believed to have found some bedbugs. But the owner brushed them away and denied their existence.

In 1938 the family took a vacation on the Baltic. The resort was then called Brunshaupten but later "rechristened" Kühlungsborn by the Nazis—a name, in a show of poor taste, kept by the East German regime after the war. (Kühlungsborn was a typical Nazi coinage reminiscent of Lebensborn, the SS "stud farms" to create more blond, blue-eyed babies for the *Führer*.)

Fig. 15.1 Ingrid, Herta, Helga, and Manfred in Brunshaupten in 1938

In the middle of our vacation my father was called back to his office because of the impending war over the *Sudetenland*. But I still have a nice snapshot from those wonderful days (Fig. 15.1).

One thing I remember was the superb banana ice cream served in a parlor between our modest apartment and the beach. My favorite toy was a small model sail boat that "always" turned around by itself when it hit the pier. But one day it didn't—no matter how long I waited. Where did it go? And where did it finally founder? Perhaps in some distant ocean.[3]

15.5 Opa and His Primus

I also remember, in fact I can still see, my grandfather haggling with a waiter who, he charged, had dared round off a half *Pfennig* to a full *Pfennig*.

A strange apparition I remember, in a park in Bad Harzburg, were two very pale (by German standards of robustness) English-speaking ladies made up with brilliant

[3] On my father's low salary we couldn't of course afford a *Pension*, let alone a hotel, right on the *Strandpromenade* ("board walk"). In fact, the whole trip was *KdF* sponsored.—The next year we took another *Kraft durch Freude* vacation in Strobl on the Wolfgangsee. We were lodged with a small-time farmer (where I caught Asthma from their moldy bedding). One day, at the end of August, I saw a long line of cars at a filling station. The reason? Impending war, I was told. In the meantime we enjoyed the lovely Wolfgangsee, the nearby mountains—especially a gorgeous view of the *Dachstein* glacier. And in Austria they were still serving strawberries and whipped cream (*Schlagsahne*, the real thing).

red lipstick—a sight I had never seen before. Lipstick! Maybe German women applied some on the sly in Berlin but not in the boondocks where we lived. (Hitler had practically outlawed lipstick as a sign of Western moral decay—but his paramour, Eva Braun, used some anyhow.)

At the end of the vacation, grandpa couldn't find his way out of the Harz mountains: "*Seesen, wo ist Seesen*?" he inquired perhaps a dozen times.

On another occasion, his car, an *Adler Primus*, wouldn't start moving. My parents knew of course the reason: Grandpa still had the handbrake applied. But they didn't say a thing, always afraid of the irascible Opa Kraemer. Then my little sister Ingrid blurted out "*Opa hat vielleicht die Bremse drin*" (Grandpa probably has the brakes applied). "*Verdammt, das Blag hat Recht*!" (Damn, the kid is right). Whereupon he released the handbrake and the car leaped forward and immediately stalled again. Yes, it was always great fun to ride with Opa in his beloved *Primus*—if it *was* running. Opa, now retired, liked to take his car apart but then always needed the help of an expert auto mechanic to put it back together again.

Once *Opa*, then about 63, took a vacation—"alone." Upon his return he always talked about "we" and "us." "Who 'we'?" my grandmother wanted to know. "Me and my Primus" was *Opa's* constant refrain. But *Oma* knew better. (And who took all the pictures of *Opa*? Always "someone who passed by," according to *Opa*.)

Going for an outing in *Opa's* car was always fun. But if there was another adult male passenger, *Opa* always made him crank the car "to save the battery." And days before an intended trip, even just to the center of the city, he would constantly consult the barometer. And if there were just a few drops of rain, the trip would be postponed for "less tempestuous" times.

I once went hunting with Opa but was afraid of shooting at a gaggle of pheasants alighting just in front of us so as not to waste any precious ammunition. Well, my grandfather didn't shoot anything either the whole day. Driving home, we stopped at a game store to buy some dead animals. But my grandmother always saw right through the ruse: the shot size in the purchased animals was never the same size as she had put into the shells in the garage. But Oma never let on. Opa's reputation as a "great hunter" was never questioned.

As a dedicated hunter, he loved dogs, but abhorred cats. One "pitch dark" night, returning from nearby Soest ("*stock-finstere Nacht, kein Weg kein Steg*"), he had encountered *eine dicke lange schwarze Katze* (a fat long black cat) crossing the road in the headlights of his car *from left to right*!—a particularly bad omen. Even so, by sheer willpower, I guess, he made it home (some 15 miles).

Of course, cats hated him right back, he was convinced. Once, with friends (the Diekmann's) on a trip to the *Sauerland*, he insisted on parking his car as far away from the street as possible so it wouldn't catch any dust. So poor Diekmann had to move his car (an *Opel* P4). The next morning *Opa's Primus* was covered in dust anyhow and showing (heaven forfend!) cat paws all over. (He was convinced *die Biester* (the beasts) knew exactly what they were doing and to whom because Diekmann's *Opel* was totally untouched.)

At one point Opa had bought a small tent that we pitched in his back yard. We wanted to spend one night in the new tent and so, in our nighties (Opa armed with a pistol), we moved in at sundown, Opa—a big man—occupied about ¾ of the floor space. Of course, neither of us could fall asleep and after about one hour of tossing we went back into the house. I can still see Opa, in the bright light of the full moon, leading the way in his nightshirt—pistol at the ready.

15.6 Bremen

After 2 years at Rotenburg I transferred to the *Lettow-Vorbeck* high school in Bremen named after the "heroic" World War I defender of German East Africa (now Kenya). In front of the school there was (and still is) a huge brick elephant standing on top of the *Ost-Afrika Denkmal* (East Africa Memorial, now closed). The monument served as a favorite playground for me and my classmates: we could hide behind the elephant's legs after having launched slingshot attacks on the bowler (derby) hats of unsuspecting Bremen burgers passing by.

Since we continued to live in Rotenburg I became a *Fahrschüler* (commuting student), always leaving Bremen at 13:26 for the 45-min return trip home. Fifty years later, the train is still running, but it takes only 25 min now. The train ride was always fun. I remember once telling my buddies that I had observed a certain pretty girl (probably not older than 16 but "old" for us 13-year olds) disappear in the direction of the woods with a *Flieger* (airman): "I guess I don't have to tell you what they were up to." Whereupon said young girl appeared from the next compartment asking: "Tell me, please, what were we up to?" I think I got very red in the face and in future always made sure there was no acoustic leakage between adjacent train compartments.

One thing we particularly enjoyed in Rotenburg were the Wümme river and its many canals. In winter they offered endless opportunities for skating (and flirting) with our classmates. In particular, we boys were always on the lookout for any "fallen angels": we were only too eager to help them back on their feet, thereby getting a first-hand feel of their physical development (or lack thereof). (We were all around 12 at the time.)

While living in Rotenburg the *Autobahn* Bremen-Hamburg was completed and our family took our first trip to Bremen on the expressway. But Bremen wasn't getting any closer and the sun was on the "wrong" side of the car; before too long we were approaching Hamburg. My father had never driven on a clover-leaf ramp and now he couldn't just turn the car around: we had to go to the next exit near Buxtehude (famous for its mentally-handicapped asylum) to turn around.

Around the same time (1938) also came my only chance to see Hitler in the flesh. My father and I had travelled to Hamburg (by train, I believe) to witness the launching of the KdF cruise ship "Wilhelm Gustloff" (or perhaps her sister ship the "Robert Ley") at the *Bloom & Voss* shipyard. KdF stands for *Kraft durch Freude* (strength through joy), the Nazi attempt to assuage the German worker after the rape of their unions on 2 May 1933. These large ships (and several smaller ones) gave working-class Germans the opportunity to visit distant shores such as Madeira and the Norwegian fjords.

Another KdF operation was the *Volkswagen*, originally known also as *KdF-Wagen* and obtainable for a mere 1,000 Reichsmark. But most buyers never saw their prewar VW: it was "high-jacked" by the Nazis and served as a German jeep, especially in Rommel's *Afrikakorps*.—As to the "Gustloff" and the "Dr. Ley," they were used, in fact designed, as *Lazarettschiffe* (hospital ships) for the war. (The "Gustloff" was sunk on 30 January 1945 by a Soviet submarine with over 5,000 refugees from East Prussia aboard.)

One thing I remember from the "Gustloff" launching were the few and isolated shouts of *Heil* when the *Führer* appeared on the platform. I was amazed because on the radio we always heard the endless *Heils* when Hitler loomed. Were we being hoodwinked? Not necessarily, because, it was explained to me, the *Bloom & Voss* workers were mostly former communists.

The actual christening of the huge vessel was performed by the widow of Wilhelm Gustloff, a Swiss Nazi "martyr" who had been assassinated by opponents.

Incidentally, Rotenburg, too, had a large asylum for the mentally handicapped: some two thousand patients—a lot in a little town of six thousand (before the air force "invasion"). Many of the patients (kept as inmates) were allowed to stroll around town on weekends: always in twos, arm-in-arm. It was here, around 1939, that I first heard the expression "*verschickt und nicht angekommen*" (sent away but never arrived), a euphemism for the Nazi program (dated September 1, 1939) to "eliminate unworthy life." Indeed, after the beginning of the war, the patient population dwindled rapidly.

Soon after the war started, British bombers appeared over northern Germany and my parents decided to move to East Germany, to Liegnitz in Silesia, where my father had just been transferred. They didn't want to get separated again—and they were afraid of bombs. Of course, the real bombing campaign didn't start until several years later but my parents never regretted their decision.

Not far from the *Luftzeugamt* Rotenburg, a sizeable sham airport was built up by the German air force. But the British were not deluded: they dropped *wooden* bombs on the make-believe operation. This was a mistake because the Germans henceforth saved a lot of money by closing down their "fairy port."

The 1 year at the Lettow-Vorbeck school in Bremen was one of my best. I befriended Dietmar Risch, one row behind me. But it was kind of friendship at a distance because he lived in Bremen and I in Rotenburg. But we met some 60 years later—and recognized each other immediately. The home-room teacher lent me his (amateur) telescope and asked me to describe to the class what I had seen in the sky.

One event at the Lettow-Vorbeck school that I remember vividly is a Reichstag speech given by Hitler at the end of April 1939, which we all had to listen to, in which he ridiculed a recent demarche by Franklin Roosevelt demanding from the German government a pledge of nonaggression toward a long list of countries. Hitler assured Roosevelt that he had no intention of attacking—and there followed another long list of mostly irrelevant nations. And his listeners got a big laugh at the expense of FDR. One of the countries Roosevelt had listed was Syria and Hitler, to great general amusement, reminded Roosevelt that Syria was a *French* protectorate and that he should address his plea to Paris.

Significantly, and no one noticed at first, Hitler left out Poland, the only country that was really threatened at the moment. This propaganda debacle (for the democracies) could have been avoided if Roosevelt had concentrated on one or two countries, say Poland and Lithuania, in which case the omission by Hitler of Poland would have been glaringly obvious.

When it came to dealing with foreign nations, Roosevelt, the superb politician, was often "out of his depths." How could he call the mass-murderer Joseph Stalin "Uncle Joe"—a term of endearment (which, by the way, the Soviet dictator, not being familiar with American mores, took as an insult). And why did he call (in May 1943, in Casablanca) for Nazi Germany's *unconditional* surrender? Churchill was aghast. For Hitler and Goebbels this was, of course, grist on their mill, giving them a superb excuse for prolonging the war. (On the other hand, Roosevelt was one of the few Americans who, early on, saw the danger of Nazi Germany. While most of his countrymen were still dreaming of peace, he supported the British, stretching neutrality to its very limits: by the Lend-Lease legislation, the donation of 22 "mothballed" destroyers, and declaring the West Atlantic a war zone in which the U.S. Navy would sink any "unfriendly" ships (i.e., U-boats).)

In those days, Hitler was at the peak of his prestige in Germany: he had just conquered Czechoslovakia, "recovered" the *Memelland* from Lithuania and was flush with hubris after his big 50th-birthday celebration on 20 April 1939, attended by diplomats and military attaches from around the world.[4]

I saw the downside of the *Führer's* policies first-hand when a (distant) relative (a customs official) showed me around the Bremen docks and warehouses where I caught sight of a lot of furniture in storage. "Belongings of Jewish emigrants," my guide explained "but they cannot be shipped on during the blockade of the German ports during the war." The loss of the Jewish citizens after 1933 created a void in German intellectual life that, 70 years later, has still not been filled, I believe. Without that deprivation, Germany would be a much more vibrant country than it actually is.

One thing that made Bremen very attractive to me as a commuting student from the boondocks, was the wonderful smell of roasted coffee—speaking to me of distant lands—that pervaded the city. Bremen was famous for its coffee roasting. *Kaffee Hag*, the first decaffeinated coffee, had its home there.

15.7 Liegnitz

When we moved to Liegnitz I had to leave the school and its director wrote a spontaneous letter to my father that, to his great regret, the best student, both academically and *im Betragen* (general comportment) was leaving. I had been an

[4] The place where Hitler stood during the mammoth parade on the *Ost-West-Achse* (later *Strasse des 17. Juni*) is now occupied by a massive Soviet monument to their victory over Germany—constructed, fittingly, with marble slabs from Hitler's *Neue Reichskanzlei*.

inspiration to the whole school, he wrote. Wow! My parents were of course delighted but the letter went to my head and in Liegnitz I became one of the worst students (especially *im Betragen*) so that the director upon my leaving *that* school said: "We are glad you are leaving." And he didn't even know some of the worst things my classmates and I had perpetrated.

For example, we broke into the school at night with a pass-key that I had illegally duplicated. Among the items we stole was potassium metal that we made to "dance" on a lake in the city's park after midnight. (When the locksmith wanted to know for what purpose I wanted the key, I said it was for our apartment door. He said it looked more like a general key for a large building. But he made the key anyhow—from a soap impression.)

We also launched for-sale ads for dogs in two local papers that brought a barrage of potential buyers to our home-room teacher's home. My ad, which was for "my dearest little lap dog," never appeared because I had to giggle too much while placing the ad by phone and the paper double-checked with the named teacher. But the ad concocted by my friend Franz Prauser for an "exceptional hunting dog," did appear and several incensed farmers, who had come from afar (Jauer, now Jawor), wanted their travel expenses back and compensation for a lost day at work. Our music teacher acted as a spy to pry the secret from us as to who had the "ingenious" idea for the ads. But we kept mum—because the affair was taking on legally actionable overtones. Nobody ever unmasked the malfeasants even though it was clear to everybody that the culprits were from class 5c (*Fünfzeher*, i.e., five-toed creatures).

At home I was manufacturing explosive powders and to accelerate the mixing process of the different chemicals used, I dissolved them in water and then evaporated the solution over the kitchen stove—boom. A great mixing invention!

At another time I ate all the crumbs from a crumb cake my mother had baked for Easter and then "plowed" up the denuded surface of the cake to make it *look* like crumbs. Of course, nobody was fooled and I was forbidden to eat any of that sorry cake (which I didn't care to, anyhow).

In my free spare time, I built a radio transmitter that fitted into an unobtrusive little box (disguised as a small suitcase). It sported a high-quality microphone and was intended for listening in on the teachers' conversation in the teachers' room. It used a single vacuum tube that served both as an audio amplifier and a radio-frequency oscillator. It never worked too well but managed to give my parents a big scare when I was testing the transmitter from a neighbor's house and then rang our doorbell to check on the signal strength. My parents thought it was the Gestapo at the door. (It was, in fact, *very* dangerous to operate illegal transmitters in Nazi Germany.)

This was also the time (1940), that I contracted my first "serious" disease: *Scharlach* (scarlet fever) which left a little scar on my heart and made me near-sighted. Four years later I fell ill with rheumatic fever which left me practically immobile for several weeks. (The nurse had to put her arm through my pyjama sleeve to pull my arm through.)

In the hospital I furtively listened to the American Forces Network (Glenn Miller!) over the BBC. While reconvalescing the SS tried to draft me but, with the help of my father, I was able to escape.

In the summer of 1940 we vacationed in Schreiberhau in the *Riesengebirge*. I remember especially July 19th, when Hitler gave his victory (over France) speech in the *Reichstag*, announcing the promotion of several generals to field marshal and Herman Göring to the unique rank of *Reichsmarschal* (a title that goes back to the Holy Roman Empire). He offered Britain peace—guaranteeing the British Empire—if he could have a free hand in continental Europe. As is well known, the British declined the preposterous gesture. In fact, they didn't respond at all.

One date that sticks in my mind from my year at Liegnitz was Monday, 23 June 1941, the day after Hitler had attacked the Soviet Union, which, true to its pact with Nazi Germany, was still shipping wheat and gold and other items in short supply in Germany the night before the attack. That morning the first course was geography, taught by Dr. Pfeiffer—a Party member (always sporting a swastika on his lapel). He unrolled a large map of the Soviet Union and said "the USSR is a very large country." Prolonged pause. Then he said—and it was obvious that he didn't believe it himself—"but I am sure that the *Führer* knows what he is doing."

During my two summers in Liegnitz I worked in the cucumber "industry." I had to sort (on a conveyor belt) the freshly-harvested produce and carry heavy sacks (40 kg) of cucumbers—no mean feat for the pencil-legs person I was at age 14. But I earned 32 *Pfennig* an hour—enough for one section of track for my future gauge-00 model railroad that, however, never came into existence because by the time I had enough money (about 250 marks for one summer, working 10 h or more per day) there were no more model railroads available. (I remember one *D-Zug* locomotive patterned on the 03-series of the *Reichsbahn* that then cost 27.50 *Reichsmarks*—which, many years later, I bought for our children for many times more *Deutschmark*.)[5]

During the cucumber harvest I also encountered several French PoWs whom I asked (since I knew little French) "Parlez-vous brittanique?" No, they didn't. But they were interested in our cucumbers (for which they would "later" give us bananas).

Another diversion was building model airplanes from plywood, parchment paper and lots of glue. This was fun. And the planes actually flew.

Since I wasn't any good at regular *Hitlerjugend* activities, they made me a *Gebietsgeldverwalter* (regional accountant) for which I had to attend a special course in Striegau near Liegnitz. I don't remember any of the math from that course, but I do recall the mid-night alarms when we had to get up and, in our nightshirts, run barefoot around the snow-covered grounds. So much for the Nazi idea of "accounting."

[5] Earlier, in Rotenburg, I was "working on my railroad" by collecting *Blaubeeren* (blueberries) for 32 *Pfennig* a (metric) pound.

Incidentally, when we first arrived in Liegnitz, my parents wanted me to attend the *Ritterakademie* (Knight's Academy), a boarding school for Silesian nobility (*Landadel*). But the school also accepted a few local students. On my first day, I was asked by the homeroom teacher my name and whether I was an *Akademiker*, meaning, in this case, whether I was a boarding student. I didn't know what *Akademiker* meant, so, to be on the safe side (and because it sounded good), I answered "yes." Then the teacher said something that became a standing expression in the family: "*Nicht mal von*?" (Not even *von*?) What's behind the teacher's comment? Well, all the *Akademiker* were at least *von* something if not a Baron or an Earl. Also the *von* Schröders were not unknown in Germany. They were bankers; one of them, Baron von Schröder from Cologne, had helped bankroll Hitler's election campaign in the early 1930s. And, of course, there is the famous London bank "J. Henry Schroder." But I was just an ordinary Schröder without title (or money, for that matter).

In any case, I lasted only for three days at the *Ritterakademie*—not because I lacked any titles but their language courses didn't match mine from Bremen.

Later, I learned that the *Ritterakademie* (now *Academia Richerska*, in Legnica, Poland) harbored an anti-Nazi-group of students—a very rare circumstance in Germany during the Third Reich. (The only other known case is the *Weisse Rose* group lead by Sophie Scholl, her brother Hans, Christoph Probst, and Professor Huber at Munich University. On the 18th of February they were betrayed by a janitor who saw them distributing "subversive" leaflets, pointing out some of the horrors committed by the *Wehrmacht* (!) in the Soviet Union. They were unceremoniously condemned to death by the infamous Roland Freissler of *Volksgerichtshof* fame, and guillotined the same day on 22 February 1943.)

Long after the war (in 1992) Anny and I visited my old school in Liegnitz. A very kind superintendent showed us my old classroom (5c) and the *Zeichensaal* ("drawing room") where I had perpetrated many pranks—some not so innocent, as when I used my limited drawing talent to inscribe some "A's" for me in the teacher's records book. This *Urkundenfälschung* (falsification of documents) was thankfully never discovered.

During the same trip we also visited Breslau (now Wrocław), Auschwitz, Birkenau, the former Jewish ghetto in Cracow, and the Polish Catholic shrine in Czestochowa (home of the Black Madonna).

The final destination of this excursion was a conference on chaos in the Royal Castle in Rydzyna (Reisen before 1919). This magnificent structure had been meticulously restored by Polish artisans after having been torched in a bonfire by the departing Soviet troops in 1945. (It was destroyed in just a few hours but it took 45 years to rebuild!)

15.8 Ballenstedt

To avoid constant school changes, my parents sent me to a boarding school in Ballenstedt in the Harz mountains: modern buildings surrounded by forests and a beautiful little lake nearby on whose shore I spent many a dreamy moment. This was in September 1941. I had just turned 15, feeling very lonely, far from family and old friends.

Other than the lake, one of the few things I remember from Ballenstedt was an assembly of the entire school in the loudspeaker-equipped mess hall on 11 December 1941. Hitler was giving a speech to the *Reichstag* declaring war on the United States. He didn't actually say as much, but instead pointed out that Roosevelt had already turned the Atlantic into a war zone by declaring that the U.S. navy would sink, without warning, any ill-intentioned submarine. Hitler also stressed that gallant Germany would of course stand by its great ally, Japan, which was "fighting for its life" since December 7th. Hitler, naturally, didn't stress the fact that Japan had been the aggressor and that Germany wasn't actually treaty-bound to join Japan in the fight. Hitler believed that the U.S. would be absorbed by its war in the Pacific. But Roosevelt, as is well known, made the decision "Europe first" because that's where he saw the greater danger.

The question has often been debated why Hitler chose to declare war on America at a time when he had just suffered his first major military defeat (in front of Moscow). The reasons, I believe, are the following:

1. He knew, as Roosevelt did, that war between the USA and Nazi-Germany was inevitable. So, in order not to lose the initiative, he preempted FDR. Otherwise it might have looked as if war with the USA had been *forced* on him.
2. He believed that America would turn to the Pacific and Asia rather than Europe, because he underestimated how much of an ogre he had become to the world outside Germany.
3. He honestly believed that America's power was much overrated. In several of his speeches he stressed the "fact" that America had only 153 million people—and he had over 300 million. (He must have counted all the captive nations) Also, he didn't think much of America's industrial prowess and banished Henrietta von Schirach, who was part-American and wife of the *Reichsjugendführer* Baldur von Schirach, from his dinner table because she had dared say some good things about the USA and the capabilities of its industry. Hitler liked to surround himself with women (Leni Riefenstahl, the legendary film-maker, Hanna Reitsch, the daredevil pilot who had first flown a helicopter *inside* a (large) building, to name just two) but Henrietta had to go. This was a pity because she could have given her *Führer* much cogent advice.

Of course Hitler was wrong on points 2 and 3, which he had to learn the hard way.

Someone who did know of Japanese intentions was Joseph Stalin who had been informed by Richard Sorge, the Nazi Party member and Soviet master spy in Tokyo. This allowed "Uncle Joe" to withdraw his crack divisions from the Far East to face the Nazis in front of Moscow and deal them the first blow in early December 1941.

Like most, if not all, boarding schools in Germany during the Nazi period, Ballenstedt was infested by ideology. On my first day, my homeroom teacher asked me whether I went to church regularly and when I answered "yes," he said that I would be free continuing to do so, but I would be the only one in the entire school.

We were also "encouraged" to study Italian. But I dropped out after a short time because I didn't like being told what to study. Later in life, when I was not forced to learn Italian, I took a great interest in the language. In 1951, I went to Italy, for a summer course *La Lingua e Cultura Italiana*, which the University of Pisa held at the Collegio Colombo in the *Pineta* (pine forest) at Viareggio. I loved it!

After 4 months, I implored my parents to find me another school. The straw that broke the camel's back for me was the following incident: I was standing by a window of our dormitory clipping my fingernails. In doing this I was of course bent over my fingers, whereupon one of my *classmates* (not even a teacher) admonished me to stand straight, without a hunchback. How do you work on your hands while standing at attention? Well, I didn't ask, and on October 16, 1941,[6] my father came through with a new school for me: Neuzelle near the Oder river.

15.9 Neuzelle

Neuzelle, like Ballenstedt, was a *Nationalpolitische Erziehungsanstalt* (national political education school). But what a difference to Ballenstedt! The official motto of these schools, the old Prussian *Mehr sein als scheinen* (Be more than appear), was inverted by us to the ungrammatical *Mehr schein als seinen*.[7]

[6] The date was also marked by one of the more important events of the "Russian campaign" (as the Germans liked to call the war on the Eastern front): October 16th saw the most serious riots in the history of the Soviet Union. In the general mayhem, the cars of party-bureaucrats fleeing Moscow were overturned by frantic Russians fearing the arrival of the *Wehrmacht* within a few days. Government departments and foreign embassies had already been evacuated to Kuybechev and a train was readied for Stalin to leave Moscow. But the Supreme Leader, after inspecting the train, made one of the most fateful decisions of the war: he decided to stay in the city and if necessary, perish with it. Soon Russian forces rallied and, only 2 months later, dealt a devastating blow to the German army just west of Moscow—the first major Nazi defeat in the war.

[7] Neuzelle was descended from Frederick the Great's famous Potsdam *Kadettenanstalt*. (The king had once offered Giacomo Casanova a position as a teacher of French (and good manners!) in Potsdam, but the great lover declined when Frederick showed him around one of the dormitories—with unemptied chamberpots. Giacomo fled Prussia forthwith for the better smells of St. Petersburg.)

I soon discovered that some students even subscribed to communism—unheard of in Nazi Germany. In other words there was much more freedom of expression. And some teachers even had a sense of humor as when one of them wanted to discourage us from using too much toilet paper: "His parents swing from chandelier to chandelier just to conserve the parquetry and what does the *Jungmann* (as we were called) do? He uses two, three or more sheets of toilet paper for a single wiping!"

Another teacher, Studienrat Blanke (our physics teacher) was perhaps an anti-Nazi because he disappeared from one day to the next without any explanation. Was he sent to a concentration camp? We never heard from him again. He had honored me by appointing me his assistant for the physics lab course. But soon after he discharged me again. Was this a precaution on the teacher's part to spare me any involvement in his fate? Blanke also impressed me with his (very un-American) credo that it takes several generations for humans to rise in society from laborer to employee to professional status to high governmental office. (This may have been a veiled allusion to Hitler who had risen from corporal to commander-in-chief in a mere 18 years.)

In Neuzelle I also operated my "secret" transmitter, which I had originally built to listen into the teacher's conferences in Liegnitz. (As a back-up, we also strung electrical wires along the wooden floors to a bug in the teacher's room. But the plan never came to full fruition.) Once, in Neuzelle, I saw an army (or SS) radio direction finder parked in the courtyard. I turned off my illegal transmitter in a hurry and never turned it on again for a long time.

I remember one student (Plokarz), 2 years older than I, who wrote in his composition assignment "What I want to become and why" that he wanted to become a German army officer because uniforms, especially officers' uniforms, carried a lot of prestige then and guaranteed a successful "love" life. As far as money was concerned: his older brother would inherit the family's shoe factory in Beuthen (Bytom) in Upper Silesia and he would get half the profit. He later did become an army officer but he had one weakness (other than girls): kleptomania. Although "rich," he would steal from gift parcels of his own troops and was thrown into the brig—from which he escaped to desert to the advancing Soviet army where he might have identified himself as Polish. (The name is Polish.) As a result, he was condemned to death (*in absentia*). I don't know what happened to my friend Plokarz; I never heard from him again. (Was he killed while crossing the lines?) The only thing I heard was from the school's director (Burow) as he broke the news to me: "*Plokarz ist zum Tode verurteilt.*"

Director Burow was one of the teachers with whom I had a close relationship: even though I was already serving as an auxiliary with an antiaircraft unit, we corresponded about mathematical problems. Once I wrote him that I could not integrate the function $\exp(x)/x$ and he wrote back that I had stumbled on an "insoluable" problem, that there was no solution in elementary functions—quite a shock to me then.

Another word about my friend Plokarz: when I first arrived at Neuzelle, shyly standing at the classroom door not knowing where to sit, a handsome, swarthy boy with beautiful, wavy long hair[8] invited me to sit next to him. After that, his grades made a "quantum" leap upward: he was always looking over my shoulder and copying what I had written. In fact, the grades in mathematics of the entire class went up: everybody was copying my homework solutions. Director Burow was very pleased with his students but in the end he became suspicious of what was going on and he assigned 12 *different* homework problems, one for each student. But the grades stayed unshakeably high. The only difference was that I now had to solve *12* problems each time.

In spite of my top grades in math and everything else (including Latin and chemistry), I was in danger of not being promoted because of an *ungenügend* (failed) grade in sports. A top school administrator was dispatched from Berlin to witness this un-German *Schlappschwanz* (weakling) called Schroeder. At shot-putting he was standing next to me while the iron ball dropped not far from our feet. In spite of his imposing presence (in the uniform of an SS colonel, no less), I didn't mind how far my shot went. And the colonel uttered a great truth about my character: "It's not that you haven't enough *strength*. The problem is that you lack the *will* to succeed." Right—as far as shot-putting and most other competitive sports were concerned, I couldn't care less—nor was I eager to prove different.

During the summer of 1942 our class was shipped to German-occupied Poland, to a village then called Nussdorf (near Środa) to help with the harvest. I was assigned to a farmer from one of the Baltic states, a dour, ethnic German who was given a former Polish farm to run which, for all I knew, may have belonged to a German before 1919. We had to get up at 5 in the morning to harvest buckwheat overgrown by weeds, especially thistle. It was very painful work and as the sun rose during the day it became hot as a furnace. One of the few good things I remember was the creamy milk from "our" cows. (By slowly lowering a glass into the milk you could skim just the cream—delicious.)

Another of my duties involved taking pigs to their male counterparts on a nearby "stud" farm. I must say, I was amazed at the amount of liquid involved in the process. Some of it was still dripping from my pigs on the way home.

Watching the impregnation of one of our cows by an eager bull was another spectacle that was new to a city boy like me. (I had never seen such a long-and-thin thing in action before.)

Some of the farm hands were Polish and they were supposed to draw their caps before members of the "master race" like us. (I am still ashamed that I once berated a Polish man for not removing his cap before me.)

During the trip to Nussdorf I saw some people in prisoner's garb working on the railroad tracks. Were they criminals or perhaps *political* prisoners, i.e., enemies of the Third Reich? I don't know, but looking at those poor men gave me a fearful feeling. Later, in Poznan (Posen), I saw a large camp, a soccer stadium surrounded

---

[8] This getup was already highly unusual in Nazi Germany where a closely cropped chevellure was *de rigeur*.

by barbed wire. After the war, from reading books about the holocaust, I learned that this was an *Arbeitslager* ("work camp"), a transshipment point for forced labor.

In a speech welcoming us to the *Warthegau*, the Nazis' name (from the river Warthe) for the annexed part of Poland, a high party official boasted that there was another *Wartegau* (waiting province), namely Switzerland, which was just waiting to become a German province. A despicable pun!

On or around 19 August German mountain troops (*Gebirgsjäger*) planted the swastika flag on Mount Elbrus, the highest peak in the Caucasus *and* Europe (5,642 m—18,510 ft.). One should think that Hitler, a lover of mountains himself, was delighted but actually he was not at all amused: he expected his troops to go after the oil in and around Baku on the Caspian Sea instead of wasting time on mountain expeditions. In the event, his armies never reached Baku, running out of gas—irony of ironies—near Mozdok at a bridge over the Terek 60 miles from Grozny (Chechnya) but still 350 miles from Baku. One bridge too many! (August 19, 1942, was also the date of the Allied raid near Dieppe on the French Channel coast.)

The longitude of Mozdok ("Dense Forest") is almost 45° East, the farthest the *Wehrmacht* ever got to the Indus river in India (around 70° East), the agreed upon demarcation line with the Empire of Japan (Tokyo is at 140° East). I forget the "boundary" between Japan and Germany in North America. (Was it the Mississippi, the Continental Divide?)[9]

This brings to mind the trick question that my friend Tom Rossing asked me: "Which of the 50 states of the USA is, respectively, the most northern, western and eastern state in terms of latitudes and longitudes." The surprising answer: Alaska, Alaska, *Alaska*! I leave it to the interested reader to ascertain this incredible-sounding result. (The southernmost state is of course Hawaii: the southern tip of the "big island" has a record latitude of just 18° North.)

Another reason I remember 19 August was that it was the day my maternal grandfather, Robert Kraemer, died of cancer at the age of 68 in distant Hamm (Westphalia). So my patriotic *Opa* never had to witness a German defeat (other than the debacle near Moscow in December 1941, which of course was kept secret from

[9] The Indian nationalist Subhas Chandra Bose (and friend of Gandhi before they parted ways) offered his services to Hitler for "liberating" India. The plan, however, fell through because the Germans failed to reach India. Bose was then shipped by submarine to Japan, where he formed a "Free India" division which actually saw action in India but was defeated by the superior weapons of the British.—The other day I asked an Indian sales clerk in a supermarket whether he knew S.C. Bose and he answered without hesitation: "Yes—a great man!" In fact, S.C.B. is still held in high regard by many Indians, because he was ready to fight the British by the force of arms—as opposed to Gandhi's philosophy of nonviolent resistance.—Another famous Bose was Satyendra Nath Bose who invented Bose-Einstein statistics and for whom *bosons* are named, such as the particles of light, the photons. S.N. Bose was also a close family friend of S.C. Bose, the nationalist leader. (From Dwarka N. Bose, University of Calcutta, in *Physics Today*, June 2007, page 12.)—(My acquaintance Amar Bose of loudspeaker renown is apparently unrelated to the two more famous Boses.)

ordinary Germans). As long as he lived, things were moving "forward" (to use a current term).

Opa Kraemer was sacked, at age 60, by the Nazis from his job in a coal mine in 1933 to make room for new workers. This was part of their solution to solve the (severe) unemployment problem in Germany. Opa always liked to drink a little beer and *Schnaps* which, however, according to my grandmother Maria, he all sweated out in the (very hot) mine.

He was also the first "acoustician" in the family: He negotiated contracts with suppliers in his personal bathroom at the mine where he kept the water filling his tub running to mask anything that was said in the privacy of his "office."

Back to Neuzelle. Another thing I remember from that school were the letters of former students from Stalingrad which were read to us in January 1943. Some letters were written in the vast Tractor Plant, one of the most hotly contested battlefields inside the city of Stalingrad. Of course, we never heard from these woeful alumni again. The last German pocket of Stalingrad capitulated on 2 February 1943, three days after the tenth anniversary of the Nazi seizure of power, and Hitler, who, like many leaders, preferred upbeat news to bad tidings, left it to his deputy Göring to give the anniversary speech. Göring, in collusion with Goebbels, turned the debacle into an historic event likening it to Greek tragedy (Leonidas, Thermopylae and Sparta).[10]

15.10 Pölitz

Two weeks after the Stalingrad debacle, on or about 16 February 1943, although I was only 16 ½ years old, I was drafted as an antiaircraft auxiliary (*Flakhelfer*) to serve with a battery of six 88-mm guns which, together with 17 other batteries, was guarding the huge hydrogenation plant in Pölitz on the Oder river (now Police,

[10] One day before the German surrender, Hitler had promoted Colonel General von Paulus to Field Marshall, hoping that von Paulus would kill himself because no German field Marshall had ever capitulated. That way, Hitler could claim that everybody had fought the Bolsheviks until their last breath—a legend Nazi-propaganda spread anyhow. It was only after the war that it became widely known that the 100,000 surviving troops (out of 300,000) had actually surrendered. To fortify their despicable lie, the Nazis, in one of their most heinous crimes against their own people, *withheld* letters and postcards from the prisoners to their families. (They were discovered long after the war in East Germany.)—The Nazis offered to exchange Yakov Stalin, whom they had taken prisoner, for Marshal Paulus (to court-martial him?). But father Stalin refused the deal. Later Yakov was killed while trying to escape the German POW camp. Stalin is said to have been relieved when the news reached him. A prisoner of war was considered a traitor and many were, in fact, shot out-of-hand when the Western Allies shipped them back (often against their will) to the motherland after they were "liberated" at the end of the war. The shootings sometimes took place in the presence of the American officers and truck drivers who had delivered the poor souls (personal communication).—(Galina Dzhugashvili, Yakov's daughter, who died in August 2007, maintained during her entire life the fiction that her father fell in battle, in spite of massive evidence to the contrary (*New York Times*, CNN, August 28, 2007.)

Poland), which was said to produce 25 % of German gasoline made from Silesian coal. When the battery commander saw me unpack a book entitled *Funkmesstechnik* (radio measuring techniques) he assigned me to the battery's fire-control radar (*Würzburg*, 50 cm) because *Funkmesstechnik* also happened to be the German word for radar. I have detailed my life as a radar operator during and after the war in the chapter "Radar" of my memoirs. But besides the radar activity, mostly in the afternoon (training) and occasionally at night (enemy planes), we had "regular" (actually somewhat curtailed) high school instruction. I think English—of all languages—(but not Latin!) was dropped. Of course we were not on speaking terms with the English-speaking countries. But why keep Latin? Did the Nazis expect a revival of the Roman empire? I remember one of our teachers asking one day whether we realized that shooting at people could mean killing human beings. Well, I must confess, we had never thought of it. But afterwards I could never forget that soldiering meant *killing* people, no matter how "justified" by self-defense or "preemption."

In August or September 1943 I, along with many other electrically interested youngsters, received an "invitation" from Hermann Göring to register for a special course in radar. This sounded interesting and I applied and was accepted. So on Saturday 23 October I decamped for the *Reichsausbildungslager* (Reich Training Camp) "Prinz Eugen" in the forsaken Westerwald near the triple point where the German states of *Westfalen*, *Hessen*, and *Rheinland* meet.

15.11 Prinz Eugen

Prinz Eugen (named after an Austrian field marshall, Prince Eugen of Savoy) was a combination high school, electronics college, and boot camp (on weekends and, occasionally, at night). The schedule of duties was so demanding that a high air force officer, General Martini, upon seeing our emaciated looks, took pity on us and sent us all to the Austrian Alps, the *Schönleiten Hütte* in the *Pinzgau*, for 2 or 3 weeks of skiing. This was in March 1944.

In addition to skiing, we had ample time for reading. My friend Jobst von Behr reminded me recently that I was reading a biography of Leibniz. The book was part fiction, part fact and I wouldn't touch it today, but it made a deep impression on Jobst and me at the time, when we were both 17. I haven't read fiction for ages, other than the "The DaVinci Code," a real page turner. But I swore not to read another fiction book for a long time. Life, *real* life, is simply too interesting to waste time on fiction, good or bad. But in my youth, I read quite a bit of fiction and enjoyed it. Like most German kids I read many volumes of Karl May's classic (American) Indian exploits (Apache Chief Winnetou, his beautiful, blushing sister *Nscho Tschi* (Fair Day), paleface Old Shatterhand and the Sioux and Mohicans). It was marvellous and of course pure fiction. (It is said that Karl May never set foot on American soil before writing these books. But where did he get the enchanting name *Nscho Tschi*?—Gesundheit!)

Another book I read, or was read to, was "Der Struwwelpeter," the hilarious depiction of the ills that befell "bad boys" (not cutting their fingernails, not eating their soup and not combing their hair—*Struwwelpeter*, in fact, means "messy-hair Peter").

And there are probably few Germans, young or old, who were not enthusiastic readers of Wilhelm Busch's captioned cartoons of assorted mischief makers, especially Max and Moritz. (Some of the funniest exploits "happened"—or, in any case, were invented—in the Ebergötzen flour mill not far from where we live, when we live in Germany.)

Once I knew enough Dutch, I read a few books in Dutch. I particularly liked the Flemish satirist Willem Elsschott (Alfons de Ridder) whose *Kaas* ("Cheese") and *Lijmen* (literally "bamboozle" but translated and made into a movie as "Soft Soap"). I cannot imagine Elsschott being translated into any other language without much being lost in translation.

The first serialized crime story I read (at age 10!) was "*Dr. Crippen an Bord*" which was later made into an equally gripping movie. The story is retold in a recent book *Thunderstruck* by E. Larson, in which the Crippen murder case is intertwined with the story of Marconi's radio experiments. (Crippen and his paramour were uncovered by the ship's captain when they tried to flee from England to Canada—thanks to radio telegrams between Scotland Yard and the ship. According to Larson "the Crippen saga did more to accelerate the acceptance of wireless as a practical tool than anything the Marconi company previously had attempted.")

Another bunch of books I loved were Hans Dominick's *Zukunftsromane* (literally "future novels," i.e., science fiction). I remember particularly his *Wettflug der Nationen* (Air Race of the Nations), describing an air race around the globe in which, of course, Germany won, but the US was a close second. I forget how the Italians fared, but the Japanese were painted as real devils, mixing sand with the engine oil of their competitors. (This book must have been written before the *Antikomintern* (Anticommunist-International) pact that allied Germany, Japan, and Italy for any future fight with the Soviet Union.)

As to more serious reading, even in high school, in Neuzelle, I once had to give a talk on an exposition by Max Planck on "free will and physical determinism." I don't remember whether I understood much (or anything) of Planck's argument. Today, 65 years later, I still don't know the answer as to how in a world governed by the deterministic laws of physics something like personal free will can exist. Yet, we all know that it does and that we are responsible for our deeds and misdeeds. The fact that quantum mechanics introduces an element of randomness looks like an easy escape hatch out of this conundrum. But is it relevant to free will? (One thing I do remember from this talk in 1942 was my near-sightedness and that I was too shy to wear glasses in class. I therefore placed my notes for the talk on top of a high stack of (learned) books, so I could read my notes without glasses.)

Another serious book that I read (or rather was read *to* by our house maid) was Erich Maria Remarque's "*Im Westen nichts Neues*" (All Quiet on the Western Front). I think she (and we children) liked the scene best where a row of soldiers are peeing "liquid silver" in the moonlight outside their dugout. The book was turned into an acclaimed movie that Goebbels and his Nazi thugs tried to boycott (because of its pacifist message) by releasing hundreds of white mice during the opening minutes of the movie. As a result, the gala premier had to be cancelled but the mice surely helped to publicize the film.

As I have said in another chapter, besides the prescribed school literature (Caesar, Goethe, Schiller, Shakespeare), I gulped English crime fiction, especially by James M(allahan) Cain (*The Postman Always Rings Twice*, *Double Indemnity*).

Chapter 16
July 1943

Manfred R. Schroeder

July 12th, 1943, was my 17th birthday. I was attending a gliding school in preparation of an intended "career" as a fighter pilot. This was ridiculous. I was already nearsighted and wearing glasses. But my father had been a combat pilot (*Kampfflieger*) in World War I so they made an exception. Of course, I never piloted a powered plane, much less a fighter. The *Luftwaffe* was literally running out of gas. But in the meantime I had great fun sailing quietly over the Pomeranian landscape. (The first time up I lost my way because the church spires I had memorized were no longer visible on the horizon: they were *below* me.)

One of our duties between flying exercises was guarding sheep. (Generally, German airfields kept the grass clipped by employing sheep.) I was given a large flock and guard dog—and promptly fell asleep in the shade under the wing of a parked airplane.[1] When I woke up, the sheep (cum dog) were gone! As far as the eye could see: no sheep. I panicked, running this way and that, all the while seeing my sheep already blocking the runway or worse. I was afraid of a court-martial. But I was lucky: after one or two false starts, I found my lost flock, all held in a tight circle by my good dog.

But in scampering across the air field, I had lost a precious birthday gift from my parents: my first "gold" (actually gold-plated) wrist watch. The entire kitchen personnel took pity on me and together we scanned the grass along the path I had run. And, lo and behold, one of the girls found my watch.

While I was amusing myself, blithely flying gliders, more momentous things happened around the world: the Allies had just begun their invasion of Sicily (July 10th) and in Russia Hitler's last offensive (July 5th to 12th) collapsed in a mammoth tank battle involving thousands of tanks (including the brand new but

Author was deceased at the time of publication.

[1] Is there a (1970) Broadway play *Sleep on the Runway* by Art Buchwald?

M.R. Schroeder (✉)
Drittes Physikalisches Institut, Georg-August-University Göttingen, Friedrich Hund-Platz 1, 37077 Göttingen, Germany

N. Xiang and G.M. Sessler (eds.), *Acoustics, Information, and Communication*, Modern Acoustics and Signal Processing, DOI 10.1007/978-3-319-05660-9_16,

untried *Tigers* and *Panthers*) outside the town of Prokhorovka near Kursk. According to latest Soviet sources, each side lost around 700 tanks in that single battle, which the USSR could quickly replace (thanks in part to American help) at a rate of up to 300 tanks *a day*.[2] Many gasoline-powered *Panthers* went up in flames because of ruptured fuel lines. Not a few tanks (on both sides) were lost to collisions because they couldn't see each other in the dirt storm kicked up by the unprecedented density of tanks on the battle field. The Germans had fielded 2,700 tanks against 7,000 Soviet tanks—including the legendary *T34*. (Another weapon of great renown was the German 88-mm gun, originally designed as a *flak* (*Flieger Abwehr Kanone*) but used most successfully as an armor-piercing gun.)

The German armed forces never recovered from this blow. Yet *before* the "Battle of Kursk," Hitler had declared: "*Zitadelle* (the code word for Kursk) would demonstrate to the entire world the enduring prowess of the *Wehrmacht*"—even after the "setback" in Stalingrad, which, after all, had happened in winter—as if the Russians didn't suffer from the cold as much as any other human being. (Most Soviet experts agree that their victory at Kursk was in fact more important than Stalingrad in winning the war.)

Once I had heard about Prokhorovka, I always wanted to visit the place to see how the Soviets commemorated this historical battlefield. Then I found out that I had, in fact, been in Prokhorovka before: in August 1959 on a night (actually 2 nights) train to Moscow from Ordzhonikidze in the Caucasus (now Vladikavkaz again as under the Czars).

From the same train I did see Izyum (south of Kharkov) which still showed the scars from the ferocious battles fought there 16 years earlier.

In August 1959, after a long drive through the Caucasus from Tblisi to Ordzhonikidze, we caught a glimpse of the beautiful Mount *Kazbek* (5,047 m). Our driver was kind enough to stop the car for about 20 min on the Georgian Military Highway when, suddenly, Kazbek was revealed, for just a few seconds, in all its towering grace.

In Ordzhonikidze (named after one of Stalin's henchmen, Sergo O.) I asked our German-speaking translator (most of our guides spoke English) whether the Germans had actually reached Ordzhonikidze. "No, but they came close." I later discovered that they had penetrated as far as Mozdok some 90 km to the north (on the Baku-Moscow rail line).

I also saw places like *Reichssender Gleiwitz* where the SS had staged a fake "Polish" attack on German soil on 31 August 1939 that gave Hitler, the next day in a speech at the Reichstag, the pretext for starting World War II: "*Seit 5 Uhr 45 wird zurückgeschossen*" (since 5:45 we are shooting back; actually it was 4:45 when the "return" fire was started—by the German battle cruiser *Schleswig Holstein*, shelling Hela near Danzig/Gdansk).

[2] When Hitler was told this number, he became very angry: "How could the Russians of all people produce that many tanks! But it was the truth.—At the same time, on the other side of the globe, in San Francisco and Sausalito, Henry Kaiser assembled one big *Liberty* ship every 24 h.

Another place I visited was the spot where the advance of the German armies was fought to standstill by the defenders of Leningrad. It had to be somewhere between Peterhof Palace and the outskirts of Leningrad. I asked my driver whether he knew and he proudly pronounced "I *fought* on the Leningrad front." He stopped at a little brook besides which there was a stone monument with a big red star and the inscription "At this point the fascist hordes drowned in their own blood" (or something to that effect).

Around July 19th I visited my father in his Blankenese apartment near his office (*Luftgaukommando* XI), just north of Hamburg, not knowing that I would never see old Hamburg again: only a week later much of Hamburg would be smoldering ruins. My father's prior assignment had been *Luftzeugamt* M (M for Moscow!). But they had to settle for Minsk, where my father was ordered to watch a mass-execution of "partisans." (This was Himmler's great idea to inure people to the "necessary measures" that had to be taken during the war.) My father was very upset. He complained to his commanding officer who answered "don't get excited, Schroeder, people here are as cheap as black berries." My father was subsequently transferred out of Russia and was never promoted again and, in fact, discharged from the *Luftwaffe* in the summer of 1944, at age 51. (One of the worst things I ever said in my life was to my father when, after the war, he complained about his low pension—a result of his not being promoted after 1941: "You lost a war and people who lose wars shouldn't get *any* pension." What a horrible thing to say, especially in view of the reason for his pension being so low!)

His next, and final, job was as chief of the Iráklion air base on Crete. There he befriended the monks of a nearby monastery, which he visited with my mother after the war.

He sometimes enriched our meagre diet by sending little boxes with raisins which however did not always arrive intact. I remember my mother opening one such box and encountering a layer of peat moss, below which another layer of peat moss appeared . . . and so forth until she was scraping the bottom: nothing but peat moss!—The theft was later exposed as a large-scale scam at Athens air base where boxes with peat moss were always waiting to be quickly switched for boxes with more valuable goods.

But Sicily and Kursk were not the only debacles that befell the Third Reich in July 1943. Beginning on July 24th parts of the great city of Hamburg were turned to ashes in a ferocious firestorm with a loss of only 17 (out of 792) British planes during the first night—thanks largely to the tinfoil strips ("chaff") tuned to the wavelength of the German fire-control radars (*Würzburg*, 50 cm) thereby completely blinding them.

Also on July 24th the Fascist Grand Council convened in Rome to depose Mussolini in the wee hours of the 25th. Undeterred, the ex-*Duce* appeared at his desk in the Palazzo Venezia the next morning, whereupon the King of Italy had him arrested and bundled off in an ambulance: destination unknown. Mussolini was located a few weeks later in a remote military reservation (*Campo Imperatore* named after Holy Roman Emperor Fredrick II (1194–1250)) near *Gran Sasso*, the highest peak in central Italy. On 12 September SS-daredevil Otto Skorzeny "liberated" the *Duce* in one of the most spectacular commando operations of the war and delivered him, now a mere puppet, into the presence of his *Führer*. (I remember the date because the next day, when the news broke, I was in Berlin for another phase of my fighter pilot career: the "courage test," meaning we had to box until all contenders were sufficiently bloodied. But how do I remember that it was September 13th when I stood in front of that news stand near the *Tempelhof* airport with the big headline "Duce liberated"? As I said on another occasion, once certain numbers enter my brain, they get stuck there "forever." I never made any notes.)

Before the month was out, Palermo was liberated, as were, shortly thereafter, on August 5, Orel and Belgorod in Russia, igniting the first of many gigantic fireworks over Red Square in Moscow in anticipation of final victory.[3]

But the Nazis, long divorced from reality, even after this disastrous July, hung on for another year and 9 months. They "stayed the course" and continued a war that was irrevocably lost on September 3rd, 1939, when Britain and France declared war on Germany following the unprovoked invasion of Poland 2 days earlier. Even the two top Nazis had queasy feelings on that day: after chief translator Paul Schmidt had finished with the British ultimatum, there was a long silence. Finally Hitler, absent his habitual bombast, said "*Was nun*?" (What now?). This question—if not an expression of utter cluelessness—must have been directed at his foreign minister, Joachim von Ribbentrop, the former Nazi ambassador to London, who had "promised" his *Führer* that Britain would not budge. But Ribbentrop kept mum. According to Schmidt's memoirs, Göring, the only other person present, looking out the window, said in an eerily prescient mood "If we lose this war, God have mercy on us." Bravo, Marshall Göring! So why didn't his Commander-in-Chief declare a "regrettable misunderstanding," recall his troops from Poland and preserve the peace? He couldn't, because he would have lost face. So, instead of losing one *face*, the world lost *millions of lives*. But that is the way of human history, driven by the inability to admit error no matter what the ultimate cost—a prospect even more frightening now in the nuclear age.

Or, as Hobbes had it: *Hell is truth seen too late*.

[3] According to only recently released Kremlin archives, "Generalissimo" Stalin was reading a *History of Greece* at the time, whence he got the idea of fireworks to celebrate great victories. (I find it hard to believe that Stalin was engrossed in ancient history at the cusp of World War II, but that's what the records say.)

Chapter 17
Radar

Manfred R. Schroeder

In mid-February 1943, around the time of Goebbel's infamous "total war" speech, I was drafted, barely 16 years old, to serve, as an air force auxiliary (*Flakhelfer*), with an anti-aircraft battery near Stettin on the Baltic Sea. This was to free the regular crews for service on the Eastern front in a futile attempt to make up for the loss of 300,000 experienced men in the battle of Stalingrad.

During unpacking, the battery commander noticed a book entitled *Funkmesstechnik* (radio measuring techniques). But *Funkmess* was also the German word for radar, so I was assigned forthwith to the fire-control radar of our battery, comprising six of the renowned 88-mm guns. There were a total of 18 batteries guarding a huge "hydrogenation" plant that converted coal into gasoline. (It was said that this single plant produced some 25 % of the German fuel that was not derived from oil—a scarce commodity.) Of course, we were "thoroughly" trained—not on a realistic large cluster of aircraft but on single slow-moving planes (*Junkers W34*) pulling air bags as targets. Our radar range data was so accurate that, combined with optical data for azimuth and elevation, we sometimes hit the bag on the first try—and were given the rest of the day off. Once or twice a week there was an alarm: approaching enemy aircraft. But they never got very close, and, after an hour or so, we could go back to our bunks. This got to the point that I never even got up during alarms.

But one night, 21–22 April 1943, matters got serious. I pulled my pants over my pyjama and wrapped myself in a greatcoat. By the time I reached my position at the radar, my screen was filled with a forest of "blips," barely 10 km away. How to pick *one* target in this mess? I selected one of the bigger blips more or less at random. But by the time my two buddies (responsible for range and elevation) had zeroed in on the selected target, it had disappeared and I had to select a new one. The same

Author was deceased at the time of publication

M.R. Schroeder (✉)
Drittes Physikalisches Institut, Georg-August-University Göttingen, Friedrich Hund-Platz 1, 37077 Göttingen, Germany

N. Xiang and G.M. Sessler (eds.), *Acoustics, Information, and Communication*, Modern Acoustics and Signal Processing, DOI 10.1007/978-3-319-05660-9_17,

scenario repeated itself over and over again during the entire night. Where did the vanishing targets go; they were not shot down: they disappeared even before our guns could fire a single round.

Pretty soon it dawned on me that the big blips on my radar screen were not really individual planes but the result of random interference of the radar reflections from *many* planes that would add up or cancel each other depending on the (slightly changing) distances between the planes. Thus, we were shooting at "thin air," random blips, all night. But what to do. I couldn't possibly run to the battery commander and advise him of the problem (and risking a court-martial). So we moved our servos as we had learned from our single-plane exercises so at least the (analog) fire-control computer wouldn't go haywire with random input data. Almost needless to say, we didn't shoot down a single plane all night. After the enemy planes had dropped their bombs (on Stettin it turned out) and left for home, our battery commander jumped on the earthen wall surrounding our radar and, arms akimbo, announced "Boys, I have never seen such smooth data!" (Never mind that we hadn't hit a single thing.) Of course, we had learned to aim our radar and "properly" track a target even without any real planes present.

The next day was our day off. But the busses weren't running. They were needed to transport wounded people, it was explained to us. Staring at our radar scopes all night, the thought had never occurred to us that it was a night of great suffering for many.

The Pölitz (now Police, in Poland) hydrogenation plant, in spite of its strategic importance, was never bombed until late in 1943. "Bomber Harris" (General Sir Arthur T. Harris of the Royal Air Force) had simply prevailed with his strategy of carpet bombing of German cities against the many Allied experts who saw a much greater potential for bringing the war to a speedy end by bombing selected targets relating to fuel supplies, rail communications, manufacture of ball bearings, and other German bottlenecks.

In early 1943, an Allied plane was downed near Rotterdam without the (navigational) radar self-destructing so that it fell into German hands nearly intact. *Telefunken*, the German electronics giant, repaired the *Rotterdam Gerät*, as the captured radar was henceforth known, and installed it on top of the Berlin Zoo bunker. Hermann Göring was taken aback when he saw a precise map of Berlin, including its lakes and rivers, on the scope of the reconstituted enemy radar. He rescinded his 1940 order of dropping radar research then and there. In his earlier opinion, the war had already been won (sounds like 2003?). Anyhow, who needed those "boxes with coils inside," his moniker for radar. It was manly courage that decided wars.

Göring then issued an order, in his capacity as president of the *Reichsforschungsrat* (Reich Research Council), for all young Germans with a background in electronics to volunteer for special radar training to catch up with the Allies. I, together with some 400 other 17-year olds, followed the call and reported for duty at the camp "Prinz Eugen" in the *Westerwald* on Saturday, 23 October 1943. In the next 6 months, we learned all about Maxwell's laws, German radars and radar receivers (preferred on U-boats), and Allied radars as far as known. (After the war I could still converse with American radar experts about the details of some of their "secret" radars, for example that the American high-μ tube 6AC7 was replaced by the similar German tube EF14.)

I once measured the wavelength, λ, of our fire-control radar with a makeshift *Lecher* setup and was immediately reprimanded for "exposing" one of the *Luftwaffe's* top secrets. This was a ridiculous charge because the Allies knew perfectly well that λ was about 50 cm—as evidenced by their tinfoil strips tuned to the "secret" wavelength.

Early during the *Westerwald* I met Jobst von Behr who became a lifelong friend. We were both interested in mathematics (and still exchange problems 64 years later) and despised military discipline. I remember chatting with Jobst during a roll call. We were standing in the back row oblivious to the goings on around us when his name was called by *Oberscharführer* Snita. Snita had to call my friend three times (Behr, *von* Behr, *Herr von* Behr) before Jobst responded with a meek "here." Whereupon the *Oberscharführer* launched into one of his favorite tirades about these "gentlemen" who won't respond until addressed with all their titles. Jobst had to step up to the front, fall down on his hands and do pushups.

Snita was also partial to wine. I remember, when we had to help him with the "tasting," we became somewhat "tipsy." In the subsequent class (*Deutsch*, if I recall correctly) my head was swaying. The teacher (Dr. Sieg) must have noticed this and he asked me something about our current reading (Hermann Löns: *Wanderer zwischen den Welten*). Being hardly able to speak, let alone give an intelligent answer, I replied in a slow voice: "*Das möchte ich gerne noch einmal lesen.*" ("I would like to read the book again.") Just the right answer! The teacher was duly impressed.

After the *Westerwald* my next military assignments were with the Navy: first boot camp and then the Radar School on the Baltic island of Fehmarn and the Radar Observation Station in St. Peter-Ording on the North Sea. The German air force wanted to get rid of me (because I had missed boot camp as a result of a foot injury contracted by stepping on a glass shard on my last outing to a local lake with my parents and sisters before being drafted). As it happened, I had also missed the *navy* boot camp so on the day of inspection I was used as a *target* for hand grenade throwing exercises. I had to climb into a fox hole while everybody was throwing live practice grenades at me. One hit the earth wall behind me and rolled into my fox hole. I was practically sitting on the sizzling thing and jumped out as quickly as I could. Everybody, the admiral included, exploded in laughter: "This is just what the enemy wants you to do. You would be mowed down by machine gun fire by now!" I didn't care. All I knew was that I was still in one piece and a potential father.

In Fehmarn we learned about German counter measures against the tinfoil strips ("chaff") which had blinded the German radars ever since the attacks on Hamburg in July 1943. The new German radars exploited the Doppler effect to distinguish between slow moving strips and fast airplanes.

We also learned about the special radar problems for submarines. To preserve "radio silence," they would only use "passive" radar, i.e. radar *receivers*. This was a good idea until it was discovered that the French-made *Metox* receiver (named after its inventor Metox Grandin) had a little wire smuggled into its circuitry that connected the local oscillator with the U-boat's search antenna. This was communicated by French resistance to the Allies which thereupon could switch off their own radars and just listen to the radiation from the German sub. This was a double blow to the subs: they were no longer forewarned by the approaching radar of Allied planes and had their own position unwittingly betrayed.[1]

Another trick, invented by the American physicist Luis Alvarez (later of dinosaur-extinction fame), exploited the relation between distance and radar intensity: The inverse square law that governs gravitation also describes the falloff of radar power with distance. This simple fact was exploited by German submarines during World War II. By measuring the increase in radar intensity, they could gauge the rate of approach of an enemy plane and dive undersea for safety before the plane could attack.

This tactic worked very well for Grand Admiral Karl Dönitz until Alvarez had a foxy vision, code-named *Vixen*. Alvarez suggested reducing the radar power so that it would be proportional to the *third* power of the range to the submarine. Thus, while the plane was approaching, the power incident on the unsuspecting U-boat was actually *decreasing*, giving thc false impression that the radar plane was flying *away*. A grand idea indeed! (For the attacking plane, however, the received radar power reflected from the boat would still increase as it closed in.)

One *German* radar invention were the highly directive dielectric finger antennas mounted on a revolving turntable under a kind of plexiglass cheese bell. These antennas fed the U-boat's radar receiver but had the bad habit of breaking off in an Atlantic storm. Then a sailor discovered that the best way to repair (and protect) these fingers were condoms. Henceforth all German submarine crews had to carry an extra supply of condoms as a regulation wartime outfit.

At the end of the war, I found myself with coastal radar in Holland, huge installations that could pick up a plane over distant London once it was 200 m above ground. The Dutch navy officers, who had missed radar while in hiding during the war, were impressed and wanted us to repair them for their use in coastal shipping. But the Allies insisted on our dismantling them being afraid (in the summer of 1945) that the Soviet Army might advance to the coast opposite Britain and use them against their former allies.

Once we had a visit of some ten higher-up Dutch officers (we counted a total of 32 golden sleeve rings) and a dispute developed as to the distance of a target, a fishing trawler: our (freshly calibrated) radar showed 7 km, but one of the older officers, a former ship's captain, insisted that it was only 3 km. Neither side would

---

[1] According to other sources, the *Metox* saga was planted by the Allies to divert attention from the real reason for the increased U-boat losses, namely, the introduction, in March 1943, of a new (10-cm) search radar (H2S) which *Metox* couldn't detect.—The *Metox* was superseded by, among other radar receivers, the *Naxos* and the *Naxos ZM*, with an antenna which rotated at 1,300 rpm and with which I had a personal encounter in January 1945.

give in until one of the Dutch officers reminded the captain that he was looking at the sea from a high dune (Noordwijk aan Zee) rather than from a shipboard and what looked to him as only 3 km away could well be 5 km or more.

Toward the end of the war, we were planning to escape from Holland to neutral territory (Spain, Portugal?) by commandeering, if need be by force of arms, the launch of the Harbor Master of Ijmuiden. To be able to leave the port, we had to make some telephone calls to have the anti-submarine barriers (chains) lowered. It worked!—The chains were, in fact, removed. But someone in the Harbor Authority must have become suspicious and, upon double-checking, was able to expose our ruse and the chains went up again. End of dream of a free post-war "cruise"—a potentially very dangerous situation (the port exit channel was also protected by two medium-caliber cannons).

Of course, everybody was happy that the war was over. The only reason we youngsters were sorry that Germany had actually lost the war was our fear that there wouldn't be any more nice German *Tanzmusik* on the radio. But it didn't take us long to discover that the Allies were broadcasting even nicer jazz ballads and swing music. I remember how we fell in love with "Don't fence me in," "Sentimental Journey," "American Patrol" and many other then very popular tunes. In the same building as ours, with walls adjacent to our living room, some elderly officers were quartered. They congratulated us on how well-behaved we were—except, occasionally, all hell seemed to break loose in our quarters. Why? Did we tell them about the American jazz that drove us crazy? Maybe. After all, the war was over and it was no longer forbidden to listen to foreign radio stations. (But we certainly didn't tell the gentlemen from next door that one of us—a navy lieutenant, no less, in his nightshirt—actually jumped on top of a wardrobe and crowed like a rooster.)

Another circumstance that lightened our burden of having lost the war was that we were sitting on endless supplies of molasses (in oil barrels!). We soon discovered how to ferment the stuff, distil it and collect it in a beaker, drop by precious drop. Needless to say, the "distilled" nights—brewing was of course illegal, so we couldn't do it during the day—were among our happiest. And what we couldn't consume on the spot, we brought to a boil, poured it over ground coffee and produced the best *Kaffee-Liquör* I have ever tasted.

We also had beautiful carpets in our "camp" quarters. One day, after a new set of guards had taken over, one of their sergeants requisitioned one of our gorgeous "groundcovers" for himself. When he came back an hour later, wondering why we hadn't delivered the carpet as ordered, the "object of his desire" had disappeared. (We were hiding it in the attic.) And the sergeant, taking a very belligerent pose, proclaimed "*Als het tapijt niet onmiddelijk ter voorschijn kommt, werd u gearresteered*!" (If the carpet doesn't reappear immediately, you will be arrested!) I was just standing there smiling. (I liked to hear Dutch.) Whereupon the sergeant turned in my direction: "*En u ook*!!" (And you too (will be arrested)!!)—The *En u ook* later became a standing expression in my family.

17.1 Outside the Geneva Conventions

Incidentally, at the end of the hostilities in May 1945, we were classified as "capitulated personnel"—outside the Geneva Conventions.[2] But in our case, in Holland, it meant Canadian army rations and even home leave, which allowed me to ascertain that my parents and sisters had survived the war unhurt and our house in heavily-bombed Hamm was still standing.

Indeed, as I approached our house, after having traversed plains of rubble, I didn't dare ring the door bell. Who would answer? Were my parents still alive and had they managed to flee from distant Silesia? So I walked around the house and saw, through a basement window, a large store of apples that maybe only my mother could have scrounged up in those desperate days. I finally entered the house through a back door into the kitchen. And I didn't have to ask who was still alive. They were all there in that one room: my parents, my sisters and even my grandmother. I was so overjoyed, I started crying. But the rest of my family, while certainly glad to see me return from the war hale and healthy, didn't seem as emotional as I. Why? They had just received a postcard that I was on my way home.

The Canadians to whom we had capitulated were boys about our age (18–20 years) and they commiserated with us that, as radar people, we had never really fired a weapon. So they returned our machine guns (and even a bazooka!) to us for one afternoon so we could have our little private war in the dunes (with tracer ammunition and the rest). (This exercise in silliness, incidentally, was not without its dangers. I remember one bazooka going off right in front of us. Nobody was hurt but all of us were blackened like chimney sweeps.)

Understandably, the higher-up brass were not amused when they got wind of the happenings.

Since we were sitting on plenty of now useless ammunition, the lure of making our own rockets was always close. And, indeed, some of our contraptions, made from wooden sticks (broomsticks) and sheet metal (from army ration tins), flew 100 ft high in the air. The only bad thing about it was that we actually *ate* some of the spaghetti-like ammo: it was sweet and tasty. But too much gave you a headache. How silly can you get?

In April 1947 I was repatriated from Holland where I had remained voluntarily since the end of the war. Originally, they had picked six older guys who didn't know much about the inner workings of radar and who were eager to be reunited with their wives and children. But three of us, all below twenty, wanted to stay in Holland and our "petition" was granted. The six older men were of course jubilant and so were we. (We were afraid of the makeshift PoW camps on the banks of the Rhine, where many unfortunate prisoners perished as a result of hunger and poor

[2] At the end of the war, General Eisenhower suddenly found himself with some three million German captives on his hands and he felt he couldn't possibly treat them all under the Geneva Conventions. So they were placed outside the Conventions.

sanitary conditions entailing typhoid and other epidemics.) Best of all, as already mentioned, we were treated by the Allies as "capitulated personnel" outside the Geneva Conventions. I know it sounds unreal after 9/11 and Guantánamo but we enjoyed many benefits including excellent food, good accommodations, and even vacations (home leaves). And of course the work was interesting: being able to demonstrate to the Dutch naval officers who had stayed in Holland during the war how radar worked. In fact, they wanted to keep some of our radars intact to monitor coastal traffic. But, as already pointed out, the British were unrelenting: all radars had to be dismantled.

After the Canadians left, a company of Nederlandse Stoottroepen (Dutch Assault Troops) became our guardians. I befriended corporal Pieter Broeks with whom (as a "prisoner"!) I travelled to Haarlem and other places. In Haarlem I became close friends with Nico Bartels who owned the radio repair shop "*De Onderduiker*" (The Underground Diver). He taught me (and gave a living example of) democracy and friendship between nations. During the war he had been a member of the Dutch underground (called *onderduiker*). This was also the time I started to study *Nederlands* (Dutch).—I visited both Pieter and Nico and their families after I was released from my Ersatz-Guantánamo on 15 April 1947.

Our next guardians were from the Palestine Regiment of the British Army. We were probably among the few Germans inside barbed wire with the Jews on the outside. We had a very fair relationship, all in English—until the last day when the Jewish sergeant-major concluded the daily roll call with some words of appreciation in perfect German.—Once, when a wristwatch of one of my buddies was stolen, captain Neumann (originally from Berlin) had his troops lined up so that the German prisoner could identify the thief (who was duly disciplined). A fine example of fairness of victor toward the vanquished! Or as Churchill wrote in 1948 in his *The Second World War* (Moral of the Work. Vol. I, *the Gathering of the Storm*):

> In War: Resolution. In Defeat: Defiance.
> In Victory: *Magnanimity*. In Peace: Good Will.

Once the radar job was completed, we were assigned to ordnance dumping duty in the North Sea—a dangerous job because the guards liked to shoot at the (punctured) ammunition canisters before they sank and sometimes there was an explosion and one or a few prisoners were killed.

My main job, though, for which I had volunteered, was to build an electrocardiograph for the Dutch Navy which I felt I could do ("slam dunk"). But to my shame, I must admit that I never got the contraption to work properly—there was always too much noise (and hum) and the EKG records were hard to read. (Now I would use tubes heated by d.c., as opposed to a.c., for the heaters in the preamplifiers.)

Buying the proper cameras for recording the EKG required some travel to different parts of Holland and pretty soon, prisoner Schroeder travelled *solo* without accompanying guards. Once, on a "private" trip through Holland, I opened the door to a railway compartment full of our camp guards. I quickly slammed the door shut before anybody could recognize me.

On another occasion I shared the compartment with several ladies who were wondering why the train wouldn't leave the station. "Had the train schedule changed?" I felt my Dutch was already so good that I could join their conversation and said: "Mij zijn helemaal geen wijzingen van de dienstregelingen bekent" (I haven't heard of any changes in the schedule). Whereupon they looked at me in great astonishment. Had they recognized my "Prins-Bernhard" accent or were they only surprised by my butting in? I never knew, but I kept my mouth shut from then on.

Another illegal activity indulged in by many camp mates was smuggling. I remember once hiding, under my greatcoat, a medium-size munitions box with ground coffee and scented soap (in the same box!) for my family in Germany. As soon as I had passed the camp guards, I panicked and started running whereupon one of the strings that held the box snapped and the box fell down. But it was caught at the last moment by a second string which almost strangled me because it went around my neck.

On my last day in camp I put all my belongings into my *Seesack* (sea bag) and had it taken to our home-bound vessel by an ambulance which some buddies were using to smuggle pots and pans and other goods that were scarce in Germany. They pretended that the patient in the ambulance was highly contagious—a ruse that worked. But when I came to the inspection point our captors were wondering why I had so few belongings. They immediately smelled a rat, arrested me on the spot and confined me to a small cell. What especially raised their suspicion were my mysterious sketches and calculations of multivibrator circuits that I had kept with me in my briefcase: they were convinced they had caught a spy. (Later these calculations came in handy when I was designing pulse generators to test early TV sets at *Grundig Radio*.)

The happenings during the next 10 min were almost beyond belief: I told one of the Dutch prison guards, whom I had befriended a bit before, to please run to the ship and warn my buddies of the impending disaster. I can still see the guard sprinting past the suspicious inspectors on their way to the boat for a thorough check. Thanks to my friend nothing was found and I was released (without apologies).

In May 1947, I was discharged from the German navy, receiving one pair of long johns and a very nice blanket for my services to the Third Reich. (The Nazis had of course promised everyone a farm in Kazakhstan, but I was never very fond of farming anyhow and the blanket came in very handy: my mother had it dyed a fashionable dark blue and turned it into a pretty overcoat for my older sister.)

Chapter 18
Lucky Breaks

Manfred R. Schroeder

18.1 The Düsseldorf Murderer

My first lucky break happened when I was about 4 years old. I was visiting my aunt Lydia in Düsseldorf and, after a downtown shopping trip, I wanted to stay behind to watch the traffic, directed by a policeman, at a busy intersection of the Grafenberger Allee. My aunt agreed if I promised not to follow the "Düsseldorf murderer," also known as the "Düsseldorf Vampire," who was then killing, mostly young boys and girls, in droves in the neighborhood. A few minutes after she had left me, a middle-aged man with a mustache, wearing a boater, approached me and invited me to go with him. (I forget where and what the enticement was-sweets, or an interesting sight?) I would have followed him because he seemed very nice, but remembering my aunt's warning I declined, telling him I was very intent on watching the traffic. Then, significantly, he turned around and walked back in the same direction from which he had approached me. He had hardly disappeared when I ran to my aunt's flat proclaiming: "I saw the Düsseldorf murderer!" But, of course, nobody believed me. Not much later, when he was caught and his picture appeared in the newspapers, I recognized him. His name was Peter Kürten. He confessed and was guillotined on 2 July 1931. In an interesting twist, he first confessed to his unsuspecting wife so that, on her reporting him to the police, she could collect the substantial reward.

When questioned by the police psychiatrist, Dr. Karl Berg, whether he had any regrets about his life of perpetual murder, Kürten's response came out almost like an epitaph... "I have no remorse. As to whether recollection of my deeds makes me feel ashamed, I will tell you. Thinking back to all the details is not at all unpleasant.

M.R. Schroeder
Author was deceased at the time of publication.

M.R. Schroeder (✉)
Drittes Physikalisches Institut, Georg-August-University Göttingen, Friedrich Hund-Platz 1, 37077 Göttingen, Germany

N. Xiang and G.M. Sessler (eds.), *Acoustics, Information, and Communication*, Modern Acoustics and Signal Processing, DOI 10.1007/978-3-319-05660-9_18,

I rather enjoy it." Kürten expressed his last earthly desire as follows: "Tell me," he asked Dr. Berg, "after my head has been chopped off, will I still be able to hear, at least for a moment, the sound of my own blood gushing from the stump of my neck?" He savored this thought for a while, then added, "that would be the pleasure to end all pleasures." (From crimelibrary.com) A severed head is, in fact, said to remain conscious for about 1½ minutes.

18.2 An SS Draft Attempt

In July 1944, I was emerging from an extended hospital stay (rheumatic fever) and ordered to convalesce for the next 6 weeks. While walking the streets of our home town, Liegnitz, seemingly hale and healthy (and 18 years old), the local SS-recruitment office caught sight of me and sent me a draft notice. Now this was outright dangerous. The war was ending and the SS needed more cannon fodder. (At the beginning of the war the SS would only take volunteers, true-believers. But in 1944, things were becoming desperate.) My parents and I decided then and there to put me on a night train to my old radar group, so that the next day my father could write the SS that I was no longer in Liegnitz and had rejoined my old air force unit.

The train which I took for my precipitous escape was so crowded with soldiers on leave from the Eastern Front that my parents literally had to push me through the door and my mother cried: "We are pushing Manfred into the war, into his grave!" But the sudden escape probably saved my life.

18.3 The Loose Hand Grenade

Toward the end of the war, February 1945, some buddies and I were transferred from Hoek van Holland to another coastal radar at IJmuiden. All ground traffic was limited to the dark night hours because at daytime the sky was completely dominated by allied fighter planes (*Jagdbomber*). In addition, gasoline was so scarce that many German military trucks ran on wood (little wood blocks that were shoveled into a stove that produced a gaseous fuel). Every so often, the truck driver had to stop for more wood to be fed into the oven. At one stop, near Leiden, the driver was checking, with a flash light, whether the oven was completely filled. And to his surprise (and ours) he was fishing a live hand grenade from the oven. "Whose is this?" he wanted to know. Upon checking, I discovered that it was one of my grenades that had fallen out from my loosened belt and got mixed up with the woodblocks. And we had all been sitting with our backs to the stove to keep warm during the cold winter night! My third escape from almost certain death—not counting my nearly botched birth when all in attendance thought I was dead on arrival and received baptism *in extremis* in the nearest available sink.

18.4 A Leisurely Trip Through Holland

Another lucky event happened to me on my very first day in Holland on the way to my new radar unit. At the end of our training with German and (captured) Allied radars on the island of *Fehmarn* in the Baltic I was picked as the "leader" of a group of six classmates who were assigned to Navy coastal radars on the Dutch North Sea coast. The first big city we hit was Hamburg and those of us who had relatives or friends in or near Hamburg wanted a day off. I said "O.k., have fun." The same thing happened 2 days later at Bremen. By the time we arrived at the first Dutch city (Groningen) we were already 4 days behind schedule. We all found Groningen so charming that we all decided to spend an extra day there. The same urge for sightseeing gripped us in Apeldoorn and Amersfoort, where we even got accidentally separated from each other—a dangerous occurrence, given the ever present military police, always on the look-out for deserters. When we finally arrived in the Hague we were a full week behind schedule. And then I made a big mistake. Upon arrival at 8 a.m. (the trains were running only during dark), I called our new headquarters ("Kommando Scherer") to tell them we had safely arrived. But then we all decided to use the day to explore the Dutch capital (none of us had ever been abroad). Two friends and I wanted to take a close look at Dutch books. Although we were "sailors" (actually Navy cadets) we were carrying rifles: Holland was considered enemy territory, infested as it was by the Dutch underground. So before going up on the ladder to reach the upper shelves, I leaned my rifle against one of the book cases and later promptly forgot about it. (We had never carried rifles before.) Only half an hour later I discovered that everybody had a rifle—except me. I was scared to death: losing your weapon in "enemy territory" was the cause for the most severe court martial prosecution for sabotage! But I was rescued by the kind book dealer. When I entered his store again, he reached under the counter near the cash register and handed me back my sorely missed weapon. (He was quite terrified too, I think, because being discovered in possession of a lethal weapon was, well, lethal to the poor occupied people.)

When we finally reached our radar headquarter in Scheveningen, all hell broke loose: why did it take us a full day to walk from the Hague to Scheveningen? In the ensuing commotion, made outright dangerous by a machine-gun wielding, drunken captain, nobody discovered that we had just "wasted" a *full week* at the fatherland's expense.

18.5 My Pants Stuffed with Cash

Another lucky escape, though less deadly but fearsome nevertheless, happened while I was a student in Göttingen. I had mentioned to a friend at the lunch table (shared with a young woman unknown to us) in the *Mensa* on Wilhelmsplatz that I had seen law student W. with yet *another very* pretty girl. During the lunch I had also mentioned the fact that I was in dire need of some extra cash, like 100 marks. When we departed from the table, the unknown woman said "Thanks for the compliment. That woman was me." A few days later, after a round of tennis, while dressing in the locker room, I found a handful of crumpled 10-mark bills in my pants pockets, a total of DM 100, just the amount I needed! But I immediately reported my find to the president of the club who happened to be in the club house. Not 10 min later a fast Porsche with W. at the wheel came roaring up proclaiming that "100 marks had been stolen from him in the locker room; could he have it searched?" The club president said that was unnecessary; Schroeder had already found the money—Can anybody imagine the situation I would have been in if they had found the money in my pockets? Expulsion from the university and worse!

18.6 An Invitation from Bell Labs

My next lucky break occurred in 1954, long after the war's end, shortly after my Ph.D. (in physics, at the University of Göttingen). I wanted to join the renowned Bell Laboratories in Murray Hill, New Jersey. But my professor, Erwin Meyer, told me in good faith that Bell wouldn't take foreigners. (He had one of his students turned down in 1938.)

But then, in March 1954, William Shockley, co-inventor of the transistor, came to Göttingen on some solid-state business and he let it be known that Bell was looking for qualified physics students. I immediately went to Meyer and the next day he wrote a letter of recommendation to one of Bell's research directors (Winston Kock). A few days later I was invited by Kock for an interview at the posh Dorchester Hotel in London. (The somewhat hilarious details of the interview are noted in my book *Computer Speech*.) The interview went well and after another 6 weeks I received an offer of permanent employment at $640 per month, which was five times as much as I could have made at Siemens in Munich (500 Deutsche Marks).

After finishing some ongoing projects at Göttingen, I traveled to Genoa to board the *Andrea Doria* (then still afloat) and arrived in New York on 30 September 1954 to begin a new life: a new country, a new language, a new job, my first car and, 3 days later, on my first foray into New York, to meet my future wife Anny, working at the Bulgarian desk of Radio Free Europe.

18.7 Kugelblitz

I was always afraid of lightning, so during a particularly violent thunderstorm in Berkeley Heights (on 26 July 1988) I repaired the basement of our home on Sutton Drive. The house had recently acquired aluminum siding, so I felt quite safe in my "Faraday Cage," especially in the basement. (As everybody knows, thunderbolts come from on high.) Then one lightning landed in a tree standing next to our house and, I couldn't believe my eyes, a globular lightning (*Kugelblitz* in German), about 15 cm across, wandered between my head and the ceiling before it disappeared with a "poof" about 3 m from where I was hiding. Even as a physicist I had never believed in *Kugelblitze* and I was convinced the deadly ball of fire I had seen was a paranormal perception. But when, after the storm, I followed the path the ball had taken, I discovered that it had run along the wires connecting the door bell with a little transformer in the basement. And that transformer was completely burned black. So that much was for real. But where did the darned *Blitz* thing come from? Well, the doorbell was of course not covered by the aluminum siding and the lightning had discovered that little opening to visit me in the basement. If my head had been in its path, it might have resembled the ex-transformer. . . .

Chapter 19
The University of Göttingen

Manfred R. Schroeder

19.1 Student Days

After having been released from my captivity "outside the Geneva Conventions," I had to face the question of what to study and where. I was always strongly attracted to mathematics but my grandmother thought I should become a physicist. Fine, but *where* to study physics in those days? In Göttingen perhaps, where several great physicists were in residence, Nobelists Max Planck, Max von Laue, and Werner Heisenberg whose quantum mechanics we youngsters were all eager to understand. We were of course duly impressed by Albert Einstein's relativity but it contained no great mysteries for us. We had accepted that space was curved, simultaneity and the "ether" were exposed as figments of the imagination and mass and energy were shown by Albert to be equivalent: $E = mc^2$.

But quantum mechanics we *didn't* understand and so I attended, in one of my first semesters, a course by Heisenberg on the shell model of the atomic nucleus—and didn't understand a word: Spin-orbit coupling! What is *spin-orbit coupling*? It's basic to the shell model, proposed by the former Göttingen physicist Maria Goeppert Meyer and two colleagues in 1949, which explained the "magic numbers" of nuclear physics, the existence of extra stable nuclei. After a few lectures, I dropped out and decided to learn some basic physics first. (Goeppert Meyer was only the second woman, after Marie Curie, to win the Nobel Prize in Physics (in 1963). Lise Meitner, who, with her nephew Otto Frisch, had explained nuclear fission, was passed over in favor of her male collaborator, Otto Hahn, who collected the 1945 Prize in Chemistry.)

Later I met Heisenberg on a more personal basis. I was asked to guard his house on Merkelstraße and his in-laws (the Schumachers) while the family were taking their annual vacation in the Alps. The house had been broken into on more than one

Author was deceased at the time of publication.

M.R. Schroeder (✉)
Drittes Physikalisches Institut, Georg-August-University Göttingen, Friedrich Hund-Platz 1, 37077 Göttingen, Germany

N. Xiang and G.M. Sessler (eds.), *Acoustics, Information, and Communication*, Modern Acoustics and Signal Processing, DOI 10.1007/978-3-319-05660-9_19,

occasion and valuable musical instruments had been stolen. (The Heisenbergs were addicted to music—see Anna Maria Heisenberg-Hirsch's collection of Heisenberg's letters to his parents *Liebe Eltern*. Their house concerts were sometimes attended by Max Planck who lived nearby and who had wanted to become a concert pianist before settling on physics.)

Another reason for my moving in with the Heisenbergs was that around 1946 suspicious characters had been observed casing the house. The border with the Soviet zone was only 20 km away and rather penetrable immediately after the war. The fear was that these guys were Russian agents preparing to capture Heisenberg and exploit his knowledge for the Soviet atomic bomb project. I never believed this cloak-and-dagger story but his son Martin, a biology professor in Würzburg, in a recent conversation, confirmed the story, except that the suspicious characters were from a *Western* secret service.

One evening, when I turned on the alarm, it immediately went off with a howling noise. In other words, one window wasn't properly closed. But I couldn't find the fault, so, to the Schumachers' dismay, I turned the alarm system off, to have it repaired the next day by the Max-Planck Institute. Needless to say, the poor Schumachers didn't sleep a wink all night, especially sincc their little dog had behaved in a funny manner all day (was he poisoned?). Anyway, the Schumachers were convinced that this night disaster would strike. But I slept like a log. This story of my alleged insouciance got around to Mrs. H. and she was not amused.

One benefit that we drew from our close association with the Heisenberg family were the CARE gift parcels from the USA that were accumulating in the attic of Merkelstrasse 18. Every so often, one of the Heisenberg maids (Marlene and Ortrud) would divert one of the parcels to us—a now scarcely imaginable joy: Hershey chocolate syrup, *white* flour, powdered eggs, and other goodies not available to the ordinary German in 1947. Erhard Scheibe, Jobst von Behr, and I baked *Topfkuchen* (pound cake) with which we could greatly impress our girl friends (Lotte Wengler among others).

On the occasion of Heisenberg's 100th birthday, I had been invited to give a talk (at the Lower Saxony State Library) reminiscing about my encounters with the great physicist (see my homepage [http://www.physik3.gwdg.de/~mrs/] for pictures of the family and the text of the talk).

Before one could even be admitted to the university, with so many young men returning from the war, one had to pass an entrance examination. I applied to the theoretical physicist Richard Becker and went to his office in early July 1947. But before I could see Becker, I was intercepted by his assistant Günther Leibfried who asked me: "What did you do during the war?" "I summed trigonometric series" (of course, he didn't want to hear any battle stories). He asked how I did it and I answered: "By using the addition rules for sin x and cos x," which is of course very cumbersome. "So, you haven't heard of the complex exponential function? Using *it*, you would have obtained a geometric series that is very easy to sum." I knew enough mathematics to know he was right, but I had missed it. How stupid can you be! I concluded that my chances of being accepted were nil when Professor Becker appeared in person. "What did you do during the war?" The exact same question, but I avoided mentioning summing series like the plague. Instead I answered "I analyzed electromagnetic circuits." "Go right ahead," Becker said and I delved into a long spiel involving, electrical fields, magnetic fields, radiation, absorption and more. After about 15 min Becker said: "Thank you. That's enough."

Scared as I was at that point, I thought he meant "that's enough nonsense." But I wanted to know for certain if I had flunked the entrance examination. So I asked, timidly, "Professor, what are my chances of being accepted?" My whole future was in the balance. And Becker answered with something that I will never forget: "If *you* are not accepted, nobody will be admitted." Thank you, Professor Becker! (Later, the high opinion in which Becker held me, went a little to my head and I became somewhat lazy.)

Before becoming fully accepted at Göttingen I had to hop over the hurdle: I couldn't produce a valid *Abitur* (high school leaving diploma)—a *sine qua non* for being registered at a German university. The administration gave me 2 months for producing some documentary evidence for my claim that I had indeed passed the *Abitur* but that the papers were lost in the final days of the war. (My family had fled from Silesia in February 1945 just before the onslaught of the Red Army.) Luckily, after a week or 2, I met an old classmate, Leo Klingen, who knew the address of our last homeroom teacher (and director) of my last high school, Dr. Sieg. He kindly provided me with a minute description of our Abitur, including grades and the names of the school district supervisor, etc. When I took his letter to the university administration, the official, after perusing the letter a long time, said: "That's enough." Again I was so timid that I asked what he meant. Enough of my baloney? No, enough to be fully accepted!

Best of all: all my buddies, who had passed the exact same type of *Not-Abitur* (wartime Abitur), had to go back to high school for a year to get a real Abitur. How lucky I was to have lost my documents (and that I had found my old teacher and that he was so kind)!

Another problem was finding a room (few dormitories then). My friend Jobst von Behr was offered a small room in Heisenberg's house, which was not under the control of the city housing department. Jobst made the room assigned to him by the city available to me and I gratefully accepted it "for the first two weeks." But, although it was very small and unheated, I kept it for the next 7 years, through my Ph.D. and emigration to America. (The rent was 12.50 marks/month, which I later "rounded off" to 15 marks.)

After just three semesters I felt I was ready for the *Vordiplom* (sort of B.S.), except for the required chemistry. So I went to Professor Becker and told him "I am ready for the Vordiplom, except that I didn't go to the chemistry lab." (It stank atrociously.) Becker said: "O.k., you can go ahead without the lab course, if you otherwise know enough chemistry." "But I have never been to a single chemistry lecture!" I countered: (I hated chemistry.) Becker: "Hmm, we never had a case like that before." "But, professor, I have attended several extra courses in mathematics," (which I liked). Becker: "O.k., why don't you do your Vordiplom in mathematics then?" "But, professor, I want to become a physicist!" "Well," Becker said, "on the day of the examinations you are a mathematician and the next day you are a physicist again." How simple things used to be in the late 1940s (with the *right* people in the *right* places)!

Becker was also amenable to our complaints about the "lecturing style" of one of our professors. Becker actually sat in on his lectures and completely agreed with our criticisms. But what could he do? For better or worse: Freedom of Teaching—good or bad—reigned supreme!

After Becker retired, he was succeeded by Friedrich Hund, one of the pioneers of quantum mechanics and its application to molecules. On his first day in office Hund wanted to see the *Schriftverkehr* (correspondence files) of the institute. But Becker's old assistant Leibfried told him "There are no files; we threw everything into the waste basket." Hund, a stickler for formalities and forms (besides mathematical formulas) had the whole *Schriftverkehr* (spanning two decades!) resurrected from the university's central files. Talk about German thoroughness! But Hund was a great teacher and, even as a professor in my own right, I attended several of his fascinating lectures on the history of physics.

19.2 Ringvorlesungen

Once, while I was in the States, I got a call from Konnie (Kaufmann): would I participate with Hund in a *Ringvorlesung* ("circle of lectures") to be called *Die Grundlagen der Physik* (the foundations of physics). I said I would if it was called *Zu den Grundlagen der Physik* (*On* the foundations of physics). Hund would cover Newton, Hubert Goenner would talk about "Einstein and Relativity," Lorenz Krüger on "Statistical Physics." But who would cover quantum mechanics? Someone said: "Schroeder." But someone else objected "Schroeder is an acoustician and not a quantum mechanic." Whereupon the great Hund said "Don't worry. Schroeder knows everything!" Yes, and Friedrich Hund knows even more.

19.3 Quantum Mechanics

This early *Ringvorlesung* was an instant success. I learned a lot of quantum mechanics (QM) while preparing my part. It is a weird subject and gets weirder and weirder the more you know about it. As Niels Bohr, the father of the atom model (1913) said: "If you sometimes don't get dizzy when you think about QM, you have not really understood it." Or, as Richard Feynman had it: "I think it is safe to say that *nobody* understands quantum mechanics." The most infamous paradoxes of QM were constructed in 1935 by Einstein, Podolsky, and Rosen. It leads to the inescapable conclusion that either physical actions can spread instantly, "ghost-like," over large distances or, worse, *there is no physical reality* as we have always thought of reality (like, say, beams in the attic on which you can bump your head or myriads of other objects that surround us).

The only way out of these fundamental difficulties that I see is the realization that our understanding of reality is the result of our daily experience with macroscopic, tangible things. For the survival of our species, it was never important to see electrons inside an atom and we should not be surprised when our images of reality fail when we try to apply them in the world of the atom.

This *Ringvorlesung* became the forerunner of many more *Ringvorlesungen* originally inspired by Kaufmann and Otto Creutzfeldt but later taken over by others. Speakers included the London art historian Sir Ernst Gombrich, Anton Zeilinger from Vienna, and Karl Friedrich von Weizsäcker ("not a bad physicist" according to a *public* pronouncement by Hund in the presence of W.).

One more word about Becker. When in April 1933 the Jewish professors were sacked except—initially—those who had distinguished themselves in the Great War (1914–1918) such as Nobelist James Franck. Franck resigned his directorship of the institute for atomic physics voluntarily, noting that he could not, in good conscience, continue to serve a state such as the Nazis were then erecting. How many German professors sympathized with Franck? In my research (for a talk on Lise Meitner and nuclear fission) I came across just one letter—by Richard Becker, expressing his sympathy with Franck's courageous stand. Nor did anybody call in by phone. (There is now a bronze plaque in the main building of the university with the words of Franck's secretary, who was hovering in vain over the phone all night "...aber niemand rief an.")

One prominent physicist who suffered from the Nazis was Lise Meitner. She was Austrian by birth but could work in Germany until Austria was annexed by the Third Reich in 1938 whereupon she became automatically a German citizen and subject to the Nazis anti-semitic policies. She fled overnight with the help of her friend Otto Hahn ("Hähnchen"—little cock—as she liked to call him). Shortly thereafter, in Berlin, Hahn (and Strassmann) discovered nuclear fission—a discovery that shook the world (and of which Hiroshima/Nagasaki are perhaps not last evil manifestations). In 1945, Hahn was awarded the Nobel Prize in Chemistry for this discovery while Lise Meitner was all but forgotten in her Stockholm laboratory. But it was she (and her nephew Otto Frisch) who first recognized what was going on and coined the term *Kernspaltung* (nuclear fission)—a realization of Einstein's famous formula $E = mc^2$. I am proud to have known Frisch personally and to have been able to pay tribute to Meitner in a talk I gave in the Berlin *Staatsbibliothek* (see my homepage http://www.physik3.gwdg.de/~mrs/).

Emmy Noether, a leading mathematician of the twentieth century, was another woman scientist who was expelled from Nazi Germany. She went to Bryn Mawr in Pennsylvania. Her brother Fritz went to Tomsk in Siberia as a professor, where he applied for Soviet citizenship in 1937. Soon after, he was arrested by the NKVD (forerunner of the KGB) as a German spy. Four years later, on 10 September 1941, he was shot at a camp in Orel as the German army was approaching the area on their way to (defeat before) Moscow.

Later, for my Ph.D., chemistry raised its ugly head again. Yet, with the help of another professor (Erwin Meyer), I was able to circumnavigate that cliff, too. Meyer cautioned me: “We have some *wandelnde Gewissen* (walking consciences) in the faculty and it may be difficult. But I will bring up your request at the end of a long faculty meeting when everybody wants to go home and I will also mention that you are leaving for the States soon.” (Never to return?) Next week Meyer told me that it had worked as predicted but that it wasn’t right to totally skip chemistry. I agreed.

Ironically, 15 years later, when I became a professor myself, quite a few of my students were working on problems that involved a lot of (physical) chemistry. I quickly got some freshman chemical texts and my handicap soon disappeared. As is well known, there is nothing like being properly motivated to learn something!

The Vordiplom was relatively easy, with chemistry substituted by mathematics. But even in math I had a deficit: I knew next to nothing about differential equations, so I asked the examiner (Dr. Lyra) to please skip differential equations. But during the actual oral examination he asked me for the solution of $dy/dt = y$. I said, “But, Dr. Lyra, that is a differential equation and we had agreed…” “But it’s a very simple one” he said apologetically. I said: “O.k., the solution is $y = \exp(t)$.” After that minor skirmish, we settled on topics more to my liking.

With the Vordiplom in hand, I registered for the obligatory advanced lab course in physics, which didn’t really stoke my interest—except when I had a chance to *umfunktionieren* (turn upside down) one of the experiments. For example, we had to measure the speed of light on a long cable (curled up in the basement of the physics building). The idea was to send a short electrical pulse through the cable and measure the delay with which it emerged at the other end. I thought, if I just connected the two ends of the cable through a feedback amplifier and cranked up the gain, the cable would start “singing” at a frequency which is simply the reciprocal of the sought-after delay.

This new approach to measuring the speed of light on a cable, which I considered almost self-evident, was rated a “major breakthrough.” The next day I was offered a position at the *Dritte Physikalische Institut* for doing a master’s thesis. Thus, while I had started out in theory, I was now an experimental physicist at a renowned institute that, 4 years later, smoothed my path to the USA and the prestigious Bell Laboratories. Some students who were much smarter than I—one even got the Nobel Prize (Horst Krömer)—who stayed behind in theory never made it to Bell Labs, because their professor wasn’t as well connected as my new boss, Erwin Meyer.

19.4 The Dritte Physikalische Institut: Historical Antecedents

Around the turn of the twentieth century, the well-known Göttingen mathematician Felix Klein proposed to promote the pursuit of applied sciences at German universities. He was further inspired in this effort by a visit to the 1893 Chicago World

Exhibition—as an official representative of Kaiser Wilhelm II—and subsequent visits to several American universities, which had a strong tradition of fostering applied sciences and engineering. But Klein's attempt to enlist the *Technische Hochschule Hannover*—let alone any university—in his endeavor failed miserably. Finally, he succeeded, with support by German industry (Böttinger), to establish an *Institut für Angewandte Mathematik und Mechanik* and an *Institut für Angewandte Elektrizität* (1905) at Göttingen.

These institutes, which had been home to such famous aerodynamicists as Ludwig Prandtl and Theodor von Karmann, morphed, in May 1947, into the *Dritte Physikalische Institut* under the leadership of Erwin Meyer, who had been professor at the *Charlottenburg Technische Hochschule* in Berlin. With Meyer, room acoustics, including concert hall acoustics, microwaves, underwater sound, and acoustics in general came to Göttingen.

The *Institut* became soon well known through several successful large projects: the *Jahrhunderthalle* in Höchst, the *Beethovensaal* in Bonn, the Hamburg *Staatsoper*, and the *Lower Saxony Landtag* (parliament) in Hanover. Especially noteworthy was the ingenious utilization of electroacoustic means utilizing the *Haas-Effekt*, related to the "precedence effect."

In fact, the *Institut* became well known internationally by the discovery, stimulated by Meyer, of the *Haas-Effekt* by Helmut Haas in 1950. This property of human auditory perception says that even an *amplified*—but suitably delayed—sound will not affect the perceived direction of the sound source. (The name Haas effect was suggested by Richard Bolt of the Massachusetts Institute of Technology.)

Around 1953, under the technical leadership of Günther Kurtze, a large *Reflexionsfreier Raum* ("free-space room") was constructed, which—unique in the world—was also designed to be nearly free from reflexions for *electromagnetic* centimeter waves to facilitate measurements with microwaves (antennas, etc.)

In general, there was a world-wide interest to supplement reverberation time by other physical parameters to characterize the acoustic quality of performance spaces. One of these criteria, favored in the early 1950s, was the so-called frequency-irregularity of the sound transmission between source (on the stage) and a listener's ears. Experimental results obtained by Heinrich Kuttruff and Rolf Thiele in the *Herkulessaal* in Munich (and other concert halls) showed that the number of maxima of the frequency response did *not* correspond to the number of normal modes (resonances)—as posited by a faulty theory by Bolt—but was actually more than a *thousand* times less. This astonishing result was explained by a statistical theory by me, then a postdoc at the institute. This theory showed that the frequency response (sound pressure and phase as functions of frequency) is, approximately, a complex Gaussian process. As a result, the average distance between maxima above a critical frequency ("Schroeder frequency") is fully determined by the (reciprocal) reverberation time—thus not affording the much sought-after new quality parameter.

Besides measurements in actual rooms, scale models were also studied, including electromagnetic models, using microwaves rather than sound waves. In my dissertation (1954), I could show that the distribution of the frequencies and

Fig. 19.1 Concert Hall Acoustics using Microwaves: Erwin Meyer explaining my microwave setup to Kultusminister Voigt

excitations of the normal modes in metallic cavities were highly irregular—even for very small deviations from the symmetry of a perfectly rectangular space, such as a cube. Thus, for all practical purposes in room acoustics, the normal modes (resonances) can be considered completely random (Fig. 19.1).

In the early 1970s, in a large study of concert hall quality, sponsored by the *Deutsche Forschungsgemeinschaft*, Dieter Gottlob, and Karl-Friedrich Siebrasse—on the basis of measurements in 22 concert halls in Europe and the USA—showed that the lack of early lateral reflections in many modern halls with low ceilings and wide (fan-shaped) ground plan was the main culprit. To counteract this deficiency, I developed sound-diffusing structures ("reflection phase gratings") based on number-theoretic principles (quadratic residues), which have found broad acceptance in room acoustics.

I also proposed a new method of accurately measuring reverberation times by "backward integration" of the (squared) impulse response of a room, using the so-called maximum-length sequences constructed from the theory of finite number fields.

Before being fully accepted at the institute, you had to design a new experiment for the institute's lab course, and I was asked to set up an exercise with 10-cm microwaves involving *round* waveguides. I said: "Professor, round guides means *Bessel* functions and I don't like Bessels." "Not to worry he said, all you have to know are the first few zeroes of those functions: 1.84, 2.40 and so forth." And Meyer was basically correct.

I soon discovered, though, that my round guides always had *two* resonances close to each other instead of just one. What was the reason? *Nobody could tell me.* I finally had to figure it out myself—drawing an analogy with the vibrations of a kettle drum. The drum in the relevant mode of oscillation always has two resonances, going with the two possible orientations at right angles to each other. Thus, my round wave guides showed what the physicist calls degenerate modes. For my master's thesis I elaborated this "discovery" into a precision tool for making very accurate measurements at microwave frequencies.[1]

Meyer was much impressed and felt I should go to the States, the microwave haven. So I applied for a 1-year Fulbright scholarship. But I flunked resoundingly—not because of my English and certainly not because of my academic credentials. The Fulbright commission felt that I was not sufficiently *politically* motivated (spreading the gospel of American democracy after my return to Europe). It was, alas, all too obvious that I was interested only in science, mathematics, and languages, to the exclusion of almost everything else: politics, sports (except tennis), etc. I remember when, in the summer of 1954 (I was just returning to the clubhouse from a few sets of tennis), the entire membership went wild, exclaiming: "We won, we won!" I thought they meant that our junior tennis team had beaten nearby Einbeck. No, Germany had just beaten Hungary (or was it Brazil?) in the finals of the soccer world championship. The whole country was in an uproar (as if Germany had belatedly won World War II) and I didn't even know that there was a championship on.

Other hobbies of mine were the theater (the famous *Deutsche Theater* under Heinz Hilpert) and movies. I fell in love with Hildegard Knef and her films (*Film ohne Titel, Die Sünderin*). I particularly liked *Les Enfant du Paradies* (with Jean-Louis Barrault), *Casablanca* (Bogart and Bergman), and *To Have and Have Not* (Lauren Bacall's first film, with Bogey) and, later, *Les Vacances de M. Hulot* (which I saw in four different languages) and other films by Jacques Tati. Another favorite of mine was Roman Polanski's *Tanz der Vampire* (*The Crazy Vampire Killers*) with his wife Sharon Tate and *A Man and a Woman* (with Jean Louis Tritignan and Anouk Aimée).

After finishing my work for the degree of *Diplomphysiker* (sort of M.S.) in just 4 years after coming to Göttingen, I had enough of university life and went into industry: *Grundig Radio*. In early 1952, they were just beginning to set up production of (black-and-white) TV sets. For their production line, they needed various testing devices and I joined that effort building a square-wave generator that could switch between its two states ("black" and "white") in just a few *nano*seconds—an unheard-of speed in the early TV manufacturing days. The price I paid was a very unusual electronic circuit that could give you painful shocks if you got too close.

[1] After I got my degree of *Diplomphysiker* on 19 December 1951, one of our co-eds appeared, out of nowhere so to speak, in my home town to spend a night with me. This kind of forwardness has happened to me only one more time—shortly after I got my Ph.D. degree. (I never realized that fresh academic degrees were that seductive. *Ius primae noctis*?)

When I told the Grundig managers that I had promised my Göttingen professor to return after 6 months in industry, they offered to increase my salary from 500 to 750 Deutschmarks and then even hinted I should name my price. But I made the right decision by honoring my word to Professor Meyer and start working on my Ph.D.

The fact that Max Grundig had said of acoustical physicist Helmut Haas (of Haas-effect fame) that all scientists should be kicked in the behind also facilitated my decision to return to Göttingen. Haas had been put in charge of developing a dictation machine and made a mess of it. It was known in the company as the *Jaulophon* (wowophone). But then Haas was not a mechanical engineer. Before this disaster, Haas was Grundig's fair-haired boy—driving around in the boss's Packard, one of the few American cars owned by Germans in Nuremberg. So perhaps I would suffer a similar fate.

When I said good-bye to my friends at the Grundig lab, the head of the storeroom called after me "rechts von den Trompeten" (to the right of the trumpets) and I knew exactly what he meant: when, on Judgement Day, I was looking for him, I could find him standing to the right of the trumpets—in other words our farewell would be for a very long time.

One of my co-workers (Gäbelein) at Grundig always came to my desk to chat in his local (Fürth) dialect—of which I understood very little. Once, he said, Herr Schroeder you look as if you were thinking of something else—which I was.

19.5 Ph.D.

I had always felt a great attraction to statistics and when I plunged my microwave antenna into a large discarded US army biscuit tin (after the last crumbs had been shaken out), I saw on my scope thousands of resonances, a veritable "forest" of spikes, all of different height and seemingly randomly distributed in frequency (around 10 GHz). This looked interesting, to say the least! But Meyer wanted me to do *acoustics* (his field) for my thesis. Yet I didn't want to part with my microwave statistics. So we compromised: I would investigate the random resonances of microwave cavities and he would interpret the results in terms of concert hall resonances. Thus, my published thesis bore the unlikely title "The distribution of resonances in large rooms: Experiments with microwaves."

I started working on my thesis in October of 1952 and was able to hand it in before the end of the following year—after just 15 months, which must have been a record for a Ph.D. thesis in experimental physics.

Our Institut, the *Drittes Physikalische Institut*, already had another Ph.D. student called Schroeder, so I became *Schroeder2*. But I soon got tired of always having to call myself *Schroeder2*, so one day, when I was asked which Schroeder I was, I answered, on a whim, *Schroeder17*—and the name stuck! For the next 50 years I remained *Schroeder17*. When the Internet came around, I chose as my e-mail address MRS17. (There is a double entendre here: some of my correspondents were eager to meet the 17-year-old "Mrs." they thought was luring behind my screen name.)[2]

I had always been (and still am) an *eye*-person strongly attracted by things visual (landscapes, flowers, people, architecture) and now I had become, willy-nilly, an *ear*-man. From an early age on, I liked photography and, while I first worked with borrowed cameras, I already was the proud owner of a precision light meter (an *Ikophot* from Zeiss). Once, when I gave a slide show at the Eggers' house, the mother of my friend Frieder Egger exclaimed "Ganz (wholly like) Caspar David Friedrich!" And I didn't even know who Caspar David Friedrich was. But I think she was right: the pictures from my "romantic" period did have the feel of C.D. Friedrich. Highly educated in the humanities as she was, Frau Eggers lacked (or pretended to lack) the most basic understanding of the physical world. Once, during a walk near Göttingen, when it started to rain, I remarked casually "No wonder, with all the clouds above us." "Herr Schroeder, you mean the rain comes from the clouds?" "Yes, Frau Eggers, the rain comes from the clouds." (I still have trouble believing this story, but it happened just as recounted.)

Another hobby of mine was dancing. We had a nice student swing band, the *Music Mixers*, that played at least once a week from 8 p.m. to long after midnight. While I was often late for lectures and other events, I was always ready to go the moment the doors of the *Mensa* (student mess hall) swung open, dressed in loosely fitting clothes and plenty of *Dextro Energen* (dextrose sugar) to keep me going until the wee hours of the next day. Another favorite venue for me was the *Taberna Academia* on Wilhelmsplatz with nightly dancing and weekly *Bockbierfeste* (bock beer fests).

Once, in Amsterdam, I went to a public dance hall with an Indonesian friend (Rudy Tian Tik Tiang) who encouraged me to ask one of the ladies at a nearby table for a dance. (I wasn't sure you could simply invite a stranger.) She was young, beautiful, and raven haired. When I asked her where she hailed from, she said Sweden. (When she noticed my surprise, she assured me that not all Swedes are blond.) What was she doing in Amsterdam? "Oh, I am on my honeymoon." And where is your husband? "He doesn't like dancing, so he stayed behind in our tent in the campground"—while she took their VW into the city. Some honeymoon!

Back to acoustics: For the distribution of the resonances, I had expected a "Poisson" process, but what I found was completely different. And, timid as I was, tried to explain it away. Now the actual distribution, named after Eugene Wigner, is recognized as a tell-tale sign for chaotic systems. But how should I have known? Wigner had introduced his distribution to describe the energy level spacings of large atomic nuclei. (It also describes the spacings of the zeroes of the

[2] The prime number 17 is a so-called *Fermat Prime*. The only known Fermat Primes are 3, 5, 17, 257, and 65,537. Fermat thought that all numbers of the form $2^m + 1$, where m is a power of 2, are prime. But Euler proved him wrong, showing that for $m = 32$ the resulting number is divisible by 641 and therefore not prime. (Try it. A modest-size pocket calculator will do! But to have "guessed" that 641 was a divisor of 4,294,967,297 required a mathematical genius like Euler.)

famous Riemann zeta-function from number theory.) It took another 30 years for people to recognize this common thread in chaotic systems, atomic nuclei and number theory.

Before delving into the actual measurements, I asked a renowned mathematician, Franz Rellich, whether a mathematical theorem existed to cover my case of "complex eigenvalues." (Hermann Weyl, Hilbert's successor, had proven the case for *real* eigenvalues, corresponding, in physics, to resonances without losses.) Rellich searched the library for maybe half an hour but found nothing relating to my case. Then Rellich said something that I will long remember: "Do you realize that you are the first physicist since the 1930s who comes to ask a mathematician for advice." He was of course referring to the fact that before Hitler there was a very close relationship between physics and mathematics in Göttingen which, along with so many other things, had been destroyed by the Nazis (Max Born, Heisenberg, Ludwig Prandtl, Peter Debye, Einstein (on visits from Berlin), Felix Klein, Hilbert, Emmy Noether, Richard Courant, Hermann Minkowski (before 1909), John von Neumann, Gustav Herglotz, Theodor Kaluza, Bartels van der Waerden (from Leipzig), Hermann Weyl, Carl-Ludwig Siegel, Wilhelm Magnus).[3]

19.6 Friends

My main contacts during my student days were Erhard Scheibe and Jobst von Behr, whom I had first met during the war in an air force radar school when we were both 17 years old. Both are still my best friends, now living in Hamburg as retired professors (of the philosophy of science and mathematical physics, respectively.). But my main support, other than my parents, was *Tante Lina*, my landlady, who had lost her entire life's savings during the currency reform in 1948. She cared and cooked for me. Sometimes, when we were both out of cash, she would produce a tasty meal with a few vegetables that I had bought with the deposit money from empty bottles. (Once, I suggested to her, in jest, that we rob a bank but she admonished me "Aber, Herr Schroeder, das ist doch verboten!" I countered: "Look, you just have to stand at the street corner and whistle when the police comes. I'll take care of the bank myself." But, no, she stood firm. We had to continue scrounging on the 85 marks/month she got from welfare and the 60 marks/month donated by my parents (who had "lost everything" during the war).

[3] I once heard it said that mathematician Emmy Noether was like a Nobelist in physics or chemistry. But that is a gross understatement. In the twentieth century some 400 physicists and chemists were awarded a Nobel Prize but Emmy clearly ranked among the top ten mathematicians of the century. She had to flee Germany in 1933 and died 2 years later as a professor at Bryn Mawr. I recently (14 February 2007) succeeded in locating one of her surviving relatives, her nephew Herman Noether, in Summit, New Jersey, from whom I could retrieve some documents of his Aunt Emmy for the Göttingen Academy Commission charged with assembling the *Nachlässe* (scientific estates) of famous mathematicians.

Later I added to our income by repairing radios and even building one or two sets from scratch. One radio I named (in sterling silver letters!) after my favorite big-band leader *Kenton*. It boasted a "dynamic expander" that made the music sound more exciting (and more disturbing to the neighbors). Once, when my *Kenton* failed, a radio repairman said: "No wonder. It says so even on the cabinet *ken Ton*—no tone." Later I added to my "income" by translating German research reports into English (at 5 deutschmarks per page). When I look at these translations today I am amazed how good my English was even then. Where did I learn it? Some in high school, no doubt, but much by reading—not Shakespeare but James M. Cain. Cain's novels (*The Postman Always Rings Twice*, *Double Indemnity*) which were on sale in Göttingen in paperback for one mark. But, as I discovered much later, I didn't know the proper pronunciation of many words: *steak*, *catastrophe*, and even the simple-looking *noise*. (I once ordered a "banana split" in a restaurant when I thought a whole banana would be too much to stomach. Was I ever surprised at the gargantuan dish—with lots of faked whipped cream—set in front of me.)

This was also the time (July 1949) that I gained my first "carnal knowledge," with a person (Christa) who was likewise a novice. She was 19 but I was already 23 years old, my somewhat late initiation being occasioned by my being caught up in the boys-only war and the "monastic" PoW camps.

I also listened a lot to AFN (the American Forces Network), which I had first encountered on D-day: 6 June 1944. I was in a hospital then (with rheumatic fever) and I had a little radio on my nightstand with which I could listen to London (very risky in those days, but I had turned the volume way down, with my ear glued to the loudspeaker). And what did I hear in the afternoon (3 p.m.) of that historic day? "This is the American Forces Network, with the Army Air Force band under Major Glenn Miller, broadcasting from London, England. Hello, boys in France." Followed by the *American Patrol*. (I wonder how many "boys in France" had the leisure to listen to AFN on those bloody beaches. But those who did must have felt a tremendous uplift.)

In 1948, after the almost total devaluation of the *Reichsmark*, life became a bit difficult for me. I therefore applied for *Gebührenerlass* (release from tuition payments), which meant I had to pass a special little exam to prove my worthiness. I selected the beginners' lab course and saw Professor Kopfermann for the test. It looked as if I knew all the answers except one which baffled me. "But this was covered in the lectures that go with the lab work," Kopfermann said. I countered, "I only went to a single lecture and then I quit. The lecture was awful." "How can you say that? I drafted the lectures myself and Dr. Paul, who actually delivered the lectures, is only a *Schallplatte* (phonograph disc)!" "But then Dr. Paul (who later won a Nobel Prize) is a very bad disc," I ventured. Understandably, this was the end of the interview. Kopfermann rose and I left his office starting to think of other ways to pay for my tuition.

19.7 Italy

What happened next is almost beyond belief. A week after the disastrous interview I got a call from the *Akademisches Hilfswerk* (the academic assistance office): Professor Kopfermann had approved my tuition request but had asked why I hadn't also applied for a *Barbeihilfe* (monthly cash support). "Yes, that would be wonderful. I could really use it." It was only 100 marks a month but, with back payments, I received some 430 marks for my first installment.

What did I do with all that money? I took the money and ran: I ran to the railroad station and bought a ticket to Italy for a summer course *La Lingua e Cultura Italiana* given by University of Pisa in the *Collegio Colombo,* situated in the beautiful *Pineta* (pine woods) near Viareggio already admired by Rainer Maria Rilke.

I'm really glad I went, following my friend Erhard who had taken the course a year earlier. I learned a lot about Italy, its people; it was the beginning of a lifelong love affair with the language. Courses (by Professor Bolelli) were given in the morning. In the afternoon—after lunch and siesta—we were free to spend on the nearby beach (*Bagno Giorgio*). They had two beautiful daughters. Fourteen years later, when I wanted to show Anny my erstwhile models, I couldn't find Silvia. I finally approached a young matron with three or four kids at her feet. "Where is Silvia?" "Silvia son' io"—that's me. I was glad to see her again and to see that her beauty at 17 had not been vain (Fig. 19.2).

I continued my Italian studies at the University of Göttingen with the *Lektor* there and read "Don Camillo e Pepone" and (with difficulty) Manzoni's "I Promessi Sposi" (The Betrothed) and Boccaccio's "Decameron."

After the summer course ended, I went with a French student into the Appuane Alps where we climbed *Monte Cavallo* (1,889 m) and spent the night in the rifugio *Aronte*. On our way up from the Mediterranean coast near Massa (think white marble and Michelangelo), we encountered a peasant who proudly pointed to a tractor, proclaiming "that's what we call a *trattrice*," as if we had never seen a tractor before.

Of course, once in Italy, I didn't want to miss Rome, the "Eternal City." I was lucky to find a room in the student dormitory of the university ("La Sapienza"). My roommate was a young Englishman who one evening returned with his hands inked in blue. He had bought, for a very low price, a "genuine" Parker pen. Except it wasn't a PARKER pen; it was a P.ARKER pen and he hadn't looked closely enough to see the little dot.

My only other memory of that dorm was that one night I woke up with a terrible toothache. I ran around the deserted neighborhood yelling "Farmáccia, Farmáccia!" But, in spite of mispronouncing the word, I soon found a drug store with a night bell. Before long two eyes appeared behind a narrow slit in the door and before I could say anything (I was still holding my aching cheek), the voice belonging to those eyes said "I denti?" "Si, Si, i denti!" He gave me some pills which I swallowed on the spot. The pain went away and miraculously never returned.

Fig. 19.2 Silvia at 17

What I found most impressive in Rome was the ancient though starkly ruined *Foro Romano*. I never had felt *History* that intensely. Luckily, I had a friend in Rome (Maria, a native of Gubbio) who showed me around the city and its sights and introduced me to some of her friends. Unfortunately, I later lost contact with Maria. She married a British diplomat and was "gone." But some 40 years later, when Anny and I visited Gubbio and bought some ceramics, the dealer, an elderly professor, remembered her: "Oh yes, Maria Moreno! But I haven't seen her in a long time" (Fig. 19.3).

Before leaving Rome, I had to pay some bills (for laundry, the return ticket?) and I had to sell my trusted *Baldinette* camera which a friend had bought for me cheap in East Germany.

On the way home, I absolutely wanted to see Milan and its *Duomo*. I put my suitcase on the train to Switzerland and asked the lady in my compartment to "watch it." I would be back shortly after some sightseeing. I had enough time to walk once around the Duomo and then I made a big mistake on the way back to the

Fig. 19.3 With Maria Moreno in Rome

Stazione Centrale: I ordered an espresso. But the espresso was so hot I couldn't drink it right away and as a near "destitute" student I couldn't just abandon it. When it was finally cool enough to drink, I resumed my way back to the station and discovered that the *Stazione Centrale* is very large. As I finally got to the right platform, I saw the red tail lights of the departing train. When the porters saw me running after the train, all yelled "Forza! Forza!" But it was too late: Train and suitcase were gone.

The next 24 h were the stuff of comic opera. I went to station master's office and told him (in halting Italian) about my *valigia gialla semiaperta*, my yellow suitcase, half open (because my tennis racket wouldn't fit in). He advised to take the next *Accelerato* (actually a slow train) to Gallarate, the next stop of the train I had missed. There the station police told me that the train stopped for only 2 min—not enough time to look for my suitcase. So I took the next slow train to Domodossola near the Swiss border. There a policeman told me he had searched the whole train (*ho visto tutto il treno*) and in fact had seen a yellow half-open suitcase in the car that I had specified, but a lady told him that it belonged to a man she knew. End of story. *La valigia é andata via*—the suitcase is gone, as the Italian policeman volunteered. Couldn't he help me further? Yes, he would take me to the representative of the Swiss Railroads, whom I told my sad story: *una valigia gialla semiaperta*... Upon which he said "Je téléphonerai avec Berne." I said, wait a minute, let me explain the whole thing in French first. Then on the phone he said "Ich hab' hier einen Herrn, der seinen Koffer..." I asked him whether I could repeat everything in German, the only language I really knew well. Then he said something that was really the highpoint of the suitcase saga. "If your suitcase was on the train as it entered Switzerland, you will have it back tomorrow. The car you indicated will be cleaned in Olten (near Basel) tomorrow, starting at 7:30 in the morning. By 8:30 the cleaning crews will have found it. The next train to Bern (where I was staying overnight) gets there some time before noon. Thus, if you go to the lost-and-found office in Bern, they will hand you your suitcase back." I thought the man was crazy. But that's exactly what happened. The next day I went to the Bern station and said (without even giving my name) "Haben Sie meinen Koffer?" The kind Swiss went backstage and returned—with my suitcase, still half open. That will be "Zwei Franken und dreißig Rappen." I hope I didn't forget to say "Thank you!"

The whole trip to Italy ended on a drole note: I lacked the two Deutschmarks to pay for the *Eilzugzuschlag* ("express" train supplemental fare). I naively thought the German railroad should take me home as quickly as possible without asking inconvenient questions. But I had evidently forgotten the pedantry of German officialdom. In fact, the conductor returned with an agent of the railroad police. One of the passengers, who had followed the whole mini-drama, quickly slipped me a two-mark piece. So when the conductor reappeared with the police in tow, I said: "What's the problem?" and gave him the extra fare.

Next year I went again to Italy with my friend Erhard and a new camera, the lovely Zeiss *Contessa*. We spent 3 weeks in a *hotel* in Malcesina on Lake Garda. Afterwards I went to Venice for another overwhelming experience. Italy had a strong mental hold on me now, and 2 years later, when emigrating to the USA, I went of course via Italy, on the *Andrea Doria* out of Genoa.

19.8 A Vacation on the Rhine

The year before my Italian adventures, I spent 3 weeks at the invitation of my maternal grandmother in Sinzig on the Rhine, between Koblenz and Bonn. I had just befriended a young French student Geneviève Hochapfel, who was spending the summer in Germany to learn German. As her name indicates, she was of Alsatian extraction. I was allowed to bring her along to a hillside cabin that my uncle Paul Dehne had purchased from his ample income as a *Heilpraktiker* (a health practitioner without a medical degree—very popular at the time). Uncle Paul was my favorite uncle, always ready with a quip or trick like pulling candy from behind his ears or money bills out of thin air or "swallowing daggers." (To add to the realism, he spent a long time in the bathroom the next day before he appeared with the "recovered" dagger.) During the war, he kept a slight wound from healing so they had to discharge him from the service.

In the summer of 1939 while most Germans were still hoping for peace and believed that Hitler (after Munich) would get away with anything, he said goodbye to his nephew with the prophetic words "Auf Wiedersehen im Massengrab" (see you in the mass grave). Actually, we both survived the war but he lost his only (step) son, Ernst-Adolf (about my age), who disappeared in the snows of Russia and was never found again. My only other male cousin, Karl-Heinz, also perished in the war, in the senseless defense of Italy (ca. 1944). But he died in a military hospital, so *his* place of burial is known. I visited my uncle and his family in Düsseldorf during my honeymoon from the USA.

One night in Sinzig, when I wanted Geneviève to stay a little longer in my room, he came to my aid by warning her that there were rattlesnakes in the stairwell. But Geneviève said she wasn't afraid of rattlesnakes and left. (But she *was* scared of "zee liddle animals"—her name for *spermatozoa*.[4])

During a walk in the woods, we came upon boar tracks and Uncle Paul explained in his English (our only common language) "wild pork has crossed here." (Did we realize that there was something wrong with the sentence? I don't remember.)

The Rhine vacation included interesting visits to the wine growing country and Rüdesheim and a few hours at a lab for dubbing movies—a "big" industry in Germany, where almost all foreign-language films are dubbed into German (with people of any description, including black people, Chinese and Eskimos, speaking perfect German). The high point was, not surprisingly, a boat trip down the Rhine from Bingen to Cologne past many of the fabled castles.

After our return to Göttingen, Geneviève helped me print my many snapshots in the darkroom of the physics institute (Fig. 19.4). We were promptly denounced by one of the professors who was jealous of his darkroom privileges. But Erwin Meyer, the director, believed *me* (rather than the accusing colleague) that the pictures we were printing were strictly personal (and not for profit, as had been alleged). I also assured Meyer that I had a "sturmfreie Bude" (undisturbed pad) and didn't need the institute's darkroom for any hanky panky. So Meyer allowed me continued use of the darkroom—alone.[5]

19.9 First Congress

It was also during my years as a graduate student that I attended my first international congress, in fact the First International Congress on Acoustics, in Delft, Holland. Since I was the only student from Göttingen who spoke Dutch, the task of introducing our group to the Dutch students who were permanent residents of our dormitory fell to me. But our attempt at politeness backfired when I concluded our visit with the words "Thanks for your hospitality, but we have to go now" and received the unexpected answer "Ja, dat is ook goed zo, want er is nog steeds hoogspanning." (Yes, that is good so, because there is still high tension). This was in 1953 (June 17th)—8 years after the end of the war. The German misdeeds during the war were long remembered, especially in Holland which had suffered disproportionately.

---

[4] Her previous boyfriend, she told me, likened protected congress to taking a shower in a rain coat.

[5] Meyer, ever the gentleman, once had every reason to be angry at me: I had "borrowed" the Institute's projector to show some slides at home. When I brought it back the next morning, he had already been waiting for it because he needed it for his lecture. Then, upon plugging the projector in, the light bulb exploded with a big bang. Meyer was obviously "boiling," but all he said was "Diese Lampe bezahlen Sie!" (You'll have to pay for *this* bulb.)

Fig. 19.4 Geneviève in Amsterdam

It was during this visit, incidentally, that we heard of the momentous events that were unfolding in East Germany on that Wednesday.[6]

While the Congress in Delft was in progress, a circus was in town and besides all the wild animals, one of the shows was a "flea circus" where "trained" fleas danced, pulled little carts, and performed other feats. During an intermission, while the flea master was outside his tent barking for new customers, my friends (Georg Schodder, Rolf Thiele) egged me on to take over the show. So I started out "Dames en Heeren, onze flooien zijn getraineerd om...." Soon the fleas started hopping and dancing and the onlookers mistook my sham show for the real thing. But it didn't last long: when the real flea master returned he was, you guessed it, hopping mad. (And that I was a German impersonating a Dutchman didn't help.)

[6] *Der 17. Juni* was soon declared a national holiday in West Germany, and called *Tag der deutschen Einheit* (Day of German Unity) implying that the East German workers had taken to the streets (and faced Russian tanks) not to protest for better working conditions but for *Wiedervereinigung* (Re-unification).—This was an obscene lie on the part of the West. If you voiced any doubt of this interpretation you were labelled unpatriotic (*Einheitsverleugner*).

At the end of the year I completed my Ph.D. thesis and on 26 February 1954 I passed the orals for my degree (*Dr.rer.nat.*, pronounced "rare nut"). Seven months later I was on my way to the USA to become a Member of the Technical Staff of what was then called Bell Telephone Laboratories.

During my thesis work, I encountered some number theory problems, which deepened my relation with the "Queen of Mathematics" (Gauss' moniker). I discovered that in my cube-shaped resonator certain resonances were missing, their absence being explained by the *3-Squares-Theorem*, which says that integers which leave a remainder of 7 upon division by 8 cannot be decomposed into the sum of three squares. For example, $14 = 3^2 + 2^2 + 1^2$, but 15 requires *four* squares: $15 = 3^2 + 2^2 + 1^2 + 1^2$. (Multiplying these "forbidden" numbers by some positive power of 4 leads to more impossible integers, e.g., 28, 60, 92, 112, ... On average every sixth number cannot be written as the sum of three squares. When I once mentioned this casually to mathematician Deuring (during a talk he gave at the Göttingen academy), he had to reflect a moment and then agreed: "Yes, asymptotically." Right!)

Much later, in 1985, I wrote a "mathematical bestseller" about number theory (*Number Theory in Science and Communication*), published by Springer and now in its fourth edition.[7]

Emboldened by this success I wrote another book that became very popular: *Fractals, Chaos, Power Laws: Minutes from an Infinite Paradise*. The subtitle is an allusion to David Hilbert's dictum when mathematicians around the world were attacking Georg Cantor's *Mengenlehre* (set theory): "From his Paradise no one shall ever evict us." (The last word in the subtitle—Minutes—is of course a triple pun.)

[7] One avid reader, Kristina Lerman, named my book as one of the few she would take with her to a coffee shop to keep her company. Some company!

Chapter 20
Bell Laboratories

Manfred R. Schroeder

Bell Telephone Laboratories in Murray Hill, New Jersey, where the Transistor (1947), Information Theory (1948) and Error-correcting codes (1949) were invented was of course a dream place to start one's career as a mathematically inclined physicist. I am not implying that people at other times and other places cannot enjoy their work as much as we did. But there is a general perception that Bell Labs was something special—a "national resource." This did not come about by accident: Bell management decided early on that freedom to pursue one's own ideas and stable long-term funding were the best well-springs of innovation.—I do hope that industrial research at some future time will see the light again and not limit research to short-range goals and immediate profit. Some of the greatest advances in the past have come from taking the long-range view.

How did I get to Bell Labs from distant Germany in the first place more than 50 years ago? Let me explain: I was studying at the University of Göttingen when a noted linguist from Bonn, Werner Meyer-Eppler, visited us around Christmas in 1951 to give a talk at the General Physics Colloquium. I don't remember the details, but he talked about Shannon and Information Theory. Perhaps this was the first time I had heard the name Shannon and I was immediately fascinated. This sounded much more interesting than the nuts-and-bolts of experimental physics. But after the talk, during the Christmas party, people professed not to have "understood a word" of the talk. The chairman of the colloquium, Richard Becker, even seemed a little miffed when I dared pronounce the lecture "interesting." How dare I! How could I find a field that was devoid of *relativity* and the *quantum* to be interesting? But undeterred, I approached my professor, Erwin Meyer, whether he could recommend me for a job at Bell Labs. Unfortunately, the answer was "no" because

M.R. Schroeder
Author was deceased at the time of publication.

M.R. Schroeder (✉)
Drittes Physikalisches Institut, Georg-August-University Göttingen, Friedrich Hund-Platz 1, 37077 Göttingen, Germany

N. Xiang and G.M. Sessler (eds.), *Acoustics, Information, and Communication*, Modern Acoustics and Signal Processing, DOI 10.1007/978-3-319-05660-9_20,

he remembered one of his students before the war (at the height of the Jew-baiting pogroms in Germany) who wanted to emigrate but Bell said they didn't hire any foreigners. End of dream to join Bell! Or so it seemed.

Two years later I learned that William Shockley, co-inventor of the transistor, was coming to Göttingen actually *looking* for bright students for Bell. Well, I flew back to my professor telling him "they *do* take foreigners ... they are even *looking* for them." Fine he said, I am corresponding with one of the research directors of Bell and I will put in a few lines for you. (Later his secretary told me that he had "put in" a whole *page* singing my praise.)

When I said good-bye to Professor Becker, he asked me where I was going and I answered: "Murray Hill." And he knew exactly what "Murray Hill" meant. In fact he responded "Hmm, Murray Hill." Now of course the name Murray Hill doesn't ring any more bells (no pun). Indeed, a street sign in neighboring New Providence directs you to Murray Hills. *Sic transit gloria mundi* (or whatever they said in ancient Rome).

Two weeks later, I received an invitation for an employment interview by said research director. I was to meet him in the lobby of the Dorchester hotel in London, no less, on April 25, 1954, at 2 p.m. Well, the interview[1] apparently went well, and after 6 weeks I received an offer of permanent employment by Bell Labs for $640 a month—a princely sum then, especially for a perennially impecunious student. One alternative would have been for me to join Siemens in Munich for the equivalent of $125. Of course, I accepted the Bell offer—in fact I would have worked at Bell for nothing, but I didn't mind getting paid for what was for me a dream come true.

So on September 30, 1954, I arrived in New York on the Italian liner "Andrea Doria"—still afloat then. I was met at the pier by my future supervisor, Ralph LaRue Miller (who turned 100 on March 11, 2007), and the director, Winston Kock, with a long black chauffeur-driven limousine. But our first stop was not the Labs but an opulent restaurant (*The Newarker*), with a Menu in French that I couldn't read, except for the word *Bratwurst*. To this day, even after Iraq, French dominates many American menus. (I just returned from a luncheon at the *Grand Summit* in Summit,

[1] After explicating my thesis work, I asked the recruiter to tell me a bit about the Bell System. Well, there was the parent company, AT&T, Western Electric, the manufacturing arm, Bell Laboratories, and 23 operating companies: New York Telephone, New Jersey Bell, Southern Bell when, in the middle of this recitation, he stopped short and, with his eyes, followed an elegant young lady (an incognito countess?) traversing the long lobby. After a minute or two, without losing a beat, he continued yes, and there is Southwestern Bell, Pacific Telephone... Twenty-five years later, to the day, on April 25, 1979, I went back to the Dorchester. At the far end of the lobby there was a kind of hat-check counter with an elderly lady behind it. I went up to her and asked "Could it be that on this day, 25 years ago, on April 25, 1954, a Sunday, a young woman might have appeared from the door behind you—it was about 2 p.m.—crossed the lobby and then exited by the revolving door?" She must have thought I was from Scotland Yard or something. But, unflustered, she answered "Oh yes, of course, at 2 p.m. we had a change of shifts then. This entrance was for service personnel—chambermaids and so forth." I asked her, "How do you know this?" And she said, "I have been here for 30 years."

New Jersey, where I "savored" the *Soup du Jour*, followed by *Chateaubriand* and chocolate *Mousse*.[2])

The following three days were perhaps the most memorable in my entire life: My first regular job, a new country on a different continent and a new language (I used to pronounce "steak" *steek* then) and, yes, a new regular companion: on my third day in the USA. I travelled to New York and met a young woman from Bulgaria, Anny Menschik, who was working for Radio Free Europe and who soon became my wife.

Incidentally, on my first outing to New York I took the Lackawanna train to Hoboken and then the ferry across the Hudson. From the ferry landing near Christopher Street I wanted to take a cab to Anny's apartment on West 85th Street. But there was only one taxi and two people were already sitting in it. Therefore, having been in a taxi only once before, I hopped in and the taxi took off—for the *East*side. Nobody said anything. The cabbie collected the fare from the couple and then, without resetting the meter, doubled back to the Westside. The fare was $8.30 (more like $50 in today's dollars). But it was worth it!

In my application for employment at Bell I wrote "My confession is Roman Catholic," as was (is?) customary in Germany (*Mein Bekenntnis ist römisch-katholisch*). My Executive Director, Bill Doherty—a Catholic himself, treasured my letter for 37 years and then gave it back to me as a memento. He took an instant liking to me.

At the Labs, Win Kock encouraged me to continue my thesis work on concert hall acoustics (the discovery of a critical frequency, now called "Schroeder frequency"), but I figured—this being the telephone company—I better do something more germane to the telephone business. So I picked "speech" as my new research field—and speech it was for the next 50 years. Nobody objected to my total ignorance of this field. Electronic Engineering, especially, was an enigma to me, but I learned a lot by "osmosis" you might say. Bell Labs was an ideal place not only for doing research but for learning. The doors were wide open and people friendly and communicative. Nobody was guarding any "trade secrets." Not getting any monetary rewards for our inventions of course helped to foster this open atmosphere. I, for one, was quite happy with the $1. I received on my first day of

[2] Some New York restaurants are actually *bona fide* French. Anny and I often lunched with the **Einsteins** (Charlene and Ernst, of carpet fame) at such establishments, Charlene, a French native, usually making the reservations (*au nom d'Einstein*). When we happened to arrive first, asking for a *table pour quartre—au nom d'Einstein*, the maitre d' invited me, in a loud voice, *Suivez moi Monsieur Einstein*—and the whole restaurant would look up but was of course disappointed when they saw little me instead of the famous physicist. But it was fun—quite apart from the good food!—While we take French for granted in New York City, I was surprised when once, in Saarbrücken, only a few miles from the French border, nobody in a French-name restaurant could handle the language.

work for all my future inventions. (This amounts to a little over 2¢ per patent for the ca. 45 patents I received while at Bell.)[3]

Just one tiny but typical example: one day I was reading a technical paper bemoaning “the metallic twang” of existing artificial reverberators for electronic music. I thought what these audio types needed was a reverberator that—when seen as an electrical filter—has an all-pass frequency response. So I swivelled around in my chair and asked my office mate, electrical engineer and country musician Ben (“Tex”) Logan, “are there any all-pass filters with an exponentially decaying impulse response?” Ben’s answer was “yes” and *colorless artificial reverberation* was born. As far as I know, Bell Labs didn’t make a penny on the patents. But you can now find colorless reverberators and artificial stereo (now called soundscape, surround sound, or virtual acoustic images) in practically all electronic music instruments, Yamaha and the rest.

Kock (whose hobby was breeding orchids) asked me to join the phono-vision research effort (known around the Labs as “phony-vision”) but, reluctantly, I declined. Phono-vision never really worked; it was too far ahead of its time. Ten years later, in 1964, AT&T introduced the Picturephone™ but there was no sufficient market for it. The real breakthrough had to await *digital* image processing. Now, of course, phone and image are nearly inseparable—think cell phones and Skype™.

After giving the required “Declaration of Intent” (to become a US citizen) on June 26, 1956, I also got a *Top Secret* clearance. But I never did much work for *Jezebel*, the code name of Bell’s Ocean Acoustics research. This was a pity, because the top secret work involved side trips to Eleuthera and Bermuda which, however, I did visit once or twice (in June 1966) to test one of my ideas (which I called “volume focussing”) for detecting the presence of submarines in a deep ocean environment. Our official host in Bermuda was the Royal Navy.

No memories of the good days at Bell would be complete without remembering some of the great people I was privileged to work with like the late John Kelly, a true genius, and Jim Flanagan, who joined my department around 1958. After fathering innumerable advances in speech and acoustics, Jim went on to the highest honors: the Marconi Medal, bestowed by the Spanish king, and the National Medal of Science, a unique honor in our field, presented by President Clinton in a White House ceremony.

[3] A patent held by one of my collaborators, **Joseph L Hall**, recently figured in a suit brought by Alcatel Lucent against Microsoft for infringing on their MP3 patent. Lucent (the former Bell Labs) was awarded $1.52 *billion* (February 22, 2007). This patent was an outgrowth of our work on perceptual speech compression, applying it to music signals.—Of course, Microsoft will appeal the decision.

20.1 John Robinson Pierce

One of the greatest leaders at Bell was the unforgettable John Pierce of satellite communication fame and the father of *Telstar*, the first transatlantic space TV (July 4, 1963). In fact, Pierce (1910–2002) was the most inspiring boss and mentor I had between 1955 and 1971 when he left Bell Laboratories to join his old alma mater, the California Institute of Technology. By his writing he inspired my own writing—although when I once told him I might call my next book "Number Theory for Almost Everyone" he suggested instead, in his usual acerbic mode, "Number Theory for Almost No One." Other famous sayings by Pierce were "A job not worth doing isn't worth doing well" (to his boss W.O. Baker) or "Artificial Intelligence is mostly real stupidity," or "In the university, no one can tell a professor what to do. But, on the other hand, in any deep sense, nobody cares." What is the main purpose of the National Academy? According to John: "To keep the riffraff out."

Pierce also coined the word *transistor*, when co-inventor Walter Brattain asked him, over lunch, to come up with a good name. (Pierce's coinage is an amalgam of "*trans*conductance," a parameter measuring the sensitivity of vacuum tubes, and res*istor*.) When Pierce suggested *transistor*, Brattain exclaimed "That's it, Pierce, TRANSISTOR!" The name also fitted in nicely with other electronic devices such as *varistor* and *thermistor*.[4]

Pierce was also not averse to risqué jokes: As when, according to John, a young bride complained to her gynecologist of "painful congress" and the kind doctor, after having explored innumerable causes, finally asks the young lady "Do you smoke afterwards?" And blushing all over she admits "I don't know, Doctor—I haven't looked."

But while often blunt, John could be quite sweet too, as when, in 1957, he called my Voice Excited Vocoder "The first speaking machine that sounds human."

---

[4] In a piece on technical neologisms in the *New York Times Magazine* (in the Fall of 2006) James Gleick, who should have known better, claimed that the name transistor was the work of a committee. Yes, a committee of two! (Where did Gleick get the idea? From *Wikipedia*? No, *Wikipedia* actually got it right.) My Letter-to-the-Editor of the *Times*, a paper that prides itself on correctness, pointing out the error, was ignored.

In fairness, I should add that the *Times* and the *International Herald Tribune* have published all my previous letters: one concerning a story (about 1980) of a "singing computer" in Dresden, then part of communist East Germany. I pointed out to the *Trib* that the first singing computer ("Daisy, Daisy...") was demonstrated by J.L. Kelly and Carol Lochbaum at the Fourth International Congress on Acoustics in Copenhagen as early as 1962. Their reporter had simply been duped by communist propaganda.

Another of my printed letters concerned the use of a script *L* by Einstein in a paper on special relativity and his famous formula $E = mc^2$.

One of the most astounding admissions of "guilt" by the *Times* concerned a (very subtle) grammatical error they had committed in an *editorial*. I wrote them that I was a recent immigrant and English was my second language but something in the construction of one of their sentences seemed amiss to me—and the *Times* published my letter in full. Long live the *New York Times*!

(The reason why modern cell phones produce intelligible speech—and in fact would sound pretty good if connected to a high-quality loudspeaker—is that they incorporate the principle of voice-excitation.)

While John and I were watching the unfolding debacle of the avant-garde spectacle “9 Evenings: Experiments in Art and Technology” at the 69th Regiment Armory in New York City, I was concerned about the bad press our friend Billy Klüver (1927–2004) and *Experiments in Art and Technology* (EAT) might get. But John reassured me “They will be written up in the New York Times, and that’s the main thing—never mind *what* they actually write.”

John always impressed me by his honesty. Every time I had something unpleasant to report, he would immediately call *his* boss, Vice-president Baker, to relay the bad tidings. I was impressed because the culture in which I had been raised, honesty—volunteering the truth—could be deadly. Lack of forthrightness, I believe, was in fact one of the root causes for the failures of Kaiser Wilhelm’s Germany and Hitler’s regime—and could yet undo other governments.

When John was mistaken for the inventor of the Traveling Wave Tube, which happened not infrequently, he would reply “Rudy Kompfner invented it—I only discovered it.” Meaning that when John went to England on a wartime mission to search for ideas for a new, more powerful microwave oscillator for radar, he found instead Austrian refugee architect Rudolph Kompfner, working on a microwave amplifier that unfortunately had a tendency to become unstable and oscillate. John exclaimed: “that’s just what we are looking for, an oscillator!” This also led to the low-noise traveling-wave amplifier without which John’s idea for an ocean-spanning satellite communications system would have been impractical. (Pierce was also the inventor of the reflex klystron which was ideal as a local oscillator in microwave receivers.—In my Göttingen Ph.D. thesis I used two such klystrons to investigate the resolved resonances of metallic cavities modeled after concert halls, something that could not be done with sound.)

20.2 Computer Music

John Pierce had a long-time interest in music. He studied the piano while a student at Caltech and later installed a pipe organ in his home near Bell Labs. John, Claude Shannon, and Shannon’s wife, Betty, who was a pianist, carried out several ingenious experiments to estimate the information content of music.

John and Max Mathews attended a piano concert in 1957, which included pieces by Schoenberg played by Schnabel. They both felt that the Schoenberg was great and Schnabel was horrible. During the concert, John said to Mathews, “Max, with the right program your equipment should be able to synthesize better music than this. Take some time and write a music program.” This sojourn into computer music was possible because to facilitate research on speech coding, Mathews with Ed David and H.S. McDonald had recently developed equipment to put digitized sound into a computer and to recover processed sound from a stream of numbers generated by the computer. John's support and inspiration led Mathews to write a series of programs, "Music 1" through "Music 5," which started and set the course of present-day synthesized music.

AT&T administrators were not enthusiastic about music programs. They asked for an explanation as to the appropriateness of the work in a telephone company laboratory. With the strong support of both John and Bill Baker, Mathews was able to show them how music synthesis grew directly out of vital speech compression research and how music synthesis techniques fed back useful technology to speech synthesis. Without the support and encouragement from John and Bill Baker, computer music would not have begun when, where, and how it did.

About 1956, Max, with Ed David, also introduced Bermuda shorts at the Labs, in the days before Bell Labs was air conditioned—thereby exposing the least attractive part of the male anatomy, but to the great relief of all male members of staff. (But it took another decade or more before women wore slacks.)

20.3 Freedom of Research

Computer music is an apt example of the freedom of research that pervaded Bell Labs.

An example of freedom at Bell, one where it backfired, occurred with a Swedish linguist, Sven Ö., we wanted to hire in the mid-1960s. His future department head, **Peter Denes**, who was to take him out for dinner, asked me what he should answer if the candidate enquired about "freedom" at the Labs. I told Denes to reassure the candidate that he had total freedom—he could do whatever he wanted to do. The next thing we knew was that Sven's office and the nearby corridor were adorned with posters of Mao Zedong, Ho Chi Min and Che Guevara—just to test what we meant when we guaranteed him complete freedom. As you might imagine—this was the height of the Vietnam War—there was quite a commotion. People came to see me saying they couldn't continue with their classified, secret work. I had to reassure them that posters couldn't actually see. Then plant security got wind of the situation. They called me and I told them I would take charge, everything was "under control"—which of course it wasn't. But at least I got the security people off my back. Then I called my boss, John Pierce, and explained the whole thing to him. When I was through, he asked me *what* I was going to do and I said "Nothing, John." He replied with a single word: "Right." Of course, in less than 48 h the whole thing had blown by.

While I may have handled the Posters Case pretty well, I made quite a few horrible mistakes. Here is just one example: Once, ca. 1964, when returning from an ultrasonics meeting in Boston, one of the return flights, an Eastern Airlines shuttle from Boston to Newark collided over Connecticut with a big transcontinental Pan Am plane enroute from San Francisco to New York. Part of a wing of the Pan Am plane was severed and fell to the ground but the rest of the plane made it safely to its destination. But the smaller plane, the shuttle, had to make an emergency landing in a potato field near Darien, Connecticut. Luckily, nobody was killed and nobody was seriously injured in the mid-air collision. But people didn't know this. They heard only on the radio about the collision and assumed the worst. One of our group from Bell who was still missing was H. (I had taken a flight an hour earlier.) Then I made the unpardonable mistake of calling his wife to see whether he was at home. He wasn't, and his poor wife started crying on the phone.—H. later returned safely to his family: he had taken a later flight.

20.4 William Oliver Baker

One person from whom I learned a lot about human relations and research administration was our Research Vicepresident Bill Baker (1915–2005). Bill would always think long and hard before making a decision—which, however, he often communicated in, to many, incomprehensible language. In other words, he made *you* think hard, too, and hopefully come to the same conclusion. This could be very frustrating but I, for one, preferred his indirect way of operating to the "methodology" of his successor, a Nobel Prize winner no less, who always seemed to act first, then talk and finally, as a last resort, start thinking.

Baker was a true "diplomat of science"—witness his many contributions as a humanist, his advocacy of the use of science for the benefit of mankind. He was a trusted personal advisor on matters dealing with science and technology and national security to Presidents Eisenhower, Kennedy, Johnson, Nixon, and Ford.—He received 27 honorary doctorates in recognition of his varied accomplishments.

Bill Baker was also instrumental in getting my main collaborator Bishnu Atal admitted to the USA. I had heard of Bishnu in the late 1950s as a very good student at the Bangalore Institute of Science in India. After a long telephone conversation I had with Bishnu, I was able to persuade our management to make an offer of employment. However, Bishnu didn't come—he couldn't get a visa. After waiting for 2 years, I asked Bill Baker to intervene, who had Bishnu declared as "crucial to our efforts in national defense." So Bishnu got a special dispensation from Washington forthwith, and he appeared at Murray Hill in no time—thanks to the excellent relations Baker had with various government departments.

Given Baker's standing, it is with great satisfaction that I learned of the high esteem in which he held me throughout my career at Bell.

Here is one of my favorite Baker quotes:

> The ideas of scientific discovery come one at a time from one person and one mind at a time. Sometimes two or three can aid each other. But scientific discovery cannot be collectivized, and it does not flourish in collectivized structures.

20.5 John Wilder Tukey

Another memorable character was John Tukey (1915–2000) who divided his time between Bell Labs and Princeton where he headed the statistics department. John, who seemed to thrive on six glasses of skim milk for lunch, is of course best known for his (re)invention (with Jim Cooley) of the Fast Fourier Transform which has revolutionized digital signal processing. (Gauss had it about 150 years earlier but, as was his wont, didn't publish it—to him it was self-evident.)

Incidentally, Tukey added some of the most important words to the English language: *bit* (ca. 1948), *software* (1958) and the less well-known *cepstrum*—and thereby hangs a story: In the late 1950s US and Soviet diplomats and scientists were negotiating a nuclear test ban treaty in Geneva, Switzerland. One crucial question was how to distinguish reliably between underground nuclear explosions and earthquakes. Tukey (with Bogert and Healy) suggested untangling the multiple reflections from the earth's mantle from the direct shock wave by the "cepstrum" method which meant taking the Fourier transform of the *logarithm* of the power spectrum of the signals received by various strategically placed geophones. (Earthquakes generate "shear waves" caused by the movements of tectonic plates. By contrast, explosions generate "compressional waves" which travel faster in the earth mantle, leading to measurably shorter delays between successive reflections.)

Being always close to Tukey, as a kindred (mathematical) spirit, I soon heard of the cepstrum idea and thought it was just what was needed for pitch detection for speech signals, namely to untangle the formant frequencies (the resonances of the human vocal tract) from the fundamental frequency, the pitch periods, the delays between successive flaps of the vocal chords. This was in the summer of 1962 and I had to attend some meetings in Europe, so I assigned the problem to one of our interns, A. Michael Noll. When I returned to Murray Hill a few weeks later, Mike had solved the whole pitch problem. The cepstrum has since found numerous other applications in speech coding and of course geophysics. Mike later became one of the pioneers in computer graphics and I was very glad to have kept him on—even without a Ph.D. degree, which was *de rigueur* under the guidance of Bill Baker. (Usually, interns are returned to the home departments after a year or so.)

20.6 Dave Slepian, Jessie MacWilliams, Ed Gilbert

Other mathematicians whose advice proved invaluable were Jessie MacWilliams, who taught me finite fields (which I needed for the design of diffusing surfaces in concert halls) and Dave Slepian, who along with his wife Jan became lifelong friends. Calculating reverberation times of acoustic enclosures (auditoria, opera houses, etc.) is notoriously difficult. According to Dave, one has to solve an "impossible" integral equation. But a day later, Dave brought mathematician Ed Gilbert along who had the brilliant idea of turning the integral equation into *two* equations which could then be solved iteratively on the computer. It worked like a dream and David Hackman, a "cooperative student" from Columbia University, obtained some beautiful results showing the importance of absorber *placement* within an enclosure. (Dave Slepian, one of the pioneers of information theory, passed away on 28 November 2007—a great loss, also for Anny and me personally.)

20.7 Ron Graham, Neil Sloane, Roger Shepard

Another example of cross-discipline interaction at Bell was the work on Hadamard coding of two-dimensional images that I did with mathematician Neil Sloane (and which earned me the Erdös-number 3, meaning that Sloane had coauthored a paper with Ron Graham, who had written many papers with mathematical legend Paul Erdös).

Bell Labs provided, in fact, a sheer inexhaustible pool of helpful talent. How could I have done our work on concert hall quality without the scaling methods of psychologists Roger Shepard, Doug Carroll, and Joe Kruskal. George Sperling, whose field was human vision, became a lifelong friend.

20.7.1 Claude Shannon and Dick Hamming

The most famous of all the Bell "mathematicians" was (actually an electrical engineer) Claude Elwood Shannon (1916–2001), inventor of "information theory." (Shannon actually called it *communication* theory.) My contacts with him were relatively sparse but we were standing daily at lunch time in the same cafeteria chow-line which gave me, the novice, not a little moral uplift. Many years later, when I was a bit better known around the Labs, he occasionally consulted me on some acoustics problem or other. (I remember he once came to my office with a funny-looking little trumpet and asked me whether I could calculate its resonances from its shape.)

Another classical Bell person from whom I learned a lot was Richard Hamming (1915–1998), father of the Hamming distance and inventor of the Hamming error-correcting codes. Dick was also a great aphorist. Here is one of my favorites:

> In science, if you know what you are doing, you should not be doing it.In engineering, if you do not know what you are doing, you should not be doing it.

20.8 Bela Julesz

Bela Julesz (1928–2003), who invented the random-dot stereogram and used it to show that human binocular depth perception does not require recognizable shapes, became one of my closest personal friends at Bell.

Bela and his wife Margit had fled their native Hungary in November 1956 during the ill-fated uprising against the brutal Soviet occupation of their country. Already in December of 1956 we were able to welcome Bela to Bell Labs where he stayed and worked for the next 32 years. Bell gave Bela the freedom—and the means—to develop and test new theories of visual perception. Later, after having declined honorable offers from the Swiss Federal Institute of Technology, the University of Zurich and Rockefeller University, Bela became the State-of-New Jersey Professor of Psychology at Rutgers University.

Bela received numerous high honors, including the MacArthur Fellow Award. He was elected to membership in several prestigious Academies, including the Göttingen Academy of Sciences, intellectual home to Gauss, Hilbert, and Heisenberg. Bela was the quintessential Bell Labs researcher who inspired us all and made it such a joy to work there.

For his friends, Bela Julesz will live on as a great teller of intelligent jokes—always delivered with a grave demeanor and a charming Hungarian accent. One of my favorites concerns the Hungarian officer at the end of a pleasant night in a Vienna hotel who is about to leave without having paid his "dues." When the young lady of the night makes the gesture of paying, the great lover waves her off and, with great modesty, says "Ungarrischer Offezerr nimmt kain Gelt."

Once Bela invited us to an "exhibition" in a basement in Greenwich Village that featured a "dead admiral" lodged in a large aquarium. On closer inspection, the body was revealed as consisting of chicken and other nonhuman parts—a gruesome sight, nevertheless. There was also a live goat present and a New York policeman guarding the goat from being "violated," as the organizers had promised. This must have been in the late 1960s—unhinged times indeed!

I am forever thankful for the many good years I spent at Bell Labs and for the superb people I was privileged to know there. Of course I was too young to have much interaction with the old icons: physicists John Bardeen and Walter Brattain (co-inventor of the transistor) and the mathematicians Hendrik Bode and Serge Schellkunoff. (I only got to know Walter Brattain personally after his retirement at an *Institut de la Vie* meeting in Paris.) But I was once engaged in a joint project with Harry Nyquist, who discovered the important (for digital communication) sampling theorem. And I thoroughly enjoyed my friendships with the next generation: the physicists Sid Millman, Conyers Herring, Phil Anderson, G.H. Wannier, Bernd Matthias, Harry Suhl, George Feher, Warren Mason, Herb McSkimin, Bob Thurston, Albert Clogston, Phil Platzman, Pierre Hohenberg, Veit Elser, Chandra Varma, Günther Wertheimer, and Horst Störmer. (Matthias, Suhl and Feher later went to San Diego to form a "Bell Labs West.")

In 1966, I introduced Wolfgang Eisenmenger from Göttingen to Bob Dayem, who together launched the field of phonon spectroscopy.

I also had friendly relations with the electrical engineer Bob Lucky, an erudite author, and the gifted mathematicians Elwyn Berlekamp, Henry Landau, Larry Shepp, Andrew Odlyzko, Ingrid Daubechies, and Henry Pollak.

In the 1960s, I joined Larry Shepp and Conyers Herring for the weekly Russian Table, presided over by Astrid Werner, Bell Labs' chief Russian translator. Ingrid

Daubechies, a Belgian native, presided at the Flemish Table where I was able to dust up my Dutch. In the early 1960s, Dave Slepian and I founded an Italian Table. Thus, I sometimes spoke five different foreign languages at lunch on the five days of the week—Mondays: French, Tuesdays: Italian, Wednesdays: German (where I was one of the "experts"), Thursdays: Russian, and Fridays: Flemish. (At work we spoke English—or some approximation thereof.)

Sid Millman was Executive Director of Physics Research and Director of University Relations. After I became a professor in Göttingen, I had lunch with Sid every time I spent a stint at Bell Labs—a mutually instructive habit. According to Bell (now Stanford) physicist Ted Geballe, Sid, who was born in Russia in 1904, was only the second Jewish scientist to become a member of Bell Labs—hired in the late 1940s by James B. Fisk, a great gentleman of science who later became president of the Labs.[5] (Sid died on November 11, 2006, aged 102.)

Other people who were of great help to me were Floyd Harvey, who was delegated to teach me the ropes when I first began working at the Labs. Bill Kefauver, Harry Hart and Al Hirsch were my capable patent attorneys. Herb Hines, assisted me in building some of my more complicated electronic circuits. Jimmy Kronmeyer was a superb Old-World style instrument maker, as was Charlie Mattke, who invented an ingenious compact rotary dial to fit the new handheld *Princess* telephone.

One of the department heads who reported to me was Warren Mason, who held over 200 US patents and who invented the sharply tuned crystal filters—crucial for carrier telephony, i.e., the transmission of many voice channels over a single copper wire (or glass fiber). His invention of highly stable quartz-crystal oscillators are used in the ubiquitous quartz watches.

Robert L. Wallace headed another of my departments, developing highly directive microphones for conference telephony. In his early days at Bell, Bob invented the *tetrode transistor* which however was never used commercially. He and his brother owned a small telephone company in Texas which provided a sturdy extra income to his family. (The original capital came from the Roosevelt administration's endeavor to provide telephone service to outlying rural regions—at 2 % interest. Ma Bell did the billing for the small fry for free.)

Throughout my career as a research "executive" at Bell, I was greatly aided by a succession of very able secretaries: Ginny (in 1958—I forget her last name), Mary Porecca (when I became a director in 1963), who stood by me in difficult times, Dorothy Caivano, and finally (when I was commuting from Germany) Esperanza Plata, a Colombian native, who once called me the best boss she ever had.

[5] Richard Feynman, one of the greatest physicists of the twentieth century (and indefatigable pranksters), couldn't even get a *summer* job at Bell on his first try in the early 1940s. But AT&T, the "mother" of Bell Labs, was by no means alone in its anti-Semitic policies among major US corporations until after World War II.

Fig. 20.1 Anny and Manfred (on his first car) (The Plymouth was manoeuvred onto the sand for this picture and got promptly stuck, but a helpful resident got his truck and pulled us out.) Long Beach, Long Island, Columbus Day 1954

My books were typed by professional typists from the Labs: Elvire Hung (*Number Theory*), Hildegard Franks (*Fractals, Chaos*), and Martina Sharp (*Computer Speech*).

Another friend was Floyd Becker. When I asked him on October 13, 1954 how he had spent Columbus Day (October 12), he said "painting my house." I got the same answer from Floyd Harvey and perhaps others and I decided never to ask again what people did on a warm, beautiful day off. Anny and I drove (in my "brand new" car) to Long Beach on Long Island for a swim and some sunbathing on an otherwise deserted beach—5 weeks after Labor Day, but still very summer-like to a recent immigrant from northern Europe (Fig. 20.1).

Besides New York State (the Hudson Valley), early excursions took us to New England for skiing and Florida for a Christmas vacation (Figs. 20.2 and 20.3).

20.9 My Early Days in the USA

Besides having to adjust to new technical fields, the social adjustments also were not always easy. I was 28 when I joined Bell and still unmarried. This was somewhat unusual at the time and my Executive Director, Bill Doherty, introduced me to one and all with "This is Dr. Schroeder. He just joined us from Germany—and he's a *bachelor*." Or, as my old math professor from Göttingen, Wilhelm

Fig. 20.2 Anny in Florida, December 1955

Magnus, then at New York University, put it: "This is not a country for bachelors. Either you will return to Europe within the year or you will get married." Well, he was right, I *did* get married 17 months after my arrival in the USA.[6] My Best Man was Ed David, who taught me how to grill steak properly (among many other American essentials), and, from 1972 to 1974, was ensconced in the White House as a presidential Science Advisor before Richard Nixon abolished the post. (Although the President had assured David of direct access to him, he actually never saw Nixon during his entire tenure—which may have been a blessing in disguise at the height of the *Watergate* scandal.)

[6] Win Kock asked me whether my bride was a "professional." I firmly swore *No*—with my still limited knowledge of English I thought *professional* was an allusion to the "worlds first profession."

Fig. 20.3 Cannon Mountain, New Hampshire, April 1955

20.10 Columbus Circle (About 1954)

Tuesday night, the regular session of the "Village Camera Club" was concluded. I left Bank Street (No. 673) to retrieve the old Plymouth from its street-side parking place: A few turns, West 4th Street (or was it 12th?), headed for Anny's place on West 85th Street. I had once made the run north as far as 72nd Street without a red light and I was determined to beat that record tonight. Would I ever reach 85th in one full swoop?

The lights on the Avenue were synchronized and the traffic not too heavy so that one could stay comfortably "in sync." But, oh my, the lights on Columbus Circle turned red just as I had crossed 58th Street, while beyond, Broadway, was all green. How could I reach these greener pastures which might carry me on to 85th? Obviously only by ignoring the ignominious red of the intervening Circle! Of course, one had to watch for other cars—who wants a mid-Circle collision?

Everything seemed to be going well with no apparent imminent collision when, suddenly, there was the traffic officer—whistling in the road. He approached the (red-faced?) offender who was ready to receive the well-earned summons. But this was New York City—Mid-town Manhattan—not Summit, New Jersey, or Zanesville, Ohio, or what other explanation would there be for the ensuing pronouncement by the "arm of the law": "Mister, if you go on driving like that, you will—one day—end up with a ticket!"

Just then the Circle lights changed to green again and the speedy traveller was on his way again.

20.11 January Night 1955

Anny had to go to the Ladies' room once more before resuming our trip to Florida. It was Christmas Eve and we had "celebrated" the evening in a roadside EAT place in Bamberg, South Carolina. Anny had given me her handbag—containing all our cash and papers—to hold while she disappeared. I wanted to take a few more pictures of the neon signs and deposited the handbag on the car! bumper . . .

Half an hour and twenty miles later, Amy needed something from her bag. Where was it? On the bumper! But the bumper was unencumbered by handbags (or anything else) . . . Luckily, the Severins had enough cash to tidy us over our week in Florida.

Upon returning to New York City, Anny found a letter in her mail: ". . . when I was driving down U.S. 301 (?) with my girl friend on Christmas night, I saw a handbag lying in the middle of the road. It contained papers with your address. What shall I do?" Keep some of the cash and return post-haste!

In the meantime, I was driving without a license (or any other identification, for that matter). Of course, one hardly ever needs to show these papers—especially, if one drives carefully, obeys all traffic lights, etc.

One very cold Saturday, we were returning from the Bronx when, watching all the lights ahead of me, I went, smack, through one right in front of me. And, would you believe it at 2 a.m., the red light came with a real live blue cop.

"License, please!" "Yes," I said, hoping the inconvenient apparition would disappear again before my bluff was called.

"Your license, please." "yes, . . . I have no license." "Show me your registration." "Yes." "Your registration, please, mister." "Yes . . . I have no registration." "Show me anything!" "Yes. . . I have nothing . . . This is my girl friend." "I see." Then I saw the Bell Lab pass with my name and picture on the car's sun visor. "I have this, officer." "That's you all right, but it doesn't mean anything. For all I know, you stole this car." "Yes, officer, that could well be. But it is my car. I lost all my papers in Florida—in my girlfriend's handbag." "Good story . . . You know how cold it is outside?" "Yes, officer, about 20 below." "Do you know that I have been on these feet for 2 hours now and that I am frozen stiff?" "I can imagine." "You are the best thing that has happened to me tonight. We'll drive to the station house and I'll get a chance to warm up, while we book you." "You are very lucky, officer. I wish I was as lucky as you tonight."

Pause. Pause. And then the voice of the officer: "Go on. It's o.k. But watch those traffic lights. Good night!" "Good night, officer. And thank you!"

20.12 One-Way Street

The show in the old Phoenix Theater was over (Had it been Marcel Marceau?) and Anny and I dashed across Second Avenue to retrieve our old Plymouth from an even older garage. It had started snowing and we were eager to reach our safe haven in New Jersey without a second to delay.

As usual, it seemed like an eternity until the familiar green vehicle was disgorged by the cavernous storage establishment. We hopped in and off we went across the Avenue and into the next side street where, one third down the block a policeman emerged from below a lantern and stopped us. What was wrong? The snow was getting thicker and we were in a hurry! The officer approached my side of the car and lowered his head (if not his voice) to speak through the rolled-down window one short sentence which explained it all: "Sir, if you would switch on your headlights, you would see that this is a one-way street—in the other direction!" Whereupon the member of New York's finest marched toward Second Avenue to hold the traffic so I could back into the Avenue without harm. Thank you so much, officer!

20.13 Spy Stories

One of my more memorable encounters with the police took place not in New York but in Hoboken, New Jersey, in November 1954. It was night and the full moon was rising over Manhattan to the east. So, being an avid photographer, I took my tripod and a very long lens and was clambering around the abandoned piers when I was stopped by a bunch of longshoremen. A citizen arrest, if you will. They alerted the Hoboken police that they had caught a Russian spy with a very long lens and a thick accent. (The McCarthy days were still in full swing.) The police soon appeared on the scene and I explained to them that I was just a "shutter bug," and it so happened that I had two boxes with me, full of pictures I had taken: landscapes, flowers and—girls. Well, the police spent the next hour or so looking at my photographs and apparently completely forgot about the serious charge levelled against me. When they had finished their "examination" of my "art," they were convinced that I was indeed just a harmless amateur and they sent me home with their best wishes for my future work.

This did not seem to please my apprehenders; they felt I had "snowed" the police. So they reported me to their management, which alerted Bell Labs, where my executive director, Bill Doherty, vouched for me. Case closed—or so you might think. *Eight* years later, during my naturalization proceedings in Newark, the immigration agent asked me: "What were you doing during the night of November 10th in 1954 on the Hoboken piers?" Well, Bill Doherty had to rescue me again and I was sworn in as an American citizen in good standing on June 10, 1963—ready to leave, on June 14th, for the Soviet Union as a US national. (But that's a different story; see my vignette on the KGB.)[7]

Lest I leave the reader with the impression that I always got away scot-free, here are a few counter-examples: I was once fined in New Providence, New Jersey, for supposedly not coming to a complete stop at a stop sign. And at another time I took an illegal left turn on 14th Street in New York City onto Park Avenue and was caught (by an unmarked police car). I felt the summons served me right because there was a large sign saying "NO LEFT TURNS."

Shortly after I arrived in the USA. I was caught speeding on the New York Thruway. I don't know how fast I was going—the needle of the speedometer was pegged at 110 miles per hour. But I am sure I wasn't going much over 100 miles; in fact, it didn't feel all that fast for a recent arrival from Europe. When a State Trooper caught up with me, he said it took him some time to catch up. But he only charged me with doing 95—the judge wouldn't believe him if he put in my actual speed. When I later appeared before a justice of the peace in New Paltz, New York, the kind man said: "The police must have been exaggerating. You couldn't have done more than 85. But I must warn you: "Speeding on the N.Y. Thruway is considered a *felony* not just a misdemeanor. And this may endanger your future application for US citizenship. So I'll put you on $5 bail and you consult with a lawyer in Summit, New Jersey, where I was living at the time, if you should plead guilty or not." But he forewarned me: "Our juries are wont to believe the police. If you plead guilty, I'll keep the $5 as a fine." My Summit lawyer (Hugh Hartlaub) felt that I should plead guilty and that there wouldn't be any long-range consequences. And there were none.

Lest some reader think that my friendly encounters with the police are exceptional, here is an excerpt from a *Dear Diary* letter by Noel Bacchus to the *New York Times* (Metropolitan Diary, October 2, 2006):

> A sunny Wednesday afternoon following a washout of tennis the previous day. We're heading for the Billie Jean King Tennis Center in Flushing Meadows-Corona Park, anticipating a day of competitive tennis. Three seniors: my brother visiting from Chicago, a friend from Mexico, and me.

---

[7] Here is what Doherty wrote to the Immigration and Naturalization Service:

As director of research in electrical communications, I was responsible for the selection of Dr. Schroeder in 1954 to work in our company as a result of strong recommendation by his professor at the University of Göttingen. My recollection of the incident in question is that I received a telephone call from the General Manager of our company, indicating that he had heard about Dr. Schroeder being observed taking pictures from the Hoboken waterfront. I immediately discussed the incident with Dr. Schroeder, and he showed me the pictures as soon as they could be developed. The pictures could not conceivably have been for anything except artistic purposes. I can well imagine that the persons who observed Dr. Schroeder might have been suspicious because of his telephoto lens. He has followed a long-time hobby of amateur photography, in which such a lens is increasingly common, and he has received awards in our company contests. I trust that this information may be of assistance in expediting Dr. Schroeder's petition. If it is possible to have this matter come up for final hearing on June 10, 1963, Dr. Schroeder would be able to commence his travel to Europe on government research business on June 14 as a United States citizen rather than a resident alien. This result would be in the public interest as well as in the interest of scientific research.

Fig. 20.4 Marion, Anny, Alexander, Manfred, and Julian in our first house in Gillette, New Jersey, September 1963

Crossing the boardwalk that extends from the No. 7 subway station to the entrance of the tennis center, we are frustrated trying to remember the words and tune to an old song, "Under the Boardwalk." Spotting a police officer—female, African-American, of a certain age—I impulsively address her.

"Good day, Officer, how are you?"

"Fine, thank you."

"Do you happen to know the song 'Under the Boardwalk,' by the Drifters?"

Looking a little surprised, she responds, "Of course!"

"Would you mind singing a bit of it for me?"

A big smile. "Are you serious?"

"Yes, Officer."

Without further ado, snapping her fingers, she sings quietly: "Under the boardwalk, out of the sun. Under the boardwalk, we'll be having some fun..."

20.14 Family Development

Not long after our wedding (in the Chapel of the Riverside Church near the Columbia Campus), Anny and I were blessed with three children: Marion Frances (born in 1956), Julian Ivan (1958), and Alexander Hilbert (1960). Marion's and Julian's middle names are based on the given names of Anny's parents (Franz and Ivanka Menschik). Alexander's middle name is that of the famous Göttingen mathematician David Hilbert (Fig. 20.4).

Here are two more snapshots from our years in Gillette (Figs. 20.5 and 20.6):

And here are our children in 1985, after we had moved to Germany (Figs. 20.7 and 20.8):

Fig. 20.5 Marion (age 7) Gillette, New Jersey

20.15 Stephen Oswald Rice

Another thing that surprised me was that people were not only very accessible but, I thought, quite humble like S.O. Rice of (mathematical) noise fame, one of my heroes even before joining Bell Labs. His office was at the old New York headquarters at 463 West Street. I had always imagined that such a renowned mathematician would reside in a palatial office with lesser people scurrying around him. But no, Steve was actually *sharing* an office with several other people who treated him like an ordinary human being. And then, a little later, we all went out together, on *foot*, for lunch in a little restaurant in Greenwich Village. I thought someone as famous as Rice would be carried by a company limousine .... Well this was just one of my European prejudices that took some time to subside.[8]

Incidentally, after lunch, when I returned to my car, which was parked illegally under the West Side Highway, I saw a cop writing a ticket. When he saw me, he apologized profusely but since he had already filled in my plate number he couldn't issue it to someone else. But he urged me to leave the car in the illegal spot for the

[8] There is now (2006) a *Stephen O. Rice Prize* by the IEEE Communications Society.

Fig. 20.6 Anny in her in first US Passport Photo, 1959

Fig. 20.7 Alexander (25) and Julian (27) in our Göttingen living room on a visit to Nikolausberg

Fig. 20.8 Marion (29) on a visit to Nikolausberg

whole day—the ticket was good for 24 hours! He was visibly disappointed when I told him I had to return to New Jersey right away.

As auspicious as my beginning at Bell may sound, it was, in retrospect, rather pedestrian. My first speech research job at the Labs, suggested by my supervisor Ralph Miller (who just turned 100 as I write this—March 10, 2007), was *inverse filtering*: you spoke a vowel sound into a microphone displayed it on an oscilloscope and adjusted tuneable filters until the formant "wiggles" disappeared and a relatively smooth waveform remained, which was a pretty good approximation of the glottal waveform. Neat! As I learned much later, as I was picking up more electrical engineering expertise, I was putting "zeroes" on top of the "poles" in the complex frequency plane. Of course, if the speech signal is the least bit noisy, inverse filtering will only lower the signal-to-noise ratio. (In order to *increase* the SNR, you have to use *matched* filters, just the opposite of inverse filters, piling poles on poles.) So inverse filtering is no good for pitch tracking for low-quality telephone signals, which was one thing we were trying to do. But it was an interesting introduction to speech signals.

My next (self-inflicted) project was a 6-channel vocoder ("voice coder"). Even though a complete novice, I felt that 10 or 16 channels for a vocoder was more than was really required, I figured that six channels would be enough. Indeed, I thought single-tuned filters would be sufficiently selective, if I added some "lateral inhibition" (as in the human eye) to my filter bank: subtracting from each filter a small fraction of the outputs of its two neighbors. It took me many months to build this contraption because I had to wind my own coils. Imagine a Ph.D. in physics, winding coils and spending his days soldering sickly circuits—the technical assistance that I had been promised *presto* did not materialize until a full year later (the very capable Tony Prestigiacomo, who later shortened his name to Presti).

Be that as it may, after all the manual labor and mental gymnastics I had invested in my vocoder project I was convinced that it produced not only *intelligible* speech, but high-quality, human-sounding speech. Thus, I alerted my supervisor, who alerted our director, Win Kock, who—if you can believe it—alerted the president of Bell Labs (Mervin Kelly). They all filed into my lab and the demonstration began.... In retrospect, I am convinced they hardly understood a word of what my vocoder said. But everybody was very polite and all left with best wishes for the future of my project, an early attempt at speech compression, which later became crucial for the Internet and mobile telephony (cell phones).

Once, a cooperative student from MIT, Tom Crystal, played me *his* speech compressor in a noisy, fan-infested, lab. He thought it sounded perfect and so did I. So I suggested that we listen in a very quiet location and I put earphones on—and it sounded horrible. Recently, Fumitada Itakura reminded me that when I first visited him in Japan in 1968, I followed the same procedure. But Fumitada's "maximum-likelihood" compression still sounded pretty good, though not as good as at the International Congress on Acoustics in Tokyo, where our two papers, Fumitada's paper and mine (on *Adaptive Predictive Coding*) were presented back to back on 28 August 1968.[9] (Little did we realize then that our two methods were actually identical. Yes, Linear Prediction and Maximum-Likelihood (or Maximum-Entropy) exploit exactly the same redundancies in a speech signal.)

Another horrible contraption I built was an *analogue* Fourier-synthesizer capable of generating periodic waveforms with 31 harmonic frequencies. The gadget, built by the able Herb Hines, comprised almost 1,000 little trimmers (variable resistors) and a number of highly unreliable transistors (this was around 1956!). Nevertheless, I got some nice waveforms that helped me solve the problem of minimizing the peak-factor of periodic waveforms by choosing the proper ("Schroeder") phases and making synthetic speech sound more pleasing.

Before I had my analytic formula, I collaborated with Vic Vyssotsky (who later became Executive Director of our Division) on a computer sorting program to find optimal phase combinations. Vic was also a great hiker and I once went on the Appalachian Trail with Vic.[10]

Manipulating just the phase angles of a periodic signal while keeping its power spectrum fixed, I could elicit some astounding perceptual effects including *flat-spectrum* intelligible speech (published with H.W. Strube). Later Schroeder phases

[9] An earlier presentation of our work on predictive coding was at the first IEEE meeting on speech communication in Boston, 1967.

[10] Vic also fled to Brazil for a couple of years around 1960 because he was afraid the Northern Hemisphere would soon become unliveable, contaminated by the radiation fall-out of an impending nuclear war.

became important in hearing research—in studying the motions of the basilar membrane in the inner ear.

I also concocted cross-correlation and autocorrelation vocoders. The autocorrelation function squares the amplitude spectrum. To undo this spectrum squaring, I employed square-rooters, one for each formant range. Only much later did I learn that the proper way to do this is via taking the *inverse* of the correlation matrix (as is done in linear predictive coding of speech and present today in every cell phone).[11]

On a more fundamental level, I was intrigued by the relationship between vocal tract shape and its resonances. Of course, a point emphasized by the Swedish mathematician Borg, who later became president of the Royal Institute of Technology (KTH) in Stockholm, the problem is ambiguous: different geometric shapes, as the ventriloquist knows well, can produce the same acoustic output. (The ventriloquist can produce intelligible speech with his lips unmoving.) But by measuring the *impedance* of the vocal tract, as seen from the lips, I was able to obtain unique area functions from acoustic data.

20.16 Max Mathews

Incidentally, the loathsome analogue-circuit building days were soon over. In 1955 Max Mathews appeared on the scene coming from MIT. He taught us that every circuit we were building could be seen as performing some mathematical operation and as such, if it could be digitized and be done by computer. In fact, John Kelly, Carol Lochbaum, and Vic Vyssotsky wrote a block-diagram compiler, called BLODI-Compiler, that could be used by computer dummies like myself to simulate almost any electronic circuit. Carol also worked with Max Mathews on the design of digital fonts—a novelty at the time.

Although I had been raised in Germany under Hamming's motto "Don't compute—think!" I was soon one of the heaviest users of digital simulation including, with Ben Logan, of a "Harmonic Compressor" that comprised 400 narrow band-pass filters and could be used to slow down natural speech. We made this available to the American Foundation for the Blind which used it in their "recorded books" program.

---

[11] To test the intelligibility of my synthetic speech, I asked people walking by in the corridor outside my lab whether they could understand it. I said "One—two—three—four—five..." over the device and, yes, they understood every word: One—two—three—four—five... Of course, people could *guess* what was said just from the very *rhythm* of the utterance. But I honestly believed my correlation vocoder produced intelligent speech! (This is reminiscent of a similarly silly procedure adopted by prisoner-scientist Rudin in Alexander Solzhenitsyn's *First Circle of Hell*, when Rudin wanted to convince himself that frequency division by a factor of 32 would produce intelligible speech—an outrageous claim, as Solzhenitsyn, an engineer himself, no doubt knew. As test material Rudin chose the first paragraph of lead articles from *Pravda*, entirely predictable boilerplate, invariably beginning with the words: "The Presidium of the Central Committee of the Communist Party of the Union of Soviet Socialist Republics ...").

Of course, the simulation didn't run in real time (this was in 1961) and some eyebrows were raised over my simulations that took several hundred times real time. Even earlier, in the late 1950s, I had started simulating frequency responses of concert halls: a complex Gaussian process in the frequency domain. To accumulate enough statistical data on some parameters that couldn't be calculated analytically, I let the computer run entire weekends. At $600 an hour, the cost was of course astronomical. But the only financial consequence to me was that my batch-processing budget was raised. Talk about "the good days" at Bell Labs! I never had to write a single research proposal in all my years at Bell—just do it and, if it made sense, the money would be provided, unless it was "additions to the plant," which were strictly budgeted. What helped of course was that, before "divestiture" (1984), AT&T—within reason—could write off Bell Labs as a business expense.

After the frequency-domain simulations, I began concert-hall simulations in the time domain, in other words: convolving the music signal with the hall's impulse response. The final output was of course reverberated music. When we first did this we had a little amplifier with a loudspeaker standing behind the digital-to-analog converter to monitor the computer output. When the people who ran the converters heard the music coming from their machines they thought some prankster was trying to fool them. But no, the music *did* come out of the computer—something entirely new in the 1950s.

Later I began simulating 16-channel vocoders, again in very "unreal" running times. Some people thought I was crazy. They had envisioned simulations for simple waveform coders, such as delta modulation or differential pulse-code modulation. But that was Bell Labs research in the years gone by: you could do "crazy" things if you so felt (and didn't mind a few raised eyebrows). A good case in point is Peter Denes's successful advocacy of *dedicated laboratory computers* for speech research—in defiance of considerable doubts by higher management who at the time (ca. 1964) were still wedded to "computing centers" with big machines.[12] In fact, we sometimes had to travel to IBM headquarters on Madison Avenue in New York City with a *trunk full* of punched cards representing a single sentence of speech (3 s). (One of the problems was the necessary double parking on 57th Street while we were unloading the digitized goods.)

The nearness to New York City also allowed us to attend many interesting meetings: the Audio Engineering Society, the Acoustical Society of America, various IEEE (Institute of Electrical and Electronics Engineers) sessions and many meetings of the American Physical Society.

I still remember with awe the "Woodstock of Physics" in March 1987 at the New York Hilton when high-temperature superconductors were first publicly presented. The session was attended by more than 2,000 people and lasted until 3:15 a.m. the next morning (I left before midnight). Many physicists followed the

[12] I remember Vice President Baker thanking me for having "insisted" on the need for small on-line computers when leaving our demonstration lab. ("Thanks for your insistence, Manny [*sic*]" etc.)

proceedings on video monitors set up around the hotel after the grand ballroom had been quickly filled. Drs. Alex Müller and J. Georg Bednorz later collected a Nobel prize for their pathbreaking work. (Müller and Bednorz had published their results first in a German physics journal to little fanfare before their paper was "discovered" by the rest of the world.)

On other occasions I saw and heard the great physicists Wolfgang Pauli ("Einstein's successor" at Princeton) and Robert Oppenheimer (leader of the wartime atomic bomb project) shortly before their untimely deaths (in 1958 and 1963, respectively).

20.17 Linear Predictive Coding

By 1966, I had become totally disenchanted by the speech quality of vocoder speech. I thought that instead of the "rigid" encoding of the speech signal practiced by vocoders, we should look for methods of speech coding that left "room for error." At this point I remembered the work of my friends in the picture-coding community, especially an engineer named Ernie Kretschmar, a refugee from Nazi-Germany—in fact the Münsterland, my native neck of the woods. The method of choice in video coding was of course prediction: point by point, line by line and frame by frame. I thought something like that should also work for speech signals, except that the predictor parameters should change with the speech sound: adapted to the different spectra of different speech sounds. That is why Bishnu Atal and I called our first paper *Adaptive Predictive Coding*, later called *linear predictive coding* (LPC[13]) and CELP (for code-excited linear prediction) which allowed the excitation signal to be transmitted by ¼ bit (!) per sample, or roughly 2 kb/s, an astonishing compression factor made possible by perceptual coding exploiting the masking properties of the human ear. J.L. Hall and I measured the masking of (quantizing) noise by the tone signals—the opposite of the usual paradigm in which masking of a signal by noise is studied. Perceptual coding is also used widely for bit-rate compression of music (MP3, etc.).

Code-excited linear prediction can also be considered a kind of *voice-excited* vocoder—hence its high speech quality. Apart from the technical breakthrough that LPC represented, its genesis is a beautiful example of the cross-fertilization that was possible, even encouraged, at Bell.

One of the more ubiquitous results of speech compression is the wide-spread use of cell phones and Internet telephony—unthinkable without compression.

[13] Correspondences with Bishnu Atal and J.L. Hall during the development of LPC are archived at the University Library Archives of the University of Göttingen.

20.18 Computer Graphics

An illustration of the freedom we enjoyed at the Labs and the good communication between researchers is our computer art work. I had already used the computer in the late 1950s to produce artificial reverberation and concert hall simulations, and generally I was very interested in any non-numerical application of the digital computer. So I was immediately attracted by the possibility, suggested by Leon Harmon and Ken Knowlton, to use the computer cum plotter to generate graphic output that no human could ever hope to draw by hand, like their rendering of a nude Debbie Hay by electronic symbols.[14]

Leon became also widely known through his *block portraits*, which looked very different from different viewing positions: from close up, you see a set of squares of different shades of gray; from afar you see a human portrait (Lincoln in Leon's most memorable work). I was likewise much interested in the idea to produce graphic art that changed appearance depending on viewing distance. One of many interests that we shared was the recognition of human faces—by man or machine and the, probably age-old question: "What exactly makes a face *beautiful*?" (Leon, one of my closest soulmates, perished by his own hand in July 1982, several years after he had left Bell Laboratories.)

At about the same time, Mike Noll, Bela Julesz, and others made their own noteworthy contributions. Mike, who joined my research area as an intern in June 1962, became prominent in the computer art world, not least through his aesthetic analysis of Piet Mondrian's *Composition with Lines* by comparing it with several statistically generated Pseudo-Mondrians. His earliest computer art was, in fact, done in 1962. I vaguely remember approving, for internal Bell circulation, his August 1962 Technical Memorandum on using the computer to draw artistic line compositions. Of course, I had no idea of the historic significance Mike's memo would later acquire.

Mike also pioneered computer-generated stereographic movies and *tactile* computer interfaces, allowing geometric shapes defined by software to be physically felt.

Bela, who joined Bell research in late 1959, came to us from the field of visual perception. His thesis, in Budapest, was on optical camouflage. At Bell he became famous for his *random dot stereograms*. Some of his creations had, in fact, the feel of Op Art.

At one of the monthly research directors meetings, instead of presenting the latest progress in speech coding, I showed my new computer graphics—and nobody complained. In 1965 there was an exhibition of Computer and Op Art, including works by Noll and Julesz, at the Howard Wise Gallery in New York City. Computer Art had arrived—as it had in Germany at the same time: Curd Alvensleben

---

[14] The nude study of Debbie was published in the New York Times on October 11, 1967. Harmon's block portrait of Lincoln was incorporated (without permission or attribution) into one of Salvadore Dali's paintings.

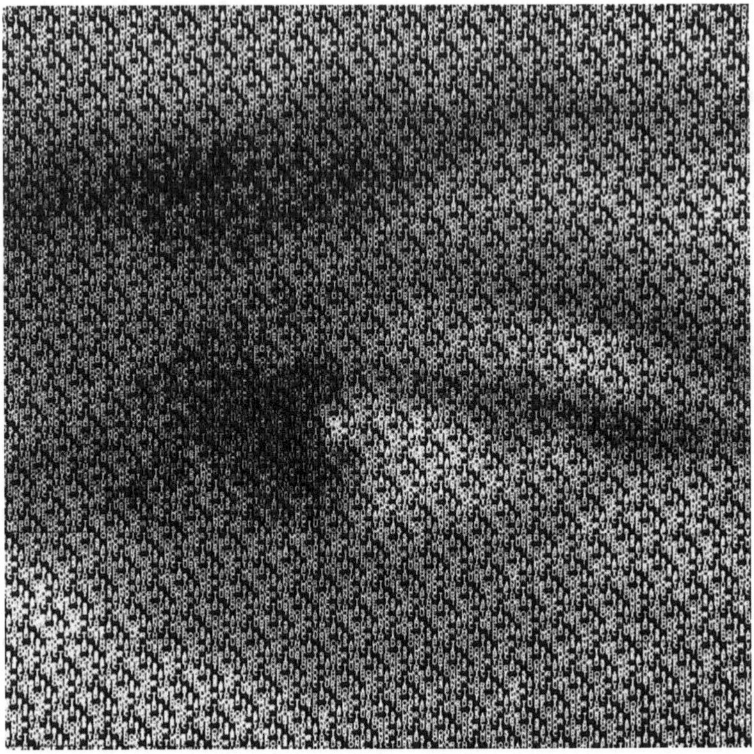

Fig. 20.9 One picture is worth a thousand words

(1962, using an analog plotter), Frieder Nake, Georg Nees, Herbert Franke and others. I even earned some money with my computer graphics—enough for a taxi ride and dinner for four in Zagreb, Croatia.

Here are a few examples of my creations. One design, programmed by Sue Hanauer, won First Prize at the Computer Art Salon in Las Vegas in 1969 (Fig. 20.9):

This computer graphic is a prime example of eliciting different perceptions depending on viewing distance: from close-up one can read the text: "One Picture is Worth a Thousand Words." From a still larger distance one perceives a kind of "weaving" pattern (caused by the periodic repetition of the sentence). And from afar one recognizes a human eye!

Another image, in the same vein, showed my initials (MRS) composed of 65,536 samples of the letter "S" with different luminosities.

Other shows at which my computer creations were exhibited included "On the Eve of Tomorrow," held in Hannover in late 1969 (organized by Käthe Schröder), "Some New Beginnings" in Brooklyn (also in 1969), "Cybernetic Serendipity: The Computer and the Arts" (Jashia Reichardt, London 1968).

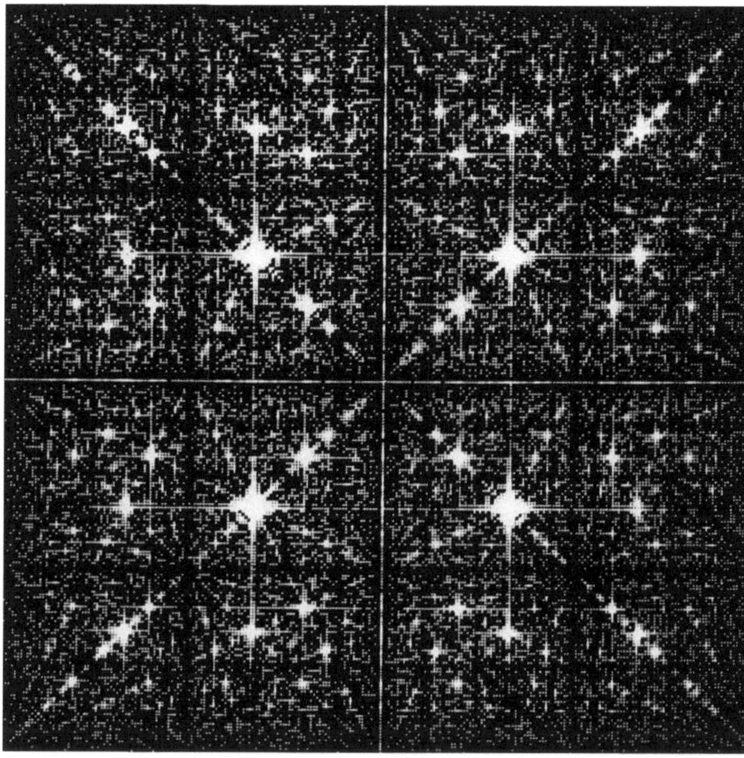

Fig. 20.10 Prime spectrum

Number theory was another inspiration for my computer-graphic endeavors. Here, as an example, is the Fourier transform of the distribution of pairs of relative-prime integers (Fig. 20.10):

Unbeknownst to me, computation of this image required shutting down the entire Bell computer center to marshal the necessary random access memory. (Fortunately, I was away on a trip at that time—or I might have prevented the effort.)

Here is another of my favorites, based on the eikonal equation of geometric optics that governs the propagation of light waves. The image (realized with the assistance of Hans Werner Strube and Wolfgang Möller) becomes recognizable as a human portrait at larger viewing distances[15] (Fig. 20.11):

With Gerhard Sessler and Mohan Sondhi I constructed an "Acoustic Camera," which was capable of reconstructing a physical object from its blurred acoustic shadow. With Mohan, I also did a study of the restoration of sharp images from

[15] I used the Eikonal Portrait as a cover of a book I edited (*Speech and Speaker Recognition*) where it was promptly misinterpreted (by G. Fant among others) as a contour spectrogram of speech.

Fig. 20.11 Eikonal portrait

pictures degraded by linear-motion blur, and of computer ray tracing in the deep ocean. This was in connection with what I called "Volume Focussing" (suggested to me by the self-steering antenna arrays keeping communication satellites focussed on the proper earth station). Volume Focussing is capable of focussing on a specified volume in the ocean, important for the defence against submarines. The surface sound channel, called SOFAR, allows sound waves in the ocean to travel over thousands of miles with little attenuation. In a peacetime application, the SOFAR channel is used from Heard Island in the Indian Ocean to Canada, Australia, Antarctica and California to monitor ocean temperatures (and the effect of greenhouse gases) over vast distances.

Other early pioneers in computer graphics were Ed Zajac and Frank Sinden. Ed made computer *movies* that illustrated the tumbling of a satellite in orbit around the earth in connection with the efforts to stabilize the Telstar® communication satellite. Frank produced many educational computer-generated movies. I will never forget his demonstration of what would happen to a planet if the gravitational law was different from Newton's $1/r^2$. Fascinating!

20.19 Among the Avant-Garde

It was also around this time (1964) that I first met Lillian Schwartz at some (IEEE sponsored?) computer graphics affair in New York City. We immediately took to each other and soon she was working for Bell Laboratories and AT&T, not as a regular employee but as a consultant.

The idea then was to acquaint *bona fide* artists like Lillian with the intricacies of the digital computer. "The computer," we told them, "is just a tool—like a paint brush, albeit more powerful—that you have to learn to use."

This thought was especially propagated by Billy Klüver who went on to found EAT. His "9 Evenings" at a New York City armory is unforgettable and the exhibition at the Brooklyn museum "Some New Beginnings—Experiments in Art and Technology" was a great success. It was co-sponsored by Marion Javitz, wife of Senator Jacob Javitz. Both Lillian and I had several works on display. I was asked to design the poster for the exhibition, an endeavor in which I was assisted by Lorinda Landgraf. (The poster showed a photograph of the Brooklyn Museum composed of letters (of different luminosities) giving the title and venue of the exhibition, the dates and opening hours.)

In my involvement with modern artists, I attended regular meetings in Robert Rauschenberg's downtown loft (planning the "9 Evenings" among other avant-garde activities), where I met David Tudor and other artists. John Cage[16] sometimes came to Bell Labs, where I had the opportunity to demonstrate some of my electronic "music" to him (generated by acoustic feedback through a reverberant hall and a frequency shifter). The composer/conductor Jim Tenney was, in fact, employed by Bell by Max Mathews, as was the French composer Jean-Claude Risset, who became a lifelong friend.

On one occasion, I was privileged to "listen to" Cage's *4'33"* with the (motionless) composer on the stage.

The same performance was also graced by the topless Charlotte Moorman playing her cello. (Later, one of her performances ended with Charlotte burning her instrument.)

In the 1960s, during the height of the Vietnam war, I also had a brief collaboration with Steve Reich when he thought he could use my speech processing equipment for one of his eerie compositions (an endless repetition of the words "Let it all come out") decrying police brutality. (Like Cage, Reich was a pioneer of minimalism, although his music has increasingly deviated from a purely minimalist style. In any case, his compositions have significantly influenced contemporary music.)

New York City was also an ideal venue to see undubbed foreign-language movies—including German films (like *Das Leben der Anderen*—The Lives of Others) which I often found hard to understand without English subtitles.

[16] Cage was perhaps best known for his "composition" *4'33,"* i.e., 4 minutes and 33 seconds of silence (divided into three movements).

Apart from languages, photography —images—have been my premier passion. The pictures that I took as a student evoked admiration: "just like Caspar David Friedrich," the German romantic painter I had never even heard of. (Yes, I was truly uncultured—nekulturnij, as the Russians say.)

When, in the 1950s, I showed my slides at the *Village Camera Club* in Manhattan (attended by such luminaries as Roman Vishniac, Weegee, and Elliot Erwitt), I was applauded for my artistry. One reason, as Norman Rothschild (editor at Popular Photography), told me was that my pictures were "totally original." And the reason for that was that I didn't know of any fashion or trends in photography because I had never been a member of a club of shutterbugs and never studied any magazines that could have given me "fresh" ideas. In fact, I didn't even know that there was a literature on artful photography. So, before joining the New York club, I was completely on my own. No wonder then that my pictures were "totally original": I had no icons to follow. (When I first arrived in the States, I joined the Summit Camera Club. But half their time was taken up by reading minutes and other trivia, so I asked members John Talpey and Paul Kennedy whether there wasn't a more stimulating club in the area and they recommended the *Village Camera Club* on Bank Street in New York City.) One of my main interests was, and still is, *unposed* human portraits—snapshots of people of all ages and genders. At the monthly slide competitions of Bell Laboratories' Murray Hill Camera Club I sometimes won First Prize. But in spite of these successes I never took photography to the heights of my friend Dankwart Köhler who has created many images of breathtaking beauty, now supplemented by digital techniques (see his Internet postings at http://tripod.com/).

In fact, my professional life was largely preempted by research in sound: speech (Linear Predictive Coding), hearing (monaural phase sensitivity, perceptual coding of sound signals), concert hall acoustics (Philharmonic Hall, New York, number-theoretic diffusors), artificial reverberation (virtual sound, soundscapes) and acoustic signal processing in general.

20.20 Philharmonic Hall, New York

In September 1962 Philharmonic Hall at Lincoln Center for the Performing Arts (located on Upper Broadway) had opened to great fanfare with a gala concert under the baton of Leonard Bernstein in the presence of the First Lady of the Realm: Jacqueline Kennedy. But audiences, not least the New York Times music critic Harold Schoenberg, were less than enthusiastic about the new hall's acoustics. For example Schoenberg complained about a persistent echo. I got his seat number and when we looked into this, we discovered that there was indeed an echo. It was easily taken care of. (Ironically, most seats were free from echoes.)

In its predicament, Lincoln Center turned to a trusted friend on *Lower* Broadway: AT&T, which in turn turned to Bell Labs where, being in charge of acoustics research at the time, it ended up in my lap.

Fig. 20.12 "Old buddies": James E. West and George W. Bush (*right*)

A Committee of Four was formed under the chairmanship of Vern Knudsen, Chancellor of UCLA and one of the founders of the Acoustical Society of America. Under the so-called Consent Decree (1956) between the USA government and AT&T, Bell Labs was actually prohibited from acoustic consulting. So it was decided that I would limit myself to making the necessary measurements to analyze the hall. And analyze we did, using computer-generated Hamming-window tone-bursts as an excitation signal and matched filtering on the acoustic output from the hall. I was assisted in this work by Gerhard Sessler, Jim West,[17] Bishnu Atal, Mike Noll, and Carol Bird McClellan, who did the complex computer programming (Fig. 20.12).

[17] Jim West, who was of African and American Indian descent, once, during a scientific talk, warmed my heart by mentioning how considerate I always was about his feeling at home in my area (in which he was the only member "of color"). In 2007 Jim was awarded, in a White House ceremony, the National Medal of Technology for his invention (with Sessler) of electret transducers now ubiquitous in cell phones and numerous other applications. Here is an excerpt from Jim's acceptance speech:

The two billion Electret Microphones made each year have transformed the world of acoustics by providing a simple, inexpensive, very linear transducer for telephones, cell phones, microphone arrays, hearing aids, professional measurements, and outer space communications. The Electret Microphone is the only transducer in mass production that delivers very linear sound reproduction universally in all applications.

It is also important to acknowledge the importance of the electret's engine, the storage of real charge in inexpensive polymers that have been used in air filtration, radiation detection, wound healing, cell development in animal and cell cultures, and piezo-active polymers.

Mr. President, Mr. Secretary, and members of the Nomination Committee thank you for awarding me The National Medal of Technology.

Before starting the measurements, I asked the ushers, students of the Julliard School of Music, whether there was one good seat in the hall and they pointed to Seat A15 on the second balcony as really good. So I decided to include A15 in our measurements. The results showed that in the center of the orchestra floor, there was a loss of almost 30 dB between middle notes (750 Hz) and low notes (125 Hz). By contrast, the loss was only 4 dB at Seat A15! I later demonstrated this effect to Cleveland maestro George Szell who agreed that A15 was indeed superior. Then, after the intermission, the legitimate ticket holder for A15 appeared and Szell and I had to sit on the floor, where it sounded even better to me. But I didn't say anything. Then Szell turned around and said: "Down on the floor here it *really* sounds good!"

What is the *cause* of this lack of low frequencies (which made the celli in *tutti* passages nearly inaudible)? By time-gating the responses on the computer, we were able to isolate different reflections and the culprit was identified as the overhead reflecting panels.

In addition to the poor base response the main problem with Philharmonic Hall (now Avery Fisher Hall) was a feeling of "detachment" from the music. To get at this fundamental difficulty, I was able to persuade the German Science Foundation to underwrite a large-scale study of concert hall quality. My collaborators (Dieter Gottlob and Karl Friedrich Siebrasse) made recordings with a specially designed "dummy-head" in 22 major halls, which were subsequently reproduced in a large anechoic space by means of an electronic filtering method. This was an early example of the now ubiquitous "virtual acoustic images."[18]

The main finding of this study, involving thousands of paired comparison tests, was that for good acoustics, there should be strong early *lateral* reflections—a difficult goal, given that most modern halls are wide and have a low ceiling which favors sound arriving in the median plane of a forward-facing listener's head. To overcome this problem I proposed diffusing surfaces based on number-theoretic principles to be incorporated in a hall's design. The most widely used reflection phase gratings are based on *quadratic residues*. I got the idea during a talk by **André Weil** (brother of **Simone Weil**) on *Gauss sums and quadratic residues* that he gave in Göttingen in 1977 celebrating the 200th anniversary of the birth of the "Prince of Mathematicians."

The accurate measurement of reverberation time also benefited from our work on Philharmonic Hall by using reverse (backward in time or "Schroeder") integration.—Ideally, concert hall measurements should be made with music as an excitation signal. This can be done by measuring the modulation transfer functions, both on the stage and in the audience area and forming their ratio.

[18] Later I became a confidant of Avery Fisher, who had donated $10 million for correcting the acoustics of the hall. Fisher introduced me to the architect **Philip Johnson** who showed me the new design, as suggested by **Cyrill Harris** of Columbia University. (Of course, I couldn't give an impartial evaluation of the design—Cyrill was a good friend.)—Avery invited me once or twice for lunch in the Century Club in New York City. Anny and her sister, Hella, had to wait in the lobby. This was in the days (1975) when the Club was a purely male preserve.

With Atal I also studied sound decays by means of computer ray simulation. The results revealed gross inaccuracies of existing reverberation time formulas and the strong dependence of reverberation time on absorber *location*.

20.21 A Call from Göttingen

In 1969 I received a call from the University of Göttingen as a professor of physics and director of the *Dritte Physikalische Institut*. Of course, Bell Labs was loath to let me go, so—thanks to Max Mathews, John Pierce and Bill Baker—a part-time arrangement was agreed upon for me to spend several months every year at Bell. This agreement held until my eventual retirement on September 30, 1987, exactly 33 years after I had first set foot on Murray Hill soil.

20.22 New York to San Francisco: With Ten Minutes to Spare

In the fall of 1960 I had to give a talk at the Audio Engineering Society at the hotel *New Yorker* in Manhattan. Five or six days later I was scheduled to speak at the Acoustical Society of America in the *Sheraton Palace* hotel in downtown San Francisco. Instead of flying, we opted for driving with our "new" Ford which we had just bought for $100. To waste no time, Anny was parked outside the *New Yorker* hotel on 34th Street—with the engine running. After my talk I hopped into the car and off we went, more or less nonstop, heading west.

All went well until Lake Tahoe when the car wouldn't start. The starter seemed to jam—or something. Anny and I were lost; there was little time to lose to arrive at my meeting in SF on time. Then, seemingly out of nowhere, a kindly "native" appeared and suggested to us recent arrivals to "rock" the car. And sure enough, the starter unjammed and we were on our way for the last leg of our transcontinental trip. After crossing the Bay Bridge, I remember making one or two right turns and, before we knew it, we had reached our hotel—with 10 min to spare before my talk.

After the San Francisco meeting we visited some of the places we had missed because of the rush across the Continent. We went up Mount Tamalpais for a grand view of the Bay, including the Golden Gate Bridge, and we stopped briefly at the historic Point Reyes Lighthouse with a grand vista of the Pacific surf and the beaches.

Of course, in the city proper, we saw many of the "required" sites, the tramway, Market Street, Chinatown, Alcatraz (from afar) and the War Memorial Opera House, which was about to be converted to a concert hall—with totally different acoustic requirements. At the invitation of Joseph Kripps, the new music director, I attended the grand opening in October or November 1963. (My plane was late because of a snowstorm on the East Coast and I had to change into black-tie getup in the men's room at the Chicago airport. Luckily my taxi driver in San Francisco knew the address, the program, the name of the new conductor and that it was Opening Night so I got there just in time—again with just 10 min to spare. How did the driver know all these things? Well, he just loved music and was building his own little organ in his basement.—Some taxi driver!)

On our way back, we stayed for a night or two at Yosemite National Park with its breathtaking views of mountains, waterfalls, rivers and trees. One morning, Anny, wrapped in just a bath towel, couldn't return to our cabin from the communal bathhouse because a big (brown) bear was blocking her way. But after a while the bear got tired of waiting for the lady and ambled away.

20.23 Baja California: A Vacation

On the West Coast, along the beautiful highway 101, we relished Monterrey, had breakfast on fisherman's wharf, drove the 17-Mile Drive, and enjoyed Mount Carmel until we arrived at our destination: Baja California, which we had chosen for our vacation because of its remoteness and low prices. After Tijuana and Ensenada (Bahia de los Todos Santos), we followed Mexico Federal Highway No. 1 toward La Paz and Cabo San Lucas, the southernmost tip of the Baja peninsula. But we didn't get very far. The "highway" became ever wider and more difficult to navigate with potholes that would arouse the envy of the moon's surface. Finally, at the end of the first day, we saw some lights in the distance: probably human beings that could help us. The lights turned out to be from a ramshackle motel with its own little generator, run by an elderly lady and her sons. Well, we spent the night there and, in fact—we liked the place so much—our entire vacation. This motel was right on the ocean and the sons turned out to be fishermen, specializing in langoustines. They were also good hunters and had an extra shotgun so that I could join them hunting rabbits, quails (*cordonices*) and other small fowl. The mother was a great cook so we persisted for the next 2 weeks on our own rich baggings. For sightseeing there was nearby Mount Kenton, which consisted entirely of volcanic cinder. We never reached the peak of this perfectly cone-shaped mount because two steps up always meant sliding back one or more steps. Since Mt. Kenton was close to the sea, it was easy to mine. The cinder, we were told, was ideal for US road construction.

The bay at which Mt. Kenton was situated was also a favored winter vacation spot for hundreds of whales arriving from farther north. Quite a spectacle: like a can of worms hugely magnified!

On our way back to the States, due to the rough road in Baja, the steering broke and the car preferred to go around in circles instead of heading north. Upon investigation, I discovered that one part of the steering mechanism was secured by three bolts, two of which had broken off. So when you turned the steering wheel, said part turned around the one remaining bolt but the front wheels didn't budge. So, although bereft of any automotive training, I inserted a little wooden peg (fashioned from a broken-off branch from a nearby bush) into one of the empty holes and—lo and behold—the wheels became steerable and we were on our way again.

Two stops that I remember were the huge asteroid crater in Arizona and the Petrified Forest. From there we called our friends, Max and Marj Mathews and learned that Anny's mother, Wanja, who had stayed behind caring for our youngest son Alexander, had fallen ill. So Alexander was transferred to the Mathews.

Since we had "seen it all," Anny and I decided to call it a day and drive nonstop back to New Jersey—2,300 miles (3,700 km). While on our trek west we had counted the tankfuls of gas, we now went by the sun: sun down—sun up—sun down—sun up for 63 h. But in Indiana, near Terre Haute, we got so sleepy that we checked into a motel to grab 5 h of sleep, a shower and something substantial to eat.

Outside Zanesville, Ohio, we were stopped for speeding by a patrol car that was in contact with an overhead "spy" plane that monitored the speeds of cars as they went from a 70-mile/h zone on US 40 to some 50 miles/h—rather unfair. But the police officer was very kind: he let me listen to the radio traffic from the plane.

We were hauled before a judge in the Zanesville court house where I was fined $20. But the trouble was we didn't have that much cash left over after the long trip and the county cashier couldn't accept checks. What to do? The alternative was going to jail for a long weekend. We called my bank in Summit but they were closed.

Then I remembered *Ma Bell*, the phone company. Bell employees had been told that, if they ever needed help, including money, they should go to the nearest phone company office and show their Bell Laboratories pass. The police officer volunteered to drive me there but I asked him to park two blocks away: I didn't want to be seen arriving in a police cruiser.

My reception at the phone company was less than cordial. Small wonder: I was unshaven and had a big hole in my sweater and of course spoke with a foreign accent. After seemingly endless questioning the Bell bureaucrat finally agreed to *endorsing* my check for $20 (no cash changed hands). I can still hear the office workers snickering behind my back because their boss "had been had."

Unfortunately, after I had secured the $20, we were back to square one, by my asking the Bell person to endorse a $100 check (to buy food for the weekend) because that really convinced him that something was amiss. Then, after about 30 min, Anny appeared and said the waiting police officer was getting impatient and feared that I had fled by the rear door. I told Anny to placate the officer—that I was making progress and would return shortly.

Thirty-four years later, in 1994, during another safari across the country, we revisited Zanesville by briefly driving off Interstate 70. The town made a desolate appearance with many stores on the main street boarded up. The only remaining industry appeared to be ceramics (pots and vases). But the County Court House was still standing gloriously. And on a side street next to it was an Automatic Teller Machine (ATM) where I could have easily drawn a lot more cash than $20 (if I had needed it).

Chapter 21
The KGB

Manfred R. Schroeder

In early 1963, I received an invitation from the "State Committee for the Coordination of Scientific Research" of the Soviet Government to "see" their work in speech coding. Bell management felt leery about my going as a German national and strongly recommended that I become a US citizen first. Before leaving for the "evil empire" (before it became known as such) I was extensively briefed by the CIA about the potential pitfalls I might encounter. Thus, if confronted with photos showing me with another man in bed, I should not doubt the veracity of the images but respond cheerfully: "Oh, how lovely! Could I have an extra print for my wife and—come to think of it—one for my boss, too?" Of course, nothing of the kind happened.

However, one day, while waiting at a busy intersection in Moscow, with my escort, Victor Ryazantsev (a KGB major—but I didn't know it at the time), I noticed several extraordinarily beautiful women on the other side of the street. Victor saw me "noticing" and said, deadpan, "Would you like one of them? They are working for us." I politely declined.

Another useful advice the CIA gave me was to never close my hands on a package that someone handed me on the street but let it drop to the pavement. This is precisely how the KGB entrapped Harvard professor Frederick Barghorn in the summer of 1963. (The parcel contained Soviet state secrets.) When President Kennedy heard about this, he blew his top: "How on earth dare the CIA turn a good Harvard professor, a friend of mine, into one of their spies!" The CIA swore that they did no such thing. Whereupon Kennedy went for the "hot-line" and berated Nikita Khrushchev. Now it was Nikita's turn to get angry at his KGB chief: "Didn't you nitwits know that Barghorn was a friend of Kennedy's? Next time pick a better target for your dirty tricks." Barghorn was released from his

---

M.R. Schroeder
Author was deceased at the time of publication.

M.R. Schroeder (✉)
Drittes Physikalisches Institut, Georg-August-University Göttingen, Friedrich Hund-Platz 1, 37077 Göttingen, Germany

N. Xiang and G.M. Sessler (eds.), *Acoustics, Information, and Communication*, Modern Acoustics and Signal Processing, DOI 10.1007/978-3-319-05660-9_21,

Moscow prison forthwith and, with some face-saving apologies, returned to the USA. He immediately went to Washington where he thanked President Kennedy personally for his intervention. (I believe the oval office encounter occurred in October 1963; a month later the president was dead. All this was covered *in extenso* in the New York Times at the time. But you can also find bits of the story by googling "professor barghorn" in the Internet.)

I was invited to give a lecture during this trip, with a translator standing next to me, translating my lecture into Russian. Thus I had to stop speaking every two or three sentences. During my introduction I stated: "I work for Bell Laboratories, a company of 14 thousand employees." The Russian translator went to work (in Russian): "...Bell, a company of 14 employees." Without hesitating, I spontaneously exclaimed *in Russian*: "Nyet! Nyet! *14 thousand*" ("Het! Het chetirnadtsat' tisyach"). The whole room broke out in roaring laughter. One of my hobbies was foreign languages and I was following every word of the translator with great interest and curiosity—to his surprise. The poor translator's face turned awfully red and he was replaced a few minutes later by another translator. (I was never quite sure if the size of Bell Labs was purposefully being down-played to the audience, but the spontaneous humor of the situation was priceless).

My 1963 visit was my second trip to the Soviet Union, after Anny and I had gone as tourists during an earlier visit. I recall waiting times of 2 h or more, and once we went without food for 28 h because we simply didn't have the necessary time it took even for a simple meal. When (after a superb puppet show) we finally got back to our hotel, the old, Victorian-style, *Astoria*, all restaurants were closed. Same luck at the other big hotel for foreigners, the *European*.

Tourists from the West were really a novelty in Russia before 1959, so people often gathered around us. Someone in the crowd outside the hotel pointed out a very beautiful, dark-haired woman to us. Could we believe that she was Russian? I said "yes" because that's what they wanted to hear, I think. (I believe she may have been Jewish or from one of the southern republics of the Soviet Union.) Someone else insisted that there was a mile-high skyscraper in the USA (he must have had a *design* by Frank Lloyd Wright in mind). But we were too hungry to argue. Food! Where is food? Yes, there was a snack bar on the fifth floor—but they were about to close. All they could serve at this late hour was caviar, white bread, butter, and champagne. Well, we didn't mind. It was a glorious finale to a very long day.

Now, instead of waiting hours for service at a restaurant, Victor always introduced me as a "guest of our government" when we arrived at a new location, and service thereafter was always near instantaneous.

One of my tasks in Moscow (other than ascertaining the state of Soviet speech research) was to advise the Russian acousticians on the sound system in the Kremlin's new *Palace of Congresses*, seating 6,000 delegates during Party congresses and being used to supplement the Bolshoi Theater for opera performances. The hall was relatively dead acoustically and the required reverberation for music was "manufactured" in the basement and piped into the main hall by a large number of (hidden) loudspeakers. I heard Glinka's "A Life for Tsar" (in Soviet times called "Ivan Susanin") there and liked it. Yet, for the final scene the audio engineers couldn't contain their enthusiasm and turned up the audio volume to an ear-splitting level. So much for *artificial reverberation* (one of my specialties)!

Once Victor and I took the night train (*Red Arrow*) from Moscow to Leningrad and upon arrival, on a beautiful June morning, he told me that, me having been in Leningrad before, I sure knew my way around and didn't need to be escorted. Would it be alright with me if he disappeared for a couple of days? He had been a student in Leningrad and, though married in Moscow, he still had an old girl friend whom he had not seen in a while. Of course, I had no objection, and we parted company right then and there on Uprising Square in front of the main train station. (I mention the name of the square because I am thrilled by its sound in Russian, as I am thrilled by the sound of so many other Russian words and phrases.)

It has been pointed out to me that Victor's disappearance may have been a ruse: that I was closely followed and observed during my entire stay in Leningrad. This is quite possible.

Shortly after arriving in Moscow around 16 June 1963, I heard the voice of Valentina Tereshkova, the first woman cosmonaut speaking from her spacecraft to Nikita Krushchev: "Dorogoy Nikita Sergeyevich..."[1]

On the last day of my state-sponsored visit in 1963, Yevgeniy Levin, head of the Foreign Department of the State Committee for Coordination invited me to dinner at Moscow's famous *Aragvi* restaurant (named for a valley near Mount Kazbek in Georgia). (Victor came along too but had to excuse himself after a while—the Stolinichnaya vodka was flowing a bit too freely.) The *Aragvi*, in addition to its ancient décor, boasted venerable sterling silver cutlery. Levin noticed my acute interest in one of the finely chiselled forks. "Take one with you, as a souvenir of Old Russia" he said. Of course, I declined the offer. After all, I had to pass exit controls by the Soviet border police the next morning. "Don't worry," Levin assured me, "tell them that Yevgeniy Levin gave you permission." Later I learned that Levin held the rank of Lieutenant General. As a polite guest, although not knowing Levin's high rank then, I finally did obey orders and took the fork with me across the Iron Curtain back to the USA. For a long time after, the *Aragvi* fork was the only sterling silver we owned, a fact that does not seem to have escaped some discriminatory thieves: One day, after returning to New Jersey after a longish sojourn in Europe, the temporary tenants had left everything in our house in place—except, alas, the treasured fork.

The Russians are traditionally a very hospitable people and I had the pleasure of experiencing their warmth on several sallies into the Soviet Union. My surprise was therefore complete when, one cold February night, upon arrival in Leningrad, nobody was at the airport to meet me—no flowers, no friendly handshakes, no nothing: nobody I knew as far as the eye could see. After pacing up and down the

[1] I recently met a salesclerk at a supermarket in La Jolla, whose first name was Valentina. Had she heard of Valentina Tereshkova? "Sure, I was *named* after her.—My parents, although Serbs, were very proud of the first woman in space, coming from the Socialist Camp."

crowded arrival hall for a while not knowing *where* to turn (I didn't even know the name of my hotel or the venue of the meeting) I finally turned to the sole "familiar" character in the thick crowd: the man I thought might be from the KGB—although in mufti, easily identifiable to the frequent (and not-so-frequent) traveler behind the Iron Curtain by his mien and demeanor. I went up to this person and said in my minimal Russian: "I, Dr. Schroeder from the Federal Republic. To attend All-Union Conference on Acoustics. Please, which hotel?" Without blinking an eye the stranger, incredibly, answered "You, *Oktyabraskaya Gostinitsa*." Two minutes later I was in a taxi headed for the Hotel October where I encountered all my friends who had expected me on a flight 2 h later.[2]

Imagine arriving at Kennedy Airport in New York City and telling a policeman you forgot the name of your hotel and being told, without a moment's hesitation exactly where to go. Yes, it's difficult to get lost in the Evil Empire—but of course, it didn't last. And I still don't know why the KGB operative at the Leningrad airport blew his cover by answering my question (correctly, as it turned out).

It is quite possible that this story, given the KGB's passion for deception, has a more sinister background too, namely that my Leningrad friends were told (falsely) that I would arrive on a later plane so that the KGB could better observe me.

One thing the Soviet secret services were particularly good at was "word spotting," now subsumed under the ubiquitous "data mining." At some time in the 1960s, Russia had suffered another disastrous wheat harvest. They had to make up the shortfall from imports, mostly from Argentina and the USA, at the lowest possible price. For this purpose they tapped into and monitored zillions of telephone calls. Their excellent word spotting program was set to recognize the words "wheat" and "price." The thus vetted calls were then examined by human listeners for clues to the US pricing strategies. This stratagem, which was unmasked only much later, is said to have saved the Soviet Union (and cost the American farmer) many millions of dollars.

How did the Russians manage to tap, without being discovered, into that many US telephone calls? Easy! They bought a hillside lot for their new Embassy in Washington which "happened" to lie directly in the path of a major telephone microwave link. Hidden antennas in the embassy did the rest—without anyone being the wiser.

21.1 Shinjuku, Shinjuku: Lost in Tokyo

When Anny and I went to Japan for the first time in 1968, we were told that, as Americans, we shouldn't try to speak Japanese. So, for our first foray into Tokyo, we dutifully asked the hotel porter to prepare a little note for the taxi chauffeur telling him that we wanted to go to the Keyo Department Store in *Shinjuku*. The

[2] This story first saw the light of day in a Letter-to-the-Editor in the Travel Section of the *New York Times* on 8 April 1984.

driver nodded and went off in some direction. But before long he doubled back and, after some more criss-crossing, we passed our hotel again. A little later, he stopped at an intersection and asked a policeman for help, but to no avail; we were still lost. Everyone he asked pointed in a different direction. Finally, I thought, it couldn't get much worse if I intervened and uttered a single word: "Shinjuku!" The driver turned around and asked "Shinjuku?" And I repeated: "Shinjuku, Shinjuku, Keyo!" And in no time we reached our intended destination, the Keyo Department Store in Shinjuku.

How did this happen? Did the hotel porter not understand us? Or was he unable to write? Or was it that the cabbie couldn't read the porter's script (and was too polite to say so)? We don't know. But what we *do* know is that it is difficult to navigate the Tokyo streets. Tokyo isn't Manhattan: its streets are like a weird warren. This was nicely brought home to me on our second day in Tokyo. Peter Denes and I had been invited to visit the *Hitachi Central Laboratories* and they had sent one of their limousines to fetch us at the *Okura* hotel in the center of town (opposite the US embassy). On the way back to Hitachi, which was after all his home base, our chauffeur seemed less and less certain as to where he actually was. He had to stop several times to ask *children playing ball in the street* for directions. These must have been very smart kids: before too long we sailed into Hitachi headquarters, happy and relieved, to be greeted by our hosts.[3]

But Tokyo taxi drivers can also be very helpful. I once left my parka in a cab at Kichijoji train station on our way to the island of Shikoku. A friend dashed after the cab, but too late: the vehicle was gone. When, a few hours later, I arrived at my hotel in Shikoku, my "lost" parka was already waiting for me at the reception. What had happened? The cabbie, on noticing the forgotten garment, drove back to the house where he had picked us up. There he encountered a secretary who was waiting for the gasman to read the meter. It so happened that her boss, too, had to fly to Shikoku and he kindly took the parka with him and delivered it at our hotel.

The zenith of Japanese service happened to me during a semi-formal dinner in Tokyo: I had spilled some soup on my pants and was asked to step behind a screen to undress. I was given a kimono and rejoined the dinner. An hour or so later I was

[3] I just bought (September 2007) a GPS (Geostationary Positioning System) navigator that I love and which would have come in very handy in Tokyo (TomTom ONE for $199 at Amazon.com). These little machines are veritable marvels. They have to use Special *and* General Relativity—or their readings would be off the true location. For one, the GPS satellites are in a weaker gravitational field, making their (atomic) clocks run faster than on earth. (So much for Heisenberg's insistence that General Relativity was just a theory with no practical consequence.)

It takes only a few seconds to connect up with up to seven geostationary satellites in the sky. And in a fraction of a second the device will consider *millions* of different road combinations. (As a test, I once asked the machine (stuck to the windshield with a suction cup) for the fastest road from our house to the Rickshaw Garage in New York's Chinatown. Answer: 28.4 miles—34 min. Right! Then I asked for a *toll-free* connection and the machine, unfazed, found one—via Albany, 7 h and 58 min.—Very fast search algorithms are in fact one of the marvels of the digital age.)

again motioned behind the screen: my suit had just returned from the dry cleaner and I was ready to partake of the proceedings in proper western dress.

Once, in the *Okura*, we asked for "sweet butter." After 20 min the waiter returned. He had searched the whole hotel, he told us: "There was no *sugared* butter anywhere in the *Okura*."

Chapter 22
Professor of Physics

Manfred R. Schroeder

Around 1966 I received some "feelers" from a physicist friend (Wolfgang Eisenmenger) whom I had invited for a summer at Bell Labs, where he invented phonon spectroscopy (in collaboration with A.H. Dayem). Was I interested in a full professorship at the University of Göttingen? I was very happy at Bell Labs and didn't know how to say "no". So I said "yes, I am interested." (I mean, how can you possibly say "no"—it would be impolite.)

When the offer was actually made, in late 1968, I was glad to accept. I had spent 14 years at the Laboratories in New Jersey and my only way "up" was, well, up: less research and more management, which I detested. Thus, in March 1969 I made a side trip to Göttingen (during a skiing vacation in Kitzbühel, Austria, where I broke a leg) to discuss the situation. They had a long list of equipment and additional personnel: seven people, including one professor and several *Assistenten* (assistant professors) that I should ask for at the state government in Hanover. One item (for 200,000 marks) on the list was "*Prozessrechner*" and I said: "What's that?" "That's what you call an *on-line computer*" (for doing real-time experiments). But I thought such a computer would really cost over a million marks. Yet I didn't want the secretary (Frau Edith Kuhfuß) to type the whole page again so asked her to put a "1" in front of the 200,000. This figure of 1,200,000 marks was accepted in Hanover (actually, the *VW Foundation* footed the bill which, in the end, came to 2.3 million).

Then came the question of my salary. Instead of answering, I showed the kind official, Ulrich Hopfe, my last monthly pay slip from Bell showing 39,000 dollars per year—then about 150,000 marks. "Well", said Herr Hopfe, "we can't compete with *that*, but we can offer you our highest rate plus all the raises until retirement", which came to about half my American salary. I said "Fine!" and accepted and I am

Author was deceased at the time of publication.

M.R. Schroeder (✉)
Drittes Physikalisches Institut, Georg-August-University Göttingen, Friedrich Hund-Platz 1, 37077 Göttingen, Germany

N. Xiang and G.M. Sessler (eds.), *Acoustics, Information, and Communication*, Modern Acoustics and Signal Processing, DOI 10.1007/978-3-319-05660-9_22,

still glad I did. Soon the dollar started dropping from 4 marks to the dollar to around 3.70 (September 1969) and it kept dropping as President Nixon cut the dollar loose from the gold standard.

Later, while already in Göttingen, I discovered that the new computer didn't fit into the space I had planned for it. Since there was no other free space available, we needed an annex to the present buildings. Of course, the *Kurator* (chief financial officer of the university) wanted to know whether I had the money for the extra building. And I lied and said (like a good groom) "yes, I do". I immediately called Ulrich Hopfe (whose telephone number I always kept on my desk) in Hanover and told him what had just happened and he said (I will never forget) "Don't worry, you *have* the money." Period. He rescued me from a very bad situation and the building was built on time.

Twenty-one years later I invited Hopfe to my farewell lecture on 5 July 1991: "*Die schönsten Experimente aus der Schwingungsphysik, vom Hören und Sehen und zur Zahlentheorie*" ("The most Beautiful Experiments from 'Swinging Physics', Hearing and Seeing, and Number Theory".) And Hopfe, long since retired, came—in a wheelchair.

Of course, Bell Labs didn't want me to leave so a part-time arrangement was reached with the state government in Hanover that allowed me to do research at Bell while I wasn't teaching in Göttingen: my transatlantic-commuter life began—still ongoing 38 years later!

Before becoming a German professor, I had to swear an oath of office. For that I went to the Rector's office to be greeted by a very friendly "*Magnifizenz*", Hans-Heinrich Voigt, an astrophysicist. When his secretary brought in the folder with the oath, he said "no, the *other* one". What was going on here? Voigt explained that, as a foreigner, I didn't have to swear allegiance to the German Federal Republic and that the "other" oath, the one I swore, made no reference to Germany. Thus, I simply promised that "I would obey the laws of the land".

There is an interesting bit of history behind these oaths. After the war, Germany decided that you could become a civil servant (like a professor) without becoming a German citizen (exceptions: federal judges, armed-forces officers, and President of the Republic). Einstein, after accepting his triple appointment in Berlin (professor of physics without teaching duties, member of the Prussian Academy of Sciences, and director of the (new) *Kaiser-Wilhelm-Institut für Physik*) had, to his dismay, become a German citizen again. (He was Swiss, after having been born German.) Hitler, before being able to run for president, had to become a German. (He lost his Austrian citizenship in 1914 when he joined a Bavarian infantry regiment to fight in World War I.) In 1932, a high-level Nazi in the Brunswick state government made him a *Landvermesser* (surveyor), which gave him Brunswick citizenship and, by the German citizenship law of 1913, German citizenship as well—with the well-known disastrous consequences.—There is an extant photo showing Hitler and two buddies

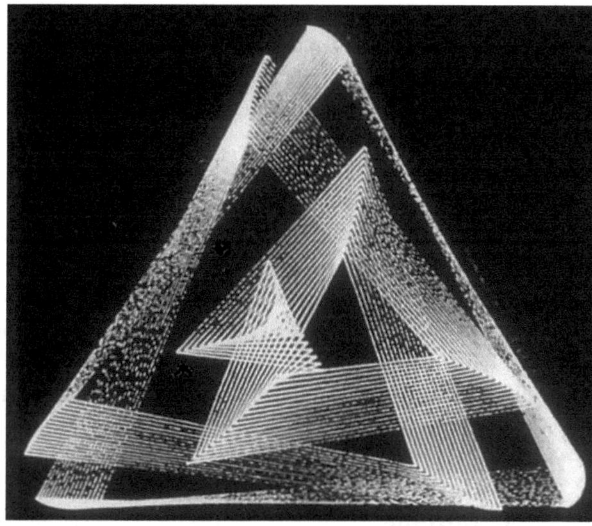

Fig. 22.1 Heinrich's "Broken-Symmetry" triangle

"on the way to the Brunswick legation in Berlin" to pick up his citizenship papers: He didn't even have to travel to *Braunschweig* to become a citizen of that state.[1]

More recently, Hitler and Einstein crossed paths again, posthumously, when *Time Magazine* ran a competition for *Man of the Century*. Einstein won; Hitler was second. (The question was "Who had affected the 20th century the most—for better or worse?") It is interesting to note that Einstein and Hitler were born within 150 miles of each other and both named Karl May (American Indian exploits) as one of their favorite authors as adolescents. (But I doubt that Hitler ever studied Baruch Spinoza, as Einstein did as a young man.)

My teaching duties at the *Institut für Schwingungsphysik* (literally "swinging physics") consisted of a weekly 2-h lecture with demonstrations on a great variety of subjects: acoustics (including speech, hearing, and concert halls), electronics, coherent optics, electromagnetics, computer graphics (one of my hobbies), and applications of number theory (my true love). I really liked preparing the numerous demonstrations. The only topic I didn't like was solid-state physics ("*squalid* state physics" as nobelist Wolfgang Pauli called it).

My lectures on computer graphics struck a particularly sympathetic chord with the students, some conceiving their own creations, like the "broken-symmetry" triangle by Gerhard Heinrich, which I find very appealing (Fig. 22.1):

Best of all, I never had to teach or test undergraduates. Our institute was what would be called a graduate school in the USA. The students were highly motivated,

[1] According to the *International Herald Tribune*, in a report dated 14 February 1932, Hitler was appointed a professor of practical pedagogy at the Technical Academy of Brunswick. But Hitler's likeness on a postage stamp of the *Reichspost* was once featured in a talk entitled *Berühmte Geometer* (Famous Geometers). Others so honored were Gauss and Eötvös, who became the prime minister of Hungary.

intelligent, and often smarter than the professor. And, of course, there were no "classes" or, perish the thought, "credits". If a student was interested in a topic, he would attend the relevant lectures. If not, he would drop out. I usually started with some 80 students at the beginning of the semester and ended up with about 40—par for the course (no pun).

When I first arrived, they gave me a little "torchlight" parade (actually lit candles): I was of course tickled pink. When my Russian friends in Leningrad (Ludmila Chistovich and Valery Kozhevnikov) asked me to characterize my new position in Germany I ventured "*Ya polubog*", which they immediately understood. (I knew the prefix "polu" from Russian champagne: *polusukhoe* = demisec. And I knew the word for god (*bog*) from Old Church Slavonic.) Yes, I had become a demigod. But while I used the term tongue-in-cheek, for some of my older colleagues it was gospel: they actually believed in it (and acted like demigods).

22.1 A Brief History of Speech Research: Early Speaking Machines

An early highlight of speech research was the work of the Hungarian count Wolfgang (Farkas) von Kempelen which resulted, in 1771, in a speaking machine that was widely exhibited at princely courts around Europe. Even Goethe in Weimar and Lichtenberg in Göttingen heard live demonstrations of von Kempelen's contraption. But while Goethe found some nice words to say about the synthetic voice (*die verschiedne kindische Worte und Töne ganz artig hervorbringt*), Lichtenberg was his usual acerbic self by pronouncing the machine incapable of saying much beyond *mama*, *papa*, and *Roma*.

The next advance came with Christian Gottlieb Kratzenstein from Wernigerode who, in 1779, won the First Prize from the Imperial Academy in St. Petersburg for designing (wooden) resonators capable of uttering the German vowels *a*, *e*, *i*, *o*, *u*. Thus, the resonance theory of vowel production was born, later elaborated by Hermann von Helmholtz and Lord Rayleigh.

One of the most useful practical applications growing out of this work was the invention of the telephone by the German high-school teacher Philipp Reis (1860) and the Scottish speech therapist Alexander Graham Bell (working in Canada and the USA). In America, the invention of the telephone led to the formation of the American Telephone and Telegraph Company (AT&T) in the late 19th century which in turn, in 1925, incorporated *Bell Telephone Laboratories* in New York City (463 West Street).

Apart from solid-state physics (transistors, solar cells) and information theory (Shannon's channel capacity, error correcting codes), Bell Laboratories was a strong supporter of fundamental research in hearing and speech, exemplified by the invention of the *Voder* and *Vocoder* (from **Vo**ice **Coder**) by Homer Dudley[2] in 1928, first demonstrated at the New York World Fair in 1939. (The vocoder played a crucial role in World War II by enabling a secret telephone link between Churchill's War Room in London and Roosevelt's White House in Washington.)

The vocoder, by analyzing and synthesizing speech signals, could be made to sound like a pipe organ or a waterfall or a VW engine (groaning for more gas). In its most recent incarnation the vocoder (having morphed into *linear predictive coding*[3]) inhabits every cell phone and most Internet speech compression.

22.2 Speech and Hearing Research in Göttingen

Some of the speech research pursued at Bell was transplanted to Göttingen in 1969 when I succeeded Erwin Meyer as director of the *Dritte Physikalische Institut*. Much of our speech research at Göttingen involved the application of measuring methods from physics to the human speaking process.

In view of the importance of human hearing for the proper encoding of speech and music signals, extensive studies of perceptual masking of one sound by another were undertaken by Sönke Mehrgardt and many others. This research was led by Birger Kollmeier and Armin Kohlrausch. Some of the work was concerned with the design of better hearing aids and tests of speech intelligibility for hearing-impaired listeners.—An important portion of the ongoing work in speech is aimed at improving diagnostic tools for voice pathologies (in collaboration with professor E. Kruse of the Medical Department of the university). Jan Lessing, my most recent Ph.D. (July 2007), developed a sophisticated acoustic method to assist in the diagnosis of various malfunctions of the human "voice box".

In December 1970 I gave my inaugural lecture "Can computers be taught to speak?" At the end I had an (American) computer recite a German poem:

Mit deinen blauen Augen
siehst du mich lieblich an.
Da wird mir so träumend zu Sinne,
daß ich nicht sprechen kann.

The Bell computer, with the help of Noriko Umeda and Cecil Coker, had been trained to properly pronounce the German *r* and *ch*. But even so, nobody in the audience understood. Then, without mentioning it, I projected a slide with the text and repeated the computer speech. Now most listeners understood. Why? Because they had (unwittingly) *read* the text.—This is a famous trick to convince sceptic

---

[2] Homer was convinced that the State Department was infested by communists and later, in 1957, that the beep–beep–beep of *Sputnik* was nothing but electronic deception—the Russians couldn't possibly be the first in space. Of course Homer was not alone in these superstitions.

[3] Correspondences with Bishnu Atal and J.L. Hall during development of LPC are archived at the University Library Archives of the University of Göttingen.

listeners that your synthetic speech is actually intelligible. (Of course, I told the August audience afterwards what was going on.)

After the lecture Frau Eggers—mother of my friend Frieder—came up to the podium: "*Sagen Sie mal, Herr Schroeder, das war doch Heinrich Heine?*" "*Ja, Frau Eggers, das war Heinrich Heine*".

Naturally, as a new professor, I was immediately condemned to some onerous committee duty: the University Senate Planning Committee. One of our tasks, under the able chairmanship of Rector (university president) Eduard Lohse, was to plan the new physics department buildings. This was in 1970, but it took until 2005 for the move to the North Campus to be completed. (First money ran out, then some of my colleagues didn't want to move, etc.)

Much later I was elected to the *Kommission für Mathematiker Nachlässe*, concerned with the unpublished papers of famous mathematicians (Gauss, Riemann, Hilbert, Noether). Since little was known about the papers of Emmy Noether, one of the greatest mathematicians of the 20th century, I was glad to have succeeded, 72 years after her death in 1935, in locating her surviving nephew Hermann Noether in Summit, New Jersey.

Emmy was forced to flee Germany in 1933 because of her Jewish descent. She became a professor at Bryn Mawr. Her brother Fritz emigrated to Tomsk, Siberia, where he became a professor at the Institute for Mathematics and Mechanics. In 1937, during the Great Purge, he was arrested as a "German spy" and sentenced to 25 years imprisonment. In 1941, he was in a Soviet camp near Orel, where he was shot on September 10th for spreading "anti-Soviet-propaganda". Some 160 other prominent political prisoners were shot the next day on Stalin's orders in a forest outside Orel. (The true reason for these hasty executions may have been the approach of the German army, which, in fact, captured Orel on October 3rd.)[4]

Another commission I served on was the *Prüfungskommission* (examinations commission) that had to settle contested cases. I remember one unfortunate student who had flunked theoretical physics in the orals twice in a row and he was not eligible for a third try. I pleaded with my colleagues on the commission to make an exception (the candidate had obviously been nervous). "No," they said, "that would set a bad precedent." I didn't see how we could nullify a student's 4 years of labor by just one flunked test. But they wouldn't budge. So the case went to the top legal eagle of the university (Herr Bahlsen) for arbitration. Bahlsen was known to be very pedantic and I saw all hope for the student fade. But Bahlsen asked a single question which saved the case: "Was the required extra witness present when the test was repeated?" No! so the second test was declared invalid and the student could take it one more time. I made sure that the examining professor was replaced by a less threatening person and the student passed and got his degree (M.S.).

One of my older colleagues actually, in the middle of the oral examination, threw out a Ph.D. candidate whose answers he didn't like (Max Delbrück

[4] Orel (pronounced arYOL) is the Russian word for eagle. Orel was liberated by the Soviet army on August 5, 1943 (see Chap. 16 of my memoirs).

(1906–1981), who later created the field of molecular biology and won a Nobel Prize for it). Talk about demigods!

When I had to examine a student myself, my first question usually was “Are you right or left handed?” (If the student was left handed, I sat on his right so his arm wouldn’t block my view of his jottings.) This was a simple question that no student had any problem with. My next question, taking a leaf from Richard Becker, was “What would you like to talk about?” which some students found difficult to answer because they were expecting the typical quiz. Then I let them talk freely, only occasionally interrupting with a question. This way it was more interesting for me, often learning something new. And it certainly was nicer for the student. Also, I could learn more about the student’s abilities than by asking questions from a “catalogue”, which the students were known to have memorized. Sometimes, at the very end of the 30 min, I would ask a pointed question or two, like “What is the spin of the photon, the electron, the graviton?”

My demonstration lectures, in which I was greatly assisted by Heinrich Henze, continued the tradition of Robert Pohl and Erwin Meyer. But my style of lecturing must have been somewhat unorthodox. Once an American student told me after a lecture that I reminded him more of an American professor than a German.

Some of my experimental demonstrations were based on number theory, such as the “2-squares theorem”, which allowed me to create circularly *polarized* sound waves. This was a small sensation because sound waves are *longitudinal* and—in contrast to light—have no polarization.

A few of my demonstrations were deliberately misleading, for example when I “showed” that microwaves traveled through a hollow plastic hose while they actually cling to its outside (depending on the wavelength and the diameter of the hose). (Of course, in the end I allowed the truth to emerge by plugging the hose shut—and the waves continued propagating.)

One of the finest accolades I received from a student was from Joachim Neumann, who wrote on his webpage, next to a picture of me, “He taught me that physics is show business”. (Some colleagues, I am sure, would be outraged by such a statement.)

One of my academic tasks other than giving lectures was selecting thesis topics. At any one time, there were on average some 20 theses, leading to the degrees of *Diplomphysiker* (M.S.) and *Dr. rer. nat.* (Ph.D.), that I had to supervise. I was much helped in this task by the untiring and conscientious Dr. Hans Werner Strube, as well as Birger Kollmeier and Armin Kohlrausch, who later became professors at prestigious institutions in their own right.

My “first professor” was Reiner T., who was in danger of not getting a job in Germany because he had a police record (participation in some leftist student demonstration). I advised him to emigrate to the USA and never mention his record of “unruliness”. I wrote a warm recommendation and after a few years he had become a *professor*. Of course, he was very smart. Before taking him on as a graduate student, I asked him “What is the numerical value of i raised to the power i?” (i is the square root of -1.) After a moment’s hesitation, he ventured “Oh, about

1/5"—which is the right answer! (I have asked some mathematics students the same question and they didn't get it.)

Shortly after I arrived in Göttingen in 1969, at the height of the Vietnam War and the worldwide student protests, the students called a *Vorlesungsboykott* (boycotting the lectures). However, the students still appeared at my lectures. So I asked them whether there wasn't a strike on. Yes, but they wanted to hear my lecture anyhow. These were graduate students eager to complete their education. Sitting on the front table, feet dangling (to signal that I didn't have any particular axe to grind), I said, "let's have a vote whether I should give the lecture". All but three students voted "yes". To the three strikers I said that I would give them a private tutorial to catch up. Talk about taking the wind out of a protest!

One of the three students, now a professor, admitted his "sin" during a recent encounter, hoping that I really wouldn't remember him. I said, "Yes, I do remember you, but I never held it against you". (His name was Gauss—no relation though to the great mathematician.) In some departments (philosophy, *Germanistik*) there were violent sit-ins and in one case the professor was locked in (or out—I forget which) of his office. Albrecht Schöne, the university's star *Germanist*, simply worked at home.

One of my Ph.D. students is Bennet Smith, an American, who later worked with Pierre Boulez at the Centre Pompidou in Paris. He didn't finish his thesis (on the peripheral auditory system) while in Göttingen, so I carried him as an external student. Whenever I visited Paris he asked me whether my offer that he could get a Ph.D. from Göttingen was still good. I advised him to closely observe the obituaries: as long as he didn't see my name there, he could still come to Göttingen and complete his thesis. (Twenty years later I am still waiting. But I guess he made his way without the degree.)—Bennet, being American, once, during colloquium, sat in the front row, traditionally reserved for professors. The second row is for assistant professors, etc. Students should not seat themselves in any of the three or four front rows—a despicable holdover from more authoritarian times. But I assured Bennet that he could sit wherever he liked.

Apart from Bennet Smith, I had several other external Ph.D. students: Jan Lessing, working at the Göttingen clinics (in phoniatrics—the subject of his thesis: telling from a recording of a voice what's wrong with the vocal chords), Olaf Schreiner working at DaimlerChrysler for a while (on automatic speech recognition in cars) and, until recently, Holger Quast working at Bosch (on spoken dialogue systems: airline reservations and the like). Holger passed the orals and got his degree on 26 January 2006. Of the five-member Ph.D. commission two colleagues had voted with me for "summa cum laude" and two had opted for "magna cum laude". So, the majority was for *summa* and I awarded Holger the "summa cum laude". (Later, one busybody colleague discovered that for *summa* the vote had to be unanimous. But I advised Quast that he could always say that his professor gave him a *summa*.)

Besides Holger, one of my star students was Tino Gramss. After his Ph.D., he could look forward to a distinguished academic career. But he was riven by self-doubt—not made better by jilted love—and chose to end his life at a very early age.

Some of my top students, like Sönke Mehrgardt, became "captains of industry". Others became professors or opened Software and Acoustics consultancies. Some rose to high academic positions, such as K.J. Ebeling, who became the president of Ulm University.

Most of my students were working in the fields of speech, hearing, architectural acoustics, and applications of number theory. Once a week we all met in a joint seminar. Some of the work in hearing was psychoacoustic (investigating special properties of human audition), but some involved animal physiology and animal hearing. The work with animals was done in a very fruitful collaboration with Otto Creutzfeldt at the Max Planck Institute for Biophysical Chemistry. I remember one Ph.D. thesis (by Rainer Koch) on guinea fowl (*Perlhühner*). Koch had electronically synthesized various "calls" by the fowl (danger, mate seeking, etc.). But his animals wouldn't budge; they only reacted to natural calls in spite of the fact that both natural and synthetic calls sounded the same to human listeners and looked exactly alike on the sound spectrum analyzer. What was going on? I thought maybe the fowl can perceive fine temporal detail that the human ear cannot perceive and the analyzer wouldn't show. After Koch modified the analysis to show the fine time-structure, he discovered that the latter was missing in his synthetic calls. After incorporating the appropriate temporal detail in his synthetic calls, the birds *would* budge as expected. Later Koch discovered that his birds were mostly sensitive to time rather than frequency detail. In other words, you could play the birds almost any melody, as long as the rhythm was right.

22.3 Underwater Sound

Research at our institute was underwritten mostly by the German Science Foundation (*DFG*) except for contracts we had with the Rome[5] Air Force Command and the British Admiralty for (unclassified) underwater sound research, supervised by Werner Lauterborn, Dirk Ronneberger and Dr. Dieter Guicking. Professor Erwin Meyer had insisted, back in the late 1940s, that we should be allowed to publish all our results, which of course precluded "secret" work. During my tenure, Leslie Kendrick was the Admiralty's representative and Leslie—and his wife Mary—became lifelong friends.

Besides the university and the Max Planck Society (of which I became a "Foreign Scientific Member" in 1972), I was elected, in 1974, to the Göttingen Academy of Sciences (founded by George II in 1751). This Academy holds *plenary* sessions every other week while the university is in session. *Plenary* means that both natural scientists, including mathematicians, and professors from the humanities attend and present short papers on, or related to, their work. Thus, someone attending regularly, as I do, gets a broad overview of much of contemporary

---

[5] Rome, N.Y.

scholarship. The Göttingen Academy has become a "second home" to me. (By contrast, many other academies have separate, specialized sessions in the humanities, sciences, etc.)

On one of my visits from Göttingen to New York I got to know Pierre Boulez, then the conductor of the New York Philharmonic. (His musical manner received mixed reviews. Harold Schoenberg (from the *Times*) once characterized his style as "the iceberg conducteth".) Pierre had been asked by the French President Pompidou, who wanted him back in Paris, whether he would like a new concert hall, but Pierre said he preferred an acoustic/music research center. The wish was granted and soon the foundation was laid for the *Institut de Recherche et Coordination Acoustique/Musique* (*IRCAM*) within the *Centre Pompidou* in Paris. During the planning, I traveled with Pierre to several existing acoustic/music research centers: to Peter Zinoviev's[6] laboratory in London and Max Mathews at Bell Labs. We also visited John Chowning at Stanford University as well as Stockholm and Göttingen (digital simulation of concert halls).

Once, during a meeting in Paris, someone pointed out that new laboratories usually run out of space after a few years—we should plan it twice as large. "No problem", said Monsieur Bordaz (Pompidou's representative), "we'll raze another street of the *Quartier Beaubourg* to make room for a bigger laboratory".

The institute's name was concocted during a lunch in the old neighborhood (then still awash with *filles de joie*). The original choice, *Institut de Recherche Acoustique/Musique* was voted down because its acronym would have been pronounced like *Iran*, where Ayatollah Khomeini had just taken over.

During my tenure I also became a "great" cyclist—both in the USA and around Göttingen. In America, Anny and I took many trips with Vermont Bicycle Touring, mostly in New England. In Germany we cycled a lot with friends (Edith and Wolfgang Schröter) and with my students (*Studienstiftler*) (Fig. 22.2).

For indoor fun, I studied (with Gerhard Sessler) the mechanics of the roulette wheel and visited several casinos in the process: Baden-Baden, Baden (near Vienna), Evian-les-Bains. Once we had ascertained that we could win considerable amounts of money (given enough starter capital), we lost interest and dropped the project—in contrast to the authors of the book *The Newtonian Casino* who pursued the idea but were of course barred from the casinos once their method became known. (But, in the meantime, they had lots of fun.)

[6] Peter was one of the pioneers of computer music. When I asked him whether he was related to the (in)famous Old Bolshevik Grigory Zinoviev, he said *No*, he was descended from a rich Russian family whose assets were confiscated after the 1917 revolution and that Grigory, whose birth name was Hirsch Apfelbaum, stole his family's name.—Grigory, who was on that sealed train through Germany with Lenin, to jump-start the revolution, first sided with Leon Trotsky and later allied himself with Stalin to denounce Trotsky (1924). Grigory himself paid the ultimate price when he became the first Old Bolshevik to be executed (August 1936) in Stalin's Great Purges. He was exonerated by Gorbachev in 1988. (For an extensive account of the murderous machinations of Stalin & Co. see Sebag Montefiori: *Stalin*.)

Fig. 22.2 Bicycling with Anny on Nantucket (1981). *Source*: In 'Sconcet (Siasconcet), Nantucket, 1981

My transatlantic commuting necessitated a permanent abode in the USA, so Anny and I decided to buy another house in New Jersey. We could no longer afford the house we had bought in 1968 in Mountainside for $60,000 and sold for $69,000 just 14 months later. Prices had gone through the roof in the meantime. However, we found a nice, smaller house in Berkeley Heights for $62,900. But we had only $700 in cash. So I went to my bank (then The Summit Trust Company) and said: "I need a personal loan for new furniture". How much? I said $20,000 would be fine. Then I went to the mortgage office (next door on the same floor) and said I had found a nice house that I wanted to buy and needed a mortgage loan. "O.k., how much cash do you have?" I said: "20,700 dollars". Fine, so they gave me a mortgage for 43,000 dollars. It was that simple (in the early 1970s) to buy a house in the USA! (Of course, the personal loan was secured by stocks as collateral—as required by law.)

22.4 The Family in Göttingen

By contrast, life in Germany was a lot more difficult. For one, there was no school bus in Nikolausberg, the suburb of Göttingen where we lived. The children had to get up very early to catch the 6:50 city bus to get to school on time. Then, not infrequently, an hour later they would return: classes had been canceled—the teacher was ill. This had never happened to us in the USA—there were always enough substitute teachers around in the New Jersey schools. (One reason may have been that in German, high-school teachers are civil servants—difficult to hire and near-impossible to fire.)

Language—the German language!—was another problem. The teachers paid little attention to the fact that our children had never been to a German-speaking school before. All three needed private tutoring in German at home—Alexander for 2 years—with many a tear shed over the abominable syntax and the orgies of compounding (see my Chap. 23). Now, of course, all our children are perfectly bilingual and, in fact, *tri*lingual because they stayed with French families during their vacations or studied at French universities (Grenoble, Julian).

Another problem was housing: in spite of spending considerable time on house hunting, we found nothing suitable. One house, owned by Percy Ernst Schramm, the keeper of the official war diaries (*Kriegstagebuch*, published in eight volumes) of the Supreme Command of the Armed Forces (*OKW*), was so run down it was disgusting just to visit. Another house, also located in the prestigious *Ostviertel* of Göttingen, was the sumptuous villa of the *Kurator* of the university. But, unfortunately, it was located near a busy intersection with the noise and fumes from accelerating cars and busses.

In our predicament the State of Lower Saxony came to our rescue by finishing two apartments ahead of schedule and combining them into one apartment with a common front door. This gave us 160 m^2, about half of our house in Mountainside, New Jersey. The bigger apartment (90 m^2) became our three children's abode while the smaller one (70 m^2) contained the master bedroom and my study. The second living room became our dining room and the second kitchen served as a nice walk-in closet. Since these apartments were subsidized for middle-level university people, our lease was limited to 4 years—ample time to build a house from scratch.

Our future neighbors (Leo and Clara deMaeyer) found some farm land (20,000 m^2) for sale at 22 marks/m^2 of which they and we took 4,000 m^2 each, the remaining acreage going to the Max Planck Society to build on.

Since we didn't want to live in a stone house (*Stein auf Stein*, huh) we looked at some 20 prefab-housing companies, of which we selected *Bien Haus* near Frankfurt because their workers drank Coca Cola (instead of the ubiquitous beer) during lunch breaks. They were also the lowest priced and the most accommodating to our wishes: we could choose our own design, as long as all measurements were multiples of 62.5 cm. In fact, we did not make use of an architect: the children and Anny designed the L-shaped layout based on our double apartment. Once the basement was finished, the house was erected in 48 h from the prefab walls that arrived on giant trucks. Curiously, the house didn't look prefab. Only our Göttingen contractor (for the basement) remarked in ill-disguised astonishment "*Herr Professor bauen fertig*?" (Mister Professor is building a *prefab* house?). "*Ja, Frau Fricke, wir bauen fertig*".

Incidentally, in all our 37 years at Göttingen, we never bought a German car. We first drove four cars imported from the USA, including an eight-cylinder Plymouth 1965 *Belvedere* with a 6-L (360 cubic inches) engine with a torque of 450 N m at 1,800 r.p.m. What a car! There was of course nothing comparable available on the German market. The car cost 2,900 dollars in 1965 (about 20,000 in 2007-dollars) and, at 14 L per 100 km (17 miles per gallon), was cheaper to run than a VW at today's gas prices.

Before we dared take the Plymouth to Europe, I asked the dealer whether we should take any spare parts. The dealer said "Take an extra fuel filter—that's all you'll need for the next couple of years". And he was right![7] After we retired the *Belvedere* (after 14 years of flawless service), its superb engine had a second life as a winch motor to pull up gliders for the local flying club to which I had donated it.

In 1983, we switched to Japanese cars—mostly Toyotas but recently some Hondas. If you want to know the reason, look at the repair statistics in *Consumer Reports* (or even the *ADAC*—the German AAA), which show Japanese cars consistently ahead of German makes.—Of course, if I was presented with a BMW as a gift, I would gladly accept it.

22.5 Talks and Trips

While a professor in Germany I continued to give talks—at various congresses and physics colloquia in Europe, the USA, Japan, and China—about applications of number theory, speech compression, hearing, and other topics. Once, at Cornell University in Ithaca, New York, I was greatly honored by the presence in the audience of Hans Bethe, one of the greatest physicists of the 20th century. On another occasion, also at Cornell, the famous former Soviet dissident, Yuri Orlov, approached me after the talk. He was honored, he said, to hear my lecture. No, I countered, *I* am honored by your presence. (Yuri was a cofounder of the Moscow Helsinki Watch Group and a close ally of Andrei Sakharov in their fight against Soviet repression. He spent 9 years in the Gulag, many weeks in solitary confinement. He was set free by Gorbachev in 1986.)

Earlier trips led me to Lisbon, from where I visited the Estoril Casino to study the predictability of the odds in roulette. I even made a little money—without any help from physics.

Four times I traveled to the Soviet Union: in 1959 (with Anny, as early tourists from America[8]), in 1963 (at the invitation of the Soviet government—see my Chap. 21), in 1971 to attend the All-Union Conference on Acoustics in Leningrad, and in 1987 to Tallin (the old Hanseatic city of Reval in Estonia) for a conference on phonetics.

One of my most momentous forays saw me in Argentina at the height of the Falkland crisis with Britain in 1982.[9] I had been invited to give a series of seminars on concert hall acoustics at the La Falda mission in the western mountains.

---

[7] In Orange, during a trip to the south of France, the car stalled in the middle of a busy intersection. What had gone wrong? I figured it must be the fuel filter—and the fuel filter it was.

[8] Apart from sightseeing in Leningrad and Moscow, we vacationed on the Black Sea (Sochi, Sukhumi) and visited Gagra and Lake Ritsa in the Caucasus (where Stalin maintained a summer home).

[9] Here is one of my favorite English headlines from those days (when Britain was ruled by Margret Thatcher): BRITISH LEFT WAFFLES ON FALKLANDS. Hmm, how very considerate!

Back in 1963 we started skiing in Switzerland, taking charter flights from New York to Zürich or Geneva (for $180 round trip). We continued this custom when we moved to Germany in 1969—except that the planes were replaced by cars or trains—until 2001. In those 32 seasons we skied all the major resorts in Switzerland (Zermatt, St. Moritz, Davos/Klosters, Arosa, Grindelwald, Verbier and the rest) and three Austrian places (St. Anton, Kitzbühel, Heiligenblut).

The triennial International Congress on Acoustics saw me in Delft, Holland (1953), Cambridge (1956), Stuttgart (1959), Copenhagen (1962), Liège (1965), Tokyo (1968), Budapest (1971), London (1974), Madrid (1977), Sidney (1980), Paris (1983), Toronto/Vancouver (1986), Beijing (1992),[10] Trondheim (1995), and Rome (2001).

It was in Tokyo in 1968, on August 28 (the same day that Warsaw Pact forces invaded Czechoslovakia to subdue Dubchek's "Prague Spring"), that I presented a joint paper with Bishnu Atal on "adaptive" predictive coding, which later morphed into Code-Excited Linear Prediction (CELP), which is at the basis of all speech compression for cell phones and the Internet.

In collaboration with Joseph L. Hall, Bishnu and I introduced *perceptual* coding that resulted in the required high quality of synthetic speech. (Yes, when you listen to a friend on a cell phone, it's not his or her original voice but a synthetic sound manufactured in real-time—with a negligible delay—inside your phone.) The reason cell phones sound so good (if connected to a high-quality loudspeaker) is that CELP is akin to "voice excitation", a principle I invented in 1957 and that, according to John Pierce, led to the first human-sounding speaking machine (see Chap. 20).

22.6 Export–Import

When we first moved to Germany, we missed many things we took for granted in the States. Of course, we couldn't do much about the (nonexistent) school bus service in Nikolausberg. (Our children had to catch a 6:50 city bus to make it to school on time!) But some things could be rectified by the "frequent flyer" in the family. Marion missed the right flour (*Bisquick*) for her pizza dough. And all of us missed the maple syrup for our waffles. Also the local lawnmowers were inadequate for our large lawn (4,000 m^2).

I once imported a Sears lawnmower from the USA and, when German customs wanted to know what I was going to use it for, I said, "Mow my lawn." Then they asked "And afterwards?" "You mean, after I mowed my lawn?" "Yes, yes." Their English was none too good and I sensed that they wanted to get off the hook, get rid of me. So I said, "After having mowed my lawn, I will take the mower back with me

[10] With side trips to Xian—the terra cotta army—and on the Li river with its sugar-cone mountains.

to the States." "Well, then it's o.k." (I can still hear them snickering behind my back about the crazy American who brings his own mower from the States.)

Another occasion, when my not speaking any German was helpful, happened when I was importing live lobsters in a cooler filled with ice cubes (replenished by a helpful flight attendant). Pointing at the cooler, the customs official said "*Was ist das*?" Of course, my not knowing German, he had to repeat in English: "Vot is in zär?" I answered truthfully "Maple syrup and lobster." Whereupon he starts poking around in the ice and first finds a bottle of maple syrup. "Vot is zis?" "Maple syrup." Sure enough, after some more poking he sees the back of a lobster. I see the front end of the animal looking at me through an ice cube (the kind with a hole in the middle). "Vot is zis?" Knowing that he wouldn't understand the English word (lobster is *Hummer* in German) I said "Oh, that's a lobster." Extended silence. Then the kind but overtaxed official said "*Ja, bitte, gehen Sie weiter*." Which, of course, I didn't understand. So he repeated "O.k., please go on."

Once, during a talk in London, on the mechanics of the inner ear, I wanted to show a film clip showing the (simulated) vibrations of the basilar membrane in the *cochlea*. But the projectionist said "The film cannot be shown". Why, this incompetent projectionist, I thought. But arriving at home in Göttingen, I discovered that some of my "smuggled" maple syrup had leaked into the film can. (What a mess it would have made of the projector! And how polite and with British understatement the projectionist had responded to the situation.)

Chapter 23
The German Language

Manfred R. Schroeder

Mark Twain has of course skewered the German language to his heart's delight (see *A Tramp Abroad*). I will comment on the language as a native speaker who is not bothered by the fact that girls in Germany are neutered (*das* Mädchen) or that a woman is not neutered but becomes *masculine* in (of all cases) the genitive form (*der* Frau). However, when you meet two or more girls they regain their femininity (*die* Mädchen). So when you encounter *Mädchen*, pay close attention to the article (no pun). Depending on *der*, *die* or *das*, *dem*, *des* or *den* you would know their number (hmm) and whether they are doing something or something is being done to them. Some language! (In Dutch, by the way, a girl is also neutered (*het* Meisje) but at least she keeps her other feminine assets intact (*haar* moeder—her mother; *haar* haar—her hair, etc.).

Why, by the way, is it *der* Euro but *die* Mark? (Of course, *der* Euro is worth about *two* Marks.)

I was once asked in chemistry class in middle school (in Rotenburg/Wümme) to elaborate on last week's topic (dimethyl-ether). So I started out "der Methyläther ist...." No, that was wrong, the teacher said. The boys behind my back were whispering "di, di, di!" So I tried again "*die* Methyläther..." Wrong again! So, as a last resort, I ventured "*das* Methyläther". Well, wrong again! So, in desperation, I said "*Der die das* Methyläther." By this time the teacher was quite angry and he called on another boy who calmly proclaimed "*der Dimethyläther*." I always knew that chemistry harbored dangerous substances but I never realized that the innocent-looking particle *di* was one of them.

But the article *die* can also be exploited to great advantage. Göttingen used to be known as "*Stadt der Wissenschaft*" (City of Science). But then someone must have remembered that Wissenschaft is actually female (*die* Wissenschaft) as it is in

Author was deceased at the time of publication.

M.R. Schroeder (✉)
Drittes Physikalisches Institut, Georg-August-University Göttingen, Friedrich Hund-Platz 1, 37077 Göttingen, Germany

N. Xiang and G.M. Sessler (eds.), *Acoustics, Information, and Communication*, Modern Acoustics and Signal Processing, DOI 10.1007/978-3-319-05660-9_23,

Latin, Italian, etc. Furthermore, by separating Wissenschaft into its two components: Wissen (knowledge) and schaft (ship), and adding a comma and an *f*, he arrived at the new slogan for our city: "*Stadt, die Wissen schafft*" (City that creates science). An ingenious pun!

The *der-die-das* business has also led to some hilarious Pennsylvania Dutch expressions like "Throw the cow over the fence some hay." Poor cow! In the German original there is of course no doubt as to what is being thrown: "Wirf *der* Kuh etwas Heu über den Zaun."

The Dutch, incidentally, did away with the distinction between the grammatical male and female a long time ago. It's always *de* instead of *der* or *die* in German. And in the genitive case they abolished the awkward *der* (*der Frau*!). Instead they write *van de* (except for the Queen, in whose "honor" they keep the *der*: *der Koningin*.

The Dutch have also repeatedly simplified their orthography, making it almost completely phonetic. Compare this to the horrors of English spelling where writing and spelling are largely divorced from each other. (A German friend of mine (Inge Spelsberg), who worked as a secretary for the U.S. Army in the Nuremberg Palace of Justice, became the supreme arbiter of spelling for her Major (in English?) and many minor officials.)

Of course, for some English words spelling and sound are actually the same: orts (leftovers).

Apart from the gender mix-ups, I am also bothered by the relentless compounding of words into veritable tapeworms of letters. A few examples may suffice: The German *Stadtabgeordnetenversammlung* is nothing but a city councilors meeting. And *Vergangenheitsbewältigung*, as Günther Grass knows best, is coming to terms with one's own past. *Waffenstillstandsverhandlungen* are simply armistice negotiations. Even a native German is often at a loss how to segment a word monster like *Seeschiffahrtsassistenzunternehmen* (tug boat assistance enterprise) to extract meaning. This of course slows down his reading and comprehension.

And some words can only be segmented from the context. Is *neuromanisch* some neuro-maniac? No, I encountered the term in an article on architecture so it probably means "New Romanesque".

What on earth is the poor reader to make of *Waldohreulen*. Using within-word capitals would surely help: *WaldOhrEulen*, i.e. we are talking about "wood ear owls" (not barn owls or owls without ears, I suppose).

In a recent article on *Stammzellenforschung* (stem cell research) I encountered a Greek philosopher unknown to me: *Ethikrates*. Was he a follower of *Sokrates*? No! After some mental gymnastics I discovered that *Ethikrates* meant "ethics council's". If it had been spelled *EthikRates* there would have been no problem. Within-word capitalization has of course been used to great advantage in computer languages, advertising, product names, and even personal names: DeGeneres, MacDonald, McGurk, O'Neal, VanDoren, etc.

On the other hand, compounding in German entails a great advantage. It allows the construction of words, even neologisms, that take entire sentences in English and other languages that don't allow compounding. How best to translate

Neiddebatte (debate engendered by jealousy); *Prozessökonomie* (cessation of judicial proceedings)? *Vertragsverletzungsverfahren* are one (or more) judicial proceedings to investigate the alleged violation of a treaty—one word in German, a longish phrase in English.

But the worst sin, as everyone knows, is the German word order. Even having grown up with German, I often have to read a sentence several times before I get its sense (if there be any). As a result, I do most of my reading in English, including German authors. (I make a few exceptions for Thomas Mann and Grass and selected others who really sound best in the German original.) It was only in the USA, at Bell Labs, composing scientific papers, that I learned to express myself understandably in English—which I hope carried over to my German writing style. As Georg Christoph Lichtenberg said in one of his more charming aphorisms: "I really went to England to learn how to write German". (He met George III, who spoke no German, several times. George I, hailing from Hanover, was of course fluent in German and George II, patron of Handel and founder of Göttingen, Princeton, Columbia and Rutgers universities, could still handle it.)

As the head of a large university institute, a lot of German missives crossed my desk, but I let my secretaries tell me what was in it. (Mostly nothing I needed to know.) One of my worst correspondents was the university president, whose long sentences may have been difficult to decode even for himself. I fondly remember one instance: the *Ehrenpromotion* (awarding of a *Dr. honoris causa*) to mathematician Fritz Hirzebruch in the festive *Aula* (ceremonial hall). The president read the *laudatio* from a sheaf of typed pages. At one point, after having turned over another page, he discovers *toward the middle of the next page* that the sentence doesn't make any sense—the proper verb was still missing or something. Then he found out that he had accidentally skipped a page! Three cheers for German syntax!

As if German words weren't long enough already, the native speaker likes to festoon them with extra garlands. Thus, coffee isn't *Kaffee* but often *Bohnenkaffee* (because during the war, coffee was made mostly from barley). And honey isn't simply *Honig* but *Bienenhonig* because the German chemical industry had contrived an ersatz product called *Kunsthonig* (which we as children—to my mother's despair—actually preferred over the real thing).

Apple juice isn't just *Apfelsaft* but, on one of our containers, *Apfelfruchtsaftgetränk* (apple fruit juice beverage) and this, I am assured, for good legal reasons. Just give me a glass of apple juice, please, and forget about apples that are not fruit and juice that isn't a beverage, or whatever other possibilities exist. Of course milk isn't just *Milch* but always *Voll*milch. To top it all, I recently discovered a label on a box that said it contained *Alpenvollmilchschokolade* while the English part of the label simply said *milk chocolate*. The poor monolingual tongue, alas, never realizes that her milk chocolate was made from the whole(some) milk of mountain-climbing cows.

On the other hand, there are a few admirable and short German word creations: e.g. *Twens* (for the awkward twenty-somethings) and *Handy* (for cell phone).

Another German weakness is to pack several, often disparate, ideas into the same sentence. I leave the patient reader with just one example in the original:

Könnte man die Artus- und Gralsdichtung als Versuch werten, die als chaotisch erfahrene Realität durch einen ‚utopischen' Gegenentwurf ethisch und ästhetisch zu bewältigen, so schienen andere zeitgenössische Werke dieses optimistische Modell, das auf Ausgleich der Gegensätze zwischen geistlicher und weltlicher Macht sowie zwischen individuellen und gesellschaftlichen Ansprüchen abzielte, unterlaufen zu wollen. Get it?

Twain also makes fun of the myriad meanings of the German *Zug* for which he lists no fewer than 29 different meanings. This reminded me of a 30th meaning: During World War II a secret telegram from Field Marshal Rommel's *Afrikakorps* was decoded (in part) as "*sie kommen in einem Zug*". One of the meanings of *Zug* is train. But there are no trains in the desert. So what did *Zug* mean in this instance? The decoding station luckily had a native speaker of German who said *Zug* in this case meant *platoon*. In other words the enemy would arrive in platoon strength. (Similar problems arose in the decoding of Japanese military messages. Just cracking the code was not enough: you needed a native speaker *and* someone familiar with Japanese military usage.)

23.1 Military Terms

One aspect of German usage that I do not particularly like is the abundance of military terms—even after two catastrophic wars. Thus, my chief secretary, was often standing *Gewehr bei Fuß* (rifle at the ready) when she wanted to signal that she was ready to take dictation. *Schützenhilfe* (help; literally: cover given by a rifleman) is another favorite, as is *Schulterschluß* (presenting a united front). Other examples are *austreten* (step out of line to pee), *Pkw* (from *Personenkraftwagen*; literally person power wagon, i.e. car), *Lkw* (from *Lastkraftwagen*; i.e. truck) and, the really horrible but thankfully disappearing, *Krad* (pronounced crud), military abbreviation for *Kraftrad* (literally: power wheel, i.e. motor bike).

My friend and comrade-in-words Eike Christian Hirsch has published a collection of military expressions still in wide use in his hilarious books. Here is a sampler:

generalstabsmässig, geballte Ladung, volle Deckung, bombensicher, Bombenstimmung, auf Vordermann bringen, Tuchfühlung, Rückendeckung und Flankenschutz, Durchbruch, Brückenkopf, Volltreffer. Etc.

Hirsch is also famous for such hilarious German word plays as "*DIE HAUT DER ARME ENTZWEI*", meaning either "the skin of one's arms..." or "he hit the poor one to pieces".

Why are even modern Germans wedded to the military terms of bygone days? Does it reflect a persisting affinity to *Disziplin und Ordnung* (discipline and order)? In a similar vein, American English abounds with words from the business world. I just read that an Olympian, who had won a second gold medal, had "doubled her holdings in gold". Who was it again, who said "America's business is business!" Charlie Wilson, the head of GM? No, he became famous for his "What's good for

General Motors is good for America." In a recent editorial in the New York Times (18 August 2006) Paul Krugman of Princeton, in a critique of conservative tax policies, coined the paraphrase "what's good for the rich is good for America".

But back to German. How can a "refined" lady, wife of a professor no less (albeit of chemistry), stoop to such abominable words as *Latrinenparolen* (rumors exchanged while sitting and doing one's business on a latrine)? I am always amazed how little attention many people pay to the origin of the words they use. Of course, in Germany, *derbe Sprache* (coarse language) is considered an asset by many native speakers; it's considered a sign of honesty. (Compare this with the prudishness of English as when one suburban housewife complains about her neighbor: "Her dog always goes to the bathroom on my front lawn.")

A pinnacle of prudishness is reached when, in the men's locker room of our local (New Jersey) swim club, men shower with their swim trunks left on. This reminds me of my mother, who was educated in a catholic convent, where students were strictly forbidden to see their own bodies. Once a week, they took a cleaning shower dressed in a one-piece bathing suit, applying the soap to the fabric rather than the naked flesh.

I was a little astonished when, not too long ago, I saw big posters all over Switzerland ironically proclaiming (in an anti-AIDS campaign) "*Ich schlafe nur mit gesunden Menschen—ich brauche keine Condoms.*" ("I sleep only with healthy people—I need no condoms."

The current German public pro-condom campaign is even more pregnant: It shows a large condom (in its virgin state) with the exhortation "*Mach's mit*" (literally: make it with, i.e. do it with condom). This is a fine pun on the frequent German plea "*Mach mit*" (come on, join in). These ads are probably more effective in preventing AIDS than calls for abstinence and "virginity pledges".

Prudishness also manifests itself when a toddler loses his or her diapers on the beach and people get very upset about such a "wardrobe malfunction", to use a current term. The original "wardrobe malfunction," displayed by Janet Jackson during intermission time of a football game at the 2004 Superbowl, of course wouldn't raise the tiniest eyebrows outside the USA, where dogs actually pee on front lawns (if not held back by a ubiquitous fence).

Another example that comes to mind is when, at the funeral of French President Mitterrand, his widow invited Mitterrand's paramour and their illegitimate child to walk behind the coffin.

23.2 Borrowed Words

Today, many sports terms in the UK and the USA, especially soccer, are German: *Eigentor* (own goal), *Abseits* (off side) and many others.

On the other hand, most languages, including German, have borrowed heavily from English (weekend, computer...). But not a few English words come from German: *Kindergarten*, *Blitzkrieg*, *Weltschmerz*, *Katzenjammer*, *Zeitgeist*,

Gemütlichkeit, *Wunderkind*, *Ersatz*, *Flak*, *Angst* and *Schadenfreude* come to mind. In fact, *über* (over) now appears in compounds such as *Übermensch* and überlobbyist. More German words entered English via Yiddish: *spiel*, *shtick*, *schmuck* and *schmo*, to name a few.

Some German words sound as if they had been taken from English but are actually German creations: *Handy* = cell phone; *Twen* = twenty-something person (a young man or woman between the ages of 20 and 29 years); *Smoking* = black-tie suit; *Beamer* = projector for digital images.

The language with perhaps the most words borrowed from German is Russian. Here are some of my favorites: *galstukh* (necktie), *bukhhalter* (bookkeeper), *shlagbaum* (road barrier), *butterbrot* (sandwich), *straf* (penalty), *strafbatalyon* (penal battalion), *feldweibel* (sergeant), *gefreiter* (corporal), *front* (army group), *pochtamt* (post office).

Bismarck, on his way to St. Petersburg as the new Prussian Minister, complained about the driver of his sled who always called the rider in front "*Verräter*" (traitor). Well, the Russian word for advance rider is the German *Vorreiter* but pronounced much like the German word for traitor (from W. Lehfeldt, Göttingen).

23.3 Gruntled

I once saw a cartoon in the *New Yorker* magazine showing two return windows in a department store, one window labeled "disgruntled" and the other one labeled "gruntled". What does "gruntled" mean? This reminded me of the many expressions in English (and German) in which only the negative form exists. *Ungeheuer* means monster. But there is no noun I know *Geheuer*.

Words like *Ungeheuer* are called "orphan negatives", of which English has quite a few other than disgruntled: inept and insipid come to mind. But actually, inept and insipid do have positive forms, albeit spelled differently: apt and sapid (having taste or intriguing, stimulating the mind) from the Latin *sapidus* (tasty).

Talking about negatives, I can't help thinking of the German prefix *un*, which can have the same meaning as the English *un* or *in*: *un*ending, *in*capable. But the German *un* has still another function besides negation, namely emphasis. Thus, *Untiefe* does not mean "undepth" (i.e. "shallow", as the innocent bather might discover to his chagrin) but great depth. (But, paradoxically, for a ship's captain looking at his charts, *Untiefe* does mean: "Attention! Shallow waters.")

Unkosten doesn't mean it's gratis, but that it costs extra. And *Unwetter* is not "unweather" but a heavy thunderstorm. An *Unfall* is an accident, not a "no-fall". I leave the reader with decoding *Ungezieferbekämpfungsmittel*, a typical German construction.

Other non-sequiturs of German are *bestimmt* (certainly) which often means "I am not sure". Or take *ganz* (wholly) which sometimes connotes "tolerable". Or look at *langsam* (slowly) which can mean "immediately" as in the phrase *Ich gehe langsam ins Bett*, i.e. I am going to bed right away (from L. Crawfords, Göttingen).

One idiosyncrasy of German, which makes it unnecessarily difficult for foreigners to comprehend, is its infatuation with the Latin neuter ending *um*, as in *Klinikum* (clinics), *Juridicum* (law school) and, perhaps worst (but perfectly legal), *Zentrum* (center). What is the lost foreign motorist, desperately seeking city center, to make of a road sign saying *Zentrum*? What kind of a tantrum is this, especially since around the world *Zen* is associated with Zen-Buddhism. So is *Zentrum* the place where Buddhists perform tantrums? Or what?

I was once berated by my colleagues in the Göttingen Academy that I had construed *Agenda* as a feminine singular in the title of a talk ("*Chaos: eine neue Agenda der Physik*"). I should have known they told me, that *Agenda* was the plural of *Agendum*, which is a neuter noun in Latin. Yes, I know. But what my friends, apparently disconnected from linguistic reality, didn't know was that the whole world, including Germany, has been using *agenda* as a singular for a long time.—The worst part of the episode was that one academician tried to console me that I didn't have to be ashamed for my linguistic lapse. Thank you, Professor W.

23.4 Du Versus Sie

Another difficulty of German (and other languages) is the strict adherence to the distinction between the familiar *Du* (thou) and the more formal *Sie* (you). Many German scientists call each other Helmut, Kurt, Karin, Erhard, Ingrid, etc. at international meetings where the *lingua franca* is English but revert to the formal *Herr Kollege*, *Frau Kollegin* when they speak German. A suitable compromise is to stay on a first-name basis in German but combine it with the formal *Sie*-form of address: Karin haben *Sie* (never Karin hast *Du*, which would carry a strong undertone of familiarity, if not intimacy).

23.5 English Versus German

I prefer English when it comes to reading (because of the simple syntax I read much faster in English than in German). However, the identity of nouns and verbs in English sometimes slows comprehension. Does "reports" mean (the) reports or (he) reports? Same with many other English noun/verb pairs. This is particularly problematic with newspaper headlines. What is meant by the headline "Ethics rules lag states." Are *rules* and *states* nouns or verbs? I had to read half the article to find out. (The proposed ethics rules in Congress lacked enough votes by the individual states to be adopted.)

During the Falkland war (before Margaret Thatcher put her foot down) the following headline in a British paper caught my attention: BRITISH LEFT WAFFLES ON FALKLANDS—hmm, how very considerate! (But my English friends never had any difficulty with this pronouncement.)

I prefer to speak German—after all my native language. And I can hear (understand) German speech better than English. In other words speaking and listening to speech were learned and programmed in the brain at a very early age. For the same reason, accents absorbed as a young child (before the age of 10 or 12) stay with us for the rest of our lives. (Audrey Hepburn, who grew up in Holland with Dutch, speaks English without much of her native accent—a rare exception to the general rule. Of course, she underwent extensive language coaching, as did Archibald Leach of Bristol while turning into Cary Grant in Hollywood.)

When it comes to writing, I do about equally well in either language—with a slight edge for English (again because of the simpler syntax, I guess).

Chapter 24
Retirement and Beyond

Manfred R. Schroeder

On September 30, 1987, I retired from Bell Laboratories 33 years after having been put on the Murray Hill payroll on September 30, 1954—the day I arrived in the USA on the *Andrea Doria*.

While my double appointment had thereby come to an end, my life as a professor of physics at the University of Göttingen continued unabated until September 30, 1991, when I became an emeritus.[1] Although *entpflichtet* (released) from many pleasant duties (lectures, chair General Physics Kolloquium, shepherding students from the *Studienstiftung*) and not-so-pleasant chores (committees, administration), my intellectual activities survived unshorn. I still offered (and offer) seminars on some of my favorite topics (speech, applications of number theory, fractals & chaos, and, lately, quantum cryptography) albeit to much smaller audiences (typically 5–15 students instead of 60–100 *Hörer* during my prime). In addition, I give the occasional talk on a variety of subjects (Bell Laboratories, Werner Heisenberg, Lise Meitner, Concert Hall Acoustics, Computer Graphics, Mathematics, and Music, The Development and Future of Physics). At the same time my attachment to the Göttingen Academy of Sciences, my second intellectual home, has grown stronger.

I also had more time to attend Manfred Eigen's *Winterseminare* in Klosters, Switzerland, where I met many famous scientists: John Wheeler, Lars Onsager, Murray Gell-Mann, Stephen Weinberg from the USA; Christopher Longuet-Higgins, Roger Penrose from the UK, and numerous Nobel laureates and near-Nobelists from around the world.

Author was deceased at the time of publication.

[1] I don't know why the date September 30 keeps cropping up in my life.—It was also the date I was kicked out of the *Luftwaffe* in 1944 (for having missed boot camp) and shoved off to the *Kriegsmarine*.

M.R. Schroeder (✉)
Drittes Physikalisches Institut, Georg-August-University Göttingen, Friedrich Hund-Platz 1, 37077 Göttingen, Germany

N. Xiang and G.M. Sessler (eds.), *Acoustics, Information, and Communication*, Modern Acoustics and Signal Processing, DOI 10.1007/978-3-319-05660-9_24,

The *Winterseminar* also allowed Anny and me to continue our "tradition" of annual skiing vacations in Switzerland, beginning in 1963 in Zermatt (three times, the first time I broke my shoulder on the downhill run to Cervinia in the Val d'Aosta, Italy[2]) and continuing in Arosa, Grindelwald (from where we took the train to the *Jungfraujoch* to ski into a distant valley along the *Aletschgletscher*—source of both the Rhine and the Rhone rivers).

Other slopes we hit included Davos, St. Moritz, Diavolazza; Seiser Alm, Italy, and Lac de Tigne (Club Med) and *Mer de Glace* (above Chamonix, France, but reached from Entrèves in Italy), and Verbier (twice; once we took a helicopter to a high glacier (Glacier du Trient) near the French border. Another time we made a very strenuous hiking tour on skis among the surrounding mountains (Rosa Blanche). And once we witnessed a large avalanche plow into the *beginner's* slope. Miraculously, although several skiers, including children, were buried under snow, they were all dug out—nobody was lost.).

We also skied in Crans-Montana, Saas-Fee and finally in Klosters with its incomparable—and incomparably long—downhill runs.

Before Switzerland we skied in New England: Stowe, Franconia Notch (Cannon Mountain), Killington (at −40 °F (−40 °C), when the snow feels like *sand*), Canada (Mont Tremblant), New York (Belleayre and Lake Placid), Squaw Valley, California, and, for the faint of heart, *New Jersey* (McAffee—only an hour's drive from New York City).

With all this skiing, from Switzerland to California, we only hit the slopes twice in Austria: St. Anton am Arlberg (including Lech and Zürs) and Kitzbühel (where I broke my leg in 1969).—Austria somehow never seemed as attractive to us as Switzerland.

We also skied on Camel Back Mountain, Pennsylvania; Snowbird and Park City, Utah; Aspen, Vail and Copper Mountain (another Club Med) in Colorado; Jackson Hole, Wyoming (at a freezing −37 °C, corresponding to −34 °F).

I also skied several times at Alta, Utah, with its superb dry powder snow and only a $7 taxi ride (in the 1960s) from the Salt Lake City airport, where I often stopped over enroute from New York to California. (The only thing I didn't like about Alta was that it was also dry in the *liquor* sense: you had to bring your own bottle of Scotch or you were out of luck—as I was during my first visit.)

With Ed David and Ann, often accompanied by their dog Gretchen, in their Studebaker Starliner we picked up Anny near the George Washington Bridge in New York on Fridays at 4 p.m. Around midnight, we reached our destination in Vermont or New Hampshire, ready to ski the next day and a half. (Ed always made sure that Anny and I occupied separate bedrooms—this was in the "chaste old days" of 1955.) There were no Interstate Highways then and once we reached New York

[2] My Zermatt doctor was a descendent of the physician who accompanied the first, ill-fated, ascent to the *Matterhorn* on July 14, 1865. (On the way down, four of the climbers fell to their deaths.) He was also responsible for keeping the outbreak of typhoid fever in 1963 under wraps.—My broken shoulder he diagnosed as "twisted" until an X-ray in the USA revealed the true state of affairs.

as late as Monday morning at 8. The next 4 days we relaxed at work for another weekend on the slopes. (How we ever got any work done is beyond me!)

Nevertheless, although not remotely of Nobel calibre, I was honored with a number of Gold Medals: The Audio Engineering Society in New York (already in 1972), the Rayleigh Gold Medal of the British Institute of Acoustics in 1987, the Helmholtz-Medal of the German Acoustical Society in 1995 and the Gold Medal of the Acoustical Society of America in 1991 ("For theoretical and practical contributions to human communication through innovative application of mathematics to speech, hearing and concert hall acoustics"). Whew!

In 1993 I received the Lower Saxony State Prize "for outstanding scientific accomplishments" (signed by then Premier Gerhard Schröder[3]).

In 2004, I was awarded the Eduard Rhein Preis (for my "scientific life work in room acoustics, psycho-acoustics, speech coding and computer graphics, and especially for the invention of linear predictive coding, the most important foundation of speech coding and speech analysis." Good citation!).

Also in 2004 the International Speech Communication Society, in a meeting on Jeju Island, Korea, gave me their Medal "For significant scientific achievement in the field of Speech Communication Science." (While in Korea, we "inspected" one of the secret tunnels the North Koreans had dug under the Demilitarized Zone.)

Apart from these "metals," I received the greatest satisfaction from the enthusiastic reception of my books. (One Kristina Lerman even recommended taking my book on Number Theory to a coffee shop to keep her company.[4]) The enthusiastic review by Philip Morrison in *Scientific American* kept me awake all night—until I could call Springer the next morning to print an extra 2,000 copies.

When I returned to Göttingen (the book was written in Murray Hill) I was amazed to learn that there was no street named of Hermann Minkowski to whom I had dedicated the book. With considerable patience (and the help of *Oberbürgermeister* Rinck) this was eventually rectified: there is now a *Minkowskiweg* (near the Frank-Born-Ring in the north of the city). At the same time the city created an Emmy-Noether-Weg, a Hermann-Weyl-Stieg, a Maria-Göppert-Weg.

---

[3] In his presentation speech the future Chancellor of Germany mentioned that he had heard that I liked good wines. He therefore thought it fitting to nominate me for membership in the *Toskana Fraktion* (Tuscany Faction). Not following German internal politics, I didn't know what he meant. Friends had to tell me that *Toskana Fraktion* designated a (fictitious) club of well-to-do *SPD* (social democratic party) bigwigs that maintained private villas in Tuscany and generally preferred the "good" life. Thank you, Gerhard!

[4] On her webpage *All the Books I have Known* she wrote: I was never terribly interested in number theory, but I regret it now. It is a fascinating subject, and Schroeder's book is a wonderful place to start. He is highly readable, if prone to corny jokes, but the best thing is his ability to show how obscure statements of mathematics can have far-reaching effects, e.g., how fundamental theorem of arithmetic leads to the well-tempered scale, and other wonderments. His other book *Chaos, Fractals and Power Laws* is a similarly engaging look at the wide ranging manifestations of chaos, and a good non-technical, though intellectually demanding introduction to the field.

Fig. 24.1 Nikki (2) on a visit to Nikolausberg

When the city asked me about Fermi, I said, of course, Göttingen should have a street named after Fermi—post-doc in Göttingen, Nobel prize winner and creator of the first "atomic pile" (now called nuclear reactor) in December 1942—something Heisenberg and his collaborators never managed before the end of the war. "Aber der Kerl hat sich nicht abgemeldet!" (But the chap never deregistered with the city!) I persuaded the city to name a street after Fermi anyhow: And now Göttingen has an *Enrico-Fermi-Eck* (Fermi corner)—not quite a street.

Later the city called about Johannes Starck. I asked: "Where did you get that name?" "Well, he was a student in Göttingen and he won a Nobel prize in physics." "And which other sources did you consult?" "The *Brockhaus Enzyklopädie*." I said: "Go to the Encyclopedia Brittanica, where it is clearly stated that Starck was the top Nazi among German physicists."

When I told Friedrich Hund this story, he stopped in mid-track, exclaiming: "Ich ziehe weg von Göttingen!" (I will move away.) I said: "No need to move away, Herr Hund." I persuaded the city *not* to name a street after Starck after all—it would have been a disaster.

Fig. 24.2 Lilly and Julia (both 7) in Bremen

In July 2001 I suffered a minor stroke at our home in Berkeley Heights, New Jersey. Luckily, our daughter Marion, a neurologist, was staying with us at the time. She could provide a quick diagnosis and arrange a speedy admission to Emergency at Overlook Hospital in Summit. This may have saved me from more serious consequences. In any case, I was discharged after a mere 9 days (I had asked Dr. Michael Suhl, who became a good friend, whether my discharge was an "honorable discharge," which he confirmed). Thus I could join family and friends for my 75th birthday celebration in Manasquan on the Jersey shore. Dr. Suhl told me that I could no longer ski which I felt was an exaggerated precaution. But in actual fact I could no longer *schuss* because I lacked the necessary balance on my feet—the only persistent residue from my mishap. In the meantime life that doesn't require physical balancing acts continues. I am 80 now but still "going strong."

The time after my second retirement (as a professor, in 1991) was further brightened by the arrival of four granddaughters: Julia in La Jolla and Lilly in Bremen, both in early 1994. Next came Nora, also in Bremen, in 1997, and then Nikki in La Jolla in 1999 (Figs. 24.1, 24.2, and 24.3).

In 2003 all 12 of us got together for Anny's 75th birthday (and my 77th) in Riva del Sole near Castiglione della Pescáia on the coast of Tuscany. A memorable vacation with breakfast on a sun drenched patio near the sea—with another cup of cappuccino and the *Herald Tribune* at my side. (I always started the day with the cartoons and the anagram.) And at the end of the day, dinner *al fresco*, lasting from dusk into the night, under the beautiful umbrella pines so typical of Italy.—I rarely felt so European.

The children availed themselves of the opportunity for side trips to Lucca, Pisa, Elba (Napoleon's first exile), Florence, and Rome. In the evening there was dancing, even for the little ones.

Fig. 24.3 Nora (8) at Schloss Hardenberg and in Göttingen (the young Lauren Bacall?)

Insanely, one of the things I liked best about the whole trip was the 17-h train ride from Germany—or rather the night-train part across the Alps from Munich to Florence: I rarely slept so well—I just love trains. (The children from Germany drove and the Californians of course flew in.)

I am closing this memoir with a letter to friends touching on some of Anny's and my activities in 2006.

Chapter 25
The Year 2006: One of Our Best (So Far)

Manfred R. Schroeder

February 25 was the 50th anniversary of our wedding in the Chapel of the Riverside Church in Manhattan. We commemorated the day—just the two of us—at the Metropolitan Opera with a performance of *Samson and Delilah* (Fig. 25.1).

On May 1 my sister Helga celebrated her 70th birthday with a day of hiking, singing, and dancing at a country resort in the *Fränkische Schweiz* north of Nuremberg.

Later in the year, my New York nephew and godson David Harkness got engaged to Leslie Johnson of Kansas—the extended family is growing, still (Fig. 25.2).

Another memorable moment was my 80th birthday. During the weekend before the great day, our children, Marion from Bremen, Julian from San Diego, Alexander from San Francisco, and my sisters Ingrid from Hamm and Helga from Erlangen—together with their husbands, children, and grandchildren—arrived in Göttingen. The weather was beautiful if a bit too warm (as it had been all summer) so our first get-togethers took place on the terrace and lawn of our house in Nikolausberg. We particularly enjoyed seeing our four granddaughters, Lilly and Nora from Bremen and Julia and Nikki (Nicola Alexandra Jazzmin!) from California. And of course they had great fun seeing each other again—great pals that they are.

For the Sunday preceding the birthday we had invited 74 of our best friends, including my oldest friend, Hermann Precking (from a time when we were both 3 years old), to lunch in the Schloßhotel Hardenberg near Göttingen. We had counted on perhaps 50 guests but actually all but two (felled by the abominable heat that engulfed most of Europe) came. Our children and my sisters entertained us with amusing reminiscences. Professor Thomssen told a story involving the great

Author was deceased at the time of publication.

M.R. Schroeder (✉)
Drittes Physikalisches Institut, Georg-August-University Göttingen, Friedrich Hund-Platz 1, 37077 Göttingen, Germany

N. Xiang and G.M. Sessler (eds.), *Acoustics, Information, and Communication*, Modern Acoustics and Signal Processing, DOI 10.1007/978-3-319-05660-9_25,

Fig. 25.1 At the Metropolitan Opera, New York, February 25, 2006

French mathematician Pierre de Fermat and Herbert Roesky, the President of the Göttingen Academy, gave a heart-warming speech about my close attachment to the Academy—My actual birthday, on July 12, was a comparatively modest affair on our terrace. But I was especially pleased by the presence of our *Ortsbürgermeisterin* Dr. Gabriele Funck who, in an earlier incarnation, had been our family physician.

Two weeks later we celebrated Anny's birthday with a most enjoyable breakfast attended by many of her friends at the patisserie *Cron & Lanz* (one of Europe's best, in our opinion).—Two days later we were off to New Jersey and our second home. Of course it was just as hot on the East Coast as in Europe, but we didn't mind because our house there is fully air-conditioned. Twice every day we went swimming, savoring a seemingly endless summer with occasional forays to the Big Apple.

Fig. 25.2 Leslie Johnson in Port Washington, New York, Easter 2007

25.1 Hawaii

On November 11 we left NJ to visit Alexander and our friends in San Francisco, where we celebrated the 80th birthday of Max Mathews with a brunch at the Sheraton Palace (where sailor Max and his future wife Marjorie had their first date some 58 years earlier). This was another chapter of our long friendship with the Mathews. Among our joint ventures were many sailing trips—from New York harbor up the East Coast to Maine, in the Caribbean and, most notably, in the Aegean among the Cycladic Islands. We visited Nauplion (Agamemnon's house) and anchored at Delos, Naxos, and Cap Sunion. I even took out a *Small Craft* license from the U.S. Coast Guard just in case the main skipper, Max, was incapacitated—which thankfully never happened.[1]

[1] In Delos (birthplace of Apollo), the aromatic smell of fresh potato pancakes wafted across the bay, emanating from a boat that flew the German flag. Starved for fresh food (after 2 weeks of canned stuff), I shouted across the bay "Haben Sie frische Kartoffeln?" "Nein." Then where does the wonderful smell come from? Answer: "Wir haben Kartoffelpuffer-Mix!" "Okay, could you spare some?" After asking permission to board our boat, which was granted, they brought a pound or two of the potato pancake mix—delicious! In Cap Sunion, one of us (Guy Mathews) had to dive into the bay to get enough cash to pay the restaurant. At another Island, I swam ashore with my clothes in a dry bucket. But when I asked (by sign language) for a private place to change, they sent me to the public area where half the village youths congregated to watch the proceedings. Since we had opted to sail without a dinghy we sometimes had to walk fully clothed into the water to reach our boat and onlookers were probably suspecting suicides.

After a delightful week in the Bay Area, we decamped for Honolulu and a cruise among the Hawaiian Islands. We had never wanted to do a cruise but a good friend, Thomas Rossing, persuaded us to join them on this—in retrospect—magnificent adventure. Our ship, the PRIDE OF HAWAII of the Norwegian Cruise Lines, was sailing under the US flag because its itinerary was completely in American waters. It was built in Germany, served by an American crew and owned by a Malaysian company—talk about globalization! (Not to mention the French, Italian, Chinese, and Japanese eateries aboard.)

After the cruise we participated in a Joint Meeting of the Acoustical Societies of America and Japan in Honolulu. The outstanding event of the meeting for us was a special session on my life in acoustics with speakers from Germany, the USA, Japan, France, and the Netherlands. I was particularly pleased to hear how some of the concepts that I had proposed a long time ago and named after me (Schroeder frequency, Schroeder integration, Schroeder phases, Schroeder diffusors) are still in current use.[2]

Also, during the meeting in Hawaii, the American Institute of Physics conducted a lengthy interview with me (conducted by Gerhard Sessler at the *Sheraton Waikiki* poolside) for their Oral History Projects. I was happy to learn that the AIP has added several photos of me and our family to their *Emilio Segrè Visual Archives*.

After the meeting (and last swim in the Pacific on December 1) we flew back to Germany via Tokyo which means you fly only during daytime without losing any sleep. (While flying over northern Siberia you see the sun, after having first set over China, rise again in the *West*.)

In Tokyo we stayed again at our favorite hotel, the International House of Japan. As on many previous occasions, Hiroya Fujisaki acted as our gracious host.

After more than a year of trying in vain, we were finally able to connect reliably to the broadband Internet so Anny could play Bridge online. I indulged two of *my* hobbies by giving a seminar at the university on *Applications Of Number Theory* and a talk in Jena on *The Early History of Computer Graphics at Bell Laboratories*.

---

[2] My friend Manfred Eigen (from my first years in Göttingen in the 1940s), who won the Nobel Prize in chemistry (actually *physical* chemistry) in 1968, once remarked to me about the several things that bear my name and that only *one* concept (from mathematical physics) was named after him: *Eigenfunktion*. This was a nice joke because eigenfunctions have nothing to do with Manfred Eigen; they are named after the German *Eigenfunktion* (proper function).

Bibliography of Manfred R. Schroeder's Publications

1. Schroeder, M.R.: Number Theory in Science and Communication: With Applications in Cryptography, Physics, Biology, Digital Information, and Computing. Springer, New York (1983)
2. Schroeder, M.R.: Computer Speech Recognition, Compression, Synthesis. Springer, Berlin (1985)
3. Schroeder, M.: Hundert Jahre Friedrich Hund: Ein Rückblick auf das Wirken eines bedeutenden Physikers. Vandenhoeck und Ruprecht, Göttingen (1996)
4. Schroeder, M.R.: Die statistischen Parameter der Frequenzkurven von großen Räumen. Acustica **4**, 594–600 (1954)
5. Schroeder, M.R.: Eigenfrequenzstatistik und Anregungsstatistik in Räumen. Modellversuche mit elektrischen Wellen. Acustica **4**, 456–468 (1954)
6. Schroeder, M.R.: Messung von Dielektrizitätskonstanten durch Aufspaltung entarteter Eigenfrequenzen. Z. Angew. Physik **7**(4), 169–174 (1955)
7. Schroeder, M.R.: On the separation and measurement of formant frequencies. J. Acoust. Soc. Am. **28**(1), 159 (A) I4 (1956)
8. Schroeder, M.R.: On the steady-state transmission in rooms. J. Acoust. Soc. Am. **29**(11), 1258(A) I5 (1957)
9. Schroeder, M.R.: New interpretation of the stereophonic effect in Lauridsen's and similar binaural experiments. J. Acoust. Soc. Am. **29**(6), 774 (A) K4 (1957)
10. Schroeder, M.R.: Spatial fluctuations of steady-state sound fields in rooms. J. Acoust. Soc. Am. **30**(7), 679 (A) I2 (1958)
11. Schroeder, M.R.: An artificial stereophonic effect obtained from a single audio signal. J. Audio Engrg. Soc. **6**(2), 74–79 (1958)
12. Schroeder, M.R.: Vocoders for military use. In: Proc. Seminar on Speech (1959)
13. Schroeder, M.R.: Methoden zur Messung der Diffusität in Hallräumen. Acustica **9**, 256–264 (1959)
14. Schroeder, M.R., Logan, B.F.: The sound of rain. Frequenz **13**, 229–234 (1959)
15. Schroeder, M.R.: New results concerning monaural phase sensitivity. J. Acoust. Soc. Am. **31** (11), 1579 (A) J5 (1959)
16. Schroeder, M.R.: Measurement of sound diffusion in reverberation chambers. J. Acoust. Soc. Am. **31**(11), 1407–1414 (1959)
17. Schroeder, M.R.: New approach to time domain analysis and synthesis of speech. J. Acoust. Soc. Am. **31**(5), 852 (A) R2 (1959)
18. Schroeder, M.R.: Improvement of acoustic feedback stability in public address systems. J. Acoust. Soc. Am. **31**(5), 851 (A) Q7 (1959)
19. Schroeder, M.R., David Jr., E.E.: Vocoder and its application to the transmission of high-quality speech over narrow-band channels. J. Acoust. Soc. Am. **31**(1), 113 (A) D5 (1959)

N. Xiang and G.M. Sessler (eds.), *Acoustics, Information, and Communication*, Modern Acoustics and Signal Processing, DOI 10.1007/978-3-319-05660-9,

20. Schroeder, M.R.: Measurement of reverberation time by counting phase coincidences. In: 3rd International Congress on Acoustics (ICA 3), Stuttgart, vol. 2, pp. 897–901 (1959)
21. Schroeder, M.R.: Improvement of acoustic feedback stability in public address systems. In: 3rd International Congress on Acoustics (ICA 3), Stuttgart, vol. 2, pp. 771–775 (1959)
22. Schroeder, M.R.: Recent progress in speech coding at bell telephone laboratories. In: 3rd International Congress on Acoustics (ICA 3), Stuttgart, vol 1, pp. 202–210 (1959)
23. Harvey, F.K., Schroeder, M.R.: Subjective evaluation of factors affecting two-channel stereophony. J. Acoust. Soc. Am. **32**(11), 1521 (A) S8 (1960)
24. Schroeder, M.R., David Jr., E.E.: A vocoder for transmitting 10 kc/s speech over a 3.5 kc/s channel. Acustica **10**(1), 35–43 (1960)
25. Schroeder, M.R., Logan, B.F.: "Colorless" artificial reverberation. J. Acoust. Soc. Am. **32** (11), 1520 (A) S3 (1960)
26. Schroeder, M.R., Crystal, T.H.: Autocorrelation vocoder. J. Acoust. Soc. Am. **32**(7), 913 (A) B2 (1960)
27. Schroeder, M.R.: Stop feedback in public address systems. Radio Electronics **31**, 40–42 (1960)
28. David Jr., E.E., Schroeder, M.R., Logan, B.F., Prestigiacomo, A.J.: New applications of voice-excitation to vocoders. J. Acoust. Soc. Am. **33**(11), 1684 (A) SP6 (1961)
29. Harvey, F.K., Schroeder, M.R.: Subjective evaluation of factors affecting two-channel stereophony. J. Audio Engrg. Soc. **9**(1), 19–28 (1961)
30. Schroeder, M.R.: Vocoders. Encyclopedia of Electronics. Reinhold, New York (1961)
31. Schroeder, M.R.: Novel uses of digital computers in room acoustics. J. Acoust. Soc. Am. **33** (11), 1669 (A) N9 (1961)
32. Schroeder, M.R.: Improved quasi-stereophony and "colorless" artificial reverberation. J. Acoust. Soc. Am. **33**(8), 1061–1064 (1961)
33. Schroeder, M.R.: Further progress with colorless artificial reverberation. J. Acoust. Soc. Am. **33**(6), 852–853 (A) Q2 (1961)
34. Schroeder, M.R., Logan, B.F.: "Colorless" artificial reverberation. J. Audio Engrg. Soc. **9**, 192–197 (1961)
35. Sessler, G., Schroeder, M.R.: Sound velocity in slightly ionized gases. J. Acoust. Soc. Am. **33** (11), 1673 (A) P11 (1961)
36. Atal, B.S., Schroeder, M.R., Kuttruff, K.H.: Perception of coloration in filtered gaussian noise – Short-time spectral analysis by the ear. In: 4th International Congress on Acoustics (ICA 4), Kopenhagen, Vortrag H31 (1962)
37. David, E.E., Jr., Schroeder, M.R., Logan, B.F., Prestigiacomo, A.J.: Voice-excited vocoders for practical speech bandwidth reduction. In: Proceedings of International Symposium on Information Theory, Brüssel (1962)
38. Flanagan, J.L., Schroeder, M.R., Bird, C.M.: Single-channel speech interpolator for 2:1 bandwidth reduction. J. Acoust. Soc. Am. **34**(12), 2003 (A) X7 (1962)
39. Harvey, F.K., Cisek, J.G., Schroeder, M.R.: Initial characteristics of impulse response for real and artificial reverberation. J. Acoust. Soc. Am. **34**(5), 726 (A) K7 (1962)
40. Prestigiacomo, A.J., Schroeder, M.R.: Practical systems for quasi-stereophony. J. Acoust. Soc. Am. **34**(5), 737 (A) T3 (1962)
41. Schroeder, M.R., Logan, B.F.: A solution to problem of compatible single-sideband transmission. In: Proceedings of International Symposium on Information Theory, Brüssel (1962)
42. Schroeder, M.R., Logan, B.F., Prestigiacomo, A.J.: Harmonic compression – A new method for speech bandwidth compression. J. Acoust. Soc. Am. **34**(12), 2003 (A) X5 (1962)
43. Schroeder, M.R.: Frequency-correlation functions of frequency responses in rooms. J. Acoust. Soc. Am. **34**(12), 1819–1823 (1962)
44. Schroeder, M.R., Atal, B.S.: Generalized short-time power spectra and autocorrelation functions. J. Acoust. Soc. Am. **34**(11), 1679–1683 (1962)
45. Schroeder, M.R., Atal, B.S., Kuttruff, K.H.: Perception of coloration in filtered gaussian noise Short-time spectral analysis by the ear. J. Acoust. Soc. Am. **34**(5), 738 (A) U1 (1962)

46. Schroeder, M.R.: Frequency correlation of sound-transmission functions in rooms. J. Acoust. Soc. Am. **34**(5), 736 (A) S2 (1962)
47. Schroeder, M.R., Kuttruff, K.H.: On frequency response curves in rooms. Comparison of experimental, theoretical, and Monte Carlo results for the average frequency spacing between maxima. J. Acoust. Soc. Am. **34**(1), 76–80 (1962)
48. Schroeder, M.R.: Natural sounding artificial reverberation. J. Audio Engrg. Soc. **10**, 219–223 (1962)
49. Schroeder, M.R.: Correlation techniques for speech bandwidth compression. J. Audio Engrg. Soc. **10**(2), 163–166 (1962)
50. Schroeder, M.R.: Improvement of feedback stability of public address systems by frequency shifting. J. Audio Engrg. Soc. **10**(2), 108–109 (1962)
51. Schroeder, M.R., Logan, B.F., Prestigiacomo, A.J.: New Methods for Speech Analysis-Synthesis and Bandwidth Compression. Speech Communication Seminar R.I.T, Stockholm (1962)
52. Schroeder, M.R., Logan, B.F., Prestigiacomo, A.J.: New methods for speech analysis-synthesis and bandwidth compression. In: 4th International Congress on Acoustics (ICA 4), Kopenhagen, Vortrag G41 (1962)
53. Schroeder, M.R., Atal, B.S., Bird, C.M.: Digital computers in room acoustics. In: 4th International Congress on Acoustics (ICA 4), Kopenhagen, Vortrag M21 (1962)
54. Schroeder, M.R., Logan, B.F., Prestigiacomo, A.J.: New applications of voice-excitation to vocoders. In: Speech Communication Seminar R.I.T., Stockholm, vol. 2, Paper G–3 (1962)
55. Sessler, G., Schroeder, M.R.: Interaction of soundwaves with a gas discharge. In: 4th International Congress on Acoustics (ICA 4), Kopenhagen, Vortrag O25 (1962)
56. David Jr., E.E., Schroeder, M.R., Logan, B.F., Prestigiacomo, A.J.: New applications of voice-excitation to vocoders. J. Acoust. Soc. Am. **35**(7), 1116 (A) (1963)
57. Harvey, F.K., Cisek, J.G., MacLean, D.J., Schroeder, M.R.: Some aspects of stereophony applicable to conference use. J. Audio Eng. Soc. 11, 212–217 (1963)
58. Schroeder, M.R., Atal, B.S.: Computer simulation of sound transmission in rooms. IEEE Int. Conv. Rec. **11**(7), 150–155 (1963)
59. Schroeder, M.R., Logan, B.F., Prestigiacomo, A.J.: New methods for speech analysis-synthesis and bandwidth compression. J. Acoust. Soc. Am. **35**(7), 1116 (A) (1963)
60. Schroeder, M.R.: Bookreview: Ungeheuer, G. (Ed.), Elemente einer akustischen Theorie der Vokalartikulation (Elements of an Acoustic Theory of Vowel Articulation). J. Acoust. Soc. Am. **35**(5), 743 (1963)
61. Atal, B.S., Schroeder, M.R.: Subjective responses of nonexponential reverberation and delayed low frequencies in concert halls. J. Acoust. Soc. Am. **36**(5), 1011 (A) C8 (1964)
62. Noll, A.M., Schroeder, M.R.: Short-time 'Cepstrum' pitch detection. J. Acoust. Soc. Am. **36** (5), 1030 (A) N5 (1964)
63. Schroeder, M.R.: Improvement of acoustic-feedback stability by frequency shifting. J. Acoust. Soc. Am. **36**(9), 1718–1724 (1964)
64. Schroeder, M.R.: An acustical journey to the soviet union. J. Acoust. Soc. Am. **36**(7), 1403–1406 (AN) (1964)
65. Schroeder, M.R.: Digital-computer method of measuring concert-hall acoustics. J. Acoust. Soc. Am. **36**(5), 1011 (A) C6 (1964)
66. Schroeder, M.R.: Bookreview: Sapozhkov, M.A. (Ed.), Rechevoy Signal v Kibernetike i Svyazi (The Speech Signal in Cybernetics and Communications). J. Acoust. Soc. Am. **36**(1), 225 (1964)
67. Schroeder, M.R., Atal, B.S., Sessler, G., West, J.E.: Acoustical measurements in philharmonic hall (New York). In: Application of Digital Computer Techniques to the Evaluation of the Acoustics of Concert Halls, vol. 3. Akustische Konferenz, Budapest, pp. 380–384 (1964)
68. Schroeder, M.R., Sessler, G., Atal, B.S., West, J.E.: Application of digital computers in the design and evaluation of concert halls. In: Nerem Record, Northeast Electronics Research and Engineering Meeting, M–4.1 (1964)

69. Sessler, G., Atal, B.S., Schroeder, M.R., West, J.E.: Acoustical measurements in philharmonic hall (New York). J. Acoust. Soc. Am. **36**(5), 1011 (A) C7 (1964)
70. Atal, B.S., Schroeder, M.R., Sessler, G., West, J.E.: Reverberation times and early energies in philharmonic hall (New York). J. Acoust. Soc. Am. **37**(6), 1188 (A) J5 (1965)
71. Atal, B.S., Schroeder, M.R., Sessler, G.: Subjective reverberation time and its relation to sound decay. In: 5th International Congress on Acoustics (ICA 5), Lüttich, 1b, Vortrag G32 (1965)
72. Mermelstein, P., Schroeder, M.R.: Determination of smoothed cross-sectional area functions of the vocal tract from formant frequencies. J. Acoust. Soc. Am. **37**(6), 1186 (A) I7 (1965)
73. Mermelstein, P., Schroeder, M.R.: Determination of smoothed cross-sectional area functions of the vocal tract from formant frequencies. In: 5th International Congress on Acoustics (ICA 5), Lüttich, 1a, Vortrag A24 (1965)
74. Schroeder, M.R.: Response to Comments on 'new method of measuring reverberation time'. J. Acoust. Soc. Am. **38**(2), 359–361 (L) (1965)
75. Schroeder, M.R.: Artificial reverberation: review and future outlook. J. Acoust. Soc. Am. **37** (6), 1192 M1 (k. Ex.) (1965)
76. Schroeder, M.R.: New method of measuring reverberation time. J. Acoust. Soc. Am. **37**(6), 1187–1188 (A) J4 (1965)
77. Schroeder, M.R.: New method of measuring reverberation time. J. Acoust. Soc. Am. **37**(3), 409–412 (1965)
78. Schroeder, M.R.: Erwin Meyer (Receives Wallace Clement S. Award). J. Acoust. Soc. Am. **37**(1), 178–180 (AN) (1965)
79. Schroeder, M.R.: Some new results in reverberation theory and measurement methods. In: 5th International Congress on Acoustics (ICA 5), Lüttich, 1b, Vortrag G31 (1965)
80. Schroeder, M.R., Noll, A.M.: Recent studies in speech research at bell telephone laboratories (I). In: 5th International Congress on Acoustics (ICA 5), Lüttich, 1a, Vortrag A21 (1965)
81. Atal, B.S., Schroeder, M.R., Sessler, G., West, J.E.: Evaluation of acoustic properties of enclosures by means of digital computers. J. Acoust. Soc. Am. **40**(2), 428–433 (1966)
82. Schroeder, M.R.: Architectural acoustics. Science **151**(3716), 1355–1359 (1966)
83. Schroeder, M.R.: Simulation with a block diagram compiler. In: Symposium on "The Human Use of Computing Maschines", Bell Telephone Laboratories (1966)
84. Schroeder, M.R., Atal, B.S.: Nachahmung der Raumakustik durch den Elektronenrechner. Gravesaner Bl **27–28**, 124–131 (1966)
85. Schroeder, M.R., Hanauer, S.L.: Nonlinear time compression and time normalization of speech. J. Acoust. Soc. Am. **40**(5), 1243 (A) 1H9 (1966)
86. Schroeder, M.R., Flanagan, J.L., Lundry, E.A.: Speech band compression by analytic-signal rooting. J. Acoust. Soc. Am. **40**(5), 1242–1243 (A) 1H8 (1966)
87. Schroeder, M.R.: Complementarity of sound buildup and decay. J. Acoust. Soc. Am. **40**(3), 549–551 (1966)
88. Schroeder, M.R., Atal, B.S., Sessler, G., West, J.E.: Acoustical measurements in philharmonic hall (New York). J. Acoust. Soc. Am. **40**(2), 434–440 (1966)
89. Schroeder, M.R.: Residue pitch: a remaining paradox and a possible explanation. J. Acoust. Soc. Am. **40**(1), 79–81 (1966)
90. Schroeder, M.R.: Reverberation time: definition, theory, and measurement. J. Acoust. Soc. Am. **39**(6), 1230 (A) 3B2 (1966)
91. Schroeder, M.R.: Articulatory information from acoustic data. J. Acoust. Soc. Am. **39**(6), 1220–1221 (A) 2B1 (1966)
92. Schroeder, M.R.: Architectural acoustics. In: McGraw-Hill Yearbook Science and Technology, pp. 123–124 (1966)
93. Schroeder, M.R.: Vocoders: analysis and synthesis of speech. A review of 30 years of applied speech research. In: Proceedings of IEEE, 54(5), M. 5184 (1966)
94. Atal, B.S., Schroeder, M.R.: Predictive coding of speech signals. In: Proceedings of IEEE Conference on Communication Processing (1967)

95. Atal, B.S., Schroeder, M.R.: Study of sound decay using ray-tracing techniques on a digital computer. J. Acoust. Soc. Am. **41**(6), 1598 (A) E/AT4 (1967)
96. Schroeder, M.R.: Speech – Man's natural communication. IEEE Spectrum, 75–86 (1967)
97. Schroeder, M.R., Hanauer, S.L.: Interpolation of data with continuous speech signals. Bell Sys. Tech. J. **46**(8), 1931–1933 (1967)
98. Schroeder, M.R.: Electronic suppression of reverberation. J. Acoust. Soc. Am. **41**(6), 1597 (A) E/AT2 (1967)
99. Schroeder, M.R.: Determination of the geometry of the human vocal tract by acoustic measurements. J. Acoust. Soc. Am. **41**(4), 1002–1010 (1967)
100. Schroeder, M.R., Flanagan, J.L., Denes, P.B.: Speech research at bell yelephone laboratories. Research Inst. Logopedics & Phoniatrics Universität Tokio: Bell Telephone Laboratories, pp. 11–12 (1967)
101. Schroeder, M.R., Flanagan, J.L., Lundry, E.A.: Comments on "Bandwidth compression of speech by analytic-signal rooting". Proc. IEEE **55**(10), 1770 (1967)
102. Schroeder, M.R., Flanagan, J.L., Lundry, E.A.: Bandwidth compression of speech by analytic-signal rooting. Proc. IEEE **55**(3), 396–401 (1967)
103. Atal, B.S., Schroeder, M.R.: Predictive coding of speech signals. In: Wescon.Tech.Pap. 12, Paper 8/2, 7 (1968)
104. Atal, B.S., Schroeder, M.R.: Predictive coding of speech signals. In: 6th International Congress on Acoustics (ICA 6), Tokio, 13–16, Vortrag C–5–4 (1968)
105. Schroeder, M.R.: Directions for speech research. IEEE Spectrum, 83–88 (1968)
106. Schroeder, M.R.: Period histogram and product spectrum: new methods for fundamental-frequency measurement. IEEE Trans. Audio **AU-16**(3), 477 (A) 200 (1968)
107. Schroeder, M.R.: Reference signal for signal quality studies. J. Acoust. Soc. Am. **44**(6), 1735–1736 (L) (1968)
108. Schroeder, M.R.: Period histogram and product spectrum: new methods for fundamental-frequency measurement. J. Acoust. Soc. Am. **43**(4), 829–834 (1968)
109. Schroeder, M.R.: Similarity measure for automatic speech and speaker recognition. J. Acoust. Soc. Am. **43**(2), 375–377 (L) (1968)
110. Schroeder, M.R.: Computers in acoustics: symbiosis of an old science and a new tool. In: 6th International Congress on Acoustics (ICA 6), Tokio, Vortrag GP–6–1 (1968)
111. Atal, B.S., Schroeder, M.R.: Predictive coding of speech signals. IEEE Trans. Audio **AU-17** (1), 58 (A) 307 (1969)
112. Lummis, R.C., Schroeder, M.R.: Technique for enhancing signals for auditory detection. J. Acoust. Soc. Am. **45**(5), 1290 (L) (1969)
113. Schroeder, M.R.: Images from computers. IEEE Spectrum, 66–78 (1969)
114. Schroeder, M.R.: Images from computers and microfilm plotters. Communications ACM **12** (2), 95–101 (1969)
115. Schroeder, M.R.: Computers in acoustics: symbiosis of an old science and a new tool. IEEE Trans. Audio **AU-17**(4), 309 (A) 527 (1969)
116. Schroeder, M.R.: Reference signal for signal quality studies. IEEE Trans. Audio **AU-17**(2), 178 (A) 379 (1969)
117. Schroeder, M.R.: Relation between critical bands in hearing and the phase characteristics of cubic difference tones. J. Acoust. Soc. Am. **46**(6), 1488–1492 (1969)
118. Schroeder, M.R.: Spatial averaging in a diffuse sound field and the equivalent number of independent measurements. J. Acoust. Soc. Am. **46**(3), 534 (L) (1969)
119. Schroeder, M.R.: Effect of frequency and space averaging on the transmission responses of multimode media. J. Acoust. Soc. Am. **46**(2), 277–283 (1969)
120. Schroeder, M.R.: Computers in acoustics: symbiosis of an old science and a new tool. J. Acoust. Soc. Am. **45**(5), 1077–1088 (1969)
121. Sessler, G., West, J.E., Schroeder, M.R.: Toroidal microphones. IEEE Trans. Audio **AU-18** (1), 69–70 (A) 583 (1969)

122. Sessler, G., West, J.E., Schroeder, M.R.: Toroidal Microphones. J. Acoust. Soc. Am. **46**(1), 28–36 (1969)
123. Atal, B.S., Schroeder, M.R.: Adaptive predictive coding of speech signals. Bell Sys. Tech. J. **49**(8), 1973–1986 (1970)
124. Schroeder, M.R.: Parameter estimation in speech: a lesson in unorthodoxy. IEEE Trans. Audio **AU-18**(4), 505–506 (A) 806 (1970)
125. Schroeder, M.R.: Synthesis of low-peak-factor signals and binary sequences with low autocorrelation. IEEE Transact. Inf. Theory **IT-16**(1), 85–89 (1970)
126. Schroeder, M.R.: Digital simulation of sound transmission in reverberant spaces. J. Acoust. Soc. Am. **47**(2), 424–431 (1970)
127. Schroeder, M.R.: Parameter estimation in speech: a lesson in unorthodoxy. Proc. IEEE **58**(5), 707–712 (1970)
128. Hall, J.L., Schroeder, M.R.: Monaural phase effects for two-tone Signals. J. Acoust. Soc. Am. **51**(6), 1882–1884 (1972)
129. Henze, H., Schroeder, M.R.: Aufsteilung der Wellenfront durch Nichtlinearitäten (Ein elektrischer Modellversuch zur Entstehung von Stoßwellen). Acustica **26**(5), 302–303 (1972)
130. Schroeder, M.R.: Neural models in autditory perception. In: Proceedings of 20th International Congress on Psychology, Tokio, 210 (A) (1972)
131. Schroeder, M.R.: An integrable model for the basilar membrane. In: Symposium on Hearing Theory IPO, Eindhoven, pp. 123–133 (1972)
132. Hall, J.L., Schroeder, M.R.: Computer simulation of a new hair cell model. J. Acoust. Soc. Am. **54**(1), 283 (A) K8 (1973)
133. Schroeder, M.R.: Computer models for concert hall acoustics. Am. J. Phys. **41**(4), 461–471 (1973)
134. Schroeder, M.R., Hall, J.L.: A model for mechanical-to-neutral transduction in the auditory receptor ('Hair Cell'). J. Acoust. Soc. Am. **54**(1), 283 (A) K7 (1973)
135. Schroeder, M.R.: An integrable model for the Basilar membrane. J. Acoust. Soc. Am. **53**, 429–434 (1973)
136. Schroeder, M.R.: Models in hearing. In: Speech and Hearing, Chelsea College, Spring Meeting, London, Session A, Paper 73SHA2 (1973)
137. Atal, B.S., Schroeder, M.R.: Recent advances in predictive coding – Applications to speech synthesis. In: Speech Communication Seminar, SCS-74, Stockholm, vol. 1, pp. 27–31 (1974)
138. Schroeder, M.R.: Speech wave processing – Some personal reminiscences and expectations. In: Speech Communication Seminar, SCS-74, Stockholm (1974)
139. Schroeder, M.R., Hall, J.L.: A model for mechanical to neural transduction in the auditory receptor. Acustica **31**(6), 362 (1974)
140. Schroeder, M.R., Gerlach, R.E.: Diffusion, room shape and absorber location – Influence on reverberation time. J. Acoust. Soc. Am. **56**(4), 1300 (T) (1974)
141. Schroeder, M.R., Gottlob, D., Siebrasse, K.-F.: Comparative study of European concert halls: correlation of subjective preference with geometric and acoustic parameters. J. Acoust. Soc. Am. **56**(4), 1195–1201 (1974)
142. Schroeder, M.R., Hall, J.L.: A model for mechanical to neural transduction in the auditory receptor. J. Acoust. Soc. Am. **55**(5), 1055–1060 (1974)
143. Schroeder, M.R., Gerlach, R.E.: Diffusion, room shape and absorber location – Influence on reverberation time. J. Acoust. Soc. Am. **55**(2), 420 (A) Q7 (1974)
144. Schroeder, M.R.: Recent advances in hearing research nonlinear mechanics and neural transduction. In: 8th International Congress on Acoustics (ICA 8), London, pp. 47–61 (1974)
145. Atal, B.S., Schroeder, M.R.: Linear prediction analysis of speech based on a pole-zero representation. J. Acoust. Soc. Am. **58**(1), S96 (A) VV4 (1975)
146. Atal, B.S., Schroeder, M.R., Stover, V.: Voice-excited predictive coding system for lowbit rate transmission of speech. In: Proceedings of International Congress on Communications, San Francisco, pp. 30–37–30–40 (1975)

147. Calavrytinos, P.G., Schroeder, M.R.: Erkennung ganzer Zahlwörter über die Energieverteilung in der Zeit-Frequenz-Ebene. In: Fortschritte der Akustik – DAGA '75, Braunschweig; Weinheim: Physik, pp. 385–388 (1975)
148. Eysholdt, U., Gottlob, Siebrasse, K.-F., Schroeder, M.R.: Räumlichkeit und Halligkeit – Untersuchung zur Auffindung korrespondierender objektiver Parameter. In: Fortschritte der Akustik – DAGA '75, Braunschweig; Weinheim : Physik, pp. 471–474 (1975)
149. Gottlob, D., Siebrasse, K.-F., Schroeder, M.R.: Neuere Ergebnisse zur Akustik von Konzertsälen. In: Fortschritte der Akustik – DAGA '75, Braunschweig; Weinheim: Physik, pp. 467–470 (1975)
150. Schroeder. M.R.: The auditory pathway. In: Proceedings of International Congress. "From Theoretical Physics to Biology", Versailles (1975)
151. Schroeder, M.R.: Amplitude behavior of the cubic difference tone. J. Acoust. Soc. Am. **58**(3), 728–732 (1975)
152. Schroeder, M.R.: Diffuse sound reflection by maximum-length sequences. J. Acoust. Soc. Am. **57**(1), 149–150 (1975)
153. Schroeder, M.R.: Computers, speech and images. Matematik och datorer i sprak-, tal- och musikforskning TR Nr. 1, Centrum för talkommunikationsforskning och musikakustik, KTH Stockholm, pp. 45–52 (1975)
154. Schroeder, M.R.: New results and ideas for architectural acoustics. In: Auditorium Acoustics. Proceedings of International Symposium on Architecture and Acoustics, Edinburgh, 1974, hrsg. v. R.Mackenzie; London: Appl. Sci. Publ., pp. 197–212 (1975)
155. Schroeder, M.R.: Models of hearing. Proc. IEEE **63**(9), 1332–1350 (1975)
156. Calavrytinos, P.G., Schroeder, M.R.: Sprecherunabh ängige Worterkennung ohne Segmentierung. Acustica **36**(3), 184–191 (1976)
157. Calavrytinos, P.G., Schroeder, M.R.: Sprecherunabhängige Ganzworterkennung ohne Segmentierung. In: Fortschritte der Akustik – DAGA '76, Heidelberg; Düsseldorf: VDI, pp. 649–652 (1976)
158. Gerlach, R.E., Schroeder, M.R.: Die Anwendung von Maximalfolgendiffusoren in einem Modellhallraum. In: Fortschritte der Akustik – DAGA '76, Heidelberg; Düsseldorf: VDI, pp. 255–258 (1976)
159. Hall, J.L., Schroeder, M.R.: Model study of nonlinear response to paired clicks. J. Acoust. Soc. Am. **60**(1), S40 (A) R6 (1976)
160. Möller, W.-J., Atal, B.S., Schroeder, M.R.: Determination of articulatory parameters of the human vocal tract from acoustic measurements. J. Acoust. Soc. Am. 60(1), S77 (A) HH10 (1976)
161. Möller, W.-J., Schroeder, M.R.: Kodierung von Sprache durch einhüllende und gewichtete Augenblicksfrequenz. In: 2. DFG-Kolloquium "Sprachverarbeitung", Michelstadt-V, pp. 7–8 (1976)
162. Schroeder, M.R.: A cocktail-party processor. In: Arden House Conference (1976)
163. Schroeder, M.R., Gerlach, R.E.: Response to "Comments on 'Diffuse sound reflection by maximum-length sequences'". J. Acoust. Soc. Am. **60**(4), 954 (L) (1976)
164. Schroeder, M.R., Möller, W.-J.: Concert hall measurements during musical performances. J. Acoust. Soc. Am. **60**(1), S9 (A) D2 (1976)
165. Schroeder, M.R.: Concert hall measurements during musical performances. Vortragsmanuskript (1976)
166. Schroeder, M.R., Gawron, H.-J., Gottlob, D., Möller, W.-J.: Messung akustischer Parameter mit Musiksignalen in besetzten Konzertsälen. In: Fortschritte der Akustik – DAGA '76, Heidelberg; Düsseldorf: VDI, pp. 237–240 (1976)
167. Steingrube, A., Schroeder, M.R.: Abänderung der Grundperiode stimmhafter Sprachsignale. In: Fortschritte der Akustik – DAGA '76, Heidelberg; Düsseldorf: VDI, pp. 645–648 (1976)
168. Strube, H.W., Schroeder, M.R.: Rauschbefreiung von Sprache bei Übertragung über mehrere unabhängig verrauschte Kanäle. In: Fortschritte der Akustik – DAGA '76, Heidelberg; Düsseldorf: VDI, pp. 665–668 (1976)

169. Wagner, J.R., Schroeder, M.R.: Untersuchung zur automatischen Sprechererkennung mit Ganzworten. In: Fortschritte der Akustik – DAGA '76, Heidelberg; Düsseldorf: VDI, pp. 653–656 (1976)
170. Calavrytinos, P.G., Wagner, J.R., Schroeder, M.R.: Automatic word and speaker recognition based on time-frequency domain processing. In: 9th International Congress on Acoustics (ICA 9), Madrid, 485, Vortrag I79 (1977)
171. Calavrytinos, P.G., Wagner, J.R., Schroeder, M.R.: Automatische Wort- und Sprecherkennung mit gleichen Parametern und Erkennungsmethoden. In: 3. DFG-Kolloquium "Sprachverarbeitung", Muggendorf, pp. 51–52 (1977)
172. Gottlob, D., Gawron, H.-J., Schroeder, M.R.: Measuring acoustical parameters of concert halls with music. In: 9th International Congress on Acoustics (ICA 9), Madrid, 3, Vortrag A1 (1977)
173. Jacobs, U., Schneider, W., Schroeder, M.R., Strube, H.W.: Vorarbeiten zur Sprachsynthese aus interpolierten Lautübergängen. In: 3. DFG-Kolloquium "Sprachverarbeitung", Muggendorf, pp. 63–64 (1977)
174. Schroeder, M.R., Gottlob, D.: New measurements for architectural acoustics. In: Proceedings of Seminar Measuremant Methods in Architectural Acoustics, Lund, Schweden (1977)
175. Schroeder, M.R.: New Viewpoints in Binaural Interaction. Academic Press, New York, pp. 455–467 (1977)
176. Schroeder, M.R.: Bookreview: Jassem, W. (Ed.), Speech analysis and synthesis. IEEE Trans. Acoust. Speech Signal Process ASSP **25**(3), 268–269 (B) (1977)
177. Schroeder, M.R.: Bookreview: Cremer, L. (Ed.), Die wissenschaftlichen Grundlagen der Raumakustik. Band II: Wellentheoretische Raumakustik. J. Acoust. Soc. Am. **62**(6), 1524–1525 (B) (1977)
178. Schroeder, M.R.: One binaural unmasking mystery unmasked. In: 9th International Congress on Acoustics (ICA 9), Madrid, 365, Vortrag H41 (1977)
179. Schroeder, M.R., Gerlach, R.E.: Diffuse sound reflection surfaces. In: 9th International Congress on Acoustics (ICA 9), Madrid, 138, Vortrag D8 (1977)
180. Schroeder, M.R.: In: Recognition of Complex Acoustic Signals, Dahlem Konferenzen, hrsg. v. T.H. Bullock: Life Sci.Res.Rep vol. 5, pp. 323–328 (1977)
181. Schroeder, M.R.: Machine processing of acoustic signals: what machines can do better then organisms (and Vice Versa). In: Recognition of Complex Acoustic Signals. Dahlem Konferenzen, hrsg. v. T.H. Bullock: Life Sci.Res. Rep, vol. 5, pp. 183–207 (1977)
182. Sondhi, M.M., Schroeder, M.R.: Acoustic camera using inverse diffraction. J. Acoust. Soc. Am. **62**(1), S22 (A) I7 (1977)
183. Steingrube, A., Schroeder, M.R.: Modification of pitch and duration of speech signals. In: 9th International Congress on Acoustics (ICA 9), Madrid, 521, Vortrag I115 (1977)
184. Strube, H.W., Schroeder, M.R.: Extraction of a speech signal from a noisy environment (Cocktail-Party Effect). In: 9th International Congress on Acoustics (ICA 9), Madrid, 435, Vortrag I29 (1977)
185. Atal, B.S., Schroeder, M.R.: Linear prediction analysis of speech based on a pole-zero representation. J. Acoust. Soc. Am. **64**(5), 1310–1318 (1978)
186. Atal, B.S., Schroeder, M.R.: Predictive coding of speech signals and subjective error criteria. In: IEEE ICASSP 78, Tulsa, vol. 18.2, pp. 573–576 (1978)
187. Calavrytinos, P.G., Wagner, J.R., Schroeder, M.R.: Kombinierte Wort- und Sprechererkennung. In: Fortschritte der Akustik – DAGA '78, Bochum, 13.-16. März; Berlin: VDE, pp. 493–496 (1978)
188. Gottlob, D., Kohlrausch, A., Schroeder, M.R.: Raumakustische Untersuchungen mit Hilfe der digitalen Schallfeldsynthese. In: Fortschritte der Akustik – DAGA '78, Bochum, 13.-16. März; Berlin: VDE, pp. 575–578 (1978)
189. Gottlob, D., Kohlrausch, A., Schroeder, M.R.: Einfluß raumakustischer Parameter auf die subjektive Beurteilung von Schallfeldern. In: Fortschritte der Akustik – DAGA '78, Bochum. Berlin: VDE-Verlag, 575–578 (1978)

190. Guicking, D., Lutze, R., Schroeder, M.R.: Messungen zur Diffusität im Halltank. In: Fortschritte der Akustik – DAGA '78, Bochum, 13.-16. März; Berlin: VDE, pp. 309–312 (1978)
191. Lewien, T., Calavrytinos, P.G., Schroeder, M.R., Strube, H.W.: Worterkennung mit Hilfe von Singulärwertzerlegung des Wortspektrogramms. In: Fortschritte der Akustik – DAGA '78, Bochum, 13.-16. März; Berlin: VDE, pp. 489–492 (1978)
192. Mehrgardt, S., Schroeder, M.R.: Die Wahrnehmbarkeitsschwelle von Unterschieden beim Paarvergleich komplexer Signale. In: Fortschritte der Akustik – DAGA '78, Bochum, 13.-16. März; Berlin: VDE, pp. 515–518 (1978)
193. Schroeder, M.R.: Modulation transfer functions: definition and measurement. IEEE Trans. Acoust. Speech Signal Process ASSP **26**(2), 180 (A) (1978)
194. Schroeder, M.R., Atal, B.S., Hall, J.L.: Optimizing digital speech coders by exploiting masking properties of the human ear. J. Acoust. Soc. Am. **64**(1), S139 (A) YY5 (1978)
195. Schroeder, M.R., Gottlob, D., Eysholdt, U.: Binaural dissimilarity: its importance and realization in room-acoustical design. J. Acoust. Soc. Am. **64**(1), S120 (A) TT1 (1978)
196. Schroeder, M.R., Gerlach, R.E., Gottlob, D.: Wall and ceiling structures for efficient sound diffusion over wide frequency bands. J. Acoust. Soc. Am. **64**(1), S120 (A) TT2 (1978)
197. Schroeder, M.R.: New results in auditorium acoustics at Göttingen. J. Acoust. Soc. Am. **64** (1), S35 (A) M1 (1978)
198. Strube, H.W., Schroeder, M.R.: A Cocktail-Party processor: noise reduction for multichannel speech signals. IEEE Trans. Acoust. Speech Signal Processing ASSP **26**(2), 180 (A) (1978)
199. Strube, H.W., Jacobs, U., Schneider, W., Schroeder, M.R.: Zur Sprachsynthese aus interpolierten Lautübergängen. In: Fortschritte der Akustik – DAGA '78, Bochum, 13.-16. März; Berlin: VDE, pp. 477–480 (1978)
200. Atal, B.S., Schroeder, M.R.: Predictive coding of speech signals and subjective error criteria. IEEE Trans. Acoust. Speech Signal Processing ASSP **27**(3), 247–254 (1979)
201. Atal, B.S., Schroeder, M.R.: Optimizing predictive coders for minimum audible noise. In: IEEE ICASSP 79, Washington, 13.2, pp. 453–455 (1979)
202. Calavrytinos, P.G., Schroeder, M.R.: Untersuchungen zur subjektiven Wahrnehmung verrauschter Signale. In: 4. DFG-Kolloquium "Sprachverarbeitung", Darmstadt, pp. 18–21 (1979)
203. Flanagan, J.L., Schroeder, M.R., Atal, B.S., Crochiere, R.E., Jayant, N.S., Tribolet, J.M.: Speech coding. IEEE Trans. Commun. **27**(4), 710–737 (1979)
204. Lewien, T., Schroeder, M.R., Strube, H.W., Calavrytinos, P.G.: Ganzworterkennung mit Hilfe einer Singulärwertzerlegung von Frequenz-Zeit-Spektrogrammen. In: 4. DFG-Kolloquium "Sprachverarbeitung", Darmstadt, pp. 121–124 (1979)
205. Schroeder, M.R., Atal, B.S., Hall, J.L.: Objective measure of certain speech signal degradations based on masking properties of human auditory perception. For Fant, G. on his 60th Birthday (1979)
206. Schroeder, M.R., Atal, B.S., Hall, J.L.: Objective measure of certain speech signal degradations based on masking properties of human auditory perception. In: Frontiers of Speech Communication Research, hrsg. v. B. Lindblom, S. Öhman: Academic Press, New York, pp. 217–229 (1979)
207. Schroeder, M.R., Atal, B.S., Hall, J.L.: Optimizing digital speech coders by exploiting masking properties of the human ear. J. Acoust. Soc. Am. **66**(6), 1647–1652 (1979)
208. Schroeder, M.R.: Integrated-impulse method measuring sound decay without using impulses. J. Acoust. Soc. Am. **66**(2), 497–500 (1979)
209. Schroeder, M.R., Gerlach, R.E., Steingrube, A., Strube, H.W.: Response to "On the theory of optimal plane diffusors" by Berkhout, A.J. et al. J. Acoust. Soc. Am. **65**(5), 1336–1337 (L) (1979)
210. Schroeder, M.R.: Binaural dissimilarity and optimum ceilings for concert halls: more lateral sound diffusion. J. Acoust. Soc. Am. **65**(4), 958–963 (1979)

211. Schroeder, M.R.: Bookreview: Cremer, L., Müller, H.A. (Eds.), Die wissenschaftlichen Grundlagen der Raumakustik. Band I. J. Acoust. Soc. Am. **65**(1), 280–281 (B) (1979)
212. Schroeder, M.R.: Progress in architectural acoustics. J. Acoust. Soc. Japan **1**(1), 71–77 (1979)
213. Schroeder, M.R., Strube, H.W.: Acoustic measurements of articulator motions. Phonetica **36**, 302–313 (1979)
214. Schroeder, M.R.: What acoustic stimulation can teach us about physiological mechanisms underlying speech perception, Repr. Experimental Brain Research, Suppl. II: Hearing Mechanisms and Speech; Springer, Berlin, pp. 247–253 (1979)
215. Schroeder, M.R.: Music perception in concert halls. In: Seminar Okt. 1978, Publ. Royal Swedish Academy of Music, Stockholm, vol. 26 (1979)
216. Atal, B.S., Schroeder, M.R.: Improved quantizer for adaptive predictive coding of speech signals at low bit rates. In: IEEE ICASSP 80, Denver, vol. 2, pp. 535–538, S7.8 (1980)
217. Atal, B.S., Schroeder, M.R.: Adaptive predictive coding of speech signals at low bit rates. In: 10th International Congress on Acoustics (ICA 10), Sydney, Vortrag A1–10.4 (1980)
218. Calavrytinos, P.G., Schroeder, M.R.: Über die Mithörschwelle von Schmalbandrauschen, verdeckt durch Sinustöne. In: Fortschritte der Akustik – DAGA '80, München, 11.-13. März; Berlin: VDE, pp. 643–646 (1980)
219. Calavrytinos, P.G., Schroeder, M.R.: Reduction of quantizing noise loudness. In: Signal Processing 1: Theories and Applications, EURASIP 80, pp. 641–646 (1980)
220. Calavrytinos, P.G., Lewien, T., Helling, D., Strube, H.W., Schroeder, M.R.: Two systems for automatic word and speaker recognition. In: 10th International Congress on Acoustics (ICA 10), Sydney, Vortrag A2–1,4 (1980)
221. Calavrytinos, P.G., Schroeder, M.R.: Subjective perception of noisy signals. In: 10th International Congress on Acoustics (ICA 10), Sydney, Vortrag A1–15,2 (1980)
222. Lewien, T., Strube, H.W., Calavrytinos, P.G., Schroeder, M.R.: Ganzworterkennung mit Hilfe einer Singulärwertzerlegung von Frequenz-Zeit-Spektrogrammen. In: Fortschritte der Akustik – DAGA '80, München, 11.-13. März; Berlin: VDE, pp. 735–738 (1980)
223. Schroeder, M.R., Hall, J.L.: Loudness of noise in the presence of tones: measurements and nonlinear model results. In: 5th International Symposium on Hearing, Noordwijkerhout (1980)
224. Schroeder, M.R., Hackman, D.: Iterative calculation of reverberation time. Acustica **45**(4), 269–273 (1980)
225. Schroeder, M.R.: Constant-amplitude antenna arrays with beam patterns whose lobes have equal magnitudes, (For Severin, H. on his 60th Birthday). AEÜ, Electronics and Communication, vol. 34, pp. 165–168 (1980)
226. Schroeder, M.R.: Acoustics in human communications: room acoustics, music, and speech. J. Acoust. Soc. Am. **68**(1), 22–28 (1980)
227. Schroeder, M.R.: Hearing mechanisms and speech. Experimental brain research supplementum. J. Acoust. Soc. Am. **46**(2), 233 (1980)
228. Schroeder, M.R.: Bookreview: Almeida, A. (Ed.), The physics of speech. Phonetica **37**, 204 (1980)
229. Schroeder, M.R.: Bookreview: Fry, D.B. (Ed.), Nasalitätsdetektion und Vokalerkennung. Phonetica **37**, 196 (1980)
230. Schroeder, M.R.: Toward better acoustics for concert halls. Phys. Today **33**(10), 24–31 (1980)
231. Schroeder, M.R.: Modulation transfer functions: definition and measurement. Acustica **49**(3), 179–182 (1981)
232. Schroeder, M.R.: Architektur und Akustik: Erst der Raum macht die Musik. Bild der Wissenschaft **9**, 106–114 (1981)
233. Schroeder, M.R.: Direct (Nonrecursive) relations between cepstrum and predictor coefficients. IEEE Trans. Acoust. Speech Signal Processing ASSP **29**(2), 297–301 (1981)
234. Schroeder, M.R.: Number theory in physics, engineering and art. Interdisciplinary Sci. Rev. **6** (3), 239–248 (1981)

235. Schroeder, M.R.: Gute Akustik – aber wie? (Good Acoustics – but how?). J. Acoust. Soc. Am. **69**(5), 1531 (1981)
236. Schroeder, M.R., Dudley, H.W.: A tribute. J. Acoust. Soc. Am. **69**(4), 1222 (N) (1981)
237. Schroeder, M.R., Atal, B.S.: Rate distortion theory and predictive coding. In: IEEE ICASSP 81, Atlanta, vol. 1, pp. 201–204 (1981)
238. Lewien, T., Strube, H.W., Schroeder, M.R.: Improvement of speech intelligibility for persons with sensorineural hearing impairment. In: Fortschritte der Akustik – FASE/DAGA '82, Göttingen, 13.-17. September; Bad Honnef: DPG-GmbH, pp. 1055–1058 (1982)
239. Rollwage, M., Ebeling, K.-J., Guicking, D., Schroeder, M.R.: Statistische Parameter zweidimensionaler monofrequenter Schallfelder. In: Fortschritte der Akustik – FASE/DAGA '82, Göttingen, 13.-17. September; Bad Honnef: DPG-GmbH, pp. 211–214 (1982)
240. Schroeder, M.R.: Bookreview: gauss: a biographical study. Interdisciplinary Sci. Rev. **7**(4), 329–330 (1982)
241. Schroeder, M.R., Mehrgardt, S., Strube, H.W.: Monaural phase effects: relase from masking. J. Acoust. Soc. Am. **72**, S90 (A) (1982)
242. Schroeder, M.R.: Concert hall acoustics and number theory? Acoustic precision measurement with music? J. Acoust. Soc. Am. **72**, S18 (A) (1982)
243. Schroeder, M.R.: Linear prediction, extremal entropy and prior information in speech signal analysis and synthesis. Speech Commun. **1**, 9–20 (1982)
244. Schroeder, M.R., Mehrgardt, S.: Auditory masking phenomena in the perception of speech. Stockholm, Mai (1982)
245. Schroeder, M.R.: A simple function and its fourier transform. The Mathematical Intelligencer **4**(3), 158–161 (1982)
246. Schroeder, M.R., Mehrgardt, S.: Auditory masking phenomena in the perception of speech. In: The Representation of Speech in the Perpheral Auditory System, hrsg. v. R. Carlson, B. Grandström; Elsevier Biomedical Press, Amsterdam, pp. 79–87 (1982)
247. Schroeder, M.R., Atal, B.S.: Speech coding using efficient block codes. In: IEEE ICASSP 82, Paris, vol. 3, pp. 1668–1671 (1982)
248. Alrutz, H., Schroeder, M.R.: A fast hadamard transform method for the evaluation of measurements using pseudorandom test signals. In: 11th International Congress on Acoustics (ICA 11), Paris, vol. 6, pp. 235–238 (1983)
249. Atal, B.S., Schroeder, M.R.: Predictive speech signal coding with reduced noise effects. J. Acoust. Soc. Am. **73**(2), 717–718 (P) (1983)
250. Mehrgardt, S., Schroeder, M.R.: Monaural phase effects in masking with multicomponent signals. In: Hearing – Physiological Bases and Psychophysics. Symp. Bad Nauheim, Apr. 1983, hrsg. v. R. Klinke, R. Hartmann; Springer, Berlin, pp. 289–295 (1983)
251. Polack, J.D., Alrutz, H., Schroeder, M.R.: Acoustique. – Une nouvelle methode pour decrire la reverberation. C.R. Acad. Sc. Paris 297, Serie II, 21–24 (1983)
252. Polack, J.D., Alrutz, H., Schroeder, M.R.: Measuring the reverberation time with music. In: 11th International Congress on Acoustics (ICA 11), Paris, vol. 6, pp. 389–392 (1983)
253. Schroeder, M.R.: Fritz E. Borgnis 1906–1982. J. Acoust. Soc. Am. **74**(1), 371–372 (N) (1983)
254. Schroeder, M.R.: Fritz E. Borgnis. Obituary: Physics Today **36**(5), 82 (1983)
255. Schroeder, M.R.: Where is the next mersenne prime hiding? The Mathematical Intelligencer **5**(3), 31–33 (1983)
256. Schroeder, M.R.: The Eikonal equation. The Mathematical Intelligencer **5**(1), 36–37 (1983)
257. Schroeder, M.R.: Speech and hearing: some important interactions. In: Proceedings of 10th International Congress on Phonetic Science, Utrecht, pp. 41–52 (1983)
258. Atal, B.S., Schroeder, M.R.: Stochastic coding of speech signals at very low bit rates. In: Links for the Future, Proceedings of International Conference on Communications ICC '84, Amsterdam, Sci. Syst. & Serv. Communications. IEEE: Elsevier Science Publishers B.V, pp. 1610–1613 (1984)
259. Polack, J.D., Alrutz, H., Schroeder, M.R.: The modulation transfer function of music signals and its applications to reverberation measurement. Acustica **54**(5), 257–265 (1984)

260. Schroeder, M.R.: Number theory in science and communication (Einige Anwendungen der Zahlentheorie). In: German-Japanese Seminar on Nonlinear Problems in Dynamical Systems – Theory and Applications. Universität Stuttgart, Inst. für Mechanik, hrsg. v. M. Hiller, H. Sorg (1984)
261. Schroeder, M.R.: Number Theory in Science and Communication – With Applications in Cryptography, Physics, Biology, Digital Information, and Computing. Springer, Berlin (1984)
262. Schroeder, M.R.: Linear prediction, entropy and signal Analysis. IEEE ASSP Mag **1**(3), 3–11 (1984)
263. Schroeder, M.R.: Number theory in music, speech, and acoustics. J. Acoust. Soc. Am. **76**(1), S58 (A) AA6 (1984)
264. Schroeder, M.R.: Progress in architectural acoustics and artificial reverberation: concert hall acoustics and number theory. J. Audio Engrg. Soc. **32**(4), 194–203 (1984)
265. Schroeder, M.R., Guicking, D., Rollwage, M.: Electronic reverberation influence by controllable wall impedance. Final Report. Pp. 88 TIB/A86-80904/GAR(No. 11,1987) Dissertation Universität Göttingen (k. Ex.) (1984)
266. Schroeder, M.R.: Speech technology for all seasons. Speech Technology (1984)
267. Smith, B.K., Sieben, U.K., Mehrgardt, S., Schroeder, M.R.: Maskierung durch strukturiertes Rauschen. In: Fortschritte der Akustik – DAGA '84, Darmstadt, 26.-29. März; Bad Honnef: DPG-GmbH, pp. 761–764 (1984)
268. Schroeder, M.R.: Linear predictive coding of speech: review and current directions. IEEE Commun Mag **23**(8), 54–61 (1985)
269. Schroeder, M.R.: The unreasonable effectiveness of number theory in acoustics. J. Acoust. Soc. Am. **78**(1), S59 (A) DD1 (1985)
270. Schroeder, M.R.: Some work on hearing at Göttingen. J. Acoust. Soc. Am. **77**(1), S62 (A) CC1 (1985)
271. Schroeder, M.R., Atal, B.S.: Stochastic coding of speech signals at very low bit rates: the importance of speech perception. Speech Commun. **4**(1–3), 155–162 (1985)
272. Schroeder, M.R.: Number theory and the real world. The Mathematical Intelligencer **7**(4), 18–26 (1985)
273. Schroeder, M.R.: Speech and speaker recognition. Basel: Karger, Bibliotheca Phonetica, vol. 12 (1985)
274. Schroeder, M.R., Atal, B.S.: Code-excited linear prediction (CELP): high-quality speech at very low bit rates. In: IEEE ICASSP 85, Tampa, vol. 3, pp. 937–940, 25.1.1.–4 (1985)
275. Strube, H.W., Helling, D., Krause, A., Schroeder, M.R.: Word and speaker recognition based on entire words without framewise analysis. Speech and speaker recognition, hrsg. v. M.R. Schroeder; Basel: Karger, Bibliotheca Phonetica, vol. 12 (1985)
276. Schroeder, M.R.: Number Theory in Science and Communication – With Applications in Cryptography, Physics, Digital Information, Computing and Self-Similarity. Springer, Berlin (1986)
277. Schroeder, M.R.: Computers in acoustics. Acoustics Australia **14**(1), 9–12 (1986)
278. Schroeder, M.R.: Some new results of hearing research. Acta Acustica **11**(1), 30–43 (1986)
279. Schroeder, M.R., Strube, H.W.: Flat-spectrum speech. J. Acoust. Soc. Am. **79**(5), 1580–1583 (Ex. S0144) (1986)
280. Schroeder, M.R.: Auditory paradox based on fractal waveform. J. Acoust. Soc. Am. **79**(1), 186–189 (L) (1986)
281. Schroeder, M.R.: Speech and speaker recognition (Buchvorlage in der Akademie- Sitzung 6. Dez. 1985). Jahrbuch der Akademie der Wissenschaften in Göttingen – 1986 (1986)
282. Schroeder, M.R., Lauterborn, W., Ronneberger, D., Pottel, R.: DPI Annual Report. Report 1 (1986)
283. Schroeder, M.R.: Predictive coding of speech: historical review and directions for future research. In: IEEE ICASSP 86, Tokio, 4, pp. 3157–3164, PL.2.1–8 (1986)

284. Smith, B.K., Sieben, U.K., Kohlrausch, A., Schroeder, M.R.: Phase effects in masking related to dispersion in the inner ear. J. Acoust. Soc. Am. **80**(6), 1631–1637 (1986)
285. Schroeder, M.R.: Cooperative and interactive models of hearing an introduction. In: 11th International Congress on Phonetic Science, Tallinn (1987)
286. Schroeder, M.R.: Das menschliche Gehör: Seine exquisiten sensorischen Leistungen und Grenzen. Akademie Vorlage (1987)
287. Schroeder, M.R., Sloane, N.J.A.: New permutation codes using Hadamard unscrambling. IEEE Trans. Inf. Theory **IT-33**(1), 144–146 (1987)
288. Schroeder, M.R.: Computers in architectural acoustics. J. Acoust. Soc. Am. **81**(1), S30 (A) M1 (1987)
289. Schroeder, M.R.: Normal frequency and excitation statistics in rooms: model experiments with electric waves. J. Audio Engrg. Soc. **35**(5), 307–316 (1987)
290. Schroeder, M.R.: Statistical parameters of the frequency response curves of large rooms. J. Audio Engrg. Soc. **35**(5), 299–306 (1987)
291. Schroeder, M.R.: Speech and speaker recognition. Karger Gazette; Basel: Karger **49**, 1–3 (1987)
292. Schroeder, M.R.: Is there such a thing as fractal music? Nature **325**(6107), 765–766 (1987)
293. Schroeder, M.R.: The unreasonable effectiveness of number theory in science and communication. Rayleigh Lecture; London (1987)
294. Simon, S., Püschel, D., Schroeder, M.R.: Bestimmung raumakustischer Eigenschaften mit Musik. Fortschritte der Akustik – DAGA '87, Aachen; Bad Honnef: DPGGmbH, pp. 721–724 (1987)
295. Schroeder, M.R.: Solved and unsolved problems in concert hall acoustics. J. Acoust. Soc. Am. **83**(1), S49–S50 (A) V3 (1988)
296. Schroeder, M.R.: Number-theoretic phase arrays and diffraction gratings with broad radiation (scattering) characteristics. Phys. Acoust. **18**, 1–20 (1988)
297. Schroeder, M.R.: Selbstähnlichkeit und Fraktale in der Physik. In: Fortschritte der Akustik – DAGA '88, Braunschweig; Bad Honnef: DPG-GmbH, pp. 91–110 (1988)
298. Schroeder, M.R.: Solved and unsolved problems in concert hall acoustics. In: ASA Meeting 88, MTG Paper (1988)
299. Schroeder, M.R.: Self-Similarity. W.H. Freeman, New York (1989)
300. Schroeder, M.R.: Number theory – The paradigm of abstract mathematics has many applications in science, engineering, even artistic design. IEEE Potentials, 14–17 (1989)
301. Schroeder, M.R.: Self-similarity and fractals in science and art. J. Audio Engrg. Soc. **37**(10), 785–808 (1989)
302. Schroeder, M.R.: Selbstähnlichkeit: Fraktale, Chaos und Potenzgesetze. Physik in unserer Zeit **20**, 144–152 (1989)
303. Schroeder, M.R.: Moderne Anwendungen Gauß'scher Konzepte aus der Zahlentheorie. Gauß-Gesellschaft Göttingen, Mitteilungen **27**, 69–73 (1990)
304. Schroeder, M.R.: Models of speech and hearing and their applications: overview and outlook. J. Acoust. Soc. Am. **87**(1), S13 (A) E1 (1990)
305. Schroeder, M.R., Kuttruff, H.: Zum Geburtstag 60. J. Acoust. Soc. Am. **71**(4), 241–242 (1990)
306. Schroeder, M.R.: Chaos, Fraktale und Selbstähnlichkeit: Eine neue Agenda der Physik. Jahrbuch der Akademie der Wissenschaften in Göttingen, 22–55 (1990)
307. Schroeder, M.R.: Fractals, Chaos, Power Laws: Minutes From an Infinite Paradise. W.H. Freeman, New York (1991)
308. Schroeder, M.R.: Circularly polarized acoustic fields. The number theory connection. Acustica **75**, 94–98 (1991)
309. Schroeder, M.R.: Selbstähnlichkeit und Chaos in der Natur. G. Sessler aus Anlaß seines 60. Geburtstages gewidmet. Acustica **73**(2), 53–57 (1991)
310. Schroeder, M.R.: Gold Medal of the Acoustical Society of America. J. Acoust. Soc. Am. **89** (4), 4 (1991)

311. Schroeder, M.R.: Chaos, Fraktale und Selbstähnlichkeit. Universitas Z. Interdiszipl. Wiss. **8**, 722–735 (1991)
312. Schroeder, M.R.: Lösung des Gaußschen Kryptogramms von 1812. In: Gauss- Gesellschaft e.V. Göttingen, Mitteilungen; Göttingen, vol. 28, pp. 87–89 (1991)
313. Schroeder, M.R.: Fractals, Chaos, Power Laws: Minutes from an Infinite Paradise. W.H. Freeman, New York (1992)
314. Schroeder, M.R.: An introduction to chaos and its control. J. Acoust. Soc. Am. **92**(4), 2402 (A) 4aMU1 (1992)
315. Schroeder, M.R.: The unreasonable effectiveness of number theory in physics, communication and music. Proc. Symp. Appl. Math. **46**, 1–19 (1992)
316. Schroeder, M.R.: From microwaves to architectural acoustics. In: 14th International Congress on Acoustics (ICA 14), Beijing, F 1–2 (1992)
317. Schroeder, M.R.: Synthetic speech: then and now. In: 14th International Congress on Acoustics (ICA 14), Beijing, P1 (1992)
318. Schroeder, M.R.: Listening with two ears. Music Perception **10**(3), 255–280 (1993)
319. Schroeder, M.R.: A brief history of synthetic speech. Speech Commun. **13**, 231–237 (1993)
320. Schroeder, M.R.: Zum Weltbild der Physik: Quantentheorie und Unbestimmtheit. Georgia Augusta; Göttingen, 35–38 (1994)
321. Schroeder, M.R.: Fraktale, Chaos und Selbstähnlichkeit – Notizen aus dem Paradies der Unendlichkeit. Heidelberg (1994)
322. Schroeder, M.R.: Zahlentheorie in der Physik. Physikalische Blätter **50**(12), 1123–1128 (1994)
323. Schroeder, M.R.: How probable is Fermat's last theorem? The Mathematical Intelligencer **16** (4), 19–20 (1994)
324. Schroeder, M.R.: Bookreview: Casti, John L. (Ed.), Searching for certainty: what scientists can know about the future. The Mathematical Intelligencer **16**(3), 72 (1994)
325. Schroeder, M.R.: How to generate thermal photons – On the computer. In: On Klauder's Path: a Field Trip, hrsg. v. G.G. Emch, G.C.Hegerfeldt, L. Streit; River Edge (NJ), pp. 197–202 (1994)
326. Schroeder, M.R.: Reverberation: theory and measurement. In: Proceedings of Wallace Clement Sabine Centennial Symposium, Cambridge, MA, pp. 75–80 (1994)
327. Schroeder, M.R.: Phase gratings with suppressed specular reflection. Acustica **81**, 364–369 (1995)
328. Schroeder, M.R.: Fractals in music. In: Fractals in the Future, hrsg. v. C.A.Pickover: St. Martin's Press, pp. 207–223 (1995)
329. Schroeder, M.R.: Objective parameters and subjective measurements related to the acoustic quality of concert halls. In: Proceedings of International Conference on Acoustic Quality of Concert Halls, hrsg. v. A.L. Sáenz, A.M. Arranz; Madrid, pp. 11–25 (1995)
330. Schroeder, M.R. (Hrsg.): Hundert Jahre Friedrich Hund. Göttingen, Nachr. d. Akad. d. Wiss. in Göttingen, II. Math.-phys. Klasse 1/1996 (1996)
331. Schroeder, M.R.: Chaos and fractals in music. Encyclopedia of Physics, ed. v. S.B. Parker; New York, 2nd edn (1996)
332. Schroeder, M.R.: The "Schroeder frequency" revisited. J. Acoust. Soc. Am. **99**, 3240–3241 (1996)
333. Schroeder, M.R.: Lucas numbers and continued fractions. Mathematics Magazine (1996)
334. Schroeder, M.R.: More on Magnus. The Mathematical Intelligencer **18**(1), 7 (1996)
335. Schroeder, M.R.: Probability and number theory in concert hall acoustics. In: Music and Concert Hall Acoustics, Conference Proceedings from MCHA 1995, hrsg. v. Y. Ando, D. Noson; London, Kap. 16, pp. 157–169 (1996)
336. Schroeder, M.R.: Number theory in science and communication – With applications in cryptography, physics, biology, digital information, and computing. Berlin: Springer, 3. erw. Auflage (1997)
337. Schroeder, M.R.: Number theory in acoustics. Curr. Topics Acoust. Res. **2**, 139–140 (1998)

338. Schroeder, M.R.: Acoustics in the audio engineering society. J. Audio Eng. Soc. **46**(1/2), 71–73 (1998)
339. Schroeder, M.R.: Concert halls: from magic to number theory. In: The Psychology of Music, hrsg. v. D. Deutsch; Academic Press, New York/London, pp. 25–46 (Kap. 2) (1998)
340. Schroeder, M.R.: Computer Speech. Springer, Berlin (1999)
341. Schroeder, M.R.: Reverberation and diffusion. In: Computational Acoustics in Architecture, hrsg. v. J.J. Sendra; Southampton: WIT Press, Chapter 1 (1999)
342. Schroeder, M.R.: Die Akustik von Konzertsälen, Physik und Psychophysik. Physikalische Blätter **55**(11), 47–50 (1999)
343. Schroeder, M.R.: Architectural acoustics. In: Wiley Encyclopedia of Electrical and Electronics Engineering, hrsg. v. J.G. Webster; Wiley, New York, vol. 1, pp. 631–635 (1999)
344. Schroeder, M.R.: Physik um die Jahrhundertwende – 1. Quanten und Unbestimmtheit. In: Europäische Jahrhundertwende, Wissenschaften, Literatur und Kunst um 1900, hrsg. v. U. Mölk; Göttingen: Wallstein, pp. 130–136 (1999)
345. Schroeder, M.R. (Hrsg.): Nachruf [auf] Friedrich Hund. In: Göttingen: Vandenhoeck & Ruprecht, Jahrbuch d. Akad. d. Wiss. zu Göttingen, pp. 193–200 (1999)
346. Strube, H.W.: Acoustic theory and modeling of the vocal tract. In: Computer speech, hrsg. v. M.R. Schroeder; Berlin/Heidelberg/New York: Springer, Appendix A, pp. 203–223 (1999)
347. Schroeder, M.R.: Erwin Meyer – Professor für Physik an der Universität Göttingen. In: Göttinger Jahrbuch 2000, hrsg. v. Geschichtsverein für Göttingen und Umgebung e.V.; Göttingen: Goltze, vol. 48, pp. 181–183 (2000)
348. Schroeder, M.R. (Hrsg.): "Computer Speech: Recognition, Compression, Synthesis" (Springer: Berlin, New York, 1999). In: Göttingen, Vandenhoeck & Ruprecht, Jahrbuch d. Akad. d. Wiss. zu Göttingen, pp. 227–241 (2001)
349. Schroeder, M.R.: Entwicklung und Zukunft der Physik. In: Wissenschaften 2001 – Diagnosen und Prognosen, hrsg. v. Akad. d. Wiss. zu Göttingen; Göttingen: Wallstein, pp. 157–178 (2001)
350. Quast, H., Schreiner, O., Schroeder, M.R.: Robust pitch tracking in the car envirement. In: Proceedings of ICASSP 2002, Orlando, FL, May 13–17, 2002, vol. 1, pp. 353–356 (on CD: 3087.pdf) (2002)
351. Schroeder, M.R.: Power laws: from Alvarez to Zipf. Glottometrics **4**, 39–44 (2002)
352. Schroeder, M.R.: Computer speech. In: Abstracts Eleventh International Colloquium on Numerical Analysis and Computer Science with Applications, Plovdiv (Bulgarien); Plovdiv, 64 [Abstract] (2002)
353. Schroeder, M.R.: Tammo's and Herman's Index. In: Past, presence and future of the Speech Transmission Index, hrsg. v. S.J. van Wijngaarden; Soesterberg: TNO Human Factors, 1–1 [Introduction] (2002)
354. Schroeder, M.R.: Sequences from number theory for physics, signal processing, and art. Acoust. Phys. **49**(1), 97–108 (2003)
355. Schroeder, M.R.: Automatyczne rozpoznawanie i synteza mowy [Automatic recognition and synthesis of speech]. Postepy Fizyki **54**, 107–112 (2003)
356. Xiang, N., Schroeder, M.R.: Reciprocal maximum-length sequence pairs for acoustical dual source measurements. J. Acoust. Soc. Am. **113**, 2754–2761 (2003)
357. Quast, H.: Speech dialogue systems and natural language processing. In: Computer Speech: Recognition, Compression, Synthesis, hrsg. v. M. R. Schroeder; Berlin/Heidelberg/New York: Springer, Chapter 4, pp. 67–106 (2004)
358. Schroeder, M.R.: Computer Speech: Recognition, Compression, Synthesis. Springer, Berlin (2004)
359. Schroeder, M.R.: Lise Meitner. Zur 125. Wiederkehr ihres Geburtstages. Jahrbuch der Akademie der Wissenschaften in Göttingen **2003**, 268–271 (2004)
360. Schroeder, M.R.: Werner Heisenberg. Erinnerungen eines ehemaligen Studenten. Jahrbuch der Akademie der Wissenschaften in Göttingen **2003**, 256–261 (2004)

361. Strube, H.W.: Acoustic theory and modeling of the vocal tract. In: Computer Speech: Recognition, Compression, Synthesis, hrsg. v. M.R. Schroeder; Springer, Berlin, Appendix A, pp. 247–268 (2004)
362. Schroeder, M.R.: Book reviews: acoustic absorbers and diffusers, theory, design and application. J. Acoust. Soc. Am. **117**, 3343 (2005)
363. Schroeder, M.R.: Music and mathematics. In: Nova Acta Leopoldina N.F. (Science and Music – The Impact of Music), hrsg. v. W. Auhagen, W. Ruf, U. Smilansky, H. Weidenmüller; Halle, 92 (341), pp. 9–15 (2005)
364. Schroeder, M.R.: Number theory in science and communication – With applications in cryptography, physics, digital information, computing and self similarity. Berlin: Springer, Springer Series in Information Sciences 7, 386 Seiten, 4. Auflage (2006)
365. Schroeder, M.R., Guicking, D., Kaatze, U.: Applied physics at the "Dritte" – Fruitful interplay of a wide range of interests. In: Oscillations, Waves, and Interactions: Sixty Years Drittes Physikalisches Institut, hrsg. v. T. Kurz, U. Parlitz, U. Kaatze; Göttingen: Universitäts, pp. 1–24 (2007)
366. Schroeder, M.R.: Number theory in science and communication – With applications in cryptography, physics, digital information, computing and self similarity. Berlin: Springer, Springer Series in Information Sciences 7, 432 Seiten, 5. Auflage (2008)
367. Schroeder, M.R.: Alltag mit der "Königin der Mathematik. Jahrbuch der Akademie der Wissenschaften in Göttingen **2008**, 157–170 (2009)
368. Schroeder, M.R.: Fractals, Chaos, Power Laws: Minutes from an Infinite Paradise. Dover, New York, 448 Seiten, 4. Auflage (2009)

Index

A

Absorbers of underwater sound, 242–245, 272
ACF. *See* Autocorrelation functions (ACF)
Acoustic echo, 3, 7, 11, 271
Acoustics, 3–11, 15–28, 33–54, 58, 78, 93–109, 114, 145–173, 178, 199, 214, 229, 241–273, 349, 365, 404, 409, 433, 440
Adaptive design, 146
Adaptive predictive coding (APC), 206, 385, 388, 420
Aesthetics, 147, 148, 153, 306, 389
Alberich, 242, 243
Alpha rhythms, 146, 167, 169
Ambisonics, 10
Analog computer, 27
APC. *See* Adaptive predictive coding (APC)
Aperiodic modulation, 193
Apparent source width (ASW), 150, 153, 160, 164
Architectural acoustics, 13, 17, 33, 35, 42, 146–149, 152, 157, 170, 171, 417
Arrivals, 59–66, 69, 71–73, 151, 200, 317, 338, 339, 376, 380, 397, 403, 404, 435
Artificial reverberation, 9, 99–102, 366, 389, 394, 403
ASW. *See* Apparent source width (ASW)
Auditory evoked potentials, 168
Auditory filter, 118, 121, 122, 124, 125, 129, 135
Auditory neurocomputation, 164
Auditory system, 128, 133, 147, 160, 164, 414
Auralization, 9, 18
Auto-correlation, 103, 106–108, 153, 160, 162–164, 166, 169, 182, 183, 189, 386
Autocorrelation functions (ACF), 128, 130, 146, 149–151, 160, 161, 163, 166, 168–170, 271, 386

B

Background information, 42, 43, 48
Background noise, 35, 37, 38, 41, 83–85, 90
Basilar membrane, 117, 118, 121–127, 386, 421
Bayes' theorem, 42–44
Bayesian evidence, 45, 50, 53
Bayesian inference, 37, 40, 42–45, 53
Bayesian information criterion, 41, 45–46
Bayesian probabilistic inference, 34, 42
Beam tracing, 26
Bell laboratories, 114, 117, 200, 204, 213–226, 340, 363–399, 402, 410, 431, 440
Binaural, 9, 18, 99–101, 115, 117, 149, 150, 153, 154, 156, 160, 163, 165, 166
Binaural hearing, 117

C

Carl Friedrich Gauss, 181
CELP. *See* Code-excited linear prediction (CELP)
Channel impulse response, 98–99
Characteristic maximum-length sequence (MLS), 95, 102
Chinese remainder theorem, 106, 107, 178, 183
Classical reverberation time formulas, 200, 202
Close-talking microphone, 218
Coded sequence, 99, 100
Coded signal, 97
Code-excited linear prediction (CELP), 116, 199, 207–209, 388, 420
Coil spring absorber, 244, 245, 256, 258
Computer graphics, 27, 371, 389–392, 409, 431, 433, 440
Computer simulation of acoustics of concert halls, 200

N. Xiang and G.M. Sessler (eds.), *Acoustics, Information, and Communication*, Modern Acoustics and Signal Processing, DOI 10.1007/978-3-319-05660-9,

Concert hall, 6, 8, 21–24, 63, 69, 73, 81, 101, 115, 116, 146–149, 152, 154, 156, 158, 159, 166, 177, 199, 200, 286, 288, 349, 350, 352, 365, 368, 371, 372, 387, 389, 394, 396, 397, 409, 416, 419, 431, 433
  design, 8, 148–153, 156
Condenser microphone, 214, 224
Cone tracing, 26
Co-prime, 106
Critical bandwidth, 129, 134, 138
Critical frequency, 266–267, 349, 365
Cross-correlation, 52, 66, 82, 96, 98–100, 102, 103, 115, 150, 153, 160, 164, 166, 254, 386
  bound, 96, 97
  function, 96, 97, 99, 102, 254
Cross-over time, 57–73
Crosstalk canceler, 9, 10
Crystallography, 180

D

Decay compensation, 68, 84, 87
Decimation, 95, 96
Decimation factor, 95, 96
Decorrelation, 8, 94, 100–105
Deep water tank, 245–246, 269, 272
Degree of maximum-length sequence (MLS), 94
Delany–Bazley model, 234
Diffractal, 105, 190, 192, 193
Diffraction pattern, 108, 178, 182, 184, 185
Diffuseness, 18–20, 148, 150, 160, 164, 259–263
Diffuse reflection, 24–26
Diffusers, 19, 20, 105–109, 159, 177, 194, 256, 270–271
Diffusion, 37, 66, 72, 148, 180, 201
Diffusion coefficient, 186, 189–193
Diffusors
  aperiodic, 192
  binary amplitude, 195
  curvilinear, 194
  fractal, 105, 190
  primitive root, 105
  quadratic residue, 105, 106, 108, 185
Digital signal processing, 4, 114, 231, 371
Directional microphone, 11, 69, 216, 223, 260
Double-slope decay, 41
Dynamical properties, 58, 73

E

Early reflections, 6, 58, 70, 71, 100, 116, 148
Effective duration, 151, 152, 159, 162, 167, 169
Electret, 214–216, 218–226, 395
Electret microphone, 8, 214–223, 395
Electroacoustic transducer, 213, 214, 220–223
Electrorheological fluids, 272
Empirical distribution, 79, 85, 87, 89
Energy room impulse response, 33–37
Envelope, 37, 100, 115, 118, 119, 121, 123, 125, 128–138, 151
Envelope fluctuation, 128, 129, 132–138
Envelopment, 99–102, 104, 105, 153, 164
Ergodic, 59, 70, 71
Evidence, 44–46, 48, 50, 52, 53, 136, 321, 331, 345
Extensible Fourier transform (XFT), 58, 61, 64–70, 72, 73

F

Fast maximum-length-sequence (MLS) transform, 99
Finite-impulse-response (FIR) filter, 99, 100, 102
First-order Markov process, 203–204
Flat plate frequencies, 193
Flicker, 146, 169, 170
Flow absorber, 252, 270
Fourier transform, 78, 82, 87, 108, 189, 192, 265, 371, 391
Frequency shifting, 7, 8, 11, 253, 254, 393

G

Galois field, 95
GAs. *See* Genetic algorithms (GAs)
Gaussian, 58, 60, 72, 77–90, 128–138, 254, 272, 349, 387
Gaussian distribution, 45, 46, 60, 61, 79, 81, 83–85, 87, 128, 261
Genetic algorithms (GAs), 146, 156–160, 272
Gold sequence, 96
Goniometer, 186, 188, 189
Gradient microphone, 216–218, 223
Grating lobes, 108, 182, 189, 192, 193
Grazing incidence tube, 233–237

H

Harmonic complex tone, 115, 118–128
Headphone, 102, 117, 206, 219
Hearing research, 113–139, 386, 411–415
Hemispheric lateralization, 168, 169
Hilbert envelope, 130–132
Histogram, 85–89
Homogeneous, 27, 58, 71
Hybrid technique, 26

I

IACC. *See* Interaural cross-correlation coefficients (IACC)
IACF. *See* Interaural cross-correlation functions (IACF)
Impedance measurement, 231, 246–250, 270
Impedance tube, 230, 232, 233, 237, 244, 246–250, 272
Impulse response, 4, 6, 7, 16, 18, 26, 33–37, 41, 49, 57–73, 77–90, 93, 94, 97–101, 124, 186, 189, 190, 219, 350, 366, 387
Impulse response matrix, 97, 98
Individual preferences, 154
Integral equation, 25, 26, 267, 371
Interaural correlation, 150, 154, 155, 166, 167
Interaural cross-correlation, 103, 115, 164
Interaural cross-correlation coefficients (IACC), 100, 103, 148, 152–161, 163, 164, 167, 168
Interaural cross-correlation functions (IACF), 146, 149, 150, 152, 160, 161, 163, 165, 166
Interaural decorrelation, 100, 101
Interaural decorrelation coefficient, 100, 101
Invariant decimation, 95
Isotropy, 18, 58, 71, 256

K

Kasami sequence, 96–100, 102
Kolmogorov–Smirnov test, 79, 80, 85, 87, 89

L

Likelihood, 43–46, 48–52
Likelihood function, 44, 45, 49, 50
Lincoln Center, 3–5, 394
Linear convolution, 99
Linear feedback register, 94
Linear regression, 34, 37–42, 69
Listener envelopment, 102, 104, 105
Listening level, 149, 150, 154,

M

Masking, 115, 116, 121, 122, 125, 129, 130, 133–136, 138, 206–208, 388, 411
Matching pursuit (MP), 58, 61–64, 69–70, 72, 73
Mathematics, 113–142, 181, 301, 319, 331, 344, 345, 348, 351, 354, 362, 412, 414, 431, 433
Maximum-length sequence (MLS), 94–97, 101–105, 350
Mean free path, 16–18, 24, 69
MEMS microphone, 223, 224
Microphone (toroidal), 8, 216, 217
Microphone array, 6, 8, 9, 11, 31, 217, 220, 223, 395
Missing fundamental (flicker), 169
Mixing, 59, 70–73, 102, 313, 323
Mixing time, 59, 70–73
Model selection, 34, 42–46, 53
Modulation filter bank, 136
Multiple-slope decay, 35, 42, 53

N

Noise, 8, 34, 58, 77, 94, 116, 161, 205, 218, 233, 254, 303, 335, 344, 368, 418
Nonlinear regression, 34, 37–42
Number theory, 6, 94, 95, 109, 180, 182, 192, 271, 354, 362, 367, 375, 391, 408, 409, 413, 415, 419, 431, 433, 440

O

Optimization, 109, 130, 157, 193
Overlap, 57, 58, 63, 71, 78, 87, 246, 249, 252, 253, 255, 269, 272

P

Paired comparison tests, 150, 153, 154, 162, 164, 396
Parameter estimation, 34, 41, 42, 44–46, 52, 53
Peak factor, 113–142
Perception, 9–10, 103, 116, 136, 138, 139, 149, 161, 162, 164, 169, 231, 341, 349, 363, 372, 373, 389, 390
Perceptual error, 206
Phase grating diffuser, 106, 107
Physics Today, 177, 179, 320
Piezoelectret, 225, 226
Piezoelectric material, 225
Pitch, 7, 115, 119, 146, 150, 160, 162–163, 169, 206, 307, 309, 371, 384

Pitch perception, 119
Plate vibration, 263, 265
Posterior probability, 43–45, 48, 49
Predictive coding, 204–206, 385
Preferences, 9, 145–159, 161, 166–171
Preferred maximum-length sequence (MLS) pair, 96, 104
Preferred pair, 96
Pressure chamber, 247–248
Prime number, 105, 106, 180, 185, 353
Primitive polynomial, 95
Primitive root sequence, 105–107, 178
Prior probability, 43–45, 49, 50
Probability, 43–45, 50, 62, 69–71, 73, 78–80, 202, 203, 256
Probability distribution, 45, 48, 49, 78, 129, 202, 203
Pseudorandom noise, 100
Pseudorandom sequence, 94, 102
Psychoacoustics, 10, 71, 99, 114, 115, 118, 127, 139

Q

Quadratic residue sequence, 94, 105, 106, 108, 148, 182, 271

R

Random, 17, 25, 26, 34, 50, 51, 66, 77–81, 85, 102, 128, 130, 163, 208, 323, 329, 330, 350, 352, 372, 389, 391
Random noise, 34, 78, 79, 81, 94, 99, 100, 102, 122, 254, 255, 258
Ray approximation, 201
Ray tracing, 15–28, 42, 99, 199, 392
  on computers, 200–204
Reciprocal maximum-length sequence (MLS) pair, 95–97, 99–105
Reflection coefficient, 107–109, 230–232, 244, 272
Reflection phase grating, 177–210, 350, 396
Reflection time, 150–152, 155, 154, 159, 167, 168
Reflectors, 18–20, 22, 189, 191, 246, 248–251, 271
Reproducible, 129
Research contract, 241, 242, 245, 263, 271
Resiliently lined water duct, 248–249, 255, 263, 270
Resonance scattering, 271–272
Reverberance, 42, 99–102
Reverberation
  tail, 99–101
  tank, 242, 251–254, 256, 259–263
  time, 5, 6, 16–18, 20, 24, 33–35, 37, 38, 53, 58, 63, 70, 72, 78, 84, 89, 99–101, 116, 151, 152, 154–157, 200, 202–204, 232, 251, 252, 254, 257–259, 349, 350, 371, 396, 397
Room acoustics, 3–11, 15, 16, 18, 25–28, 33–54, 58–60, 62, 63, 70, 71, 78, 94, 99, 114, 229, 231, 232, 235, 237, 270, 271, 350, 433
Room impulse response, 6, 7, 33–37, 41, 49, 57–73, 81, 94, 99–101

S

Sabine and Eyring formulas, 203
Schroeder, M.R., 3–6, 9, 11, 15, 39, 42, 77, 94, 114, 122, 128, 131, 139, 146–148, 153, 160, 164, 171, 173, 177–179, 181, 190, 192–196, 199, 200, 205, 213–226, 229–237, 241, 242, 245, 252, 260, 270, 273
Schroeder decay curve, 36–38, 40, 41, 47
Schroeder decay function, 33–35, 37, 38, 41–43, 53
Schroeder decay model, 35, 36, 41, 46, 53
Schroeder diffusers, 270–271
Schroeder frequency, 57, 59, 60, 349, 365, 440
Schroeder integration, 33–35, 37, 53, 396, 440
Schroeder phase, 118–127, 385, 440
SCOTTs theory, 234
Seat effect, 5, 232, 237
Secure voice communication, 205
Self-imaging, 270
Self-similar maximum-length sequence (MLS), 95
Shallow water basin, 242, 250–251, 256, 265, 271
Shannon, C.E., 205, 368, 372
Shape optimizer, 194
Shell vibration, 263, 267–269
Signal phase, 125, 127
Silicon microphone, 224
Single-slope decay, 35, 42
Sound absorption measurement, 19, 256–258
Sound decay in two-dimensional rooms, 201–202
Sound field
  channel with absorber, 234
  empty channel, 234, 235
Sound power radiation, 251
Sound rays, 58, 59, 69, 71–73
Sound-soft water duct, 248–249
Sound transmission in rooms, 200

Source size effect, 235
Sparse signals, 61
Spatial audio, 3–11, 94, 102
Spatial envelopment, 99, 100
Spatial hearing, 10, 114, 117, 148
Spectral compensation, 87
Spectral estimator, 87
Specular reflection, 17, 19, 24, 26, 27
Speech coding, 115, 116, 204, 205, 207, 368, 371, 388, 389, 433
Speech perception, 138
Speech processing, 114, 231, 393
Squeeze film absorber, 244, 272
Statistical sound field theory, 252–255
Statistical test, 60, 79, 85, 87, 90
Statistics, 18, 20, 66, 68, 78, 128–138, 301, 320, 352, 371, 419
Steady-state energy decay, 34, 35
Subjective loudness, 206
Switch-microphones calibration, 233

T
TEF analyzer, 186–188
Teleconferencing, 3, 7, 10, 11, 223
Telephony, 7–9, 214, 216, 374, 385, 388
Temporal design, 146, 171
Texture, 146, 163, 169
Timbre, 102, 103, 115, 146, 152, 160, 163, 169
Toroidal microphone, 8, 216, 217
Touch dial, 219
Transducer, 8, 213–226, 246, 247, 249, 269, 270, 272, 395
Transfer function, 10, 78, 80–82, 85–89, 93, 161, 168, 230, 233, 237, 256, 260, 396
Tree search, 207, 208
Two-circuit resonance absorber, 243, 244
Two-microphone method, 217, 219, 220

U
Ultrasonic transducer, 214, 219
Underground sound, 180
Unidirectional microphone, 217, 218
Upper limit of the integration, 35, 37, 38, 40, 41

V
Variant decimation, 95
Vector base amplitude panning, 10
Viscometer, 270
Vocal tract
  geometry, 231–232, 234
  impedance, 229, 231, 234, 386
Vocoders, 115, 199, 204, 205, 367, 384–388, 411

W
Wave field synthesis, 10
Weyl, H., 57, 59, 354, 433
Wide impedance tube, 249–250
Wiener, N., 205

X
XFT. *See* Extensible Fourier transform (XFT)

Z
Zero-order Markov process, 203

MIX
Papier aus verantwortungsvollen Quellen
Paper from responsible sources
FSC® C105338

If you have any concerns about our products,
you can contact us on
ProductSafety@springernature.com

In case Publisher is established outside the EU,
the EU authorized representative is:
Springer Nature Customer Service Center GmbH
Europaplatz 3, 69115 Heidelberg, Germany

Printed by Libri Plureos GmbH
in Hamburg, Germany